Sustainable Industrial Water Use: Perspectives, Incentives, and Tools

Sustainable Industrial Water Use: Perspectives, Incentives, and Tools

Edited by

Cheryl Davis and Eric Rosenblum

IWA PUBLISHING

Published by

IWA Publishing
Republic – Export Building, 1st Floor
2 Clove Crescent
London E14 2BE, UK
Telephone: +44 (0)20 7654 5500
Fax: +44 (0)20 7654 5555
Email: publications@iwap.co.uk
Web: www.iwapublishing.com

First published 2021

British Library Cataloguing in Publication Data
A CIP catalogue record for this book is available from the British Library

ISBN: 9781789060669 (Paperback)
ISBN: 9781789060676 (eBook)

This eBook was made Open Access in February 2021

Dedicated to the memory of John Anderson (1947–2018), 'the messiah of water recycling in Australia,' whose commitment to sustainable water use continues to inspire us all.

Contents

Sponsor statements

ARCADIS

Sustainable water use by industry is a dynamic and multifaceted journey molded by diverse perspectives offered from within and outside. Frequently, focus is placed upon final solutions without fully embracing the myriad opportunities and challenges posed to balance business strength, steward resources and improve quality of life. As a first of its kind, this book describes the entire journey in equally measured detail, empowering readers to seize upon opportunity and surmount challenges.

I thank and congratulate every author who contributed to this unique book; together we have created a holistic guide to facilitate industrial water use in balance with our aspirations for a sustainable future.

Dr. Brian C. Moore
Vice President – Industrial Water Use Lead
Arcadis

DEG

Water-related risks have consistently emerged near the top of the World Economic Forum's Global Risk Report. At the same time they are one of the most underrated challenges faced by companies. Water issues impact companies and investors around the globe. The Carbon Disclosure Project has reported growing corporate losses driven by water risks, year-over-year, amounting to over US$30 billion in 2018 alone. DEG supports this book and the Open Access publication to highlight the importance of sustainable water use by industry to turn a financially material risk for companies and investors into an opportunity.

Jens Hoenerhoff
Vice President Sustainability and Corporate Governance
DEG – Deutsche Investitions- und Entwicklungsgesellschaft mbH

DELFT UNIVERSITY OF TECHNOLOGY

Water demand is increasing world-wide due to increasing industrialization, urbanization and population growth. A sustainable use of water is key to avoid water scarcity and competition between sectors. This book gives an excellent overview of the challenges and perspectives for such sustainable water use in an industrial setting, not only in industrialized countries but also in emerging economies. The contributions provide both technological and governance perspectives, making it a highly valuable book for students to learn about technological and conceptual tools. I think this book will especially inspire young engineers to contribute to a better world in general and a more healthy environment in particular.

Luuk Rietveld
Professor in Drinking Water and Urban Water Technology
Delft University of Technology
The Netherlands

ECOLAB

The 2020s are a decade of exceptional importance. If we don't take decisive action, the world will see a freshwater deficit greater than 50% by 2030. Science tells us that we must also halve our carbon emissions by that same year to keep climate change within manageable limits. These tightly intertwined challenges are huge, but they also afford us an extraordinary privilege. We are the generation that is tasked with building the platform for truly sustainable economic growth. If we succeed – and there's no reason we shouldn't – we'll help create a world with opportunities to thrive for all.

Emilio Tenuta
Senior Vice President and Chief Sustainability Officer
Ecolab, Inc.

HELVETAS SWISS INTERCOOPERATION

This is a unique book that compiles different cases on industrial use of water. Recognizing the perspective of industry is crucial to understand how their decisions impact shared water resources and what is needed to remove barriers in bringing different users closer for promoting multi-use of water. This is key for ensuring sustainability of water resources and addressing intricate issues such as competition among different uses and users. As a people-centered NGO, Helvetas contributes to securing rights of communities to access water in developing countries. This is only possible with the participation of stakeholders committed to good water governance.

Bernita Doornbos
Head Water, Environment & Climate Change
Helvetas Swiss Intercooperation

LUNDIN GOLD

This publication is of special interest because it provides the vision of different stakeholders regarding the responsible and sustainable management of water in diverse environments and industries. For the large-scale mining industry, the management of this resource is of vital importance both in terms of our industrial activities and community engagement. Importantly, water is core to the worldview of many indigenous communities, an interesting aspect that is also addressed in this book.

Modern mining technology promotes water recovery processes, reduces water use, and minimizes the need for freshwater collection. Additionally, the integrated environmental management of a mining project must incorporate trust-based community relations.

This book compiles a series of valuable tools and good practices. I invite you to read it.

Ron Hochstein
President and CEO
Lundin Gold

NESTLE

This book is a great collection of different perspectives and insights from a wide variety of water practitioners.

It has the merit of clearly explaining with practical examples how industry can sustainably manage water resources focusing on both internal practices and external stakeholder engagement. Strongly suggested as a key reading for young water experts starting their career in industry and also recommended for anyone interested in understanding better the industry challenges and solutions in the field of water stewardship. I wish I could have had the pleasure of such an interesting reading when I started my water career 30 years ago!

Carlo C. Galli
Technical Director – Water Resources
Nestlé

SUSTAINABLE RICE PLATFORM (SRP)

The growing challenge of global freshwater scarcity cannot be resolved by individual stakeholders. Concerted efforts are needed to deliver sustainable solutions for water stewardship. Whilst such calls for collaborative action are frequently voiced, finding practical examples remains a challenge. This encouraging book fills exactly this gap, providing a valuable collection of contemporary examples, tools and innovative partnerships that help contribute to realizing sustainable water stewardship.

As a global multi-stakeholder partnership committed to sustainability, I am proud to see the WAPRO project, where SRP plays a pivotal role, profiled as one of the many cases in this valuable collection.

Dr. Wyn William Ellis
Executive Director, Sustainable Rice Platform

VLIR-UOS BIODIVERSITY NETWORK ECUADOR (BIONET ECUADOR)

We are very grateful to the authors and editors for this excellent collection of valuable articles on innovation in the water sector. We are glad that our project could contribute to bringing many of the authors together in an inspiring atmosphere during the IWA-IDB Innovation Conference on Sustainable Use of Water in 2019 (Guayaquil, Ecuador). We are sure that these materials will be a key source of inspiration for many, including the students and professors in the diverse water training programs in Ecuador developed and organized in the context of this project.

Peter Goethals
Professor of Ghent University (Belgium)
General Coordinator BioNet Ecuador

Luis Dominguez
Director CADS-ESPOL (Ecuador)
Water Coordinator BioNet Ecuador

WEST BASIN MUNICIPAL WATER DISTRICT

As with any field, the water industry has evolved through the development of new technologies. Whether it be advanced water and wastewater treatment, water quality testing, or water use efficient fixtures, communities are now able to make existing water supplies more sustainable. The authors in this book provide unique experiences and strategies to advance and promote responsible water supply management. This book can and will make a difference in how communities view water and will open the door to new possibilities for communities that want to achieve a more sustainable environment. I want to thank everyone for including West Basin Municipal Water District in this effort, and congratulate all those involved.

Edward (E.J.) Caldwell, Esq.
Water Policy and Resources Development Manager
West Basin Municipal Water District

WWF

Freshwater resources are facing unprecedented threats: the 84% collapse in freshwater species populations since 1970 is the clearest possible warning sign. The same pollution, over-exploitation and mismanagement of our freshwater resources that are pushing biodiversity to the brink are also putting people and businesses at risk. WWF supports this book as it provides companies with valuable information on tools and good practices so they can become better water stewards and turn risk into opportunities – helping to create a sustainable water-secure future for all.

Ariane Laporte-Bisquit
WWF Water Risk Filter Project Manager

List of contributors

Kelly Alexander (Kelly.Alexander@care.org) and **Andrés Córdova** (Andres.Cordova@care.org) *Alleviating poverty through sustainable industrial water use: A watersheds perspective* (p. 25).

Kelly Alexander has a Master's degree from Emory University in Public Health, global infectious disease. She is currently a Senior Technical Advisor at CARE for the Water+ team, Food and Water Systems unit. A self-proclaimed eco-feminist, she believes the world's systems would have greater equity and environmental sustainability if women had more leadership roles.

Andrés Córdova from Quito, Ecuador, has a Master's degree in Sustainable Agriculture and is trained as an engineer in Renewal Natural Resources. Andres has worked at CARE Ecuador since 2010 where he is the Head of the Entrepreneurial, Resilient, and Violence-Free Societies Program and is the national lead on water and climate change.

Dave Archambault II (davearshambo@gmail.com) *The indigenous perspective on water: A source of life, not a resource* (p. 15).

Dave Archambault II is an independent consultant for Brave Fox, Inc. Dave was the 45th Chairman for the Standing Rock Sioux Tribe. As Tribal Leader, Archambault's focus has been on education, economic development, renewable energy, and government reform. Dave earned a Bachelor's of Science in Business Administration at North Dakota State University and a Master's Degree in Management from the University of Mary. Recently Chairman Archambault emerged as a global leader for Indigenous Peoples' rights as he led the opposition to the construction of the Dakota Access Pipeline. He strives to find ways to help tribal nations, being a voice for tribal sovereignty in championing protection of tribal inherent rights.

Zineb Benjelloun (z.benjelloun@ocpgroup.ma) and **Karim Saoud** (k.saoud@ocpgroup.ma) *Phosphate mining and the circular economy: Morocco's OCP group's approach to sustainable water use* (p. 63).

Zineb Benjelloun from Casablanca, Morocco, has worked for OCP Group since 2008 as Water and Energy Manager. She was responsible for steering the implementation of the water and energy strategies set up by the company to support its important industrial development program. In 2018, she became a core member of the company's Circular Economy Program and is currently Sustainability External Engagement Lead.

Karim Saoud from Casablanca, Morocco, has nearly three decades of experience in utilities, first as an operator of private water and electricity networks at the OCP site in Khouribga and then as project manager for the construction of wastewater treatment plants, water pipelines, and high voltage electrical networks. He is currently Director of the Water & Energy portfolio in the Sustainability platform at OCP Group.

Troels Kærgaard Bjerre (tkb@vcsdenmark.com) *Applying the UN sustainable development goals as a framework for corporate sustainability* (p. 351).

Troels Kærgaard Bjerre, from the island of Funen in Denmark, is a project director at VCS Denmark, one of the largest Danish water utilities where he is responsible for sustainable groundwater resources management and protection. Since 2015 he has also been spearheading VCS Denmark' effort towards corporate sustainability.

E.J. Caldwell *Designer water: One utility's unique approach to industrial sustainability* (p. 255).

E.J. Caldwell is Manager of Water Policy and Resources Development for the West Basin Municipal Water District, responsible for local, state, and federal advocacy efforts as well as planning and water conservation programs. He previously practiced law, advising local governments on financial and legal issues after serving as a member of the Speaker's Policy Unit for the California Assembly. E.J. has a BA in Political Science from the University of Arizona, a JD from Willamette University, and is a member of the California State Bar Association.

Thomas Chesnutt (tom@antechserv.com) *The economics of sustainable industrial water use, reuse, and the value of water* (p. 199).

Thomas Chesnutt Ph.D. is CEO of A & N Technical Services, Inc. in Encinitas, California and has extensive experience in water rates, stochastic simulation and forecasting, and financing. He was the Principal Investigator on the recent (2020) study of *Probability Management for Water* sponsored by the Water Research Foundation (WRF Project 4742).

Cheryl Davis (ckd@cherylkdavis.com) *Sustainable water use by industry: What can we do?; Staff training: An integral component of sustainable water use by industry* (p. 1, 181, 379).

Co-Editor Cheryl Davis, from the San Francisco Bay Area of California, worked for over three decades as a manager of the Bay Area's largest water utility (San Francisco Public Utilities Commission). Cheryl chairs IWA's Specialist Group on Sustainability in the Water Sector as well as the IWA Task Group on Sustainable Use of Water by Industry. She also consults on workforce reliability and sustainable use of water by industry.

Leandra Diaz (leandradiazrios@gmail.com) *Water footprint: A sustainability tool for industries* (p. 333).

Leandra Díaz, from La Paz, Bolivia, has ten years of experience in planning, implementation, and monitoring of climate change adaptation and mitigation projects in 15 cities in 5 countries in Latin America. She was director of the Adaptation Mechanism of the Plurinational Authority of Mother Earth. She specializes in greenhouse gas inventories and water footprint analysis, and is a member of a water footprint practice community in Latin America and the Caribbean (LAC). In that capacity she has evaluated the International Standards Organization standard ISO14046 and Life Cycle Assessment products in the LAC beverage industry and is part of the IPCC Roster of Experts.

Peter Easton (p.easton@watersustain.com) *Water stewardship* (p. 303).

Peter Easton, from the UK, is a hydrogeologist and water stewardship specialist with 30 years international experience across Europe, the Middle East, and Africa in a range of industry sectors including food and drink, sustainable agriculture, mining and wetland restoration. Peter has contributed to the development of international water stewardship standards. Peter is currently a Senior Advisor on Water and Sustainability to Revolve Media, a Brussels based international media agency and regularly publishes on water and related environmental issues (https://revolve.media).

Mutshiene Deogratias Ekwanzala (ekwanzala.md@gmail.com), **Phumudzo Budeli** (bphumu@gmail.com), and **Maggy Momba** (mombamnb@tut.ac.za; momba1958@gmail.com) *Sustainable solutions to the impact of industrial water pollution on the environment and community health* (p. 127).

Mutshiene Deogratias Ekwanzala, residing in Seoul, South Korea, is a qualified biotechnologist and trained microbiologist in water quality and wastewater management. He holds a Ph.D. degree in Water Science and Technology, and is currently a Postdoctoral Researcher at Kyung Hee University exploring the RNA virome of sewage.

Phumudzo Budeli, from South Africa, a trained microbiologist with expertise in wastewater management and drinking water treatment, holds a Master's degree in Water Science and Technology. Phumudzo is currently a Ph.D. candidate at Tshwane University of Technology focusing on bacterial and algal immobilised enzymes for the degradation of endocrine disrupting compounds.

Maggy Momba, from South Africa, trained as a microbiologist and has extensive expertise in drinking water and wastewater treatment, bioremediation, and the use of nanotechnology for water treatment. Prof. Momba has a Ph.D. in environmental microbiology, and began her career as a research professor in 2007. Since 2013 she has been South African National Chair at Tshwane University of Technology for water quality and wastewater management.

John Etgen (john.etgen@projectwet.org), **Nicole Rosenleaf Ritter** (rosenleaf@aya.yale.edu), and **Will Sarni** (will@waterfoundry.com) *The business value of water education to corporations* (p. 385).

Project WET CEO John Etgen has 35 years of experience delivering conservation education around the world and a passion for advancing water education to understand global challenges and inspire local solutions. He is the Chair of the Alliance for Water Stewardship Technical Committee and a credentialed AWS Professional Consultant.

Until 2020, Nicole Rosenleaf Ritter was the VP of Communications for the Project WET Foundation, championing the importance of water education as a solution. She is now an editor and writing mentor for Articulate, an e-learning software company.

Will Sarni is Founder and CEO of Water Foundry. An internationally recognized thought leader on water strategy and innovation, Will has written extensively on the value of water, innovations in digital water technology, the circular economy, and the energy-water-food nexus. He helps multinationals and startups quantify and mitigate water-related risks.

Carlo C. Galli and **Jonathan Sottas** *Nestlé: 'Caring for Water' through people, farmers, and communities* (p. 43).

Carlo C. Galli has worked at Nestlé since 1991 in various positions, where he has promoted the use of leading practices for sustainable management of water resources. He is currently Technical Director for Water Resources at Nestle Group level and Head of Sustainability for Nestlé Waters, the Group's bottled water category. For many years he has been a key actor at the Alliance for Water Stewardship (AWS), which awarded him the AWS Prize in 2017.

Jonathan Sottas is a hydrogeologist for Nestlé working in corporate water stewardship strategy and sustainable water management in business operations and supply chains.

Barry Greig barry.greig@gov.scot and **Jon Rathjen** jon.rathjen@gov.scot *Promoting sustainable industrial water use: Scotland's 'Hydro Nation' at home and abroad* (p. 395).

Barry Greig works for the Scottish Government Directorate for Environment and Climate Change: Water Industry Division. Formerly a lawyer and commercial arbitrator, Barry leads the Scottish Government's innovative 'Hydro Nation' agenda. In that capacity, he helps to develop the economic, social, and environmental value of Scotland's water resources while ensuring Scotland manages its water environment responsibly and sustainably, and deploys its knowledge and expertise effectively both domestically and internationally.

Jon Rathjen works for the Scottish Government Directorate for Environment and Climate Change: Water Industry Division. He leads the Scottish Government Water Industry Team with policy responsibility for industry, including sponsorship of the public corporation, Scottish Water, and the Hydro Nation agenda to develop the value of Scotland's water resources. Jon is a board (or equivalent) member of Aqua Publica Europea, UK Water Partnership, and UK-Unesco IHP.

Cornelis (Niels) Groot ckgroot@dow.com *Triple water reuse at Dow; Challenges in regional collaboration; Staff training: An integral component of sustainable water use by industry* (p. 89, 277, 379)

Cornelis (Niels) Groot, of Dow Benelux in the Netherlands, supports Dow's assets regarding water and wastewater technology with special focus on developing water sourcing strategies in water stressed areas. Niels has a part time assignment to HZ University of Applied Sciences as a professor in water technology.

Brent Haddad bhaddad@ucsc.edu *Severe water crises: Industry's role and response* (97).

Brent Haddad, MBA, Ph.D., is Professor of Environmental Studies at University of California, Santa Cruz. He studies sustainable water systems, policies, economics, and communications. In addition to his teaching and research, he advises water agencies, companies, and regulators.

Juan José Herrera juan.herrera@lundingold.com, **M. J. Pozo** maria.pozo@lundingold.com, **María Cristina Acosta** macristina.acosta@lundingold.com, and **Nathan Monash** nathan.monash@lundingold.com *Ecuador's Fruta del Norte: Early engagement as a tool to build trust* (p. 51).

Juan José Herrera, an economist from Quito, Ecuador, joined Lundin Gold in March 2018. He previously worked at Grupo FARO, an Ecuadorian Think Tank, as well as other academic and research institutions. Juan José has a Master's in Energy Governance and a Specialization in Extractive Industries and Sustainable Development with experience in mining, oil, energy and sustainable development.

M. J. Pozo, from Quito, Ecuador, has been involved for more than ten years in research, consulting, and environmental management. She has participated in research and projects on water management and has worked in the public sector in water and environment, in the oil industry, and for four years at Lundin Gold.

María Cristina Acosta, from Quito, Ecuador, has been involved for more than nine years in the Fruta del Norte gold project, leading the Environment and Permitting department. She is an Environmental Engineer with a Master's degree in Quality, Safety and Environment, as well as a Professional Coach. Prior to FDN she worked in the oil industry and in consultancy for different industrial sectors.

Nathan Monash, based in Quito, Ecuador is the VP of Business Sustainability for Lundin Gold. He has approximately 20 years of experience in sustainability, community relations, and external affairs. He previously worked for a number of companies and institutions, including AngloGold Ashanti, the International Finance Corporation (IFC), Rio Tinto Alcan, and the World Economic Forum.

Paula Kehoe pkehoe@sfwater.org and **Taylor Chang** TaChang@sfwater.org *Onsite water reuse: A collaborative strategy to manage water* (p. 271).

Paula Kehoe is Director of Water Resources with the San Francisco Public Utilities Commission. She is responsible for diversifying San Francisco's local water supply portfolio through the implementation of conservation, groundwater, recycled water, and innovations programs. Paula spearheaded San Francisco's landmark legislation allowing for reuse of alternate water sources in buildings and is Chair of the National Blue Ribbon Commission for Onsite Non-potable Water Systems.

Taylor Chang is a Water Resources Analyst at the San Francisco Public Utilities Commission. She manages the Onsite Water Reuse Program, which allows buildings in San Francisco to treat and reuse water onsite for non-potable applications. Taylor graduated from UC Berkeley in Environmental Studies, and has a Masters in Public Administration from the University of San Francisco.

Susanne Knøchel skn@food.ku.dk *Fit-for-purpose water reuse in the food processing industry* (p. 139).

Susanne Knøchel has worked with microbiological safety and quality in food and water for almost 40 years. She is professor and Section Head for Food Microbiology and Fermentation at the Department of Food Science, University of Copenhagen, Denmark.

Ariane Laporte-Bisquit arianeLaporte-Bisquit@wwf.de *WWF water risk filter: Assess, respond & value water risks* (p. 321).

Ariane Laporte-Bisquit has been the WWF Water Risk Filter Project Manager since 2018. She is responsible for overseeing the ongoing evolution of the Water Risk Filter tool's development and strategy. She also engages with investors and corporate users of the tool, in collaboration with the wider WWF network, to drive water stewardship awareness and action. Before joining WWF, she was the Project Officer for CDP's Water Security Program in Europe and worked previously on several water-related projects in South Africa, Bangladesh, and French Guiana.

Oliver Maennicke oliver@maennicke.com and **Jens Hoenerhoff** *Development finance: Encouraging sustainable water use by industry* (p. 207).

German-Australian Oliver Maennicke trained as an engineer for water management at Dresden University of Technology. His experience includes six years of work for WWF as a Water Stewardship Specialist and Corporate Partnership Manager; since 2018 he has worked as an independent consultant. Oliver is a member of the Technical Committee of the Alliance Water Stewardship, and has extensive experience working with private sector organizations and NGOs to promote sustainable use of water by industry. He also serves as Secretary on the management committee of the International Water Association's Specialist Group on Sustainability in the Water Sector, and is a member of the International Water Association's Task Group on Sustainable Use of Water by Industry.

Jens Hoenerhoff, degreed Energy and Process Engineer at the TU Berlin, works as senior environmental specialist for the DEG, the private sector development financing branch of the German KfW-Group. As vice president for sustainability, he is inter alia responsible for the development of the WWF-DEG 'Water Risk Filter', the mapping tool for water risks.

Felicia Marcus feliciaamarcus@gmail.com *Musings of a former regulator, or 'how can we do better?'* (p. 151).

Felicia Marcus has spent her career in and out of government at the local, state, and federal levels, most recently as the Chair of the California State Water Resources Control Board. Previously, she served as Regional Administrator for the US Environmental Protection Agency and President of the City of Los Angeles' Board of Public Works in addition to senior management roles in national environmental NGOs. She is currently a Landreth Fellow at Stanford University and a Member of the Water Policy Group (http://waterpolicygroup.com).

Juan Pablo Mariluz Silva jmariluz@ana.gob.pe and **Ursula Antunez de Mayolo** ursula.antunez@sgs.com *The Certificado Azul: Peru's innovation for encouraging sustainable use of water by industry* (p. 235).

Juan Pablo Mariluz, from Lima, Perú has 15 years of experience in water resources management for the National Institute of Natural Resources and tha National Water Authority of the Ministry of Agriculture and Irrigation. He is a member of the evaluation committee of the Water Footprint Program and participated in the preparation of the study of the water footprint of the agricultural sector in Peru.

Ursula Antunez de Mayolo, from Lima, Peru, is a registered Specialist in Water Stewardship and conducts verification assessments of the Water Footprint with a focus in auditing and teaching courses. She has a background of Industrial Engineering from the University of Lima (Peru), a Masters in Environmental Science and Technology from IHE Delft (the Netherlands), and a post-graduate degree in Resource Studies from Lincoln University (New Zealand). She worked in SGS Peru (eight years) and SGS Australia (five years). In 2019 Ursula was elected as a member of the Board of AWS (Alliance for Water Stewardship).

Brian Moore Brian.Moore@arcadis.com *Implementing sustainable water use: A roadmap for industry* (p. 363).

Brian Moore, Ph.D., is based in New York (USA) where he serves as Industrial Water Use Practice Leader for Arcadis. Prior to joining Arcadis, he was the technical water and wastewater expert for GE where he developed cost-effective projects to reduce their global water consumption by nerarly 50%. Through his experiences with GE, Brian learned the importance of return on investment (ROI), and developed a unique talent for identifying and driving water conservation opportunities that meet business financial needs.

Domenica Mosca Angelucci (mosca@irsa.cnr.it) *Biological wastewater treatment as an opportunity for energy and resource recovery* (p. 407).

Domenica Mosca Angelucci from Rome Italy has a Ph.D. in Chemical and Process Engineering, and is an Environmental Engineer working as a researcher at the Water Research Institute of the Italian National Research Council. Her research topics include the development of innovative technologies for the treatment of urban and industrial wastewater and sludge and for the bioremediation of contaminated soils.

Arjumand Nizami arjumand.nizami@helvetas.org, **Jawad Ali** jawad.ali@helvetas.org, and **Sophie Nguyen-Khoa** Sophie.NguyenKhoa@helvetas.org *Government-industry partnership for sustainable water use: Insights from Pakistan* (p. 115).

Arjumand Nizami has been engaged in development for 27 years. Currently she works for Helvetas Swiss Intercooperation to promote climate resilient agriculture and water management systems leading to improved productivity, equity, nutrition, and diversification in Pakistan. She has written numerous journal articles, reviews, and chapters on these topics. Her work involves multi-partner collaboration with low-income communities, government, development actors, the private sector, and academia.

Jawad Ali has worked in the development sector for 25 years. His experience includes teaching and research at the Universities of Life Sciences Norway and Agriculture University Peshawar, Pakistan.

Currently working with Helvetas, his interests include natural resource management, climate change, and water. Ali has supervised several developmental projects and has published numerous articles on topics related to deforestation, water management, and climate change.

Sophie Nguyen-Khoa is currently a senior advisor on the water-food-climate nexus at HELVETAS. She has over 25 years of international experience in integrated water management and holds an interdisciplinary Ph.D. from Imperial College London. In earlier work she served as a researcher at the International Water Management Institute (IWMI) and the WorldFish Center and co-directed the CGIAR Program on Water and Food.

Karla Raettig karlaraettig@nwa.org and **David Muth** muthd@nwf.org *Recovering from disaster: Holding industry accountable for restoration* (p. 165).

Karla Raettig is the Executive Director of the National Wildlife Federation Action Fund, where she leads the political and electoral work of the organization. She previously served as the National Wildlife Federation's Program Director for Mississippi River Delta restoration and practiced public interest environmental law for a decade.

David Muth lives in the Mississippi River delta, studying its geology, ecology, plants, wildlife, history, and culture. After 30 years with the National Park Service, he joined the National Wildlife Federation and is the Director of the Gulf Restoration Program in New Orleans.

Olivier Crespi Reghizzi crespio@afd.fr *Reducing pollution from industrial wastewater in developing and emerging countries* (p. 215).

Both civil engineer (MSc., Politecnico di Milano and Centrale Nantes) and economist (Ph.D., Bocconi University and AgroParisTech), Olivier Crespi Reghizzi works in the water and sanitation division of the French Development Agency in Paris. In the past Olivier worked for AFD in Senegal and Mexico and for Eau de Paris, the Paris municipal water utility.

Eric Rosenblum eric@envirospectives.com *Incentivizing sustainability: How utilities can support industrial water conservation and reuse*; *Implementing sustainable water use: A roadmap for industry* (p. 261, 363).

Co-Editor Eric Rosenblum PE is an environmental engineer based in San Jose, California. After managing public water and wasterwater utilties in the Silicon Valley area of northern California, he now specializes in helping cities and companies recycle their water. He also researches and writes about sustainability and environmental ethics.

Guatimozin Santos guatimozin.santos@pmi.com *Business transformation as the gateway to sustainability: A tobacco company's perspective* (p. 35).

Guatimozin Santos from São Paulo, Brazil is the Institutional Relations Manager of Philip Morris International (PMI) affiliate in the country. Social scientist with a master's degree in Political Science, is currently leading the sustainability strategy in Brazil and helping to advance PMI's transformation goal of becoming a smoke-free company.

Petra Schneider petra.schneider@h2.de and **Christian Wolkersdorfer** christian@wolkersdorfer.info *Dimensions of water management in the extractive industries* (p. 73).

Petra Schneider, from Magdeburg, Germany, holds a Ph.D. in Hydrogeology. Her professional experience includes 20 years in the private sector, followed by 5 years at Magdeburg-Stendal University of Applied Sciences. Since 2015, Petra has been a professor for International Water Management and leads the master course 'ecological engineering'. She has a focus on environmental rehabilitation.

Christian Wolkersdorfer is a full research professor and has 30 years of professional experience. He is a mining and geothermal hydrogeologist specialized on mine water tracer tests, mine water geochemistry, and remediation. Prof. Wolkersdorfer is also the Technical Editor for the journal *Mine Water and the Environment* as well as the President of the International Mine Water Association (IMWA), the Industry-Academia coordinator for the Mine Water Division of WISA (Water Institute of Southern Africa), and he is a member of the Global Alliance.

Dr. Henri Spanjers h.lf.m.spanjers@tudelft.nl *Extreme industrial effluents: Opportunities for reuse* (p. 417).

Environmental Engineer Henri Spanjers has worked with a number of research institutes in various EU countries and was project director with a consultancy firm. He is currently a professor at Delft University of Technology and a member of the faculty of Civil Engineering and Geosciences where he leads the Industry Water group. He is Editor-in-Chief of the journal *Water Resources and Industry*.

Michael Spencer michael.spencer@monash.edu *The culture of water needs to change* (p. 243).

Michael Spencer is a Research Fellow at Monash Business School in Melbourne Australia. He has held leadership positions with the Alliance for Water Stewardship and Forest Stewardship Council following a successful career in business (BHP Billiton, National Australia Bank and BlueScope Steel), government (senior advisor, Premier of Victoria), and journalism.

Alex Stefanakis stefanakis.alexandros@gmail.com *Promoting sustainability in the oil industry: The benefits of using constructed wetlands for oily wastewater treatment* (p. 427).

Alex Stefanakis Ph.D., from Crete, is Assistant Professor at the Technical University of Crete and a professional Environmental Engineer with 15 years of experience in the design and management of constructed wetlands for wastewater treatment. He participates in various international research projects and has published several scientific articles, books, chapters, and conference papers.

Adrian Sym and **Sarah Wade** *The AWS Standard: A common language for the global water stewardship community* (p. 313).

As Chief Executive, Adrian Sym leads the pioneering work of the Alliance for Water Stewardship (AWS). Since joining AWS in 2011, Adrian has overseen the development of AWS into the a truly global network that ignites and nurtures leadership in credible water stewardship.

Sarah Wade joined AWS as Strategic Programs Manager in 2018, having previously worked in WWF-UK's water stewardship team. She works with partners from civil society, the public sector, and private sector to catalyze water stewardship activity around the world.

Emilio Tenuta etenuta@ecolab.com *Closing the execution gap: How industry can lower its water use and help tackle global water scarcity* (p. 295).

Emilio Tenuta is Senior Vice President of Corporate Sustainability for Ecolab where he is responsible for linking Ecolab's market strategy with its sustainable solutions delivered through customer partnerships. He holds a Bachelor of Science degree from the University of Wisconsin-Eau Claire and an MBA from Northwestern University's Kellogg School.

Maria Concetta Tomei tomei@irsa.cnr.it *Best available techniques as a sustainability tool for industrial water management and treatment*; *Biological wastewater treatment as an opportunity for energy and resource recovery* (p. 343, 407).

Chemical engineer Maria Concetta Tomei is Senior Scientist at the Water Research Institute of the Italian National Research Council in Rome. Her research activity is focused on urban and industrial wastewater treatment involving both theoretical (process modelling) and practical (technology development) aspects. She has authored more than 200 publications in the field.

Arthur Umble arthur.umble@stantec.com *The role of the consultant in supporting sustainable industrial water use* (p. 105).

Arthur Umble Ph.D. leads the Global Wastewater Treatment Sector for Stantec Consulting, focusing on converting waste streams to value streams. He provides technical leadership for new and rehabilitated wastewater treatment plants, with emphases on nutrient removal and recovery, process optimization for capacity and energy management, and water reuse.

Santiago David Vicencio, Chico. vicencios@produbanco.com *The banking sector as an intermediary in supporting sustainable use of water by industry* (p. 225).

Santiago Vicencio, from Quito, Ecuador, studied management, finance, and economics and has an Executive MBA and several international certifications in Sustainable Finance and Risk Management. Santiago leads the Green Lines Program in Produbanco GP, which funds the development of sustainable processes, projects and operations in all enterprise segments.

Introduction

Cheryl Davis[1] *and Eric Rosenblum*[2]
[1]*CKD Consulting, Piedmont, CA, USA*
[2]*Water Resource Consultant, San Jose, CA, USA*

This anthology contains the work of multiple authors, each of whom has a unique and important perspective on the subject of sustainable use of water by industry. These authors come from a variety of backgrounds and disciplines, providing an unusual opportunity to listen to the views of individuals not always represented in publications on this topic. In this book we hear not only from industry, government, and academia, but also from water and wastewater utilities, financial institutions, consultants, environmental and social organizations, and underserved and indigenous populations. This diversity reflects the fact that before the idea of 'sustainable water use' can be realized in practice, it must pass through many hands.

While comprising less than 20% of global water use, industry impacts local water supplies not only by its demand but also by its discharge of pollutants into rivers and streams. As this volume demonstrates in detail, companies can now purify and even reuse water through a variety of advanced water treatment technologies, including micro-, ultra-, and nanofilters, microbial fuel cells, and wetlands designed to accelerate natural treatment processes. Public experience with water resource management has generated many innovative and effective policies, incentives, and regulations that stimulate the adoption of sustainable practices. Even so, the road to implementation is often long and complicated, with detours and dead ends. A promising new technology can achieve its potential only if companies invest in it and then train their workers to operate and maintain it properly. Similarly, the most innovative policy depends for its success on public advocacy and the development of regulations endorsed by technical experts, monitored by inspectors, and enforced by the courts.

DOI: 10.2166/9781789060676_0001

A COMMON PURPOSE

In our opinion, the goal of sustainable industrial water use can only be achieved through collaborative effort on the part of individuals from multiple sectors. Despite their many differences, the authors represented here share several attitudes. First, they agree that people matter, and so does the planet we live on. While each author may use the term 'sustainability' somewhat differently, there is general agreement that economic tools should be employed to promote sustainable water use. Or to quote the economist Herman Daley, '*The market is not the end of society, and is not the right instrument through which the ends of society should be set*' (Daley & Cobb, 1994).

This shared understanding is consistent with the 'Triple Bottom Line' framework, which expands the metaphor of the financial spreadsheet to include social and environmental costs and benefits. Monetization of social and economic impacts is a useful exercise when it allows the damage caused by industrial operations to be included in the cost of goods and services. Many impacts are difficult to monetize, however, resulting in a growing awareness that the Triple Bottom Line framework itself could be more comprehensive. As an example, our colleague Bruce Beck, past Chair of the IWA Specialist Group on Sustainability in the Water Sector, identifies no fewer than 14 dimensions of the Triple Bottom Line that gauge the progress of integrated water resource management (Beck, 2011).

Another shared assumption is the importance of a perspective that extends beyond the next quarter, beyond the next fiscal year, and beyond our lifetimes. Since human beings throughout history have been only too willing to inflict pain and suffering on others in the present, with their victims standing right before them, it may seem unduly optimistic – even naïve – to imagine widespread adoption of such a long-term and inclusive frame of reference. Nevertheless, we are unlikely to make significant investments in sustainable programs unless we can cultivate imagination and compassion enough to respect the value of humans not yet conceived and the future plant and animal inhabitants of the earth. A realistic conversation about sustainability needs to address both the planet's carrying capacity and our own 'caring capacity.'

ORGANIZATION

We have arranged the chapters into three sections – *Perspectives, Incentives, and Tools*. The first section highlights perspectives because we believe it will be easier for us to achieve the collaboration needed among different sectors if we can begin to hear and understand each other more clearly. The section on *Incentives and Barriers* exists because we believe businesses respond to the potential rewards and penalties they perceive, and make their choices accordingly. Further, we think it is unrealistic to expect far-sighted altruism to be sufficient motivation for modification of industrial practices at the level and speed needed to protect water, people, and the environment. The section on *Tools* provides a glimpse into the variety of analytical, technological, and management tools available to guide and support more sustainable water use by industry.

At the beginning of each section, the reader will find a synopsis of the articles. In addition, we asked each prospective author to complete a brief survey to assess their views on the key elements and barriers to sustainable water including how they thought others might help industry use water more sustainably. Their responses to our survey are reported collectively (and anonymously) at the end of the introduction to each section, further highlighting their different points of view.

TRENDS AND OBSERVATIONS

In the process of assembling the articles in this book, we have been privileged to work with dozens of authors who have provided insights into industrial water use across the planet. The anthology reflects what can be accomplished in the most affluent parts of the globe, with advanced technology and an established regulatory framework, as well as the achievements in less-affluent countries, where economic, political, and educational challenges must be overcome to use water sustainably.

Looking at this issue broadly, we can make a few general observations about the direction we are going and the challenges ahead. Technologically, many promising trends support reduced water use within industrial facilities. Sophisticated membrane-based processes, including microbial fuel cells and aerobic and anaerobic sequencing batch reactors, provide very high levels of treatment in a small footprint. Where sufficient land is available, natural treatment methods can be adapted for industrial use, while refinements in conventional treatment continue to lower the overall cost of industrial reuse. Plant management, too, has been enhanced through computerized control that enables operators to remotely supervise equipment with advanced diagnostics that help prevent problems before they occur.

As a result, many companies (like those in the high tech sector) are treating and reusing their wastewater onsite, reducing their dependence on drinking water supplied by public utilities. Further up the supply chain, companies that embrace the principle of 'the circular economy' have begun to substitute products and redesign processes to reduce pollution and waste (McDonough and Braungart, 2002). Water footprint analysis and similar techniques have proved useful in guiding product selection and process design decisions, and standards like the UN Sustainable Development Goals also help frame the choice of industrial water options with clear global and local objectives.

On the other hand, thousands of industrial water users around the world, small and large, continue to discharge waste into local waterways, endangering the lives of people and the health of the environment, often for generations. In addition to the damage they inflict on the community, when they duck cleanup costs these companies also undercut the ability of responsible competitors to invest in water improvements. Sustainable water use technology may be getting cheaper, but it is not free, and even sympathetic executives must wrestle with the cost of modifying plant operations when every number on the 'expense' side of the ledger is scrutinized with an eye towards reducing costs and boosting profits.

For this reason, some of the burden of supporting sustainable use of water by industry must fall to others, principally government agencies and non-governmental advocates. These groups, too, are challenged to influence industry in a meaningful, measurable way. For instance, environmental advocacy organizations often encourage companies to minimize their impact on the environment in order to attract ecologically minded customers. But while NGOs frequently publicize the offenses of polluters, occasionally organizing boycotts to punish them, they are less successful at consistently driving trade toward good companies.

Ultimately, it falls to government to protect public watersheds, but government performance has been inconsistent at best. Environmental regulations vary from country to country. In some places, they are effectively nonexistent, while in others they stiffen and relax with each shift of the political winds. On a national level, this reduces industry's incentive to invest in sustainable water use; globally, it allows companies to site their factories wherever they face the fewest requirements, putting pressure on their competitors to do the same. On the local level, effective enforcement of industrial water regulations is frequently hampered by a lack of trained inspectors, and sometimes undercut by outright corruption. And like private companies, public utilities can overlook the value of watershed improvements by focusing too narrowly on revenues, as when they fail to encourage wastewater reuse out of concern for its impact on rates.

So although we may be more protective of water than we were a half century ago, with many more tools to remove pollutants and conserve water, globally it will take more than partial solutions by individual companies to achieve sustainability. As company managers more fully appreciate the economic value of water, weak local and national governance must be bolstered by vigorous public advocacy, while consumers will ultimately have to support the cost of more sustainable approaches. In short, it will take consistent, coordinated action by all those who decide how water is used, as well as those who must live with the consequences of its misuse.

CONCLUSION

In summary, if industry is to use water sustainably in a way that is consistent with overall watershed stewardship, it will take active engagement by all parties. As noted elsewhere by one of our authors (Sarni, 2011), 'Solutions to water scarcity and sustainability will ultimately require close collaboration with those who make public policy, NGOs, investors and the local communities in which these companies operate.' Reflecting on how the diversity of viewpoints represented in this book might be harnessed to achieve sustainable use of water suggests to us that all concerned must improve their skills in three areas: conflict and collaboration, communication, and change.

Conflict and collaboration

As the saying goes, 'The opposite of love isn't hate, it's apathy.' Similarly, when it comes to industrial water use, the opposite of collaboration is not conflict, it's isolation. Conflict is to be expected when people engage in problems from different perspectives, and can be a beneficial part of the process when it draws attention to problems. If customers continue to buy cheap products from companies that leave polluted water in their wake, people will continue to get sick, and wildlife will continue to decline. However, if those same customers stage a boycott demanding that industry upgrade their practices, things will change. If governments are reluctant to police industry for fear of losing jobs, companies will have little incentive to improve. But if utilities increase rates, regulators monitor discharges, and law-makers mandate environmentally protective practices, competitive companies will install new technologies to conserve water and reduce pollution.

Industry depends on the resources of the community – water, raw materials, social structures, labor – and can't become sustainable simply by perfecting the water-intensive processes within its own facilities. Sometimes the greatest gains lie 'outside the fenceline.' This necessarily involves working with government agencies, utilities, environmental and social organizations, and even competitors operating within the same watershed. This, in turn, may involve professional associations, consultants, academicians, and researchers located half a world away.

It is neither helpful nor realistic to pretend that all these parties share the same goals from the start. For their part, industry can only be expected to use water in a manner rewarded by the market, which means they must experience significant negative financial penalties for unsustainable water use. This will only happen if pressure is brought to bear in ways that threaten to disrupt the status quo. But as we get to know each other, we may identify shared values that motivate us all. In the meantime, we can speak openly about our differences, recognizing that respectful conflict will bring us closer toward the goal of protecting water.

Communication

The purpose of this book is to help industry put the concept of sustainability into practice. We hope that by bringing together the various sectors whose voices are heard here we can create a platform for ongoing communication among industry representatives, water professionals, advocacy groups, researchers, regulators government policy-makers and regulators, finance institutions, and the general public. This is challenging, because the language we water professionals use to talk amongst ourselves is often indecipherable to other audiences. Water professionals need to translate technical information into understandable terms in order to reveal the meaning embedded in graphs and charts. Those satisfied with our current modes of communication might ask themselves, 'How is it that advertisers convince people to buy poor-quality products they quickly discard, while we struggle to rally support for water, which is essential to life?' We may know a lot about water, but clearly have a lot to learn about communication.

Change

Sometimes working for change feels like leaning against a locked door. But as illustrated by the massive and unexpected impact of the coronavirus pandemic, established realities can transform overnight. The world changes in fits and starts: our job is to push, and be prepared to walk through when the doors open.

How do we prepare for change, not knowing what the change will be? We first must work to advance the sustainable use of water by industry as we recognize it today, making the business case for sustainability, lobbying for more protective regulations, supporting the work of environmental organizations and others who pressure businesses to take responsibility for the long-term impacts of their actions. At the same time, we need to consider how the value of sustainable water use might be most effectively expressed in the future. What laws would best protect water as a public good, and ensure its availability for communities and the environment? What treatment methods should we develop today so that we can better remove contaminants and recover nutrients and energy from industrial wastewater tomorrow? By asking the right questions now, we will be able to make the most of future opportunities when they arise.

Sometimes the work of preparing for change can feel both hopeless and foolish. However, some questions are answered based on who you want to be, your stance toward life, and not measures of visible productivity (This may be the point of a sign glimpsed in a garden at a Buddhist retreat: 'Work diligently, achieve nothing.') We can't predict the future, but we know with certainty that sustainable use of water by industry is critical to our welfare and the survival of those who come after us. Water is precious, and those of us who recognize its value are committed to its conservation and protection.

We have attempted to collect the best ideas about how together we can protect our communities and the environment against the effects of unsustainable water use by industry. Our authors have brought an abundance of information, from broad concepts to specific, practical suggestions.

In the final analysis, the question is not 'What have we learned?' It is 'What will we do?'

REFERENCES

Beck M. B. (2011). Cities as Forces for Good in the Environment: Sustainability in the Water Sector, Warnell School of Forestry and Natural Resources. University of Georgia, Athens, Georgia, p. 107.

Daly H. and Cobb J. (1994). For the Common Good. Beacon Press, Boston, p. 14.

McDonough W. and Braungart M. (2002). Cradle to Cradle: Remaking the Way We Make Things. North Point Press, San Francisco.

Sarni W. (2011). Corporate Water Strategies. Earthscan, New York, p. 3.

ACKNOWLEDGEMENTS

We would like to extend our deepest thanks to the many authors who generously donated their time in contributing their perspectives and expertise to this book. In a very real sense, our aim as editors has been to serve as a vehicle for their voices, to share their knowledge, their vision, and their values. We recruited these authors precisely because each one shoulders important responsibilities with respect to the use of water by industry. To their immense credit – and our unending gratitude – they devoted considerable time and attention to prepare these articles for publication, from our initial hour-long video conversations to the completion of the editorial review process.

We would also like to acknowledge the assistance of the members of the specialist groups of the International Water Association (IWA), as well as the IWA Task Group on Sustainable Use of Water by Industry who highlighted the importance of our effort in their 2019 work plan. Jamie Workman, Tom Chesnutt, and Brian Moore each provided advice and encouragement, and Mark Hammond, Books Commissioning Editor for IWA Publishing, helped us coordinate with dozens of authors on six continents during a global pandemic. Finally, on a personal note we would like to thank our respective spouses, Mike Dadik and Lisa Ginsberg Rosenblum, for sharing our commitment to sustainable use of water and supporting the work associated with this project.

Cheryl K. Davis
Piedmont, California

Eric Rosenblum
San Jose, California

Part 1

Perspectives

In '*Perspectives*', authors involved in making decisions about industrial water use and those impacted by their decisions describe their goals and challenges and share their experiences. They are a geographically diverse group representing many different industries, companies, governments and non-governmental organizations, as well as a variety of technical specialties and academic disciplines. As each author discusses a facet of industrial water use, the focus of the articles shifts from the meaning of water to indigenous people and its importance in raising the standard of living in developing countries to the ways in which various multinational corporations work to ensure their use of water is consistent with community needs. The need to prevent industrial water pollution is explained, and the different ways governments can regulate industry as well as how NGOs help restore ecosystems injured by industry.

Although they have different priorities – and in some cases little history of productive interaction – within the covers of this book they openly share their perspectives, achievements and frustrations, as well as their suggestions as to what their counterparts in other sectors might do differently. We hope these articles serve as a springboard for continuing conversations about what each of us can do on our own and in collaboration with others to promote sustainable use of water by industry.

1. **Archambault** *The indigenous perspective on water: A source of life, not a resource*
   To begin, Archambault offers a view of water not as a resource, much less as a commodity, but as described by the Lakota phrase *mni wiconi* – 'the source of life'. To Archambault, industrial misuse of water (and in fact, many of the ills of industrialized society itself) stem from a short-sighted vision that fails to see the reality of the larger world we share with nature and other people. By contrast, he shares the perspective common to many indigenous people, in which humans as individuals are part of a community, and as a community are part of a living network that encompasses 'everything with a spirit or soul, including the stars, plants, animals, air, land, clouds, sun, and water'. In these societies, he claims, individuals mature not by amassing more wealth but by broadening their concern for others, from the individual, to the clan, to the

DOI: 10.2166/9781789060676_0007

community, and ultimately to all. According to Archambault, this fundamental clash between the holistic indigenous outlook and individualistic Western culture often erupts in struggles between industry and native communities over water, as recently expressed in the Standing Rock Sioux protest against the Dakota Access Pipeline.

2. **Alexander and Córdova** *Alleviating poverty through sustainable industrial water use: A watersheds perspective*
   Alexander and Córdova call attention to the role of water as key to a better life for economically challenged communities in developing countries. They point out that water and hygiene are essential to human dignity, warning that, where industries are permitted to use water without constraint, the poor, vulnerable, and marginalized are most likely to bear the brunt. They also note that girls and women suffer most from inadequate access to domestic water supplies, both physically and by virtue of their role as 'water gatherers' where indoor plumbing is not available. CARE partners with several global companies to improve conditions for women through better water management. Examples given include CARE's Water Smart Agriculture program which helps women farmers in Malawi and Ghana increase crop yield while using less water (Coca-Cola); She Feeds the World program which teaches efficient irrigation to farmers in Egypt and Uganda (PepsiCo); and the Water and Women Alliance program which educates women cotton farmers in India at 'the bottom of the supply chain' on water conservation techniques, community leadership, and business skills (Gap, Inc.). They urge industries to consider water stress when siting facilities, minimize water consumption, maximize reuse and modify production to avoid discharging harmful chemicals to the environment.

3. **Santos** *Business transformation as the gateway to sustainability: A tobacco company's perspective*
   Tobacco company Phillip Morris International (PMI) has adopted many of these practices, as **Santos** describes in his article on that company's aspiration to become 'water stewards' on a global level. Adopting the standards promulgated by the Alliance for Water Stewardship (AWS), PMI considers water availability in siting new facilities and in making structural changes to its 38 existing processing facilities. The greatest part of PMI's water footprint, however, is attributed to the 350,000 tobacco farmers that supply its raw material worldwide. In Brazil, for example, which produces 25% of the world's tobacco, PMI collaborates with Hydrographic Basin Committees to identify farmers with the greatest impact on local waterways and encourages them to abandon destructive agricultural practices. PMI helps improve the watershed by annually compensating farmers for reforestation, spring protection and other environmental services. PMI also sponsors a mobile training unit that visits rural areas and trains farmers on good agricultural practices, health and safety issues, controlled use of pesticides, and water sanitation and hygiene (WASH).

4. **Galli and Sottas** *Nestlé: 'Caring for Water' through people, farmers, and communities*
   Food and beverage giant Nestlé also uses the AWS tools to evaluate water risk; like PMI they have discovered that their greatest water use is in their supply chain. According to **Galli and Sottas**, Nestlé pays close attention to water in their food production facilities. For example, their ZerEau technology extracts water from fresh milk for reuse, reducing the overall demand for fresh water in milk processing operations. However, only about 120 million m^3/yr of water is used in processing facilities compared to an estimated 4 billion m^3/yr of irrigation water used by farmers in its agricultural supply chain. Nestlé has focused on three main goals to reduce its impact on water availability in the regions it operates: (1) optimize water efficiency ('More Crop per Drop'); (2)

reduce water withdrawals; and (3) protect and replenish area watersheds. To implement these goals Nestlé's team of 675 agronomists and 5000 extension workers provides education and training assistance to over 100, 00 farmers in 21 countries. They also maintain an ongoing dialogue with suppliers, governments and NGO partners through the 'Farmer Connect' organization.

5. **Herrera *et al.*** *Ecuador's Fruta Del Norte: Early engagement as a tool to build trust*
   Lundin Gold, a Canadian-based mining concern, is similarly committed to sustainable water use. As described by **Herrera et al.**, they developed a unique approach to identifying risks to meet the needs of local communities. When they arrived in Ecuador in 2015 as that country's first large-scale goldmining operation, Lundin Gold faced high expectations from local stakeholders regarding employment, procurement, and environmental management. They initiated a 'comprehensive participatory engagement process' to identify and map community perceptions of the social and environmental risks. Using tools developed by the company, they created a picture of the relationships between stakeholder groups at the local level, and identified a portfolio of concerns related to their operation. This list was translated into a series of 'thematic roundtables' where issues were discussed and agreements reached about how Lundin Gold could mitigate identified risks. Responses enacted to date include water monitoring programs and a biodiversity management program in which flora and fauna were relocated during construction.

6. **Benjelloun and Saoud** *Phosphate mining and the circular economy: Morocco's OCP group's approach to sustainable water use*
   Morocco's nationally owned phosphate mining and processing company – the world's largest – has also worked to ensure that its activities don't deplete or degrade its country's scarce water supplies. As **Benjelloun** and **Saoud** describe, OCP (formerly Office Chérifien des Phosphates) went through a major restructuring in 2006 and took the opportunity to align its operations with the principles of a 'circular economy'. Review of its phosphate mining and purification process, as well as the operation of its fertilizer manufacturing facilities, led the company to focus on three strategies: (1) resource conservation in mining, enrichment, and processing; (2) customization and appropriate use of fertilizers; and (3) processing waste so that it can be used as a resource by OCP and its ecosystem. They constructed a 187 km (113 mi) slurry pipeline to transport crushed rock from inland mines to coastal processing plants, replacing their rail transport operation, reusing the slurry water in the manufacturing process and saving 3 million m^3 of water per year. The company built municipal wastewater treatment plants for three communities to recover and reuse over 10 million m^3/yr of water in adjacent mining and processing facilities, and manufactures an additional 25 million m^3/yr of water at its seawater desalination facility which also utilizes surplus energy from the phosphate manufacturing process.

7. **Schneider and Wolkersdorfer** *Dimensions of water management in the extractive industries*
   Schneider and Wolkersdorfer prescribe a range of actions to mining companies intent on using water sustainably. From exploration and mining design to mine tailings management, mine closure and ecosystem restoration, they provide numerous detailed examples demonstrating how mining companies can include sustainable water use at each stage of the process. Regions can derive long-term economic and social benefits from mining, they argue, but only when mining operators are socially responsible. They add that while European Union countries require companies to guarantee they have access to sufficient funds for post-closure rehabilitation, in practice this

requirement is not always enforced. They observe that the water needs of the public are often subordinated to the demands of the mining industry, and recommend that public health and the long-term health of the ecosystem should be a priority when mining projects are considered, licensed, and monitored.

8. **Groot** *Triple water reuse at Dow*
Chemical processing also takes significant quantities of water, and Dow Chemical demonstrates a range of techniques to 'reduce, reuse and recycle', water at its chemical plants. Groot notes that while some conservation practices (like reuse of steam condensate) are common in the chemical industry, others, like Dow's use of the Global Water (WBCSD) and Aqueduct (WRI) tools to assess the water risk, require additional commitment. Using these risk assessment techniques, Dow identified six locations worldwide that were particularly vulnerable to water-related stress, and focused its efforts on improving water use at those facilities. Improvements included the installation of an online monitoring system that increased cycles of concentration and reduced chemical addition in evaporative cooling towers, and application of advanced treatment technologies (RO, UF, EDR) to treat water for reuse within the plants. Significant water savings were also achieved by developing new chemical manufacturing processes; by oxidizing propylene with peroxide instead of chlorine, Dow reduced fresh water use by 70%. After reducing, reusing and recycling water internally, Dow looked 'outside the fenceline' for opportunities to save water. As a result, Dow co-funded design and construction of a membrane bioreactor (MBR) at the City of Terneuzen's (the Netherlands) wastewater treatment plant and built a 13 km (8 mi) pipeline to convey effluent to meet its chemical plant for reuse, reducing its demand on fresh water by 15%.

9. **Haddad** *Severe water crises: Industry's role and response*
Dow's collaboration with the local utility exemplifies the positive way in which industry can help communities deal with water shortages, especially during times of drought. Haddad suggests that instead of seeing industry only as a water consumer, utilities should consider the role businesses can play in responding to the increasing global scarcity of water, which he traces to increases in infrastructure, technology, population, greenhouse gases, and demand for environmental flows. In addition to its role in keeping the local economy afloat by paying wages to employees, industry can participate in planning exercises, share its water technology expertise, and provide mutual aid to communities in water emergencies. With its technical and financial resources, industry can also participate in public-private partnerships that help communities build water projects which would otherwise be financially out of reach for government-owned utilities. It was in this context, Haddad claims, that the City of Santa Cruz, California (USA) decided to allocate water during its recent drought in a way that had the least impact on commercial and industrial operations.

10. **Umble** *The role of the consultant in supporting sustainable industrial water use*
Umble adds that consulting engineers can help companies use water more sustainably by analyzing industrial processes and helping them design and build treatment technologies that maximize water conservation, recycling, and reuse, He encourages consultants to look at problems holistically in order to identify and deliver effective, sustainable technology solutions, and to design wastewater treatment systems so they produce water that can be reused. Umble envisions the consultant's role as the 'trusted advisor' who performs four basic functions: (1) delivering technology solutions; (2)

minimizing environmental impact; (3) navigating regulatory complexities; and (4) brokering key stakeholder partnerships. In Umble's view, the consultant bridges the gap between company management and regulatory authorities by supplying the business case for sustainable water solutions and proving the appropriateness of new approaches to meet permit limits. He notes that in certain circumstances, the consultant must be willing to lose a client (and fees) rather than provide outmoded solutions that don't contribute to a sustainable future.

11. **Nizami *et al.*** *Government-industry partnership for sustainable water use: Insights from Pakistan*
Based on their experience in Pakistan, Nizami, *et al.* emphasize the role that government plays in protecting public water supplies. The authors report that when industrial water consumption and discharges are not monitored and regulations are enforced weakly (or not at all), industries act like water consumers 'with few responsibilities'. In that case, industrial users behave as economic 'free riders', transferring the cost of their water use to other stakeholders, that is, the public. The authors suggest that first steps to improve this situation should include identifying, licensing, and regulating industrial facilities and metering all industrial water use. They estimate that at present only about 1% of industrial wastewater in Pakistan is treated before being discharged into rivers and drains, even though wastewater from many of Pakistan's industries contains pollutants like detergents, dyes, acids, soda, salts, and heavy metals. At the same time, the authors are mindful of the jobs and other economic benefits industry provides (of particular value in developing countries), and recommend that government work with industry to create 'win–win' solutions.

12. **Ekwanzala *et al.*** *Sustainable solutions to the impact of industrial water pollution on the environment and community health*
Individually and in aggregate, industrial pollutants can have a devastating effect on human health and the environment. Ekwanzala *et al.* identify dozens of inorganic and organic contaminants frequently found in wastewater from twenty different industries, and explain the harm they do to different organs of the human body. Some pollutants pose a threat to health not only when ingested directly but also when present in water used to irrigate food crops. They also note that even relatively non-toxic industrial discharges high in biodegradable organic matter can destroy aquatic life by depleting the oxygen in receiving streams. In addition, heated wastewater from power stations and cooling towers can exceed the temperature range in which aquatic life can survive. On an optimistic note, the authors also provide information on several 'green' technologies that that effectively remove pollutants from industrial waste, including membrane filtration, photo-catalysis, and enzyme-based bioremediation.

13. **Knøchel** *Fit-for-purpose water reuse in the food processing industry*
The importance of clear, appropriate regulations is also discussed by Knøchel, who points out that while many food processors would like to reuse process water to reduce their water footprint, they are stymied by uncertainty about applicable rules. Given the interconnectedness of global trade, the problem is complicated by the fact that food production is regulated by most national governments as well as some international bodies. She maintains that a more streamlined regulatory process would help overcome reluctance on the part of the food industry to invest in water reuse. To support this streamlining, Knøchel suggests the use of a Hazard Assessment and Critical Control Point (HACCP) process to identify and mitigate microbiological risks associated with water reuse in the food industry.

14. **Marcus** *Musings of a former regulator, 'how can we do better?'*
Marcus agrees that governments need strong, enforceable rules to protect the environment and public health from industrial water pollution. At the same time, she observes that some companies see advantages in sustainable water use, and asserts that government regulators should be able to 'incentivize' sustainable practices by developing more flexible regulations based on achieving better overall performance instead of meeting minimum requirements. The proposed trade-off might be, for example, faster permitting of facilities in exchange for enhanced environmental protection. She notes that successful implementation of such an approach would require transparency in monitoring, reporting, and public review, as well as a shared agreement between industry and regulators on what constitutes 'superior performance'. Possible approaches could include reducing emissions significantly below regulatory thresholds, prioritizing important community needs, and developing multiple-benefit projects not prescribed by regulations. Based on her experience at the local, state, and federal levels, Marcus suggests that countries looking for a model regulatory program should consider flexible outcome-based regulations from the outset rather than relying exclusively on regulation of specific contaminants.

15. **Raettig and Muth** *Recovering from disaster: Holding industry accountable for restoration*
When, despite regulations, industrial activities damage the environment, all available remedies must be used to restore degraded ecosystems. **Raettig and Muth** relate the role of the National Wildlife Federation – a major US environmental advocacy organization – in making sure that the billions of dollars of penalties BP paid were applied in ways that actually mitigated the extreme damage caused by the catastrophic failure of the Deepwater Horizon drilling rig that spilled 5 million barrels of oil (780,000 m^{3})) into the Gulf of Mexico. Their first step was to document the extent of the disaster, and to assist media in relaying that information to the American public. Over the subsequent decade, NWF persisted in carefully assessing the impacts of the spill on the entire Gulf food chain, including whales, dolphins, sea turtles, pelicans, and fish. Only then could they motivate the environmental community to lobby legislators to hold BP accountable by levying appropriate fines and penalties, and pass legislation that included a restoration plan providing long-term habitat enhancement.

CONTRIBUTOR SURVEY RESULTS: PERSPECTIVES

As part of the process of preparing this anthology, we also asked each author to complete a survey reflecting their views on sustainable use of water by industry. The survey gave authors the opportunity to report on their own activities, achievements, and challenges in relation to sustainable use of water, as well as their perspectives on useful actions that could be taken by organizations in different sectors to promote sustainable water use. Survey results relating to the topics raised in each section are discussed following the summary descriptions.

We reviewed the survey results by sector to better understand the differences and similarities among the various groups. For example, all industry contributors were keenly aware of their impact on water resources, both globally and locally, directly or through their supply chains. Like **Santos** and **Galli and Sottas,** several respondents reported providing suppliers with training and support in sustainable water management techniques. As one company manager described it, 'sustainable water use is a three-legged stool, resting on the awareness, commitment, and capability of the company and its suppliers'. Another described their company's goal as 'turning a strategic risk into an opportunity to create value' by minimizing their water footprint and adopting an integrated water management strategy throughout their supply chain. In general, survey respondents from industry were proud of their accomplishments, especially their careful assessment of regional water risk, their willingness to engage with stakeholders in their communities, and their ability to apply technical solutions to reduce water use. When asked to identify the biggest problems related to industrial water use, water scarcity topped the list, followed by pollution, which they attributed to a lack of competent wastewater treatment and uneven regulation in many countries.

By contrast, government contributors expressed the view that water seemed an afterthought for many companies, trailing other higher value (and higher cost) concerns. Notwithstanding the importance of water to communities whose supplies are dwindling due to climate change, these respondents alleged that most companies lacked commitment to pollution reduction, water recycling, and reuse. Compared to energy conservation, for example, they saw industry reluctant to invest in water improvements. By the same token, they claimed their own agencies were underfunded and unable to adequately monitor industrial waste discharge. This claim was echoed by environmental and social organizations, who agreed that government rarely seemed to hold industry accountable. As they saw it, not only were regulations lax and enforcement haphazard, they also found the role of government to be ambiguous. One respondent commented that government agencies were more likely to celebrate a company's contribution to the economy than to penalize it for polluting a river or draining a reservoir. Their common complaint was that regulators failed to enforce existing laws. This suspicion touches on the assertion by **Archambault** that Western society is hard-pressed to value water above material gain. Interestingly, utility contributors also blamed a lack of enforcement for problems they experienced with industrial waste discharge, citing the cost of treatment and the lack of appropriate incentives.

Academic respondents proposed the need for a paradigm shift to a circular economy, where industry closes production cycles by extracting water and energy for reuse and eliminating waste altogether. Sustainable water use should be practiced at the catchment level, they suggest, as industry engages with the community, motivated by appropriate regulations and incentives. They cited research that promises new, more efficient solutions to water treatment problems. Consultants represented in the survey took a different tack, some arguing that corporations often 'spend too much time looking at the big picture, while the real risks are local'. They held that industry undervalues water and fails to take advantage of existing water conservation technologies, reinforcing the position of **Umble** that consultants must not only present their clients with sustainable solutions, they must also explain their value to the company. Survey participants from all sectors recognized the need to maintain adequate supplies of clean, safe water in the face of climate change and population growth. (**Haddad** offers a more comprehensive list of threats to the global water supply.) They also agreed that while much work remained to be done, industry can't do it alone – that it will take the combined efforts of government, environmental, and community advocates, and consumers to motivate industry to use the tools now available to help them use water more sustainably.

Chapter 1

The indigenous perspective on water: A source of life, not a resource

Dave Archambault

Brave Fox, Inc., Cannon Ball, ND, USA

Keywords: culture, environmentalist, Grand Mother Earth, indigenous, infrastructure, not a resource, respect, source of life, Standing Rock, water

1.1 INTRODUCTION

It's natural for human beings to take what they have for granted. In my neighborhood, we wake up and turn on the lights, go to the bathroom, wash our faces, brush our teeth, shower, turn on the stove, make coffee. As residents of the developed world, we consider it a human right to have instant access to clean water and electricity. They bring us comfort, and play a role in everything we consume, including food and clothing. We know we could live without electricity, although it would mean the end of our comfortable lifestyles. But without water, there would be no life.

Is it possible to run out of water on a planet whose surface is 71% water? The Earth has a way of recycling water, giving the impression that we will have enough water forever. However, when you observe the changes happening with our water supplies and climate, it is obvious that nothing is guaranteed. We must be mindful of how we sustain water.

My name is Dave Archambault; my Lakota name is *Tokala Ohitika* which means 'warrior from the kit fox society'. I grew up on two Lakota reservations in the United States: Pine Ridge or Oglala Sioux Reservation and Standing Rock Sioux Reservation. In 2016, when I was Chairman of the Standing Rock Sioux Tribe, I found myself in the middle of a movement. The Tribe opposed the Dakota Access Pipeline because it crossed our treaty lands and went under Lake Oahe, one of our main sources of drinking water. The movement drew thousands of water protectors to our community and gained global media attention. As I reflect on the movement, I realize it was a moment of unity in Indian country. It was the first time in modern history that there was a coming together of tribes from over 300 nations (Figure 1.1) to stand up for our traditional way of seeing the world around us.

DOI: 10.2166/9781789060676_0015

Figure 1.1 Indigenous people representing 300 tribes gathered on the Standing Rock Sioux reservation with others to protest construction of the Dakota Access Pipeline. (*Credit: Frank White Bull*)

In my early years on Pine Ridge Reservation I was fortunate to live with family members who practiced our ancestral ways, so I learned to pray. I became familiar with prayer, and the connection with all beings in prayer – not taking anything for granted, completely respecting all. Today I realize that my specific ancestral way of thinking and believing is shared by many indigenous communities throughout the world. Whenever I meet Indigenous People, I hear a lot that sounds familiar. Where I visit indigenous communities – the Kayapo in Brazil, the Mapuche in Chile, the Andean in Peru, the Wangan and Jagalingou in Australia, the Maori in New Zealand, First Nations in Canada, Gwich'in Alaska, Swahili in Africa, and Sawau in Fiji – I find areas rich not in monetary wealth but in cultural wealth, with a deep understanding of the universe, of the world, of community, family, and youth. Some common beliefs I find particularly interesting, like the idea that you have two parts to yourself, your physical body, and within that body your spirit, so that even when your body no longer lives in this world, your spirit continues. Another common belief is the notion that water is more than a resource: it is a source of life.

1.2 INDIGENOUS AND MAINSTREAM THINKING

Despite subtle differences among indigenous cultures, there are a number of important similarities. It's easy to distinguish Indigenous People from those who follow the Western or European influenced 'mainstream' way of life. Indigenous Peoples are connected to the land and continue to pass down the teachings from their ancestors on how to care for the land. Where Westerners typically think in terms of self or individualism, Indigenous Peoples think in terms of community. Perhaps the most important divergence from mainstream thought is the indigenous belief that Nature – everything living and non-living in the surrounding environment – transcends all. This means that the role of humans is not to rule Nature, but to be a part of it. This and other differences are illustrated in Table 1.1, below, which compares some of

Table 1.1 Comparison of indigenous peoples' and mainstream way of thinking.

Indigenous Way of Thinking	Mainstream Way of Thinking
Communal	Individualistic
Circular	Linear
Nature	Humans
Past Present Future	Future
Connection, related to all things in universe	Life is focus, advance to top
Oral Stories	Written history or documents
Eye contact is overly assertive	Eye contact is part of the conversation
Mystical	Scientific
Praise the group	Praise the Individual

the core beliefs of indigenous communities and mainstream populations, which also influences how these two groups think about water.

All human beings are built the same. We all share the same patterns of physical and mental growth, and we go through these stages of life for the most part not realizing what we are doing until the time we return to Mother Earth. We live much of the time in our minds, hearing our own thoughts, and we long for acceptance by others. In the mainstream thought process, if we do well for ourselves, we will be accepted. In the indigenous thought process, we must do well for others, and acceptance will come. So even though we all crave for our egos to be stroked, the outcomes and impacts are quite different when we strive for our own gain rather than for the benefit of all. Over time, as mainstream thinking continues to expand around the world, so does its impact: individual convenience has become both the norm and the priority.

In the same vein, the word 'all' to the indigenous person does not mean mankind alone. The word 'all' applies to everything with a spirit or soul, including the stars, plants, animals, air, land, clouds, sun, and water. To be connected to 'all' means to be part of something that is greater – greater than oneself, greater even than human beings. Wherever mankind has focused too narrowly on its own perceived self-interest, other beings begin to die off, so the more we take and destroy for our own gratification, the more, over time, we threaten our own survival. To develop for mankind alone necessarily damages and disrupts all other beings. By contrast, the indigenous way of thinking offers a broader perspective that's consistent with both self-gratification and with preserving the environment that we depend on for existence.

In the culture I grew up in, our ceremonies were performed in a circle, reflecting the four stages of life:

- Starting at the bottom, the first quadrant represents our first stage: the baby, the toddler, the child. Immature socially and physically, we think only of ourselves. In this stage, it might be said, we exhibit all the characteristics of the mainstream way of thinking.
- The next quadrant represents the second stage: the young adult. At this stage, we begin to think beyond ourselves, as we also care about a pet, a boy/girlfriend, a spouse, or a child. In this stage of maturity, we recognize that someone or something is more important than our individual selves.
- The third quadrant represents adulthood. At this time you think not only about those closest to you but also the environment or community that you are a part of. In this stage, you want to make a better place for all.

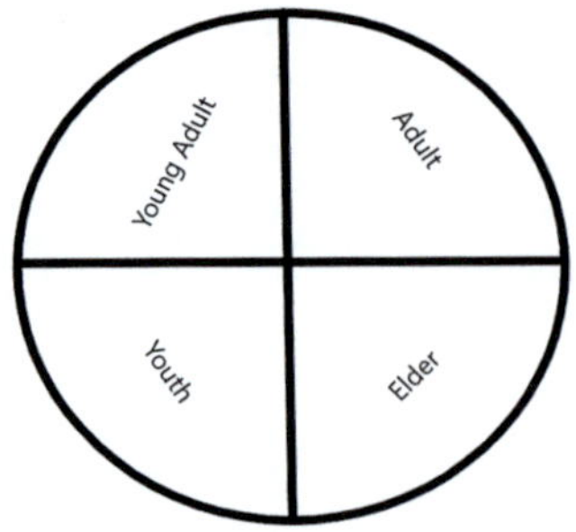

Figure 1.2 Indigenous 'circular' thinking about human society.

- The last quadrant represents the elder. In this stage, you seek world peace. You think not just of the environment but the universe, and the spiritual connection that binds all together. At this point, as we look beyond ourselves we perceive that connection with the same immediacy with which we experienced the world as an infant. (The baby and the elder are said to be closest to the spirit world.)

Our progress, as we mature through life, is illustrated by the diagram in Figure 1.2.

In the mainstream way of thinking, life is not a circle but a straight line, as illustrated in Figure 1.3. This way of thinking views time as linear, and the eyes are focused on some idealized future of the self. The focus is on me and (perhaps) mankind, rather than us all or the universe. The problem is that when you only think of benefits for mankind and self, all others pay the cost.

From the indigenous perspective, water is not a just a resource: it is a source of life. There are four sources of life, and without any one of them, life will not exist on this planet: Water, Air, Land, and Sun. In Lakota, the phrase '*Mni Wiconi*' means 'Water is a source of life'; or to put it even more briefly, 'Water is life!' Many freshwater deposits are starting to vanish – glaciers, lakes, and aquifers. If land developers viewed water as a source of life rather than a resource to exploit, perhaps their projects would plan for sustainability rather than simply consumption. With proper planning and appropriate use of technology, we can secure the source of life for future generations, but only if we are mindful of our actions. If we continue to take this source of life for granted, it will continue to diminish, and the next wars will be fought over water rather than oil.

1.3 A HISTORY OF TAKING

The difference between the indigenous and the mainstream view of water is not just philosophical; it has real world implications that can result in serious conflict. This conflict has been acted out repeatedly on the national stage in the United States, throughout its history and up to the present day. As an example, the recent and highly visible conflict over the construction of the Dakota Access Pipeline (DAPL) can be shown to stem directly from the historic conflict between the mainstream Western view of the world and the more holistic perspective of the Indigenous People in this country.

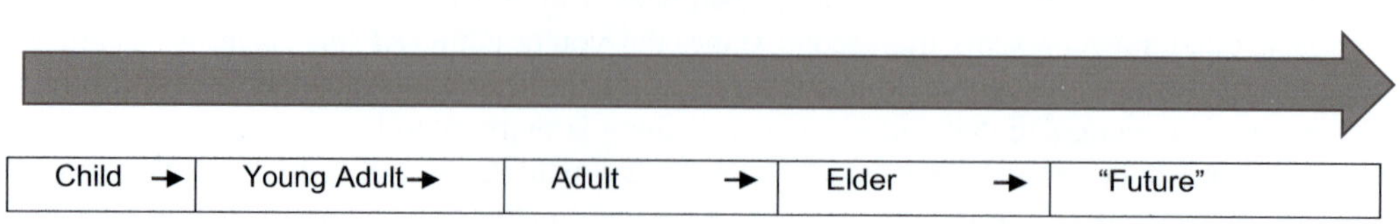

Figure 1.3 Mainstream 'linear' thinking about human society.

For over 500 years, the Standing Rock Sioux Tribe of the Dakota/Lakota people and all the other Great Sioux Nation tribes have warned that the benefits associated with mainstream thinking came with a cost borne by the Tribal Nation, altering the indigenous way of life and endangering the natural world and all that entails. Our creation story includes the buffalo: 'A man changed himself into a buffalo to help our people survive'. In 1803, it was estimated that 70 million buffalo roamed the great plains of America and Canada. The buffalo was our economy, our food, shelter, tools: our everything. Then the railroads arrived – the first infrastructure project to encroach on our human rights – and nearly exterminated the species. Hunters rode railcars across the plains, shooting buffalo for sport, and as a strategy to eliminate the Indians and take their land. By 1889, there were fewer than 100 buffalo roaming the plains. When the railroad came, our way of life changed forever.

After killing off the buffalo, in 1851 the US government ratified the Horse Creek Treaty, securing for the Great Sioux Nation the rights to 60 million acres to reside upon forever, without any more interference. Before the ink was dry, however, that treaty was broken when gold was discovered in South Dakota's Black Hills, the heart of the Great Sioux Nation. To get the gold to back its currency, the US Congress passed new laws that carved the Black Hills out of the Great Sioux National Lands, eventually placing the indigenous population on reservations. Not US citizens, their treaty rights revoked at will, my ancestors, the Indigenous People, at the time were considered less than human. Mainstream society failed to understand the dignity of our cultures, our spiritual beliefs, and our interconnection with the natural world. And despite our courage and strength, the physical power of the indigenous population was further weakened by breaking the Great Sioux Nation into individual tribes, like Standing Rock.

Along with the treasured gold, land was also considered a resource to exploit. Throughout the late 19th and early 20th centuries the US government routinely opened up the best land on the reservations to farming by non-indigenous homesteaders. Out of the 2.3 million acres on the Standing Rock reservation, fewer than 1.3 million acres remain reserved in government trust for the tribe and members of the tribe. (See Figure 1.4.) Out of the desire to exploit 'land resources' for the sake of monetary profit, the US government continually reneged on its promises and took land it had promised to the Tribes 'forever'.

And the taking continues, this time inspired by another '*resource*': water. In 1944, Congress approved a massive infrastructure project which resulted in construction of seven dams on the Missouri River. At the time considered the largest dam project in the world, the dams and reservoirs were strategically placed to maximize the flooding of Indian reservation lands. They were created for hydroelectricity, for flood control, for recreation – or more broadly, for money and wealth. This was done without regard to the great and enduring costs to those of us connected to the land, and to our ancestors who, when forced onto the reservation, had settled on the river bottoms where the water, game, fruits, and wood were most plentiful. By 1958 this land was under water, a result of the Oahe Dam, and once more we were forced to move, abandoning the fertile and protected bottomlands of our ancestors for the harsh and windswept uplands. Once more the clash of the mainstream and indigenous mindsets forced us to alter our way of life.

1.4 THE LAST STRAW: DAKOTA ACCESS PIPELINE

The history of infrastructure projects is a history of transferring benefits from indigenous populations to the nation, with the US government as the perpetrator that abuses Indigenous Peoples. This history has repeated itself time and again. But recently, a point came when the members of Standing Rock were no longer willing to accept wrongs committed against not just us but against all, including the ones that don't speak our language: the ones that crawl or fly, the ones with four legs, the ones that blossom or bloom, the ones that swim or slither, the ones that sway in the wind. From that perspective, perhaps what happened at Standing Rock was inevitable. At some point, there was going to be a stance taken by a concerned

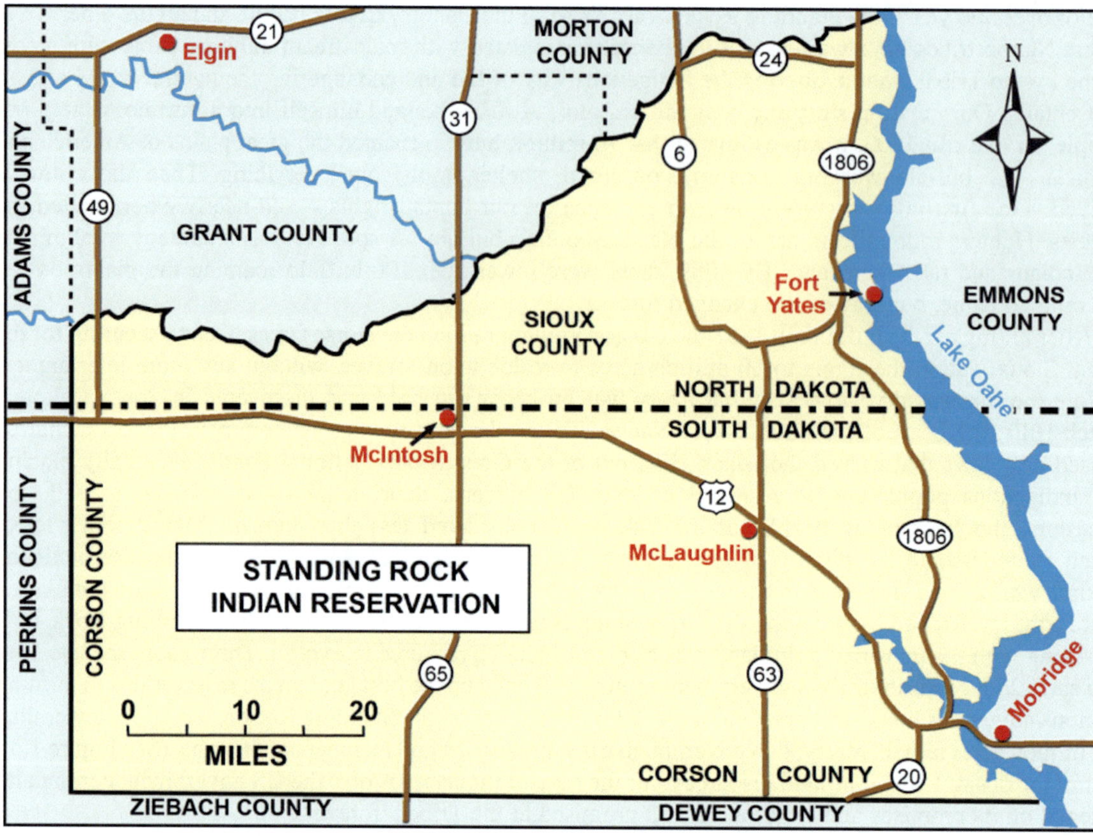

Figure 1.4 Standing Rock Sioux Reservation, North and South Dakota, USA. (*Credit: State Historical Society of North Dakota-North Dakota Studies. https://www.ndstudies.gov*)

indigenous population against the government and corporations; the resistance to this one particular pipeline just happened to take place at Standing Rock, where an oil company planned to construct the DAPL across our land and under our water.

The Tribe opposed the DAPL because it crossed our treaty lands and went under Lake Oahe, one of our main sources of drinking water, and the very same lake whose construction by the US government 70 years earlier had displaced our grandparents from their homes. The movement drew thousands of water protectors to our community and gained global media attention. As I reflect on the movement, I realize it was a moment of unity in Indian country when tribes from over 300 nations came together to stand up for their rights, and the rights of the water and the land. We received prayers and letters of support from Indigenous Peoples across the country and around the world. Our message of *Mni Wiconi* – water is life – resonated with other Indigenous Peoples who were also fighting to protect our Mother Earth (Figure 1.5).

We had campaigned against the project from the time we first heard about it, even though in June 2014, when Energy Transfer Partners announced their plans to build it, the planning and scoping of the project was already complete. According to federal law, the process for massive infrastructure projects requires that the owner first get permission from all the federal agencies and the states that the project will impact. After those initial approvals are granted, the exact route is determined, and additional permits and approvals are

Figure 1.5 A campground at Standing Rock during the 2016 demonstration against the Dakota Access Pipeline. (*Credit: Frank White Bull*)

obtained but nowhere are Tribes/Indigenous Peoples contacted or consulted, much less asked for their consent. Only after all agencies and states approve the project does the permitting federal agency consult with the impacted Tribes, and these consultations are designed to be *pro forma* – an insignificant item on a procedural checklist.

Nevertheless, we wanted to have a say, and to be given the opportunity to decide whether or not this infrastructure project could happen. We asked for in-depth study of the impact DAPL might have on our way of life, not the conclusory environmental assessment written for the company by a former general of the US Army Corps of Engineers that barely glanced at the facts and determined there would be 'no impact to environment'. We knew that this was not true, and that a deeper study needed to be conducted. We knew it was not true because in the past, every infrastructure project has had a negative impact on our way of life.

We argued the laws in federal court, and the federal judge ruled against us, just as the court has ruled against us for centuries. Just as the court ruled against the heirs of Thomas Johnson in 1823, when they contested a government land grant to William M'Intosh on the basis that Johnson had previously purchased the same land from the Piankeshaw tribe. In that classic case (Johnson v M'Intosh) the US Supreme Court upheld M'Intosh's grant by invoking the 'Doctrine of Discovery' which declared that all 'discovered' lands belonged to the government in whose name they were claimed. (It's worth noting that this doctrine, a joint invention of the Catholic Church and European kingdoms, was itself based on the premise that since the Indigenous Peoples who lived on this 'discovered' land did not believe in Christ, the conquistadors were authorized to take their land – and to maim, mutilate, and kill these sub-humans – in the interest of wealth.)

Despite this legal history, we at Standing Rock felt compelled to pursue every possible angle to try and stop the Dakota Access project. We didn't want our descendants to be able to say, 100 years from now, that we didn't try so when our youth spoke out and said 'We don't want that pipeline to go under our precious

Figure 1.6 Authorities block a road on the Standing Rock Sioux Reservation during demonstrations against the Dakota Access Pipeline. (*Credit: Frank White Bull*)

river', our elders held a ceremony. In the ceremony, the spirits' guidance was this: With peace and prayer you will stop the pipeline; with violence and arrest, the pipeline will go under the river.

We had no idea when we began our peaceful opposition to the DAPL that people from around the world would come in support. When people came, we asked them to come in peace and in prayer; no alcohol, no drugs, no violence, and no weapons. The spirit of the movement was most powerful when those that came adhered to our request. We had the power of prayer on our side; the voices of our ancestors permeated through to today's generation. Like many things grounded in a spiritual understanding, the indigenous opposition to the Pipeline was more than words – it was a deeply felt and abiding feeling. Imagine an oppressed population being liberated for the first time in over 200 years and finally understanding the teachings, the way of life, the way of thinking that was passed down for centuries.

Despite the Tribe's opposition, the DAPL was built. Unfortunately, the Tribe will be the ones impacted when there is a spill and so we continue the fight. Tribes across the country continue to face development projects that impact their treaty, land, and water rights. These projects include Keystone XL, Line 3, and the Liberty Pipeline. Our concerns about these projects are not going away (Figure 1.6).

1.5 THE INDIGENOUS PERSPECTIVE ON WATER: A FAMILY MATTER

I get tired of people saying, 'get over it!' meaning that we should 'get over' the past. They don't understand that the past continues to repeat itself – that federal law continues to be an impediment to the interests of Indian tribes, and our ability to protect our lands, waters, and way of life The major difference from the time of Johnson v. M'Intosh until today is the identity of those now partnering for greed and wealth to our detriment. The laws that are enacted by the US Congress are supposed to be in the interest of the people, but when corporations lobby and fund the campaigns of politicians, federal laws favor business and wealth over the universe, planet, environment, and human rights.

The Lakota teach that water is a source of life and that we are all related. From the business perspective, on the other hand, water is a resource. Yet there may be a path forward that can bring the indigenous view into business and development decisions.

As I look out onto the prairie, I see and hear movement – the grasses sway, the trees dance, a deer flips her tail in alertness, and a bird chirps as the clouds float by. There is movement all around me. In the Lakota way, when we see movement, we say *taku skan skan*, which means there is something out there. If something moves, then it has a *nagi*, a life force or spirit. If something has a *nagi*, then in our way, we say *mitakuyasin*, or all my relatives. We are related to it. And if we are related to it, then we should treat it with respect, just like we would treat our own mother or child.

We also learn that there are four sources of life: water, air, sun, and earth. Not only do we demonstrate respect and love for our relatives, but we do the same for these sources of life, with the understanding that they too have a *nagi*. These sources nourish life so that our *nagi* can grow and mature within our bodies. They are part of *Tunkasila*, or the Creator. *Tunkasila* is not one being. *Tunkasila* is all around you. With every prayer and every ceremony, we give thanks to these four elements. We treat them with respect and understand that we are all connected and rely on each other. In this way, our ancestors understood sustainability before the term 'sustainable development' ever existed. Today's environmental problems and the mismanagement of water arise from the mistreatment of our relatives. Corporations and governments exploit our relatives because they do not recognize how we are all connected. They extract and transport oil, take fossil fuels from Mother Earth, manufacture and create products that pollute the air, and exploit water, treating it as a resource rather than a source of life.

Industry is blind to the environmental and social impacts of their projects because their decisions are based on money. Money does not move. Money does not have a *nagi*. We are not related to money, and it does not provide a source of life. Money is fictitious, and yet people make decisions based on whether they will increase or decrease the money they can obtain. People are so focused on money that they forget to respect the very things that give us life. They continue to drift, even as there is growing recognition that our environment is dying because of our actions.

This money-focused perspective was evident in the DAPL controversy. In September 2014, a few months after the announcement of the DAPL, the Standing Rock Sioux Tribal Council met with representatives from Energy Transfer Partners, the principal company behind the project. At this meeting, we expressed our concerns about the pipeline because it crossed our treaty lands and had potential impacts on our relatives, including our water, our burial grounds, and other sacred sites. The representatives listened but they did not internalize what we told them. I wonder if the investors behind the DAPL would have made the same decisions if they had fully listened to the Tribe and meaningfully brought our perspective into the process. Would they have more deeply considered our concerns about the crossing of our treaty lands and rerouted around the lake? If they had, they would have saved money. Their failure to fully consider the environmental and social impacts of the project led to a six-month delay the resulted in numerous costs. We estimate that the project cost 70% more than originally anticipated – over $7.5 billion. In addition, the banks financing the project lost at least $4.4 billion in account closures, and taxpayers and local stakeholders suffered costs valued at least $38 million. These are very real costs that stemmed from poor project management and a failure to account for the project's social risk.

1.6 BUILDING PROJECTS, BUILDING COMMUNITY

I am not against development, I am not anti-pipeline, and I recognize the need for water projects. However, when governments and companies give Indigenous People a voice at the table, they will only strengthen their projects, mitigate social risks, and prevent unnecessary costs. My hope is to see more water management projects that are led by Indigenous People and centered on community needs. This is consistent with human rights principles found in international instruments, including Article 32 of the United Nations Declaration on the Rights of Indigenous Peoples, which states that 'states shall consult and cooperate in good faith with the Indigenous Peoples … in order to obtain their free and informed consent prior to the approval of any project affecting their lands or territories and other resources, particularly in connection with the development, utilization or exploitation of mineral, water or other resources'. This is also aligned with Goal Six of the UN Sustainable Development Goals to ensure access to water and sanitation for all.

The Bell Waterline Project is a great example of a water project that incorporated the indigenous perspectives. It was pioneered by Wilma Mankiller, the first female chief of the Cherokee Nation. Prior to becoming Chief, she worked with the community of Bell, a small village on the Cherokee reservation in the 1980s. She asked the community what they most wanted, and their response was 'access to running water'. She helped the community build an 18-mile-long water distribution system and required that they contribute their own time and labor to build the project. Her work is a model for how water development projects can be successful when they are consensus-driven and community-led.

It is going to take a generation who cares to save our existence. *Tunkasila* and all our relatives will recover in time. Mother Earth will always find a way to heal and to recover. However, if we continue down this path of exploitation and destruction, we will continue to hurt our sources of life and to threaten our existence as a species. Instead of seeing this opposition as a problem, I hope companies will begin to see it as an opportunity: an opportunity to learn more about how their projects are impacting our way of life and an opportunity to bring indigenous values into their planning, which will save money and prevent the negative environmental, social, and historical impacts to Indian communities. We are all related, after all.

1.7 AFTERWORD

Over 300 tribes came to Standing Rock to pray for Grandmother Earth. The federal government and the corporation ignored us, giving Grandmother Earth no choice but to protect herself. At the time of this writing, we now must self-isolate, create social distance, and quarantine. We are forced to spend time with our loved ones if we want to protect them. In short, people all over the world have an opportunity to re-learn what family is, and that family is more important than money. From this perspective, COVID-19 can be seen both as a cleansing and a warning – Grandmother Earth speaking again: '*You can no longer just take, you must give back*'. Indigenous People around the world have for generations faced extinction from epidemics and armies, but we are resilient. We will make it through this pandemic, too, for we have always understood that we are part of something greater than ourselves. Now everyone, for the moment, is forced to acknowledge the need to reciprocate the nurturing of our Grandmother Earth and to explore renewables and sustainable living.

At the same time, in March 2020 the District Court of the United States for the District of Columbia ordered the Army Corps of Engineers to prepare a full environmental impact statement on the DAPL. Just as we had argued, the Court ruled that the original DAPL environmental review did not adequately consider how an oil spill would affect our tribe's water, our fishing and hunting rights, and would further impact the environment. Oil began flowing through the pipeline in June 2017; a decision whether to shut down the pipeline while an environmental study is being completed is expected sometime later in 2020.

Chapter 2

Alleviating poverty through sustainable industrial water use: A watersheds perspective

Kelly Alexander[1] and Andrés Córdova[2]

[1]CARE USA, Food and Water Systems, Atlanta, USA
[2]CARE Ecuador, Project Coordinator, Quito, Ecuador

Keywords: conservation, equity, gender, partnerships, sustainable, vulnerability, women

2.1 INTRODUCTION

CARE's mission is to save lives, fight global poverty, and promote social justice. Since clean water and sanitation are necessary for both human and economic health, many CARE programs throughout the world include a focus on water. This chapter focuses on the water-related aspects of CARE's Food and Water Systems program, which works with communities, industry, and government to address water supply and service provision, water quantity and quality, provision of sanitation and hygiene infrastructure, preservation of watersheds, and water resource governance. In particular, this article highlights the ways in which CARE's holistic watershed approach supports improved governance to protect vulnerable communities from uncontrolled use of water resources by industry. Particularly in the context of a changing climate, where water shortages are increasing, there is a need for a focus on climate justice and adaptation to ensure that vulnerable populations, and particularly women and girls can continue to have access to safe water.

2.2 CARE'S COMMITMENT TO REDUCING GLOBAL POVERTY: WHY WATER?

2.2.1 History and mission

Cooperative Assistance for Relief Everywhere, known simply as 'CARE', is one of the largest and oldest humanitarian aid organizations focused on fighting global poverty. CARE was founded in 1945 as an

DOI: 10.2166/9781789060676_0025

impartial, non-governmental organization (NGO) that distributed CARE packages to post-war Europe. Today, CARE works to tackle the root causes of poverty, with a particular focus on empowering women and girls to lead their communities out of poverty. CARE takes a holistic and multi-sectoral approach to poverty reduction that includes improving nutrition and food security; increasing access to safe, sustainable water supplies; strengthening health systems, particularly through sexual and reproductive health; expanding educational opportunities; and challenging negative social norms by not only empowering women and girls, but also engaging men and boys as partners and champions. In 2019, CARE had programs in 100 countries, measurably impacting the lives of over 100 million people.

As a non-profit organization, CARE receives funding from a wide variety of donors, with government agencies, such as USAID, being the largest source of funding.

2.2.2 Water, gender, and vulnerability

We all use water every day; it is essential for our health, our hygiene, and our basic dignity. We use water to cook nutritious food as well as to clean our bodies, our clothes, our homes, and the faces of our children. Water is used in every aspect of our lives, from farming to industrial production. However, when access to clean, safe water is curtailed due to diminishing freshwater resources, mismanagement or unfair distribution, water can become a barrier to social progress and a major contributor to inequality.

CARE places an emphasis on women in its water programming because impoverished women are often excluded from decisions regarding water allocation, use, and management. Like people with disabilities and other vulnerable populations, women and girls suffer disproportionately when water is scarce or hard to access. As the task of collecting water generally falls to them, they risk gender-based violence and sexual assault when water sources are remote or unreliable (Figure 2.1). Without access to safe, clean water, even normal occurrences like menstruation and childbirth can be harmful or fatal to women (Cheng *et al.*, 2012; Thompson *et al.*, 2011). This risk is only heightened by the twin forces of climate

Figure 2.1 Asmarech, mother of six, collects water in the highlands of Northern Ethiopia. She and her daughters have to make several trips a day to carry the water needed in their home. (*Credit: CARE, Josh Estey 2013*)

change and rampant environmental degradation that stress already-fragile ecosystems in many parts of the developing world, further reducing the availability of groundwater and rainwater and exacerbating the challenges that women and girls face. CARE's programs work to increase access to high quality water supplies, so that women and girls spend less time finding, treating, and hauling clean water, less time caring for family members sickened by water-borne disease, and more time participating in income-generating activities. Access to clean water also means that instead of collecting water, girls can attend school. Girls are more likely to stay in school or participate in collective action and advocacy when water – and clean, private latrines – are reliably provided.

CARE encourages equal participation in decision-making for women by including them in discussions on the establishment and management of water resources at all levels. If they are to serve the entire population, governments can no longer listen only to educated male leaders. Women, the poor, and other vulnerable populations must also be able to exercise their voices and influence policy. In fact, involving women in community decision-making about environmental resources leads to more sustainable solutions that are more likely to conserve water for future generations and to share water resources with others in the watershed (Elwell N. & Williams Y. 2018). For example, CARE has observed that when women engage in water resource decisions, water pollution decreases, water efficiency increases, and water-related ecosystems are more likely to be restored. A recent World Bank report described how inequalities in access and use of water and sanitation facilities can represent inequalities in other aspects of a society and that, conversely, improving gender equality can result in improved equality of water use and other water, sanitation, and hygiene (WASH) outcomes (Das, 2017). The report suggests that if women and disadvantaged groups become more involved in decision-making and public community committees, such as water committees, they will better articulate and demand their rights to water and other social services.

Developing and maintaining reliable water supply not only increases economic and educational opportunities for women and girls, it also reduces their risk of experiencing violence or assault, and leads to more sustainable and equitable management of water resources for everyone (Figure 2.4). At the same time, good governance of water resources can only be achieved by providing women the opportunity to make their voices heard, by hiring them, placing them in positions of influence, and listening to their advice.

2.3 DEVELOPMENT OF CARE'S WATER+ PROGRAM

With respect to water, most NGOs have traditionally focused on WASH programs dedicated to providing clean water for drinking and domestic use; installing latrines or toilets; and educating the community about proper handwashing, menstrual hygiene, and food safety. Access to water and sanitation is a human right, but one that far too many people cannot realize. Indeed, the Human Right to Water and Sanitation was recognized by the UN General Assembly in 2010 (UNGA, 2010).

Without disputing the critical importance of WASH programs, CARE has become increasingly aware of the need to address water issues from a watershed perspective. To this end, in addition to the quantity and quality of water available, CARE considers sustainability, productivity, equity, and the resilience of the political and socio-ecological environment in which our communities live. Looking at water this way changes the kinds of questions asked, which in turn changes the kinds of problems solved. House by house, for instance, people may continue to worry, '*Will I have enough water for my family today?*' and '*Will my children get diarrhea from drinking it?*' But from a watershed perspective they ask, '*Will our community have enough water 15 years from now to irrigate our farms and feed ourselves?*' More pointedly, they will also ask: '*Who else is drawing water from the river or ground in this area?*', '*How much water are they using?*' and '*How is climate variability impacting water availability?*'

Figure 2.2 Niva (center) sits with her husband Agrepino (left) looking at the glacier that is the main water source of the village Vilcacoto. Niva, president of her community, explains that as the glacier melts the community will lose their main source of water for agriculture. *(Credit: CARE, Zak Bennett 2014)*

One concrete example of this is demonstrated by the complex situation in Peruvian communities that survive off the glacier-melt from the mountains (Figure 2.2). The melting of glaciers is seen from two perspectives: as a risk when rocks and ice fall and devastate towns (and diminish water reserves), and also as an opportunity for hydroelectric power and increased use of water for irrigation and fish farming. The management of the water resource, which is a multi-sectoral process, must recognize the benefits, challenges, and varying perspectives and effects involved in the watershed, and the quantity and timing of the water flowing through it.

In Latin America, CARE's programs focus on water conservation, watershed protection, and restoration to maintain water sources into the future. For more than 20 years CARE has viewed domestic water supply from a watershed perspective, understanding that to build domestic access to water, we must also strengthen watershed protection and water conservation. This beginning-to-end awareness may have been inspired by the fact that mountainous areas of Latin America often create smaller, well-defined watersheds (*cuencas*) where it is easier to identify partners who all share an interest in the same water source. By contrast, identification of stakeholders can be more complicated in areas of Africa and Asia where watersheds are larger and more widely shared.

It is in this broader context of water resource sustainability that CARE works. CARE uses the term 'Water+' to recognize that WASH, Water Smart Agriculture, integrated programming and improved water governance are all essential aspects of ensuring the availability of safe and adequate water for a community. To provide water for a community, it is not enough to drill a single borehole. The community also needs to be able to operate and maintain the well, with access to technical staff to repair and maintain the whole system. In addition, the community and government both need to be able to protect the watershed from contamination so that each borehole can continue to serve as a reliable water supply. Within this wider view, CARE's work on water is not just about digging wells, but also about ensuring the welfare of the community, the involvement of all stakeholders (including private and public sector actors), and the environmental health of the surrounding watershed over time.

2.3.1 Water smart agriculture

CARE began its work to promote 'Water Smart Agriculture' in 2010. The techniques promoted in CARE's Water Smart Agriculture Program are not new, but instead draw on simple, proven agricultural practices that support healthier soils and agro-ecological methods that require less water. As an example, in Mali, Water Smart techniques – mulching, composting, digging zai pits (to catch water and concentrate composting) and intercropping – have resulted in a range of social, economic, and environmental benefits, including reduced need for watering with higher crop yield. Where small-scale women farmers used to water their plots twice daily, they can now produce a comparable yield by watering only once every other day. By saving water, they also avoid the 30-minute trip to collect water, giving them more time to care for their children, generate income, and rest from their labors.

In Malawi and Ghana, women farmers in CARE's Water Smart Agriculture program have experienced similar increases in yield and income on fields where less water is used, and where no artificial fertilizer is applied. In each of these countries, CARE works closely with government representatives to ensure alignment with local policies, and to expand the program through government-run agricultural extension systems and wider natural resource management policies and approaches. Water Smart Agriculture has even enabled some women to farm in areas no longer thought to be productive. CARE is now working with Sahel countries to reclaim degraded land and transform it to provide critical ecosystem services such as increasing biodiversity. Local governments all across this semi-arid region of north-central Africa, south of the Sahara Desert, are witnessing first-hand the regeneration of degraded lands, and are beginning to understand the value of women farmers who use Water Smart Agriculture techniques. As farmers report less consistent rainfall patterns, increasing droughts, and rising temperatures, Water Smart Agriculture helps communities adapt to a future with an unstable climate by increasing the productivity of land in the midst of water scarcity.

2.3.2 Industrial water use and water governance

CARE works to raise awareness about the impact of water management on impoverished communities, working closely with government partners. It takes the authority of government to create and enforce regulations that ensure everyone in a watershed is using water wisely and disposing of it properly. This is particularly important when working with industry and large corporations. Industries often purchase or rent land for the sole purpose of extracting and using water. When companies are granted rights over land where indigenous peoples live, whole communities are put at risk. In Ecuador, for example, CARE has worked to protect the rights of vulnerable groups in the Amazon by facilitating discussions with governments, industry, and local indigenous peoples to make sure their voices are heard. Although CARE itself does not generally engage directly with industry to influence their practices, our teams seek to shape the policy environment in which they operate through advocacy with government and collaboration with vulnerable populations.

In addition to working with governments to promote fairer allocation of water resources used by industry, CARE works with a number of multi-national companies to improve their use of water. Recognizing that economic success can come at an enormous cost to the regions where they operate, these companies, discussed further below, are paying more attention to the watersheds they draw from, and the communities that depend on them. In some cases, industries are setting aside 'water recharge zones' within a watershed in order to decrease their impact on water availability.

2.4 CARE'S INDUSTRIAL PARTNERSHIPS

CARE's Food and Water Systems program receives significant funding support from corporate sponsors. These companies are concerned about water use and waste disposal, and they work with CARE (and other NGOs) to improve the lives of the people working in or near their factories. Water is the primary ingredient in soft drinks, and three times more water can be used in manufacturing and bottling the product – so international beverage corporations know they must do more to protect the water supplies in the countries where they do business. Fashion brands who make and sell cotton clothes also understand the enormous levels of water used in agricultural and factory production of their products. Programs sponsored by three CARE partners – Coca-Cola, PepsiCo, and Gap, Inc. – reflect this awareness in ways that are important to the communities CARE serves, as shown below.

2.4.1 Coca Cola: Replenish Africa Initiative (RAIN)

The goal of Coca-Cola's RAIN program is to replenish every drop of water used in their product and provide 6 million people in 35 African countries access to safe water. Along with other NGOs helping the company to save water, CARE is working with Coca-Cola to help accomplish one specific RAIN goal:

> 'Economically empower up to 250,000 women and youth; promote health and hygiene in thousands of communities, schools, and health centers; and replenish up to 18.5 billion liters of water to nature and communities every year.'

CARE's Water Smart Agriculture program's activities in Mali, Malawi, and Ghana, described above, all fit within this goal, where CARE and local government staff increase the capacity and knowledge of small-scale women farmers through trainings, farmer collectives, demonstration plots, and by encouraging 'last-mile' agro-dealers to operate closer to communities. Over three years, (2016–2019), with support from RAIN, the Water Smart Agriculture program reached over 30,000 people with improved techniques to increase crop yield using less water.

2.4.2 PepsiCo: She feeds the world

In six of the countries where PepsiCo operations are located, CARE received funding for a program designed to help women farmers produce crops more sustainably. As of 2020, this program was running in three countries (Egypt, Uganda, and Peru), training women farmers through CARE's 'Farmer Field Business School' approach: a hands-on learning-by-doing approach through which farmers are trained on the adoptions of sustainable agriculture practices in order to improve yields, increase income, and improve their household nutrition. Through these farmer groups, CARE works to educate farmers about gender equity, climate change and ways to conserve water use on their fields through drip irrigation and other Water Smart Agriculture methods to increase moisture retention in the soils.

In Uganda, CARE is promoting the use of mulching and practices that limit soil tillage and other agroforestry and integrated crop and livestock practices that promote efficient use of water (Figure 2.3). PepsiCo credits the water conserved by the women they train against the magnitude of their own water footprint, used by their local processing and bottling facilities. In the Minya region of Egypt, CARE and PepsiCo are working with farmers to increase efficiency of irrigation practices, and increase the moisture holding capacity of the soil to decrease irrigation requirements, in an effort to replenish or conserve a total of 167 million liters of water per year.

Figure 2.3 Efulazia and Peter have eight children and live in the village of Kigando in Uganda. They are subsistence farmers and participate in PepsiCo Foundation funded She Feeds the World program. Farmers learn methods to increase yields, improve access to markets, change gender power dynamics and empower women while helping change the attitudes of men. *(Credit: CARE, Josh Estey 2019)*

2.4.3 Gap, Inc.: Water and women alliance

The Water and Women Alliance program is co-funded by Gap, Inc. with assistance from USAID's mission in India. As a leader in the garment industry, Gap recognizes the huge amount of water required to make its clothes (at least 1,000 gallons, or 3,785 liters, of water per pair of jeans) and seeks to empower rural women who are often at the bottom of the cotton supply chain. Through the Water and Women Alliance (originally a five-year program, starting in 2017), CARE works with Indian women in cotton growing communities, to teach not only water conservation techniques, but also leadership and business skills. In the Water and Women Alliance, rural Indian women engage with the Personal Advancement & Career Enhancement (P.A.C.E.) curriculum program for six months, supplemented with modules on their rights to water and sanitation and techniques for proper hygiene.

In addition to the P.A.C.E. instruction, one Water and Women Alliance Program partner is working to increase access to capital so that rural communities can take out loans for water and sanitation improvements, while another is facilitating cooperation between the community and local government to monitor water quality and develop water safety plans. On an industry-wide level, one ambitious Water and Women Alliance partner is piloting methods to use less water for growing and processing cotton.

2.5 ADDITIONAL OPPORTUNITIES FOR ADVANCING SUSTAINABLE WATER BY CORPORATIONS

CARE's awareness of the role that water and reducing gender inequality plays in maintaining human health and prosperity underlies its support of sustainable water use practices by industry. These practices include the following:

(1) Take water availability into account in placement of industrial facilities.

When extraction and processing facilities are located in geographical areas where water resources are limited, substantial extractions for industrial purposes may endanger the ability of communities to meet their own agricultural and personal needs. This inevitably puts more strain on women and girls (and increased risks of violence towards them), as they are most often charged with gathering and ensuring food and clean water for their families.

(2) Analyze water use in all industrial processes, so that water consumption can be minimized, and water reuse maximized.

Analyze water use of all components of the supply chain, both in terms of water consumption and provision of adequate water and sanitation for workers (keeping in mind the unique needs of female workers) in factories and throughout the supply chain.

(3) Implement production processes that ensure that the quality of water effluent discharged from industrial processes is not damaging to humans or the environment.

Responsibility for sustainable water use doesn't end with conservation. It includes making every effort to return water to the environment in a condition as good as or better than it was when originally extracted. Poor water quality affects women and girls and other vulnerable groups in unique ways due to hygiene needs, domestic duties, and familial roles.

In response to an awareness of the need for industries to do more in the area of sustainable water use, some organizations have developed systems for certifying 'water friendly' companies. Certified B Corporations like Patagonia that commit to a standard of social and environmental performance and public transparency, also use social media to encourage companies to become better stewards of the environment (B Corps, 2020).

2.6 NEED FOR GOVERNMENT ACTION

2.6.1 Inadequate regulation means no incentive for sustainable industrial water use

Inadequate regulation means there is a lack of negative consequences for unsustainable and destructive industrial water use. Industries have little (monetary) motivation to use and dispose of water in a socially and environmentally responsible way. One challenge faced by communities is that water utilities often set tariffs lower for industries. These rates are designed to attract water-intensive industrial facilities that governments perceive as advantageous to the economy because of the jobs created. Too often, more effort goes into recruiting the company than consulting local communities prior to locating a water intensive-industry or monitoring its impact on surrounding communities and the environment. This problem is exacerbated when there is a lack of coordination within government and information is not shared between departments or ministries. The department that approves the factory and collects the taxes is generally divorced from the departments and other agencies that monitor community health and well-being, and environmental impact.

When the government charges low fees for water, places no limits on industrial water use, and fails to punish companies for over-extraction or degradation of water quality, industry has no financial incentive to adopt sustainable water use nor are they held accountable for unsustainable use. Sadly, it is not uncommon for a bottling company to set up shop in a region, extract most of the water in the water table, then sell the water to the people who live there after their traditional water supply has been depleted.

2.6.2 CARE facilitates the strengthening of government capacity

Governments are the duty bearers of water services and must ensure that communities have access to water. It is through advocacy for underserved communities and strengthening the capacity of government that CARE works to ensure provision of sustainable water services to vulnerable communities. As an NGO, CARE tries to balance its need to fulfill donor expectations who want to see 'measurable progress' in concrete terms (like higher incomes, number of latrines constructed, and an overall increase in average child height), with the long-term benefits that may be gained from less obvious changes like the strengthening of inclusive governance systems that prioritize community equality and environmental health. CARE puts considerable effort into working with governments to increase their organizational capacity. Below are a few examples:

- In **Madagascar**, CARE is helping federal and regional governments expand local water systems through contracts with private companies that operate government-owned water utilities. People are willing to pay for water, but not for an unreliable, poor-quality water supply, and so, CARE is working to increase the capacity of service providers.
- In **Ethiopia**, CARE works with zone (district-level) government partners to map waterpoints and water resources and detail zonal budgets for maintenance and repair of waterpoints, as well as conservation of key recharge areas. Mapping has influenced where and how the Government of Ethiopia invests in water supply.
- In **Somaliland**, CARE has partnered with the government to establish a training center that equips technicians with skills to fix and maintain rural water systems – previously a major barrier to delivering consistent water to communities.
- In **Guatemala,** CARE advocates for establishing Municipal Offices for Water and Sanitation. To date, half of Guatemala's municipalities have set up water and sanitation offices to support sustained water and improved community health.

2.7 FUTURE CHALLENGES

As climate variability increases, the availability and accessibility of water will become more unreliable. It is essential for governments to understand that despite the economic benefits of industry, industrial use of water needs strict regulation, and the impact of local industries on communities and the environment needs close monitoring. When industries are permitted to use water resources at their discretion, and to dispose of wastewater without quality improvements, it is the health and livelihoods of the poor, vulnerable and marginalized that are most affected.

However, the impacts of unequal access to water are not limited to the poor, as shown by the recent spread of COVID-19. The difficulty of halting a global pandemic in communities where WASH services are inadequate or non-existent reveals the deep connection between universal access to water and worldwide public health (Figure 2.4). As climate change increases the threat and spread of disease, vulnerable communities must have clean water to adapt and to thrive, highlighting the urgency of CARE's Water+ work.

Figure 2.4 Pamela and her younger sister Monica depend on the generosity of their neighbors for food, and live in perpetual fear of abuse, since they have no adult to look after, or protect them. Their only options for water are an open well filled with dirty water 200 m away, or a cleaner source 4 km out of the way where Pamela can walk to collect water after school. *(Credit: CARE, Timothy Buckley 2017)*

REFERENCES

B Corps (2020). About Certified B Corporations. https://bcorporation.net/about-b-corps (accessed 30 April 2020).

Cheng J. J., Schuster-Wallace C. J., Watt S., Newbold B. K. and Mente A. (2012). An ecological quantification of the relationships between water, sanitation and infant, child, and maternal mortality. *Environmental Health*, **11**, 4. doi: 10.1186/1476-069X-11-4

Das M. (2017). The Rising Tide: A New Look at Water and Gender. World Bank, Washington, DC. https://openknowledge.worldbank.org/handle/10986/27949 (accessed 9 February 2020).

Elwell N. and Williams Y. (2018, September 26). If You Care About the Environment, You Should Care About Gender. Retrieved December 01, 2020, from https://www.wri.org/blog/2016/03/if-you-care-about-environment-you-should-care-about-gender

Thompson J. A., Folifac F. and Gaskin S. J. (2011). Fetching water in the unholy hours of the night. *Girlhood Studies*, **4** (2), 111–129. https://www.berghahnjournals.com/view/journals/girlhood-studies/4/2/ghs040208.xml. https://bcorporation.net/about-b-corps (accessed 30 April 2020).

United Nations General Assembly (2010). Resolution adopted by the General Assembly on 28 July 2010: 64/292. The human right to water and sanitation. https://www.un.org/ga/search/view_doc.asp?symbol=A/RES/64/292 (accessed 9 February 2020).

Chapter 3

Business transformation as the gateway to sustainability: A tobacco company's perspective

Guatimozin Santos

Philip Morris International, External Affairs, São Paulo, Brazil

Keywords: Alliance for Water Stewardship, business transformation, payment for environmental services, supplier engagement, sustainable agriculture, sustainable water management, tobacco growing

3.1 OUR APPROACH TO SUSTAINABILITY

For Philip Morris International (PMI), sustainability is an organizational value that affects every aspect of its operations. For a tobacco company, sustainability must begin with addressing the negative health impacts of its products. For around 20 years, PMI has been working to develop and commercialize better alternatives to cigarettes. Experts including many regulatory bodies such as the US FDA, indicate that the primary cause of smoking-related diseases is not nicotine but the inhalation of harmful and potentially harmful constituents (HPHCs) formed as a result of burning tobacco. To leave combustion behind, PMI has developed a portfolio of smoke-free products which are likely to lead to reduced risk of harm when compared with cigarettes, with the ultimate goal of eliminating cigarettes and offering better alternatives as soon as possible to smokers who otherwise would continue smoking.

While replacing cigarettes is the biggest overall contribution we can make towards achieving the United Nations Sustainable Development Goals, PMI can also make significant progress by managing both its social impact and its environmental footprint. Effective environmental management across our operations and value chain goes beyond compliance with applicable laws and regulation; we're committed to continuous improvement of our business activities in order to achieve the highest standards of environmental sustainability (PMI, 2018). To that end, we have reduced environmental impact and improved performance of our manufacturing operations.

The majority of our environmental footprint is generated elsewhere in our value chain, however, especially in tobacco growing and curing. This chapter describes how the implementation of

DOI: 10.2166/9781789060676_0035

international water management standards in our operations have supported our collaborative work with tobacco growers, suppliers, NGOs, and governments to reduce PMI's overall water footprint through enhanced agricultural and industrial water management.

In addition, PMI has set science-based targets for greenhouse gas emission levels, aiming to achieve carbon neutrality by 2030. By 2017, PMI had already achieved a 31% reduction of all scope emissions, and by 2019 PMI had reached the A-list of global disclosure system CDP, in both greenhouse emissions and sustainable water management (PMI, 2019a). PMI also works with farmers and suppliers across its supply chain with the goal of reducing greenhouse gas emissions during tobacco curing by 70% by 2020, and eliminating net deforestation by 2025 (PMI, 2019c).

3.2 ACTING AS A WATER STEWARD

Globally, the organization sources tobacco from 335.000 farmers and processes it in 38 manufacturing facilities that supply approximately 150 million consumers of PMI products. Water is used in our manufacturing facilities, our agricultural supply chain, and in the production of other raw materials and supplies, such as cellulose acetate filters, paper, electronic devices, and packaging materials. (Although water is not a major input in conventional cigarette production, manufacturing smoke-free products is more water intensive.) To manage an operation of this scale, PMI takes decisive actions in areas where its work can have the greatest impact to minimize the amount of water used per unit manufactured.

Being part of such a large integrated value chain represents a great opportunity, as we participate in each step of the production cycle. In our sustainability programs, we seek not only to reduce what we must use, but also to act as a water steward. This means responsibly using water, without negatively affecting the needs of others, consistent with the following principles (PMI, 2019d):

- Protect the quality of water resources;
- Provide access to water, sanitation, and hygiene at all PMI facilities, and foster availability of these services to tobacco farmers;
- Respect the vulnerability of water and its importance to the community;
- Seek partnerships to improve our water management practices.

PMI works to continuously improve these practices through our ongoing program to educate, encourage, and enable farmers to increase sustainability in the tobacco supply chain. Launched in 2002, PMI's Good Agricultural Practices (GAP) program aims is to maintain the quality of water resources and ensure access to safe water and sanitation for tobacco-growing communities (PMI, 2019b).

In addition to its focus on farmers, PMI also invests in improvements in the use of water in the production of both cigarettes and smoke-free products. While a Global Water Footprint analysis conducted by PMI showed that water consumption in manufacturing operations accounted for only around 6% of total water consumed across our value chain, that amount still translated to 4.4 million cubic meters per year. As we ramp up production capacity to meet the increased water demand, since the manufacturing of smoke-free alternatives is more water intensive, PMI considers both local and regional water availability in siting new factories or making any structural changes to existing facilities.

Our water strategy considers the risks we face from water stress and pollution, and the significance of those risks in catchments relevant to operations up and down the supply chains. This analysis drives action to ensure that water is managed sustainably, as a shared resource. By systematically adopting an integrated water management strategy along our value chain, we can go beyond optimizing our manufacturing process and positively influence water quality in the communities where we operate, improving the lives of thousands of people.

We believe that communities neighboring our factories should not experience any negative impact on water resources due to our operations. We regularly consult with local communities on environmental and community issues, and recognize the importance of engagement with external stakeholders.

3.3 AWS STANDARDS GUIDE PMI'S SUSTAINABLE WATER MANAGEMENT

Beginning in 2017, PMI began working with the Alliance for Water Stewardship (AWS), a leading organization dedicated to better water catchment management. By adopting the AWS International Water Stewardship Standard, PMI has been able to better understand and manage its own water use and to work collaboratively and transparently with others to increase the sustainability and resilience of the communities where we operate. The AWS framework (AWS, 2019) takes into account the needs of local communities by addressing five key areas:

- *Water governance*: analyzing the procedures and rules that govern industrial water use, and respecting local customary rights and applicable regulatory aspects;
- *Water balance*: ensuring that water use is compatible with naturally available volumes through mitigation of water risk and adverse impacts on water availability;
- *Water quality*: mitigating physical risk and reducing the adverse impact of poor water quality on the economic, environmental, and social dimensions;
- *Water-related areas*: assessing the condition of relevant areas of the water basin which, if damaged or lost, would adversely impact environmental, social, cultural or economic benefits;
- *Water, Sanitation and Hygiene (WASH)*: ensure access to safe and sufficient water for drinking, food preparation, and other basic human water needs (including washing and toilet facilities), and provide hygiene education to combat the spread of water-related illness.

AWS not only requires excellent management of water at an industrial site, it also requires an understanding of the local watershed and extensive engagement with local stakeholders on water-related topics. PMI's factory in Brazil, certified by AWS in March 2018, offers a good example of the AWS Standard in action. The first factory in Latin America to be AWS-certified – and first in the world to be certified under the newest (2.0) standard – PMI is now building on the experience at its Rio Grande do Sul facility to apply the AWS Standard across its operations to certify all PMI factories by 2025, with 10 certifications during 2020 alone.

3.4 THE EXPERIENCE IN BRAZIL

Brazil is a global leader in tobacco production and accounts for 25% of global sales. More than 150,000 families grow tobacco in Brazil, mostly in small farms in three states the South Region of the country. Directly or indirectly, over 2.1 million Brazilians are involved in tobacco production, exporting some 549,000 tons in 2019 with an estimated value of $2.14 billion USD. Nearly half of Brazil's tobacco output is destined for EU markets.

Philip Morris Brazil operation is based in the state of Rio Grande do Sul, where tobacco output accounts for 9.62% of the region's total exports. PMI's Rio Grande do Sul operation is a vertically integrated industrial facility, engaged in every step of the process, from development of tobacco seeds to manufacturing cigarettes for domestic and foreign markets, with tobacco supplied directly or indirectly by 47,000 farmers in the region. It was in this facility that PMI piloted implementation of the AWS Standard (Figure 3.1).

Figure 3.1 Philip Morris International factory in Santa Cruz do Sul, Brazil.

The factory had already achieved ISO14001 certification, and because it already had strong environmental management policies and record-keeping practices in place it was able to internalize the AWS Standard in only four months. The combined effect of on-site initiatives implemented throughout the years have cut water use in half (57% compared to 2010 usage). These initiatives included the following:

- Collection of information regarding the source and use of all process water, including storage and reuse;
- Preparation and maintenance of a water balance for the facility, updated annually;
- Mitigating operational risks by monitoring water consumption in all production areas, identifying incidents and taking immediate corrective actions;
- Development of a water contingency and resilience plans with clear definition of roles and responsibilities;
- Reusing water from cooling towers and capturing rainwater for reuse;
- Engaging all employees by soliciting suggestions to improve water use efficiency, and establishing an expert panel to assess the feasibility of new initiatives.

To date (2020), eight recommendations from employees have been successfully implemented, including the development of a system that combines a water compressor, heater, and pressurizer to reduce up to 50% water used to clean the factory's production lines.

3.4.1 Working in collaboration with tobacco farmers and the community

Sustainable water management requires an inclusive approach, taking into consideration the priorities of the river basins and the communities where companies are present. It goes beyond requiring excellence in managing water-related issues within the manufacturing process, requiring deep understanding of the local watershed and extensive engagement on water-related topics with local communities, suppliers, and stakeholders. Industries do not usually have enough understanding of the water challenges beyond their own fence-line, so the AWS Standard helps them look for solutions in collaboration with others, including governments, organized civil society, and academia.

3.4.2 Engaging suppliers on sustainable practices

PMI works with tobacco farmers through our GAP program to support the sustainability of family farming in Brazil. Key GAP initiatives include (PMI, 2019b):

- *Improve working conditions on the farms*: systematically adopt international labor and human rights standards to improve employment conditions all through our value chain. Unique in terms of scale, scope, and level of transparency, PMI's Agricultural Labor Practices (ALP) program, provides specific initiatives training, monitoring, and other areas, resulting in tangible improvements 'on the ground';
- *Address the impact of tobacco farming on the environment*: reduce carbon emissions, promote water stewardship, conserve biodiversity, and combat deforestation;
- *Make tobacco farming profitable and sustainable*: monitor and support farmers implementing good agriculture practices to improve yields and quality and reduce labor requirements, resulting in higher income for farmers. Help them maximize tobacco production to allow more land for alternative crops, to increase food security, and generate additional sources of income.

3.4.3 Engaging with the community and external stakeholders

The Hydrographic Basin Committees are community groups representing certain hydrographic basins and part of the Water Resources National Management System, the governing body in charge of implementing the National Policy for Water Management, the regulatory framework set by the Brazilian Government to manage water resources in the country. With membership including both citizens and government authorities, these committees periodically debate and decide issues regarding water management in Brazil and play a key role in shaping public policies, especially in regions subject to floods, droughts, and water quality problems. PMI has actively engaged with the River Basin Committees surrounding its operations to better understand basin-risks and identify opportunities to support collaborative water management.

As an example, the Water Guardian Project grew out of an effort by the Pardo River Hydrographic Basin Management Committee to restore the degraded river margins of the Andreas Creek. The city of Vera Cruz's main water source, Andreas Creek is within the Pardo River Basin, home to Philip Morris Brazil. The initiative aims to protect springs around Andreas Creek by incentivizing farmers to adhere to good water and soil conservation practices, financially compensating them for their environmental services. Farmers' adherence to the program is voluntary, and the amount paid to each farmer is calculated based on technical criteria, taking into account the crops grown at each farm and the size of the protected area.

The Water Guardian Project began by collecting data from area farms along with information on existing springs. This provided a basis for creating action plans for recovery and protection of damaged areas. The River Basin Committee initially identified 66 farms that negatively impacted Andreas Creek with their agricultural practices, which PMI was able to improve by annually compensating them for reforestation, spring protection, and other environmental services. The environmental work of each farmer is evaluated and certified once a year by the University of Santa Cruz do Sul (UNISC), prior to payment. PMI is accountable for the financial support to both farmers enrolled at the program and the technical support provided by the University of Santa Cruz do Sul, being one of the few initiatives on this model financed by the private sector in Brazil.

Since its implementation, there has been a significant improvement in the water quality in the Andreas Creek. When the project has first started, less than half (45%) of the water analyzed by the program was considered fit for consumption. Based on recent current analysis, according to the National Water Agency criteria some 90% of the water in the creek is now considered suitable for consumption after simple treatment (filtration, disinfection, or pH correction). This upgrade in water quality has improved the lives of more than 18,000 people in the urban area of Vera Cruz and has been recognized by Brazil's National Water Agency (ANA) as a benchmark achievement. In addition to the direct benefit to the population, the initiative allowed significant savings for municipal government, reducing costs for water treatment. The National Water Agency's recognition of the benefits of this program also made the municipality eligible for federal funding for investments in infrastructure.

Based on the experience acquired through this initiative, in 2019 we expanded our efforts to another area, located in Sinimbú, a town of 10,000 with a strong presence of tobacco growers. Also in partnership with the University of Santa Cruz do Sul (UNISC), PMI is now evaluating the quality of water from springs that currently have no treatment or protection. From the diagnosis made, an individualized intervention project will begin at the springs of each farmer selected for participation in the project. To date environmental improvements have been completed at 20 farms and there are still 6 to be evaluated. Water quality is also being performed (Figure 3.2).

PMI has also implemented a program called +Campo, which consists of an itinerant training unit that visits rural areas and trains farmers on our GAP principles. The program emerged from the challenge of reaching out to farmers and workers for training sessions, since the vast majority of farms are located on remote areas. Rather than inviting them to specific locations (such as community centers), our

Figure 3.2 Philip Morris Brazil field technician, UNISC environmental specialists and a tobacco farmer in Sinimbú, Rio Grande do Sul.

experience demonstrated that the most effective way to provide those trainings was visiting them at their properties. Since 2015, dedicated instructors traveling in two special vehicles have trained 8,000 people on topics such as health and safety on the farm, the importance of controlled use of pesticides, and proper usage of crop protection agents (CPA) to prevent water contamination. In 2018, a new module including WASH, among other subjects, was launched.

Bringing transparency and making our efforts on water management public are also important recommendations from the AWS Standard. Since 2019, in addition to the data disclosed in our Annual Sustainability Report and platforms such as CDP, PMI is also disclosing its water performance locally. Philip Morris Brazil is part of the Brazilian Business Commitment on Water Security, a business coalition focused on sustainable water management practices led by the Brazilian Business Council for Sustainable Development (CEBDS), a leading sustainability organization in Brazil, where companies are encouraged to disclose their commitments and goals for the upcoming years and have their progress monitored. The platform also allows companies to exchange experiences, best practices and develop joint initiatives.

3.5 KEY LEARNINGS AND STRATEGY MOVING FORWARD

A key lesson from the broad approach that the AWS Standard requires is that water issues are always more complex than anticipated, but that engagement with stakeholders can help to resolve problems that individually may be viewed as insurmountable. The standard fosters an active involvement of the private sector in water-related discussions, especially at the local level, promoting collaboration and the development of innovative solutions by different stakeholders (farmers, NGOs, public and private sector, academia, and others).

In terms of challenges, our experience demonstrated that the most critical factor in successful implementation of a long-term sustainable water management strategy is the development of a diverse network of stakeholders with clear roles and accountabilities. In line with our commitment to certify all factories by 2025, we have collected our practical experience and learnings from the AWS implementation in Brazil into a toolbox that has been shared with PMI factories across the world. In addition to the PMI factory in Brazil, six other PMI factories are already certified on the standard, in Italy, Indonesia, Turkey, Portugal, Russia and Mexico.

As a matter of principle, we take a collaborative approach and seek out partnerships. Although the controversies surrounding the tobacco industry make it difficult for many stakeholders to even consider collaborating with us, we know that if we are transparent about our intentions and the challenges we face, and are willing to make clear commitments that we consistently honor, there are stakeholders willing to hear about the concrete contributions we can make.

REFERENCES

Alliance for Water Stewardship (2019). International Water Stewardship Standard Version 2.0 22.03.2019. https://a4ws.org/download-standard-2/ (accessed 12 January 2020).

Philip Morris International (2018). PMI's Environmental Commitment. https://www.pmi.com/resources/docs/default-source/sustainability-policies-commitments-and-positions/pmi's-environmental-commitment.pdf?sfvrsn=6bd595b5_4 (accessed 12 January 2020).

Philip Morris International (2019a). CDP Climate Change Submission. https://www.pmi.com/resources/docs/default-source/sustainability-cdp-submissions/cdp-climate-change-2019-submission.pdf?sfvrsn=8f235cb4_0 (accessed 12 January 2020).

Philip Morris International (2019b). GAP – Good Agriculture Practices. https://www.pmi.com/resources/docs/default-source/sustainability-policies-commitments-and-positions/good-agricultural-practices-gap.pdf?sfvrsn=22d595b5_4 (accessed 12 January 2020).

Philip Morris International (2019c). Sustainability Report 2018. https://www.pmi.com/resources/docs/ default-source/sustainability-reports-and-publications/pmi-sustainability-report-2018-high-res-may-2019.pdf?sfvrsn=43ef95b5_4 (accessed 12 January 2020).

Philip Morris International (2019d). Water Stewardship Policy. https://www.pmi.com/resources/docs/default-source/sustainability-policies-commitments-and-positions/pmi-water-stewardship-policy.pdf?sfvrsn=379272b4_2 (accessed 12 January 2020).

Chapter 4

Nestlé: 'Caring for Water' through people, farmers, and communities

Carlo C. Galli and Jonathan Sottas

Nestlé, Vevey, Switzerland

Keywords: industrial water management, manufacturing facilities, supply chain, water for society, water stewardship

4.1 INTRODUCTION

Water has been a 'natural' management priority for Nestlé since its beginnings, more than 150 years ago. Farmers need water to grow food; factories need water to operate; and consumers need water to prepare their dishes. Water touches every part of Nestlé's value chain. This chapter describes how Nestlé's management has made water a priority for the company. Nestlé's initial goal was to become a leading company in the field of industrial water management within its own manufacturing facilities. However, over time, Nestlé realized that the impact of a water strategy built solely on internal excellence is limited, both from an operational and from a reputational perspective. Therefore, Nestlé has implemented a new strategy – the 'Caring for Water' initiative – which builds upon the principles of water stewardship, focusing on collective action at the level of the watershed.

4.1.1 Importance of water for Nestlé

The results of Nestlé's 2018 materiality analysis highlights water among the top seven societal company issues – based on importance to stakeholders and on impact on Nestlé's success (Figure 4.1).

Nestlé needs water to grow source materials, run its factories, and manufacture its products; Nestlé employees need water to be productive; and Nestlé's consumers need water to prepare and use many food and beverage products such as Nescafé, Maggi, or Milo. In all, Nestlé factories use about 120 million m^3 of water per year, while customers use an additional 100 to 200 million m^3 per year to consume its products.

DOI: 10.2166/9781789060676_0043

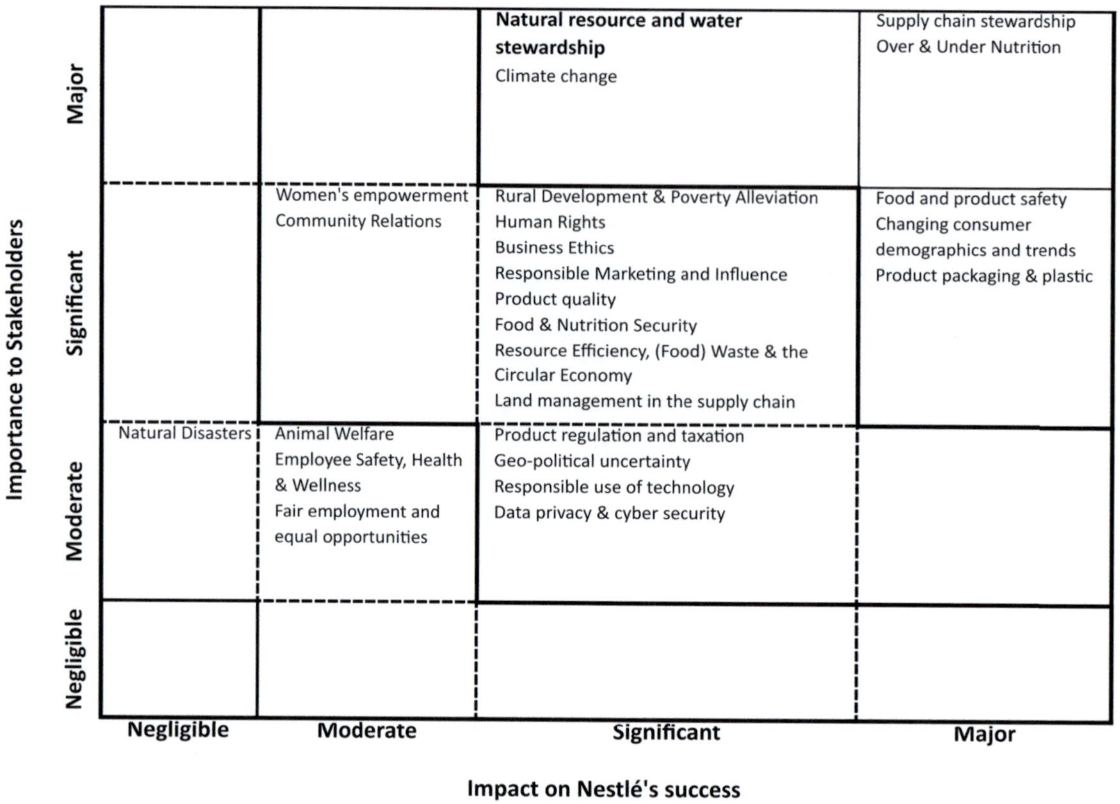

Figure 4.1 2018 Nestlé materiality analysis. (*Credit: Nestlé, 2019*)

In Nestlé's agricultural supply chain, water use is estimated to be around 63.5 billion m^3 of water per year: this is a combination of rainwater (green water) and irrigation water (blue water). Out of this 63.5 billion m^3/y, only around 4.3 billion m^3/y is blue water. The ingredients with the highest blue water footprint in Nestlé's supply chain are dairy, sugar, cereals, and meat/poultry. Even though coffee and cocoa rely traditionally on green water, both are increasingly using irrigation water in many parts of the world (e.g. Vietnam, Brazil).

About 30% of Nestlé's production sites, and a significant amount of the sourcing areas around those factories, are located in regions of water stress. Without water, crops are not able to grow, livestock is not able to survive, factories are not able to function, and there is no business for Nestle. To ensure the sustainability of both sourcing and manufacturing activities, and to avoid serious financial implications from permanent or temporary business disruptions, Nestlé initiated, in the early 2000s, the design and the implementation of a new corporate water strategy – Water Stewardship.

4.2 WATER STEWARDSHIP AT NESTLÉ

4.2.1 A look back

Beginning in the 21st century, Nestlé' has emphasized the importance of water as a key driver for sustainability and endorsing 'the human right to water'. Nestlé approaches water management strategically, starting with its production sites where the company optimizes its water withdrawals, water processing, and water effluents to use less water, have less impact on local water resources, and at the same time become more resilient to local water stress. In 2008, water became one of the three pillars of Nestlé's Creating Shared Value (CSV) program, designed to take the company *'a step beyond corporate social responsibility to create value through our core business both for our shareholders and society'*. In 2011, Nestlé was awarded the Stockholm Industry Water Award as recognition of driving water efficiency across its operations. In 2013, Nestlé published its first set of CSV commitments, which included an objective to reduce the direct water withdrawal per ton of product by 40% between 2005 and 2015. In 2014, this objective became a firm part of Nestlé's commitment on Water Stewardship.

4.2.2 Going outside

Following significant improvement in water efficiency in our production sites, the company began taking a wider approach, considering how water is managed by all users across the watershed.

While a dairy factory may become less dependent on local water supplies and more resilient to water stress through the implementation of our ZerEau technology (extracting water from fresh milk), the milk delivered to the factory comes from cows that consume no less water than before. A broader approach to water management allows the company to look beyond its factory gates and explore water strategies with all stakeholders that use the same water resource within a watershed.

In catchments where Nestlé operates, water is a shared natural resource and sustainability can only be ensured by an active engagement of all stakeholders, within a clearly defined policy framework. In changing its water management strategy from an introverted to an extroverted approach, Nestlé strives to address water scarcity and water quality issues collectively, and at a pre-competitive level, through an Integrated Water Resource Management (IWRM).

4.3 THE CARING FOR WATER INITIATIVE

The Caring for Water initiative (C4 W) was launched in 2018 as the flagship initiative for the 'Planet' pillar of Nestlé's 'Purpose and Value Framework' (Figure 4.2). Its implementation is mandatory everywhere Nestlé operates, and responds to a need to have a consistent approach in tackling such a critical issue. It focuses on four pillars: Factories, Watersheds, Agriculture, and Communities. It takes Nestlé's factories as the starting point to engage in collective action at the level of a water catchment (or 'Watershed'), recognizing the importance of agriculture for both the water quantity and water quality in the watershed where Nestlé's activities take place. The 'Communities' pillar represents the positive social impact that Nestlé can play around its operations and within its supply chain.

To operationalize the Caring for Water initiative, the company currently uses the principles and approach set in the Alliance for Water Stewardship (AWS) Standard. With its standard, AWS provides a common language for those willing to be pro-active in Water Stewardship activities through a thorough implementation approach and a credible external auditing/certification process. AWS guides companies in assessing risks and needs, and in focusing efforts where it matters most, for the benefit of all.

Nestlé Waters (NW), our bottled water business, has committed to have all its factories AWS certified by 2025. NW organizes internal trainings and 'train the trainers' sessions to strengthen our operational teams' capacity. These capacity-building sessions on AWS start by focusing on the concept of Water Stewardship, before going into the details of all criteria and indicators of the standard itself.

Regarding external stakeholder engagement, a key element in the AWS standard, training sessions cover the sociological mechanisms driving local engagement ('how to grasp concerns and expectations') before moving to technical training on our internal stakeholder engagement tool (CRP), learning about

Figure 4.2 Caring for Water is the flagship initiative of the Planet pillar in Nestlé's Purpose and Value framework.

stakeholder mapping and action plans. On top of that, the company takes advantage of its worldwide commitment to share experiences between sites going through the AWS certification process.

Finally, as part of the AWS process, NW frequently organizes awareness-raising sessions for their employees to ensure an understanding of the AWS certification and its purpose. These meetings may also provide an opportunity to present and discuss shared local water challenges with local employees.

4.3.1 Factories

Reduction of water consumption and improvement of Nestlé's internal Water Use Efficiency (WUE) have been part of management practices at the plant level for many years. WUE (expressed in m^3/t) refers to the volume of water used in Nestlé's factories to produce 1 ton of product. At global level, it is the ratio between the total volume of water coming into the factories versus the total volume of production. To ensure the constant improvement of this ratio, all factories apply the 'Reduce, Reuse, Recycle' approach, going as far as necessary. By the end of 2018, Nestlé had improved its global WUE by 30% since 2010 and by 78% since 1997 (Figure 4.3).

Moving forward, we have designated a group of 'Where it Matters' (WIM) factories which, based on their annual water withdrawal and exposure to local water stress, will be our focus for further WUE improvement in coming years. We also define specific WUE thresholds for key product categories (e.g. water, coffee, and dairy) in our manufacturing facilities. To support these efforts, we host webinars on the most recent technological advances for our technical community throughout the world, and we have published a 'Do-it-yourself' catalogue with 375 water management solutions available to help factories identify energy and water savings opportunities and build long-term action plans.

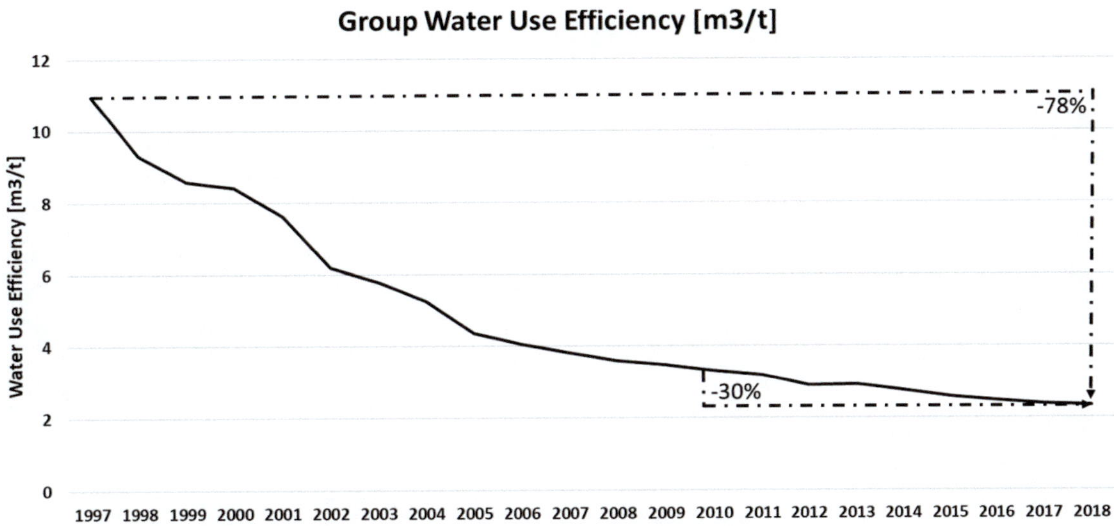

Figure 4.3 Evolution of the Group Water Use Efficiency at Nestlé [m^3/t].

4.3.2 Watersheds

Achieving water sustainability at catchment level requires, on the one hand, a solid technical and environmental understanding of the resources available and the associated demand; on the other hand, it means sharing this information with local stakeholders (such as other water users) and engaging with them in collective actions addressing local water challenges. The success of such a collaborative approach relies on full transparency from all stakeholders.

Nestlé has initiated water stewardship projects in many parts of the world. In 2009 Nestlé launched the ECO-Broye program to facilitate and coordinate efforts to preserve natural resources and sustain economic development in the Swiss district of Broye-Vully around the water source of its Henniez brand mineral water. This project includes ecosystem restoration and sustainable forest management on the infiltration (recharge) zone of Henniez springs as well as non-intensive cultivation of 120 hectares of farm land and 2,000 hectares of ecological corridors. A biogas production project adjacent to the Nestlé Waters plant transforms organic waste from local farms and industry into clean energy, reducing greenhouse gas production by 1,750 tons of CO_2 equivalent and saving the factory around CHF 60,000 per year.

In Lebanon, Nestlé joined the Shouf Biosphere Reserve to conduct a two-year monitoring study of the reserve's groundwater. The study highlighted important impacts on local water resources due to seasonality effects, climate change, and human activities. With the rainy season extending from October to April, there is a need to better manage water resources to avoid droughts caused by seasonal shifts in the flow of springs and rivers in the area during the dry period of the year (Figure 4.4). Through the AWS certification process, our employees are offered training sessions to learn about this process; by applying the Standard, they become water stewards and ambassadors of good water management practices within their communities.

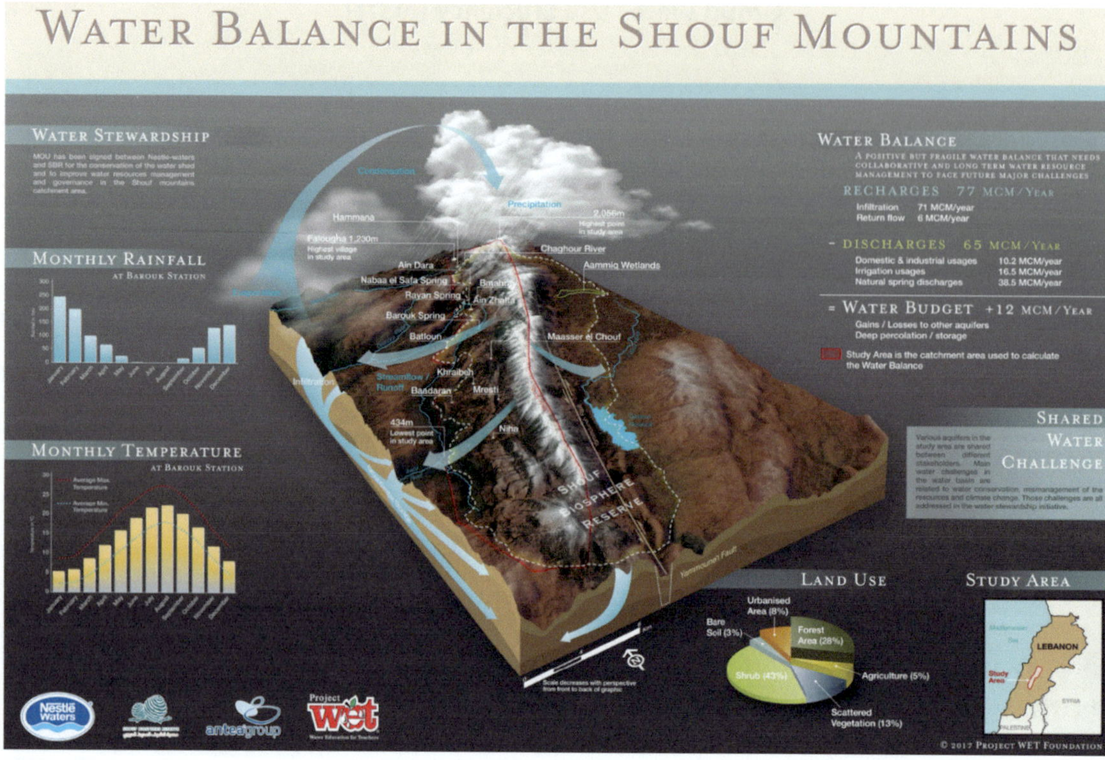

Figure 4.4 A watershed delimits a portion of land where all users share a common water resource. The water balance (water availability vs water demand) is critical to characterize the level of water stress.

4.3.3 Agricultural supply chain

Addressing water use in Nestlé's agricultural supply chains poses the greatest water stewardship challenge but also offers the best opportunities to make a positive difference. Agricultural activities use 70% of the world's freshwater withdrawals, and up to 90% in some developing countries. Quantitative water stress from agriculture can be caused by overexploitation of local water bodies, inefficient irrigation practices and surface water run-off. Qualitative stress results from contamination of water bodies with pollutants of agricultural origin and soil erosion. Nestlé focuses on three areas for actions in Nestlé's agricultural supply chain: (1) Optimize water efficiency – 'More Crop per Drop', (2) Reduce water withdrawal – 'Use Less', and (3) Protect and replenish watersheds – 'Protect'.

Nestlé's team of 675 agronomists and approximately 5,000 extension workers and contractors offer support and training assistance to 111,800 farmers via 28 projects in 21 countries. They also maintain an ongoing dialogue with traders, governments, and NGO partners through our 'Farmer Connect' organization. There are no conditions attached to the assistance Nestlé gives, and farmers are not obliged to sell their produce to Nestlé. Such projects secure supplies of higher-grade raw materials, which ensures Nestlé can maintain the production of high-quality products. Nestlé also invests in research and development projects and breeding programs to supply cocoa and coffee farmers with millions of drought-resistant plantlets every year.

By sharing good agricultural practices with Nestlé's suppliers in water-stressed locations, suppliers can become more productive and resilient to climate change. In Vietnam, the 'More Coffee with Less Water' project delivers water efficiency trainings, covering topics such as coffee irrigation, pruning and grafting techniques, fertilization and composting, and harvesting techniques and post-harvest technology. Jointly sponsored by Nestlé and the Swiss Agency for Development and Cooperation (SDC), in 2018 it provided training sessions to 45,287 farmers in four provinces (Dak Lak, Dak Nong, Lam Dong, and Gia Lai). Nestlé also launched a free app, available in Vietnamese and English, which provides farmers with short-term (3.5 days) rainfall and temperature forecasts to help them manage water usage on their farms. Completed in 2018, the 'Manos al Agua' project (supported by Nestlé, the Colombian Coffee Growers Federation, Cenicafé, and the Netherlands Enterprise Agency) empowers Colombian coffee farmers to tackle climate change through regional solutions for water management, The program was centered around community participation training groups, which trained farmers, and worked on reforestation, bioengineering, and climate-monitoring projects in several river basins.

Looking beyond the 'Farmer Connect' landscape, it is estimated that around 5 million farmers supply agricultural raw material to Nestlé on a yearly basis. Many products are purchased as a raw commodity in globalized markets, supplied by key trading organizations with whom we work via our Responsible Sourcing Standard. For example, in 2018 Nestlé Purina, The Nature Conservancy and Cargill launched a three-year water project to improve the sustainability of the US beef supply chain by reducing the environmental impact of row crop irrigation in Nebraska. More than 50 per cent of water used in U.S. beef production is dedicated to irrigating the raw crops that become feed for cattle. By putting first-of-its-kind, cost-effective, irrigation technology in the hands of farmers, the amount of water needed for raw crop irrigation is greatly reduced, as is the environmental impact of the beef supply chain.

The Nebraska project enables farmers to make more informed irrigation decisions, by installing smart weather sensors in crop fields and using Internet of Things (IoT) technology on sprinklers connected to a smartphone app. By using smart weather sensor technology, this program could help save 2.4 billion gallons of irrigation water over three years, which is equivalent to roughly the amount of water used by 7,200 households over that time period. The reduction of pumping also means less energy used and less labor expense for farmers. Furthermore, these solutions are scalable and can be adopted by farmers across the United States.

4.3.4 Communities

Access to water, sanitation, and hygiene (represented by the acronym 'WASH') is key to maintaining healthy communities. Supporting access to WASH among employees, their families, and in priority communities around our factories and our supply chain regions has a positive impact on health, food safety, and labor productivity; and it also improves relations with local stakeholders. Nestlé is a signatory of the World Business Council for Sustainable Development's (WBCSD) WASH Pledge, guaranteeing adequate access to WASH services to all employees. Using the WBCSD self-assessment tool to identify and close WASH-related gaps, in 2015, more than 90% of employees had access to WASH services in our factories. This figure rose to an estimated 100% in 2016, and we continue our self-assessments across our facilities, identifying and correcting gaps through action plans. Our employees receive training on hygiene and WASH upon hiring and are provided with refresher training annually. We also extend this training to communities.

In 2018, Nestlé rolled out Guidelines on Respecting the Human Rights to Water and Sanitation, to make sure that Nestlé's operations and upstream supply chain do not have a negative impact on the human right to water. It is built upon guidance from the CEO Water Mandate and leverages existing audit/assessment tools within Nestlé (CRP tool, Water Resource Reviews, Human Right Impact Assessments, Farm Assessments, Tier-1 Audits, etc.). Countries where Nestlé operates are split into high, medium- and low-risk based on UNICEF/WHO data, and different due diligence requirements apply to the three risk categories. Additionally, Nestlé works with the International Federation of Red Cross and Red Crescent Societies (IFRC) and local stakeholders to improve WASH access in cocoa-farming communities in Côte d'Ivoire and Ghana. A total of 1,140 staff and volunteers received training in various skills, including 38 people from all 12 project districts trained as trainers in Community Led Total Sanitation (CLTS) and Community Based Health and First Aid (CBFA). At the community level, volunteers from all 50 target communities received training in community-based health and care (CBHFA) focused on public health and social mobilization skills.

As a result of these capacity-building efforts, Ghana Red Cross became more competent in the provision of WASH service delivery; cooperation between Nestle and the ministries of Water and Sanitation and Community Water Agencies has significantly improved; and standardized quality control mechanisms have been established with increased community engagement and accountability. Trained community-based volunteers conducted a total 9,846 house-to-house sessions on health and hygiene practices, encouraging positive behavioral change at the household level. Since 2007, over 300,000 people have benefited from this partnership, and globally Nestlé's WASH projects have reached 750,000 beneficiaries, including hot spots in West Africa and South Asia (Nestlé, 2019).

Chapter 5

Ecuador's Fruta Del Norte: Early engagement as a tool to build trust

Juan José Herrera[1], M. J. Pozo[2], Nathan Monash[1] and María Cristina Acosta[2]

[1]Lundin Gold Ecuador, Business Sustainability, Quito, Ecuador
[2]Lundin Gold Ecuador, Environment and Permitting, Quito, Ecuador

Keywords: biodiversity conservation, community-based water monitoring, community engagement, community roundtables, Fruta del Norte, water management, water monitoring

5.1 INTRODUCTION

Lundin Gold is a Canadian mining company that operates Fruta del Norte (FDN), a high-grade gold mine located in the southeastern part of Ecuador's Amazon basin (Figure 5.1), in the Zamora Chinchipe province, Yantzaza municipality, Los Encuentros parish. The 'province' is the first level of Ecuador's administrative division, and the 'municipality' is the second. The 'municipality' is subdivided into 'parishes', which are classified as urban and rural. The Los Encuentros parish is located on FDN's indirect area of influence.

Lundin Gold arrived in Ecuador in 2015, after acquiring mining concessions from Kinross Gold Corporation. As the first large-scale gold mining operation in the country, Lundin Gold faced high expectations from local stakeholders regarding local employment, procurement, and environmental management (Hilson, 2012). To address this, soon after its arrival Lundin Gold conducted a comprehensive participatory engagement process to identify and map community perceptions of the social and environmental risks associated with the mine's construction and operation. Out of this process emerged the 'thematic roundtables' which have served as the forum for community input into the company's activities and have become a model for other mining companies throughout the country.

This case-study presents the engagement strategy implemented by by Lundin Gold to define risks and priorities through a comprehensive participative process with local stakeholders. The chapter reviews

DOI: 10.2166/9781789060676_0051

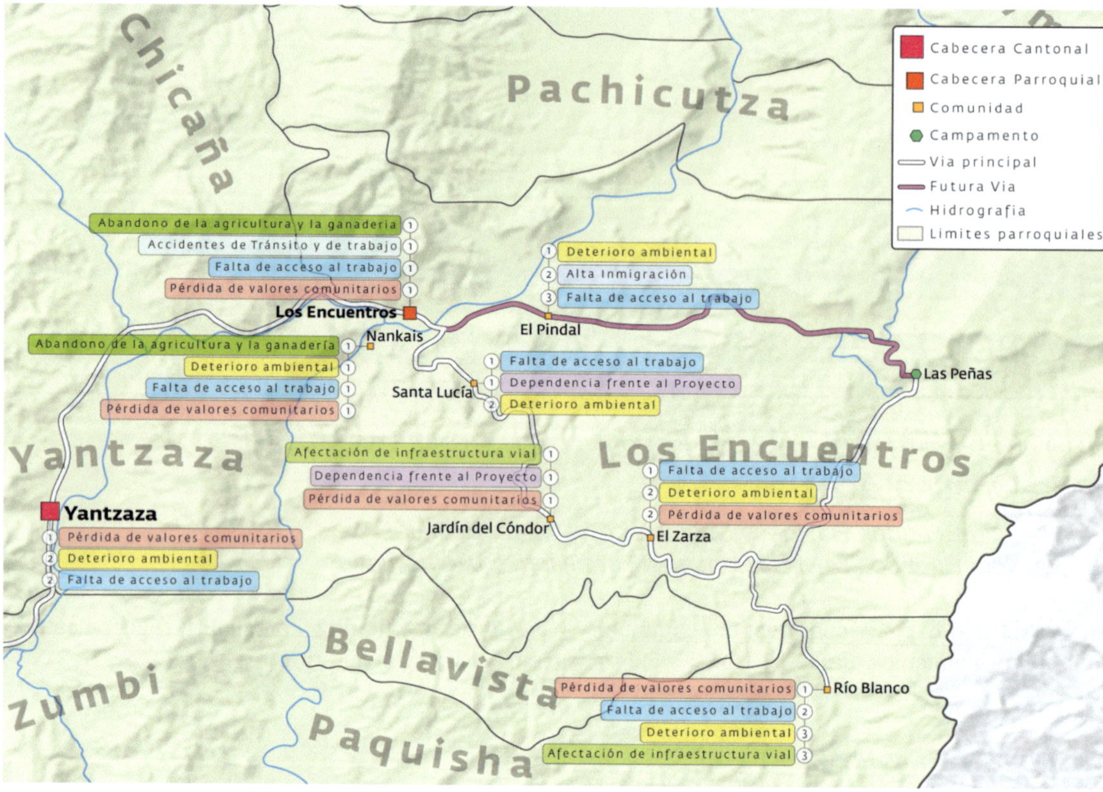

Figure 5.1 Location of Fruta del Norte mine site and surrounding communities. (*Credit: Lundin Gold and INSUCO*)

how the process created trust among stakeholders and resulted in a strong environmental management policy (Bowen, 2010). In addition, it describes how local stakeholders may prioritize economic opportunities over environmental management, and how a company can help balance these priorities.

5.2 ENAGEMENT PROCESS

As illustrated in Figure 5.2, Lundin Gold's approach to community engagement is to establish a dialogue that shifts the focus from the company as the center of all relationships to one where all stakeholders within the geographic area identify and share common objectives. This approach, which depends upon active stakeholder engagement, emphasizes common interest and opportunities and results in a vision of shared risk management. This strategy represents a clear shift away from the more traditional 'transactional' approach, and promotes collaborative management of natural resources, where the mining industry is an equally engaged 'co-responsible' stakeholder in the governance of the region (Crowson, 2009).

5.2.1 Methodology

Upon acquiring the FDN asset, Lundin Gold made the decision to promote this approach to promote social sustainability. By engaging stakeholders in the evaluation of the social risks associated with the construction

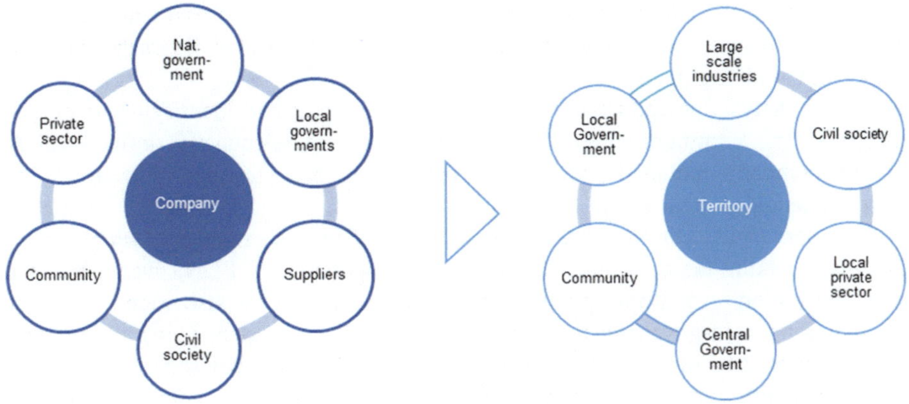

Figure 5.2 From a conventional Corporate Social Responsibility structure to a Territorial (Regional) approach. (*Credit: Lundin Gold and INSUCO*)

and operation of the mine, the company set out to develop and implement a risk management strategy jointly with local communities (Devonshire Initiative, 2016). As a first step, a roadmap of sequentially implemented activities was established that promoted ongoing stakeholder participation. These included: (1) stakeholder mapping; (2) social risk analysis; (3) strategies for social risk prevention and mitigation; and (4) monitoring and evaluation of regional impacts (Insuco and Lundin Gold, 2016).

The engagement process began in early 2016 when the company hired an apolitical third-party facilitator to ensure the neutrality of the dialogue process. A neutral facilitator was important in order to promote engagement, develop partnerships, and create an institution capable of understanding local interests as expressed by each of the different participants in the process.

5.2.1.1 Stakeholder mapping

The objective of the first phase of this process was to identify and characterize the stakeholders who should participate in the development of collective community actions associated with the construction and operation of the mine. Throughout this process, efforts were made to ensure the engagement of a demographically representative range of stakeholders, including women, youth, the elderly, and indigenous peoples.

Lundin Gold developed its own tool for this stakeholder mapping process in order to collect strategic information about the area. The tool considers several different aspects of any given stakeholder: position with respect to the mine, but also flexibility and vulnerability, amongst others. It also considers both the current situation and the desired status within one year to facilitate the design of stakeholder-specific strategies.

This tool was used in several stakeholder workshops to produce a dynamic picture of the social and governance relationships between stakeholder groups at the local level. This analysis enabled the identification of priority stakeholders.

5.2.1.2 Diagnosis and evaluation of socio-economic risk

The first participatory exercise sought to understand the local assessment of potential social risks and impacts, their causes, the stakeholders affected, and the point in the life cycle of the FDN project at which such risks might occur. This exercise involved a multi-stakeholder assessment not only of the social impacts that the company could cause directly, but also of any impacts stakeholders thought might possibly result indirectly from a mining operation (Insuco and Lundin Gold, 2016).

The rigorous risk analysis procedure involved defining each risk and identifying its potential impacts in personal, family, and community terms. Once each risk was adequately defined, all stakeholders were named (the company included) who might proactively prevent such risks from materializing, including those stakeholders that could help mitigate potential impacts. This information was shared with all stakeholders and communicated in a straightforward manner, which reinforced the understanding that Lundin Gold, while the owner and operator of FDN, was only one of many stakeholders in the region. Additional potential strategic partners were also identified as stakeholders at this time, including non-government organizations (NGOs), academic institutions, national and local public institutions, and other companies in the area with their own social investment strategies. By building partnerships and strengthening existing programs, the company sought to avoid duplication of effort (Insuco and Lundin Gold, 2016).

Workshops were held with eight different communities or clusters of communities, and approximately 170 individual stakeholders participated. In addition to identifying the most pertinent risks, the workshops uncovered a variety of public perceptions regarding large-scale mining, and opinions and perceptions that generated the greatest interest were prioritized.

In all, local community stakeholders identified a total of 446 perceptions of risk, 185 causes and 200 potential impacts were identified. In a subsequent step, local stakeholders prioritized these risks according to probability and impacts, thus calling out the following as their main concerns:

- Lack of employment opportunities
- Possible environmental degradation, including impact on water sources
- Loss of moral/community values
- Abandonment of farming and ranching
- Traffic and work accidents, and related impacts on road infrastructure
- Lack of local procurement
- Use of mining royalties
- Coexistence with artisanal mining activities.

The findings of the risk identification workshops were then shared and validated with local communities.

5.2.1.3 Measures for the prevention and mitigation of priority social risks

The next phase served to identify measures to prevent priority risks entirely or mitigate them, should they occur. These measures were not limited to actions by Lundin Gold; the exercise commenced with a discussion of community strengths in order to help participants consider actions that they could implement themselves to avoid undesired outcomes. The exercise then addressed the previously identified causes of each risk, as well as community strengths to identify mitigation measures. This stage of the process allowed local communities to identify short-term (up to 6 months) and medium-term (6–18 months) actions that could prevent the causes of a risk or mitigate their impacts.

Subsequent to this work, follow-up roundtables were set up to: (1) promote multi-stakeholder coordination; (2) facilitate group formulation of prevention/mitigation actions; (3) identify and mobilize relevant stakeholders; (4) promote capacity-building in local communities; and (5) promote coordination between stakeholders. This last action sought to align the company's social investment strategy with local government plans and actions prioritized by the local communities.

5.2.1.4 Thematic roundtables

The next step involved organizing thematic roundtables (TRs) around each of the priority issues to develop proposals for addressing them. Since the process of risk identification naturally tended to frame these topics in a negative light, stakeholders early on decided to change the focus of the TRs from one fixed strictly on risk management to one that included the development of opportunities for the local communities. As a result, the names of the initial TRs were changed to mark this more positive direction. Additionally, participants decided that some of the risks were not suited for this type of engagement or were not as important as initially perceived (e.g. artisanal mining activities and use of mining royalties), while some new challenges and opportunities were identified. Table 5.1 lists the original and modified titles of the TRs, illustrating the dynamic and adaptive nature of the process.

Operating guidelines were set up for the roundtables, in order to facilitate an understanding of the process, the role of each TR, and the way that new members could join. Additionally, an internal decision was made on the importance of the engagement of Lundin Gold senior management staff (with decision-making capacity) during each TR to legitimize the process (Lundin Gold, 2018).

A specific roadmap was created for each TR to document commitments and progress toward meeting those commitments. Each commitment is registered with a date and the TR session at which it was made. There is a description of actions and activities required together with timeframes, background information, and updates on progress in fulfilling the commitment and current status of the commitment (achieved, in process, or pending implementation). As noted, this tool serves to strengthen the process through the systematization of operational procedures, practices, commitments, and projects.

Between October 2016 and February of 2020, 24 TR sessions were held with nearly 2,000 participants, representing local authorities; state institutions; local communities; education institutions (schools and universities); Lundin Gold representatives; local private sector; and civil society organizations. These categories that were established in the Monitoring and Evaluation Protocol for the TRs. Local stakeholders presented 67 proposals for projects or initiatives aligned with the TR topics. A total of

Table 5.1 Development of thematic roundtable topics.

Original Theme	Revised Theme
Lack of employment opportunities	Employment and capacity building
Lack of local procurement	Opportunities for local businesses
Loss of moral/community values	Promotion of community ethics and cultural values
Abandonment of farming and ranching	Agro-economic development
Traffic and work accidents, and related impacts on road infrastructure	Road safety and infrastructure
Possible environmental degradation, including impact on water sources	Environmental responsibility
Use of mining royalties	Eliminated
Artisanal mining activities	Eliminated
–	New: Tourism development
–	New: Inter-institutional coordination

37 projects worth almost $1.5 million USD were accepted and implemented, most co-financed by different stakeholders, mostly through 'in-kind' contributions. Ongoing efforts to evaluate this process annually included local stakeholders' recommendations to improve its dynamic, and after more than two years of implementation the first results of this innovative approach could be observed (Insuco and Lundin Gold, 2018).

5.3 ENVIRONMENTAL RESPONSIBILITY ROUNDTABLE

One of main concerns identified by local stakeholders at the beginning of the process was the environmental management of the FDN project, especially concerns raised by local stakeholders about the potential effects that FDN project could have on biodiversity and water. Over time, however, stakeholder and community interest in this issue decreased, and the roundtable was sparsely attended compared with others such as Employment and Capacity Building, Opportunities for Local Businesses, and Agro-economic Development. This dynamic shows that, although recognized as priority, environmental issues were ultimately overshadowed by the critical economic needs of local stakeholders who live in a rural area that registers high rates of poverty and unemployment.

To generate more interest, the company made further efforts to engage the community by explaining the Lundin Gold approach to environmental management. Since environmental stewardship is one of Lundin Gold's fundamental principles, the company also promoted active participation in this TR by implementing a community-based water monitoring program involving local stakeholders, academia, and NGOs.

5.3.1 Lundin Gold's environmental management approach

The FDN project is located in a highly biodiverse area, and as a leading responsible mining company in Ecuador, Lundin Gold is committed to world-class environmental standards, especially as regards rescue and monitoring of flora and fauna and ecological restoration. In addition, as detailed further below, Lundin Gold's environmental management program also embraces water management, erosion control, waste management, and chemical management (including cyanide), covering all aspects of mine activity, from construction and operation through closure and restoration. As it will be also described below, the Environmental Responsibility Roundtable has been a key channel between the company and local stakeholders to exchange information and concerns on this issue.

5.3.1.1 Biodiversity

Currently, Lundin Gold implements four different environmental management plans, each designed according to international best practices, including International Finance Corporation standards (IFC, 2012). These plans include specific measures for each phase: construction, as well as operations and closure. As an example, during the construction phase, 16 rescue and relocation of flora and fauna campaigns were undertaken. Biodiversity management is based on the landscapes and ecosystems of the project area (Figure 5.3). The objective is that all activities related to the biotic component are compatible with the life zones and the ecosystems. There is a permanent staff of 14 employees and one biotic coordinator in charge of biotic monitoring; this team carries out permanent monitoring, ecological restoration, and nursery management activities at a time investment of approximately 54,000 hours/year. Two biotic monitoring campaigns are undertaken each year, which analyze ecological variables, such as abundance and diversity of species, as well as changes in the ecosystem due to deforestation. Another important aspect is the presence of archaeological, biotic, and environmental monitors that were present in the different work fronts, in order to prevent damage to any sensitive environmental or cultural

Figure 5.3 (a) The Coatí (*Nasua nasua*) was found during the pre-clearance rescue campaigns and the bird. (b) Tororito Pechiocráceo (*Grallaricula flavirostris*) was identified during the biannual biotic monitoring. These species are just two examples of the great biodiversity that exists in and around FDN's area of operations. (*Credit: Lundin Gold)*

heritage sites. The company also monitors noise, air quality, vibration, and other factors, including water quality and flow.

5.3.1.2 Water management and monitoring

In Ecuador, the use of water is regulated by the State through the issuance of use authorizations, which are obtained prior to the analysis of the environmental flows of the source of interest, and the review of the water use priority. The State prioritizes the use of water for human consumption and for food sovereignty. Observing these precepts, Lundin Gold requested authorization from the State to use a flow of 10 L/s, required from the beginning of the operational phase. During the construction of the project's infrastructures, it was not necessary to collect water from surface sources, since the water obtained from the construction of the access ramps to the mine met the needs of the project. The State issued the water permit (10 L/s) to the Company; however, its use has not been necessary so far, even though the operational phase began in November 2019.

Water treatment and recirculation or reuse are the main water quantity and quality mitigation and protection measures applied by the company. Water from the mine and water collected in site sedimentation ponds is used in surface and mine operations, or for road irrigation. A portion of the water stored in the Tailing Storage Facility (TSF) as a result of precipitation and from the process, is reused in the process plant. Excess water coming from rainfall to the operations areas is collected, treated, and recirculated, or discharged in compliance with current regulations.

The water discharged to the environment always comes from the Main Water Treatment Plant, where water coming from the management ponds system undergoes physical-chemical treatment. These ponds are operated according to the operations water balance to ensure the quality and quantity of water. Through this treatment, discharge complies with the permissible limits under Ecuadorian regulations and other international standards such as those of the IFC (2007a) and the International Cyanide Management Code (ICMC, 2016).

The company's water management includes a permanent monitoring program, part of which is carried out by an external laboratory accredited by the Ecuadorian Accreditation Service (SAE), and which also complies with the processes and methods of the Annual Book of ASTM Standards (American Society for Testing and Materials, 2019), the Standard Methods for the Examination of Water and Waste water (American Public Health Association, Several editions) and the United States Environmental

Protection Agency (EPA). Quarterly monitoring is carried out in a network of 34 points for surface water in rivers and streams; and 12 for groundwater. This network covers the project's micro-basin and serves to evaluate and verify the water quality upstream (before operations), in the area of mining operations and camps, and downstream (after operations) over time. The accredited laboratory monitors discharges on a monthly basis (Figure 5.4). In addition, an internal control of critical parameters and basic (field) parameters, to ensure the necessary adjustments in the treatment, is carried out daily by the project's on-site chemical laboratory.

During the construction phase, several erosion and sediment control structures were put in place, including containment barriers built along the road trenches connecting the project, silt fences, and main slope stabilization techniques. Accelerated revegetation was also applied to slopes, such as hydroseeding and manual revegetation of native species, to promote stabilization.

Regarding domestic water management, three sewage water treatment systems are in place in the camp and office area. Through this treatment, discharges comply with the permissible limits under Ecuadorian regulations and other international standards, such as those of the IFC (2007b).

Figure 5.4 Water monitoring by Lundin Gold staff on an unnamed stream, tributary of the Machinaza River, downstream of the operational area, in Eastern Ecuador. (*Credit: Lundin Gold)*

5.3.1.3 Waste management

The company's waste management program focuses on preventive measures, such as the reduction and classification of waste at the source and its correct management until final disposal. The responsible consumption of resources allows us to minimize the amount of waste generated; as an example, through the 'Empty Plate' campaign, Lundin Gold reduced employee food waste by 40%.

5.3.1.4 Training

Finally, continuous training is a cross-cutting axis in environmental management and involves all project participants and visitors. There is a program of on-boarding workshops, daily talks, and weekly and monthly trainings. There are internal environmental campaigns that address issues such as: cyanide management, biotic rescue, environmental and archaeological monitoring, management of waste, responsible consumption, and water care, among others. As an important part of environmental management, the company supports research projects on biodiversity conservation and archeology.

5.3.2 Environmental responsibility thematic roundtable

The TR has been a key channel to share information on Lundin Gold's environmental management activities described above. At each TR, senior staff from Lundin Gold's Environment team share detailed information with local stakeholders on aspects of the companýs environmental campaigns described above. This has included a site visit by local communities to experience at firsthand how Lundin Gold is implementing its environmental policies. The TR has also been an important platform to share information on the advances of the 'Biodiversity conservation program in Fruta del Norte' project, implemented by Conservation International Ecuador (CI-Ecuador). This project focuses on four pillars which are good governance, natural heritage conservation, sustainable production, and communication and environmental education. The initiative uses a 'Sustainable Landscape Approach' (CI, 2018) that allows the effective integration of nature conservation and human well-being as a benchmark for responsible mining in the country.

5.3.2.1 Community-based water monitoring

As a result of a request made by local stakeholders through this TR, Lundin Gold agreed on the development of a community-based water monitoring program. As a first step, a group of 23 volunteers of different communities of the area of influence of the project were trained in sampling, analysis, interpretation, and dissemination of results. The theoretical-practical training was delivered by the Private Technical University of Loja (UTPL by its Spanish acronym), through a cooperation agreement between the Company and the University.

As a second step, a new cooperation agreement was put in place with the Catholic University of Cuenca to develop a methodological proposal for the execution of the water monitoring program. This effort sought to define, together with academia, the technical aspects of community-based monitoring; that is, the establishment of monitoring points and parameters, the frequency of sampling and the methodology for analyzing and interpreting results with a participative approach between the local communities and the academia.

The first phase of this methodological proposal includes the monitoring and analysis of the water quality of the Machinaza River (in the wastewater discharge zone of FDN). The parameters considered in this phase correspond to those related to wastewater discharges, which are also regulated in Ecuadorian environmental legislation. The second phase seeks to establish community monitoring and the analysis of water quality in the area of mining operations (upstream and downstream). The parameters correspond to those typically evaluated for discharges from the mining industry, including heavy metals and cyanide.

Community participation is expected throughout the process, including the collection of samples, interpretation, publication, and discussion of results in their communities with the support of the academy. This program is intended to start its full implementation in 2020 with the support of the academia. The final goal is to build capacities and create a water monitoring culture among local communities. Having empowered communities will contribute to the sustainability of the process and the creation of trust and accountability between the company, local communities, and other stakeholders.

Academia and NGOs have been key actors in encouraging the interest of local stakeholders in environmental and water issues. To date, academic institutions and environmental NGOs, such as CI-Ecuador have been considered as neutral and reliable stakeholders by local communities; therefore, the engagement of these parties in the process has been fundamental. The promotion by academia of initiatives such as joint projects, research activities, case studies, and capacity building certainly generate interest among local stakeholders and brings renewed focus to environmental issues as a priority topic for a geographical area that will support a large mining project.

5.4 CONCLUSIONS AND LESSONS

- Early involvement and continuous participation of stakeholders is key to an accurate evaluation of social and environmental risks. Their engagement contributes to creating trust and commitment to the process of multi-stakeholder dialogue and collective action. It also facilitates the sharing of responsibilities and the management of coordinated actions for territorial development.
- Community dialog is a dynamic process. When the social risk management strategy is implemented at an early stage, potential risks may not yet be evident. A long-term, participatory risk management process adds extra confidence to the multi-stakeholder dialogue and collective action approach.
- The company's senior management staff must be involved to sustain the dialogue over time.
- Companies that operate in highly sensitive ecosystems must develop and implement comprehensive environmental management policies early in the process.
- In a local context that faces challenges such as poverty and underemployment, environmental issues may be perceived as a secondary priority. It is important for companies and other stakeholders to generate interest and promote an active participation on these issues. Information sharing, community monitoring programs, and the involvement of academia and environmental NGOs can be powerful strategies to encourage local interest.
- Additionally, building capacity within communities creates a strong basis of trust and accountability, allowing local stakeholders to confirm the data provided by the company.
- The involvement of other stakeholders, such as academia, NGOs, local government, international cooperation agencies, private sector, is fundamental to carrying out processes of participative engagement such as the thematic roundtables or a water monitoring program.

REFERENCES

American Society for Testing and Materials (2019). Annual Book of ASTM Standards. USA. https://www.astm.org/BOOKSTORE/BOS/index.html

Bowen F., Newenham-Kahindi A. and Herremans I. (2010). When suits meet roots: The antecedents and consequences of community engagement strategy. *Journal of Business Ethics*, **95**(2), 297–318. http://dx.doi.org/10.1007/s10551-009-0360-1

Conservation International (2018). Sustainable Landscape Approach. Implementation Guidebook. Arlington, VA, USA. https://www.conservation.org/docs/default-source/publication-pdfs/ci_laf-sustainable-landscape-approach-implementation-guidebook.pdf?Status=Master&sfvrsn=b772ba44_2 (accessed 2 October 2020).

Crowson P. (2009). Adding public value: The limits of corporate responsibility. *Resources Policy*, **34**(3), 105–111. http://dx.doi.org/10.1016/j.resourpol.2008.10.001

Devonshire Initiative (2016). Beyond zero harm framework A Participatory Process for Measuring Community Well-Being. https://www.devonshireinitiative.org/beyond-zero-harm (accessed 2 October 2020).

Environmental Protection Agency (U.S. EPA) (Several editions). Methods to Determine the Physical-Chemical Characteristics of Water, Discharge Water, Drinking Water, Soil, Solids and Solid Waste. USA.

Hilson G. (2012). Corporate Social Responsibility in the extractive industries: Experiences from developing countries. *Resources Policy*, **37**(2), 131–137. http://dx.doi.org/10.1016/j.resourpol.2012.01.002

IFC (International Finance Corporation) (2007a). Environmental, Health and Safety Guidelines for Mining. Washington, DC, USA. https://www.ifc.org/ehsguidelines (accessed 2 October 2020).

IFC (International Finance Corporation) (2007b). Environmental, Health, and Safety General Guidelines. Washington, DC, USA. https://www.ifc.org/ehsguidelines (accessed 2 October 2020).

IFC (International Finance Corporation) (2012).Performance Standards on Environmental and Social Sustainability. Washington, DC, USA. https://www.ifc.org/wps/wcm/connect/Topics_Ext_Content/IFC_External_Corporate_Site/Sustainability-At-IFC/Policies-Standards/Performance-Standards (accessed 24 November 2020).

International Cyanide Management Institute (2016). The International Cyanide Management Code for the Gold Mining Industry. Washington, DC, USA. https://www.cyanidecode.org (accessed 2 October 2020).

Insuco and Lundin Gold (2016). Proceso de Identificación y gestión territorial de riesgos. Proyecto Fruta del Norte. Insuco, Yantzaza, Ecuador.

Insuco and Lundin Gold (2018). Resultados del monitoreo y evaluación del proceso de mesas temáticas. Insuco, Yantzaza, Ecuador.

Lundin Gold (2018). Lundin Gold Sustainability Report 2018. Quito, Ecuador. https://www.lundingold.com/site/assets/files/16806/2018_sustainability_report_lug-eng.pdf (accessed 2 October 2020).

Standard Methods for The Examination of Water and Waste Water (Several editions). American Public Health Association, USA.

Chapter 6

Phosphate mining and the circular economy: Morocco's OCP Group's approach to sustainable water use

Zineb Benjelloun and Karim Saoud

OCP Group, Sustainability Platform, Casablanca, Morocco

Keywords: circular economy, desalination, innovation, mineral economics, non-conventional water, phosphate rock mining, policy-making, resource efficiency, resource stewardship, sustainability, sustainable mining, SWRO, wastewater treatment, water management system, wastewater treatment plant, WWTP

6.1 INTRODUCTION

Phosphorus is a unique mineral – an element of nature (P) and a product for which there is no substitute. It is fundamental to all living things, including humans, animals and plants, where it plays an important role in the germination, growth and development of crops and is a critical element in agriculture. In fact, the stability of the world's food supply has been linked to the continued availability of phosphorus to farmers everywhere. Morocco is home to 70% of the world's phosphate reserves, which have been mined, processed and marketed by the OCP Group (OCP) for a century. As custodian of this incredible natural resource, OCP is committed to the principles of a regenerative economy, including water conservation, recycling and reuse before, during and after extraction.

In 2008, OCP set out a major 20-year industrial development program to double mining operations and triple processing capacity in order to meet the growing global demand for food, while reducing the freshwater resources' part in its specific water consumption. Now, OCP's ambition is to go 100% non-conventional water resources by 2030.

This chapter describes how OCP increased its phosphorus production, bolstering Morocco's economy while reducing its impact on the environment and making significant social contributions at home and abroad. This transformation was accomplished through a deliberate culture change within the organization, and collaboration with partners and stakeholders within and outside the company. In water

DOI: 10.2166/9781789060676_0063

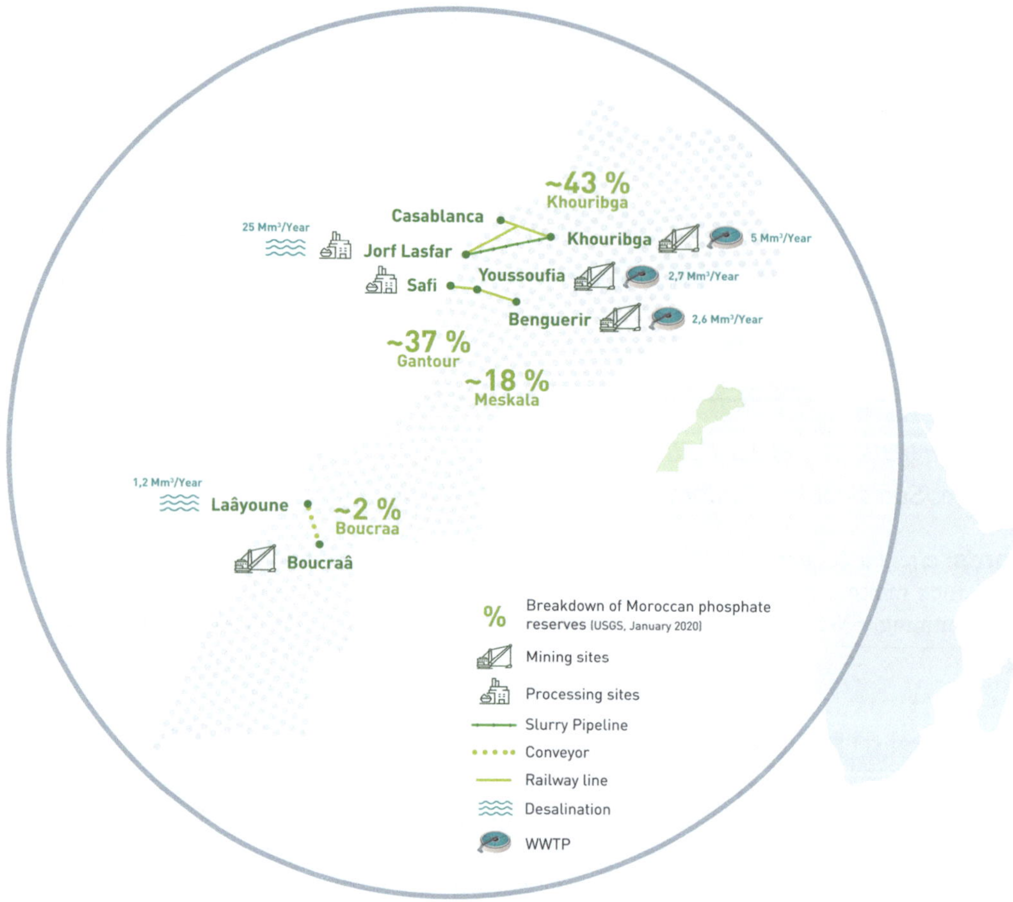

Figure 6.1 OCP's water treatment units in Morocco. (*Credit. OCP*).

stewardship, OCP's main achievements to date include (1) saving water and energy by bringing phosphate rock from inland mines to coastal processing facilities through a 235 km (146 mile) slurry pipeline; (2) desalinating seawater using self-produced clean energy to respond to its further need for water for phosphate processing and (3) reusing industrial wastewater and recycling municipal effluents within its mining areas (Figure 6.1). In addition, OCP has worked with farmers around the world, promoting water-saving techniques and customizing fertilizer products to meet local needs and maximize the benefits of phosphorus in agriculture.

6.2 PATHWAY TO A CIRCULAR ECONOMY

Founded in 1920, OCP plays an important role in feeding a growing global population, by providing essential elements for soil fertility and plant growth. It covers the entire phosphate value chain, from mining to processing. The company produces phosphate rock, phosphoric acid and a wide range of customized fertilizers, the bulk of which are exported worldwide. OCP's activities are concentrated in three regions in Morocco: the mines of Khouribga and the processing hub and port of Jorf Lasfar in central Morocco; the mines of Gantour and the processing hub and port of Safi, and in central Morocco; and the mines of Boucraâ with the port of Laayoune in Morocco's southern region (Figure 6.1). OCP is the Kingdom's largest company generating nearly 6% of the country's GDP (Trading Economics, 2020) employing around 21,000 people and supporting an additional 40,000 indirect jobs in Morocco, across Africa and around the world.

As a major contributor to the global fertilizer market, OCP supports the transition towards a more prosperous, sustainable and resilient agriculture. The Group is implementing a strategy that strengthens its status as one of the most sustainable fertilizer producers in the world and thus contributes to achieving the UN Sustainable Development Goals (SDGs).

According to the Global Footprint Network, people are consuming nature's resources 1.7 times faster than our planet's ecosystems can regenerate them (UNFCC, 2018). If nothing changes, this rate may double as a result of population and economic growth. The issue then is to meet growing consumption needs while preserving resources as much as possible, a major challenge that the world must take up today, especially industrial companies. OCP has responded to this challenge by creating the 'Circular Economy program' in the phosphate and fertilizer industry. The objective was to take OCP's long-standing commitment to the environment and sustainability one step further. As stated by Chairman and Chief Executive Officer Mostafa Terrab, in OCP's 2018 Sustainability Report, 'the challenge is to feed a growing world population while using resources responsibly and continually striving to reduce its environmental footprint' (OCP, 2018).

6.2.1 Transitioning from Linear to cyclic production

In 2018, OCP set ambitious goals by launching the 'Circular Economy program' whose aim is to create a green dynamic, as well as to foster symbiosis with the industrial ecosystem and communities.

This new approach consists of moving from a linear pattern of resource consumption to a circular model, optimizing products from their design to their end of life, including their production, use and reuse. It is implemented through four components: preservation of resources, sustainable production, smart consumption and valorization of waste into resource (Figure 6.2). The 'Circular Economy program' contributes to better value creation for the Group's customers and partners; it also benefits its employees and the people living on its sites.

OCP's focus on the circular economy is demonstrated, in part, by its implementation of technologies that reduce pollution. For instance, air control technologies have reduced emissions of SO_2 in sulfuric acid production by up to 98% (ten times lower than the World Bank threshold), while 86% of the power used by the Group comes from clean energy, with an objective of self-sufficiency by 2030. In 2019, more than 30% of OCP's water needs were met through non-conventional water resources and the firm's new ambition is to reach 100% by 2030 thanks to its water program.

6.2.2 Creating a movement

OCP's ability to implement the 'Circular Economy program' has been based on an organizational culture and agile management that actively supports creativity, collaboration and innovation. This ambitious program is the result, like so many others within the company, of an original program called the 'Movement.' Launched in 2016, it allows any OCP employee to develop a project to address the challenges faced and to be backed with the proper resources to implement their idea. The Movement's goal is to 'positively disrupt the company' by allowing the 9000 employees involved in the program today to make proposals through a non-hierarchical process and according to their own areas of interest.

Figure 6.2 OCP's Circular Economy framework. (*Credit. OCP*).

OCP also invests heavily in professional development and encourages employees to volunteer in their surrounding communities through its 'Act4community program', designed to boost citizenship initiatives. In 2019, 3307 OCP employees volunteered almost 11,110 hours to various associations and communities (OCP, 2019) (Figure 6.3).

The company's support for building expertise and supporting innovation extends beyond its own employees. OCP's support for education and skills training is reflected in all layers of its business ecosystem. In 2012, OCP developed the Mohammed VI Polytechnic University (UM6P), which aims to offer cutting-edge world-class education in applied research and development in science and technology, providing the tools and skills for future generations of African leaders. Besides, the company provides skills training to its suppliers through four Industrial Expertise Centers. In 2018, OCP provided training on improved agricultural practices in seven African countries outside Morocco, as well as 10,000 farmers in India. The Group collaborates also with NGOs and civil society to promote business skills, job creation and access to education, health and culture.

6.3 WATER FROM THE ROCK

OCP's $20 billion 20-year capital investment program to increase its industrial capacity began with a full engineering review of the phosphate value chain: mining, phosphoric acid processing and production of finished fertilizers. Knowing that the use of water occurs at each stage of its value chain (mining, transport and transformation), the company invested heavily in a system that integrated the reuse and sustainable preservation of water throughout its entire production process, starting with extraction.

Figure 6.3 OCP human capital in one of Khouribga's phosphate washing plants. (*Credit. OCP*).

Phosphate mining involves spraying water for dust control along mining trucks' roads. Once extracted, the phosphorus-bearing rock is washed in the beneficiation plants, which recycle 80% of the used water (Figure 6.3). Then, the washed phosphate is either dried for transport by train (from Benguerir and Youssoufia in the Gantour basin to Safi) or sent as a pulp down a slurry pipeline (from Khouribga to Jorf Lasfar) for processing. Once in the processing units and depending on their requirements, more water is added to the pulp from the slurry pipeline or to the dried phosphate that has been transported by train to produce either phosphoric acid or fertilizers (Figure 6.4).

OCP's challenge was to meet its growing mining and industrial water needs – which will rise from 63 million cubic meters (46 mgd) in 2010 to 180 million cubic meters per year (130 mgd) by 2030 – without either withdrawing more water from nearby dams or tapping underground aquifers and draining the Kingdom's strategic water reserves.

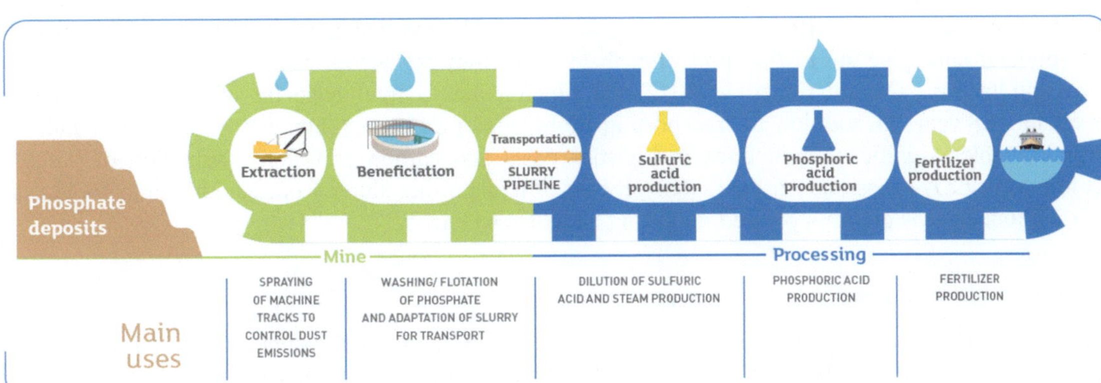

Figure 6.4 Water chain configuration for slurry pipeline transportation mode. (*Credit. OCP*).

Figure 6.5 Slurry Pipeline's terminal station in Jorf Lasfar. (*Credit. OCP*).

Facing the increasing global demand for fertilizers and Morocco's increasing water scarcity, OCP started by asking its engineers and scientists to reconsider water use in every aspect of its production in terms of what would minimize impact on the environment and what would most benefit its multiple stakeholders. Taking into consideration the National Water Strategy, their analysis produced a dedicated integrated water program based on circular economy principles. In response, the company launched many projects to optimize water use across the entire value chain. OCP decided to stop relying on groundwater and to reduce surface water use while promoting use of non-conventional water resources: treated (domestic) wastewater and desalinated seawater.

For instance, OCP realized an annual saving of nearly 3 million cubic meters of water per year (2 mgd) by constructing the world's largest phosphate slurry pipeline to transport phosphate rock from the mine to the processing facility (Figure 6.5). This mode of hydraulic transport eliminated the drying step, which was necessary for rail transport, making it possible to retain the natural humidity of phosphate rock which can be recovered and reused at the processing facility. OCP invested $500 million in a 235 km (146 mile) underground pipeline to transport phosphate rock from the Khouribga mines to the Jorf Lasfar processing facility. Commissioned in 2014, the pipeline has a total capacity of 38 million tons of phosphate per year. In addition to transporting higher volumes than previously possible via rail, the pipeline enables significant reductions in logistics costs and carbon emissions (reduction of 930,000 metric tons of CO_2 emissions per year). The process of transporting slurry from Khouribga to Jorf Lasfar takes 24 hours and is gravity-powered, requiring almost zero energy.

6.4 CIRCULAR DESALINATION

Once the phosphate slurry arrives at the processing hub, additional water is needed to produce phosphate-based products. To supply this water, OCP has invested in seawater reverse osmosis (SWRO) desalination facilities, including the Jorf Lasfar unit, the largest desalination plant in Morocco, supplying up to 25 MCM per year (18 mgd) with plans to expand by 2022 to 40 MCM per year (29 mgd). This

Figure 6.6 Ultrafiltration unit of OCP's desalination plant in Jorf Lasfar. (*Credit. OCP*).

follows the SWRO plant built in 2006 at Laayoune site with a capacity of 1.2 MCM. As an example of the circular economy, the Jorf Lasfar unit (Figure 6.6) runs on surplus clean energy created by the phosphate manufacturing process, where the generation of steam to produce sulfuric acid is transformed into electrical energy.

Taking into account the constraints of seawater quality variation on the costal intake, the plant has been designed with an advanced pretreatment unit (Dissolved Air Flotation + Ultrafiltration) to insure a high pretreated water quality and maximum availability. OCP has also adopted the best-in-place energy recovery device's technology, to produce the desalinated water with the most efficient manner at that time. The water then is re-mineralized and sent to water tanks in order to be distributed to the different consumers. In addition to the on-site effluent treatment, the brine generated by the plant (less than 3.5% of the overall water used by the platform) is diluted into the pumped cooling water and then reused in the hub's processing units.

Additional SWRO desalination plants are under study in several other areas to respond to OCP's industrial growth and the resulting increasing water demand. OCP is tracking the most innovative technologies that allow reductions to the cost of desalinated water, including renewable energy.

6.5 FROM WASTE TO RESOURCE

In addition to seawater desalination plants, OCP has also invested in urban wastewater recycling facilities (wastewater treatment plant; WWTP). In fact, OCP has helped several municipalities treat their domestic effluent, to preserve the environment as well as the national freshwater resources. This service helps protect communities and the environment while producing highly treated wastewater that can be used in OCP's production facilities.

Over the last decade, OCP has developed municipal treatment plants in the mining cities of Khouribga, Benguerir and Youssoufia in order to recover and reuse over 10 million cubic meters per year (7 mgd) of urban wastewater (Figure 6.7). These plants have been designed to treat the wastewater with

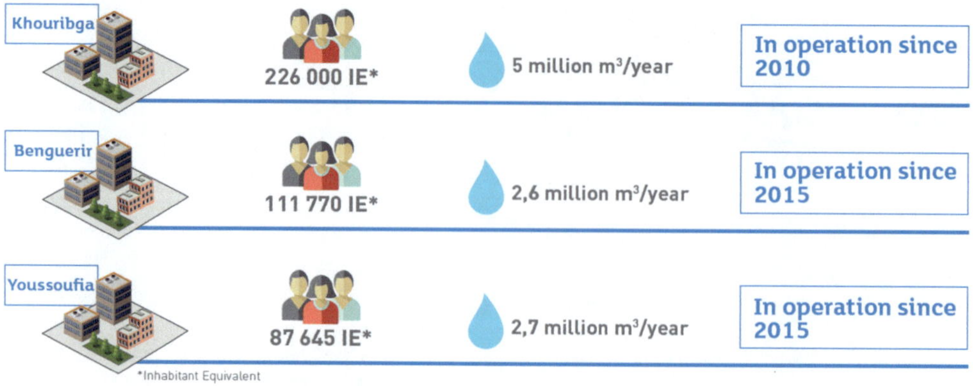

Figure 6.7 The wastewater treatment plants built in OCP's mining cities. (*Credit. OCP*).

cutting-edge techniques, providing excellent overall performance (activated sludge technology). Khouribga's WWTP, the largest plant, has been in operation since 2010 and enabled the first experience worldwide of reusing treated wastewater in phosphate mining.

Benguerir is located in the central region of Morocco, between Casablanca and Marrakesh, and is home to the recently built Mohammed VI Polytechnic University (UM6P). The region suffers from a scarcity of water resources, and is a good example of collaboration between OCP and the municipality, where the WWTP generates about 2.6 million cubic meters per year (2 mgd) of treated domestic wastewater (Figure 6.8).

Figure 6.8 Benguerir's wastewater treatment plant. (*Credit. OCP*).

The human waste initially travels from homes to a pumping station, 4 km outside of Benguerir, where it is pumped to the WWTP. There it undergoes pretreatment, including screening and a process to remove oils and sand followed by primary sedimentation. The secondary treatment process begins in a biological tank that provides an environment that encourages bacteria to grow with a sufficient and controlled supply of oxygen that varies with the quality of the wastewater received. This is followed by tertiary treatment, a three-stage process: (1) microfiltration, which eliminates residual dirt and suspended solids in the water; (2) granular activated carbon filtration, which removes many organics and produces a high quality effluent; and (3) disinfection with hypochlorite, which kills bacteria, viruses and other potential pathogens. After treatment, the majority of water from the plant is sent to the Benguerir mine where it is used primarily for dust control on mining roads. It is also used to water the green spaces in the Mohammed VI Green City of Benguerir. The remaining water is piped to Youssoufia 60 km (36 miles) away, where it is used in place of groundwater to wash phosphate.

The sludge generated by the wastewater treatment is sent to the thickeners and the digester where it produces biogas. The recovery of biogas emitted during the wastewater treatment process covers more than 30% of the electrical and thermal energy needed to operate the WWTP. After that, the stabilized sludge is dewatered by centrifuge machines and allowed to dry to 70% solids in solar drying units. Reuse of sludge in the mining areas is currently being tested.

Still in accordance with the circular economy vision, the industrial reuse of treated wastewater from other existing or new WWTP is being studied.

6.6 LOOKING TO THE FUTURE

As an essential input for its industrial processes, OCP recognizes, respects and values Morocco's most precious natural resource: water.

Over the past decade, the water program implemented by OCP has enabled the company to sustainably ramp up production and ensure food security while preserving national freshwater resources. 'Thinking

Figure 6.9 OCP agronomist supporting a Moroccan farmer (Credit: OCP).

outside the reservoir' has led OCP to identify every possible drop of water to be saved throughout its value chain and to make the voluntary decision to shift from conventional to non-conventional water resources, even at a higher cost (Figure 6.9). OCP's initiative has attracted international attention as a model of corporate sustainable water use and financial institutions, including the French Development Agency (AFD) and the German development fund (KFW), have granted OCP a $500 million USD loan to implement the first wave of its water program. In addition to its technical innovations, OCP's continuous improvement efforts show how a state-owned company's commitment to sustainability can achieve environmental and social objectives that support domestic and foreign policies.

The adventure is not over. The company has set even more ambitious goals and is constantly looking for new ideas, cutting-edge tools and revolutionary models to improve its performance and increase its innovation capacities. This would not be possible without collaboration with the other stakeholders of its business ecosystem, increasing joint-ventures in different fields and inclusive partnerships with world-class companies and universities and working hand-in-hand with authorities and civil society to create a more sustainable future for business, the community and the environment.

REFERENCES

OCP. (2018). Sustainability Report. https://www.ocpgroup.ma/en/sustainability-report-2018, p. 4 (accessed 17 May 2020).

OCP. (2019). Sustainability Report. https://www.ocpgroup.ma/index.php/en/sustainability/reporting-performance-data, p. 162 (accessed 12 August 2020).

Trading Economics. (2020). Morocco GDP. Website. https://tradingeconomics.com/morocco/gdp (accessed 17 May 2020).

UNFCC. (2018). Earth's Annual Resources Budget Consumed in Just 7 Months, July, 2018. https://unfccc.int/news/earth-s-annual-resources-budget-consumed-in-just-7-months (accessed 2 October 2020).

Chapter 7

Dimensions of water management in the extractive industries

Petra Schneider[1] and Christian Wolkersdorfer[2]

[1]Department of Water, Environment, Civil Engineering, and Safety, Magdeburg-Stendal University of Applied Sciences, Magdeburg, Germany, ORCID 0000-0001-7489-9192

[2]Department of Environmental, Water and Earth Sciences, Tshwane University of Technology (TUT), Pretoria, South Africa, ORCID 0000-0003-2035-1863

Keywords: extractive industries, mine water management, mining and post mining landscapes

7.1 INTRODUCTION

The aim of this chapter is to provide the readers with limited technical knowledge about the field of mining with an overview of the dimensions of water management in mining. It describes the range of water management tasks required and provides information on how water management challenges can be addressed.

Mining is the extraction of raw materials from the Earth's crust by construction of surface openings (open pits, shafts, adits, inclines), the extraction of raw materials from the host rock by means of treatment plants, the dumping of residues in waste rock heaps or tailings ponds, the drainage and lowering of groundwater in large areas and all related transport activities. Mining activities require the management of water resources in the surface and underground watersheds. Commonly, mine site rehabilitation is also considered a mining activity, which includes the revitalization of mining areas after the extraction of raw materials. Technical measures for this are the groundwater rebound after switching off the pumps (mine flooding), the closure and dismantling of the mining facilities, the covering of heaps and tailings facilities and the reclamation of the land used for mining and their landscaping.

Water management plays a substantial role in all phases of mining and in all types of mining activities (Grünewald, 2001; Tiwary, 2001; Wolkersdorfer *et al.*, 2020). During the exploration phase, extensive hydrogeological studies must be carried out to allow comprehensive planning for

DOI: 10.2166/9781789060676_0073

the subsequent mine drainage including flood protection, water supply and disposal, disposal of liquid processing residues as well as to initiate the respective water management approval procedures within the overall permitting procedure framework for the mining site. As a rule, a hydrological-hydrogeological site model is created using the results of the site exploration, which provides information on the hydraulic properties of the subsoil in order to predict future dewatering scenarios (Rapantova *et al.*, 2007).

In addition to the geotechnical and hydrogeological site exploration of the extraction site, the water management activities during the exploration and site preparation include the planning and implementation of processes for (1) lowering the groundwater level, (2) disposing of groundwater and wastewater, and (3) planning of development of facilities for processing wastes from the site. Both liquid and solid residues need to be addressed (Dold, 2008; Fields, 2003; Pepper *et al.*, 2014; Younger & Wolkersdorfer, 2004). While solid residues are deposited on waste rock dumps, processing residues are dumped into sedimentation basins called tailings ponds. Depending on the rock composition and target raw mineral, these tailings ponds may contain potentially toxic substances that are used to process the rocks to extract the target minerals. For example, in the case of gold mining, cyanide is used as a solvent; in the case of uranium mining, acid or alkaline solutions are used. The pollutant-containing suspension is deposited into the tailings ponds, so that the sediment can settle and thus forms a sediment layer with an excess water level. Furthermore, flood protection measures must be planned for all mining facilities.

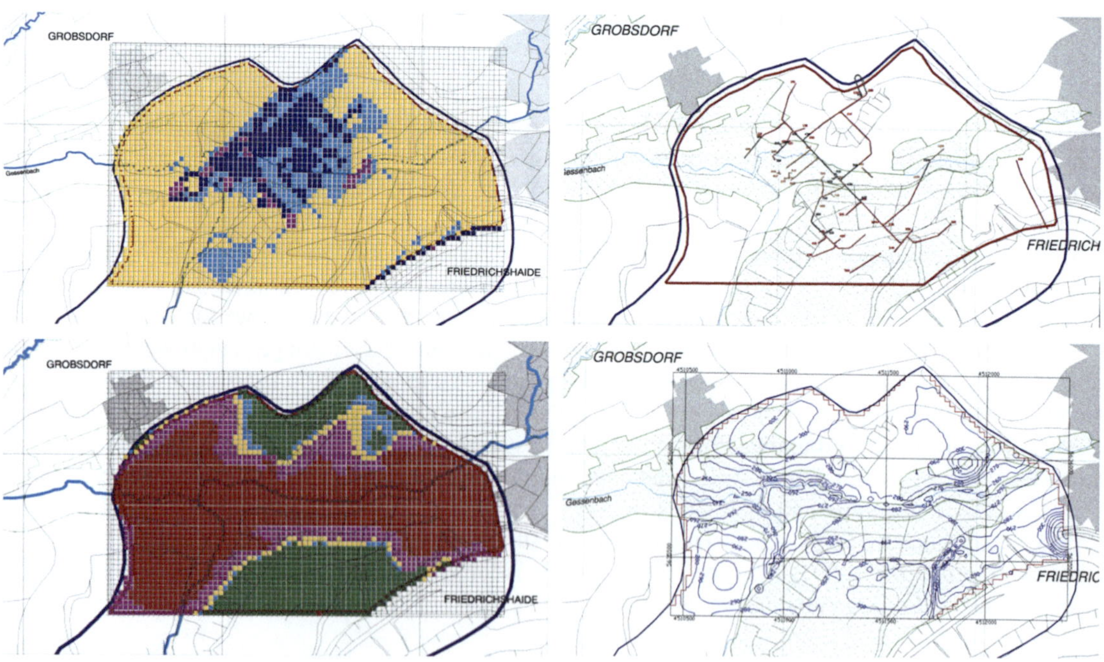

Figure 7.1 Exemplified preparation of extraction by site exploration; top left: rock permeability with consideration of tectonics, top right: map with the routes implemented in the model, bottom left: calculated groundwater corridor distances, bottom right: groundwater balance plans as the basis for the scenario calculation.

Thorough water management preparation during the exploration and site preparation phases is key for successful water management during the extraction phase (Dold, 2008). For planning the water management of the extraction site, methodical baseline data collection is carried out, to determine the water balance (natural and anthropogenic groundwater recharge, marginal inflows, groundwater abstraction), geological setting (drilling data, geological maps), hydrogeological situation (hydraulic parameters in coarse rock and bedrock) and the planned mining (mining crack, the location where the ore will be extracted). These data are used to build the hydrogeological and hydraulic model for the mine site (Figure 7.1).

Basic preparation steps for commissioning of a mine can be quite extensive (Figure 7.2), and the dimension of water management in mining is illustrated by numerous photos hereafter.

7.2 WATER MANAGEMENT DURING MINE OPERATION

During the extraction of raw materials, liquid and solid waste (such as treatment residues, wastewater) is generated for which disposal facilities must be planned. In some cases, the mining residues might be toxic if the treatment has been carried out with chemicals. For economic reasons, these disposal facilities are close to the mining activities and must be equipped with appropriate safeguards against potential environmental emissions. The planning of waste disposal facilities for liquid waste concerns industrial landfills (tailings ponds) and solid waste rock dumps (Ritcey, 1989). In addition, water treatment plants might be necessary, in particular when mine water with elevated concentrations of potentially toxic elements or acidity is formed during the excavation of the deposit. Usually, this occurs when geogenic iron (di)-sulphides (pyrite, marcasite, pyrrhotite) are oxidized by water and oxygen; in this case, sulphuric acid reacts with the host rocks, forming Acid Mine Drainage (AMD) (Blowes *et al.*, 2014; Evangelou & Zhang, 1995; Wolkersdorfer, 2008). Water management activities during the extraction process include groundwater and mine water management, the operation of the water supply and disposal, the disposal of liquid processing residues in tailings dams (Figures 7.3 and 7.4) and mine flooding protection.

The planning of tailings dams has to be carried out according to the recommendations of the International Commission on Large Dams (ICOLD), which issues bulletins on dam safety management. ICOLD is an international non-governmental organization aiming to share professional information and knowledge of the design, construction, maintenance, and effects of large dams (any dam above 15 m in height), and leads in setting standards and guidelines to ensure that dams are built and operated. ICOLD presently has 30 Technical Committees, which issue bulletins with 'state of the art' recommendations for engineers to ensure long-term geotechnical and environmental stability for tailings pond dams (Bowles *et al.*, 2007). Depending on the dimension of the mining site and extension of the processing activities, tailings ponds can cover as much as several hectares and be several tens of meters deep. They contain the processing sludge that is undergoing sedimentation and thus gets separated into a solid and a liquid phase, which is called pore water or interstitial water and, usually, is contaminated with the processing residues.

The term mine water describes all water that circulates in the mine, comes into contact with the underground or open pit host rocks or raw material. It does not include the processing water, and it is not necessarily polluted. Usually, mine water with the highest environmental relevance is AMD, which is mine water with a pH below 5.6. It forms when iron (di)-sulphides in the rock are oxidized and not enough buffering minerals exist. The acidic environment in the mine water provides optimal conditions for the solution of metals and metalloids from the surrounding rock. As a result, acid mine water is usually highly mineralized. If it is not treated and is not prevented from spreading, large amounts of acid

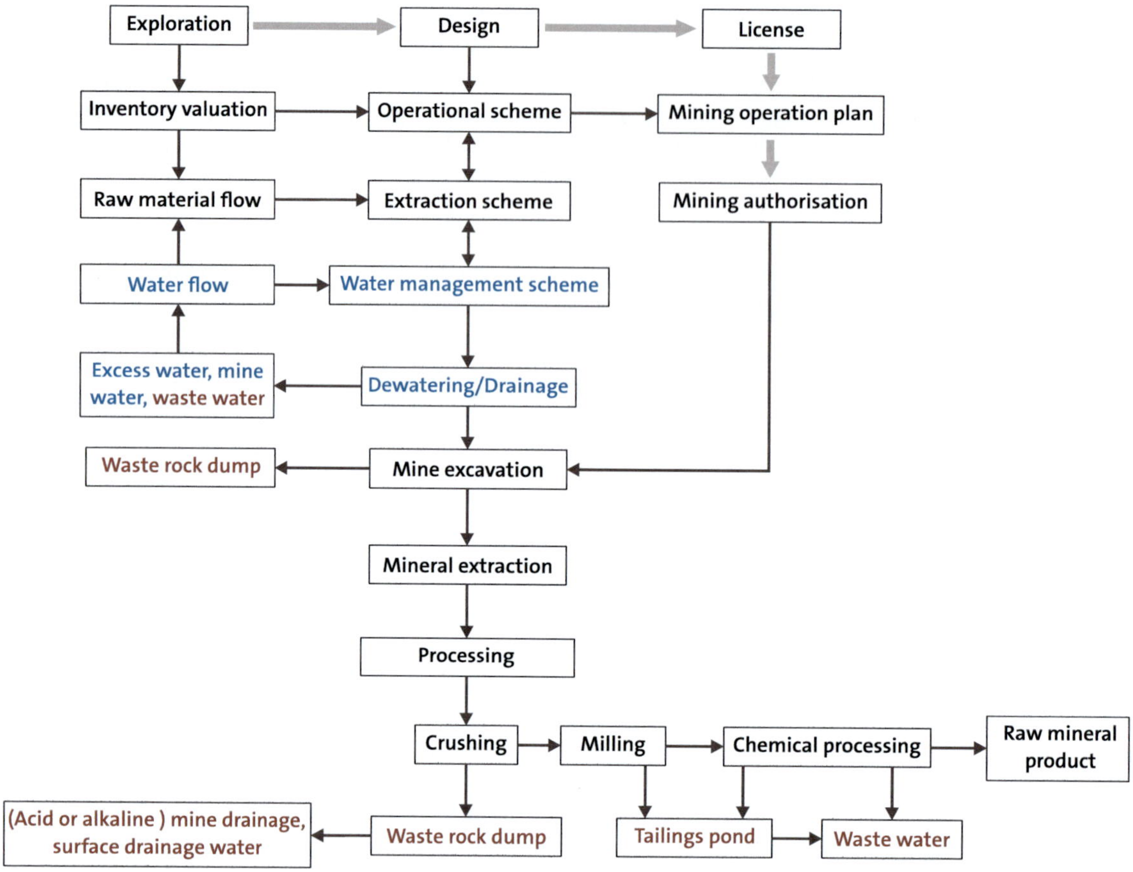

Figure 7.2 Exemplified scheme of mine design and mining process. (*Key.* Black font: activities related to mine preparation and operation; blue font: water-related flows and activities; brown font: waste-related flows or activities)

Figure 7.3 Example of an operating mine tailings pond in South Africa (left) and Chile (right). (Photographs: *left, Petra Schneider; right, Christian Wolkersdorfer*)

Figure 7.4 Example of the dimension of an operating mine tailings pond in South Africa. (Photographs: *Petra Schneider*)

mine water can flow into nearby rivers or pollute the groundwater, as can be seen in many regions of the world like South Africa and Spain. (Olías *et al.*, 2004; Sarmiento *et al.*, 2009) (Figure 7.5).

AMD represents a major challenge in mining practice, because (1) it is heavily polluted and (2) there are mining sites where it is produced in large quantities. A classic method for treating AMD is liming (i.e. neutralization by means of calcium carbonate, calcium oxide, calcium hydroxide or sodium hydroxide). This method of treatment results in potentially dangerous residues that must be disposed of appropriately. Currently, there is no industrial-scale method for obtaining economic benefit from mine water; as a result, treatment and disposal still represent the best option available for handling of mine water. The level of toxicity of the material flushed into tailings ponds may also be affected by treatment that has occurred in the mining process. Metal and energy minerals are often treated with chemicals; the highly polluted wastewater resulting from this treatment process is flushed into tailings ponds as well.

Aggregate mining typically has less effects on water quality, as this type of mining is carried out without chemicals, and the target raw materials rarely contain pyrite or marcasite. Nonetheless, aggregates mining also requires sound water management, with a focus on water supply (such as dust control and sanitation) and flood control. Aggregate quarries operators usually maintain a sedimentation basin for rain and flood

Figure 7.5 Examples of Acid Mine Drainage (AMD) from South Africa (left) Rio Tinto mine, Spain (right). (Photographs: *left, Christian Wolkersdorfer; right, Petra Schneider*)

Figure 7.6 Example of an aggregates mine and the respective extraction and processing installations in the Sevilla region, Spain (left) and increased sedimentation due to aggregates mining in the Nakku Khola River, Kathmandu Valley, Nepal. (Photographs: *left, Petra Schneider; right, Christian Wolkersdorfer*)

retention, but also for the provision of service and extinguishing water (Figure 7.6). Another type of aggregates mining that has a substantial effect on watersheds is sand dredging from rivers. Shortage of sand to support urban construction is a pressing problem in many areas, such as South East Asia; unsustainable sand mining is often accompanied by massive geotechnical and ecological hazards (Kondolf, 1997).

7.3 WATER MANAGEMENT DURING MINE CLOSURE AND REHABILITATION

When the economically viable potential of the raw material is exhausted, mining rehabilitation should begin. Most mining sites that require landscape rehabilitation are large (Figure 7.7). Usually, the use of the area for mining has had a substantial effect on the regional water balance system, as water regulating vegetation may have been removed and natural watersheds drained. The stage of mine closure and rehabilitation provides an

Figure 7.7 Dimensions of mining: active lignite mining site Jänschwalde (Germany, left); former iron mining area Río Tinto (Spain, right). (Photographs: *Petra Schneider*)

Figure 7.8 Example of a flooded open cast mine in the Lusatian area of Germany. The embankments are equipped with stabilising materials. (Photographs: *Petra Schneider*)

opportunity to restore ecosystems and aquifers (Heikkinen *et al.*, 2008; McHaina, 2001; McKenna, 2002; Sheoran *et al.*, 2010; Slingerland & Wilson, 2015), though rehabilitation of most mining sites requires decades of restoration work (Sheoran *et al.*, 2010; Slingerland & Wilson, 2015).

In order to insure that funding will be available for the rehabilitation of mining sites, financial reserves must be built up during the operational phase. In addition to mine closure and dismantling of the facilities, backfilling and flooding of the mines may be required. In this case, groundwater recovery management is necessary, probably including water treatment. In most instances, the first step in the reclamation of the mining affected environment is to drain the pore water from the tailings and to cover the waste and tailings storage facilities. Polluted pore water is treated in a wastewater treatment plant.

Open pits, particularly those resulting from lignite mining, will often be flooded. In the frame of mine site rehabilitation, flooding refers to the cut-off of the drainage systems in order to allow the natural surface and groundwater to enter the mine through shafts and adits or to fill an open cast excavation (Melchers *et al.*, 2019). Mine flooding is a rehabilitation activity that can take several years or even decades. Large volumes of water are necessary that would then be unavailable for other purposes during the flooding period (Figure 7.8). Flooding measures are closely connected to other geotechnical measures that are necessary (e.g. the stabilisation of the embankments of the future lakes). In addition, a chemical stabilisation might be required in case AMD is generated during the flooding process (Johnson & Hallberg, 2005).

Water management during closure and rehabilitation work in the extractive industry typically requires groundwater retention (Figure 7.9, left) as well as the construction of water diversion channels (Figure 7.9, right). This is especially true, when several mines on a mining site are undergoing rehabilitation at the same time, as in the case of the 'Lusatian Lakeland' in Germany (Figure 7.10). In this area, water was diverted to flood open cast mines, developing these lakes into a tourist area. The diversion channels used to flood the mines were also used to support recreational use of small boats by tourists (Lintz *et al.*, 2012).

Tailings pond rehabilitation is a challenge during the whole mine closure process, as the fine sediments are not always settled after mining ceases due to strong pore water pressures. To ensure long term safety of tailings ponds, a methodology using verticals drains has been developed (Wismut GmbH, 2019). Usually, the supernatant water is pumped off to take off free pore water from the surface. Pore water from the

Figure 7.9 Rehabilitation of lignite mines: the case of Lusatia (Germany). Left: groundwater management, right: flooding water channel to fill an open cast mine chain. (Photographs: *Petra Schneider*)

remaining sedimentation layers is drained off through vertical drains. Holes are drilled into the processing sludge (typically 3 m deep) to introduce vertical drains. Stabilisation of the complete system is performed through a geogrid, underlain by geofabric and drain mats (Figure 7.11). In general, the layered system is covered with 1 to 2 m of mineral soil or waste rock material to force tailings consolidation.

Both the extracted pore water and the supernatant water usually need treatment (Wismut GmbH, 2019). In principle, the approaches used entail mechanical, chemical or biological methods or a combination of these (Wolkersdorfer, 2008). Due to the size of the sites and the lengthy time required for the mine water treatment, the extractive industry often employs passive water treatment methods which are based on a combination of aeration and biological treatment e.g. with microorganisms (Baker & Banfield, 2003; Espana *et al.*, 2005; Gazea *et al.*, 1996; Kleinmann *et al.*, 1981; Martínez *et al.*, 2019; Neculita *et al.*, 2007). These mine water treatment methods are designed to occur in open basins (e.g. constructed

Figure 7.10 Flooding water channels for transport of water during the rehabilitation process. After rehabilitation, these channels serve as waterway for small touristic boats. (Photographs: *Petra Schneider*)

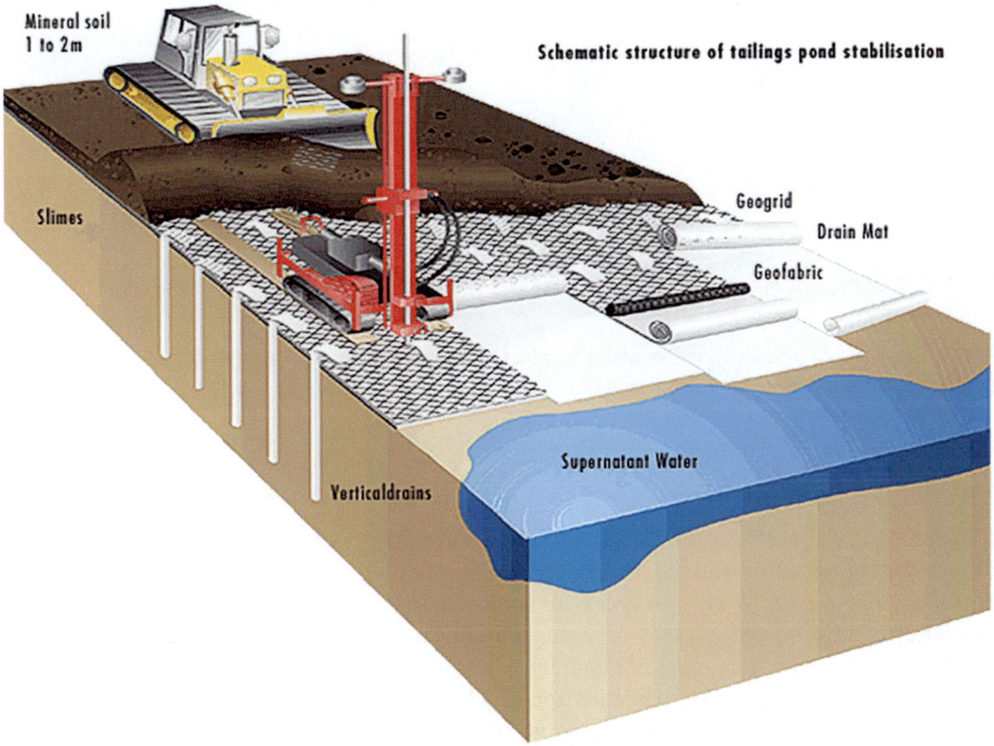

Figure 7.11 Stabilisation technology for tailings ponds: vertical drains in combination with a stabilisation package made of geofabric, drain mats, and geogrid. The system is covered with 1 to 2 m thick mineral soil or waste rock material to force consolidation. (*Modified from Wismut GmbH, 2019*)

wetlands) and are located at the bottom of the mine water outlet (Figure 7.12). Further chemical treatment options for metals include the use of reactive materials (Schneider *et al.*, 2001).

Typically, the remaining waste rock dumps must be rehabilitated as well. In practice, the remediation approach depends on the pollution potential. While sites with lower pollutant potential are often not surface sealed (Ludwig *et al.*, 2003) and, as a preferred solution, are often left to be remediated by natural plant succession (Sweigard *et al.*, 2017), waste rock dumps with higher pollutant potential or in close proximity to protected areas usually need to be covered (Ludwig *et al.*, 2003). A further after-use for biomass production might also be taken into consideration for sites with a small pollution potential that are left to natural plant succession (Bungart *et al.*, 2000). For sites that require a mineral sealing, this sealing may involve a hydraulic barrier (mineral sealing layer with a permeability coefficient of $<1 \times 10^{-9}$ m/s and a thickness of about 50 cm), or pure mineral layers, which form a gas and dust barrier and serve as root space for plants. A special mitigation solution for waste rock dumps with a very high pollutant potential can be reactive horizontal barriers (Schneider *et al.*, 2002).

Aggregates mining quarries are usually reused (Oswald *et al.*, 2018) with different approaches, including industrial heritage tourism (Edwards *et al.*, 1996). After appropriate geotechnical safeguarding, potential uses for reuse of quarries include swimming, fishing, as semi-natural habitat; or as a soil deposit or waste landfill.

Figure 7.12 Example of a passive mine water treatment system in South Africa (left), and the Río Tinto area, Spain (right). (Photographs: *Petra Schneider*)

7.4 ENVIRONMENTAL IMPACTS OF MINING ACTIVITIES

By nature, mining usually causes substantial environmental damage; sometimes this damage is irreversible. In addition to damaging nature and the landscape, mining often affects both the quantity and quality of the water in the catchment area. Water resources are subject to competing usage claims as a result of mining activities. Strict control on the amount of water withdrawn for mining should be enforced by a water management authority. The municipal community is also often faced with major challenges from mining in its environment, as mining unleashes emissions such as dust and noise, in addition to the damage to nature and landscape. The licensing procedure for mining should require environmental impact assessment in the preparation phase, as well as an environmental management plan in the operation phase. In some countries, legislation requires an environmental and social impact assessment.

Assessment of the effects of past mining and treatment activities on the environment is critical for defining the precise requirements of the rehabilitation work that will be needed at the conclusion of the project, in order to justify the spending of funds for this purpose and to monitor the success of the rehabilitation measures. As a result of the long-term nature of the monitoring programs, an optimized approach is required for monitoring equipment and procedures, as well as for the frequency, dimension and accuracy of the monitoring network installed during or after the rehabilitation work. A financial plan and an institutional framework are also required for implementing the monitoring plan.

7.5 SOCIAL IMPACTS OF MINING ACTIVITIES

In many regions of the world, extraction of resources through mining is a way to greatly enhance the economic performance of a region. Plans to mine often create a polarizing effect on the local population. On the one hand, the economic benefits that the region can derive from active mining are seen when the mining operator takes its social responsibility. However, this also presupposes that its corresponding benefit sharing takes place with the municipalities concerned and their inhabitants. Implemented in this way, mining activities can bring about a societal upswing associated with job creation and increasing prosperity. There are numerous examples on a global scale for this. On the other hand, resistance to mining activities is increasing in many regions, as has been the case in recent years, for example in Romania and Ecuador. At the forefront of resistance to mining activities is an awareness of the need to protect natural resources and biodiversity, coupled with a recognition that mining has not always been done in a manner that protects either nature or communities. For these reasons, mining activities should reflect a consensus decision between all protected goods and stakeholders, and should only advance when a social license to operate has been secured.

The Lusatian mining area in East Germany provides a well-defined example of the social effects of mining and its closure in a region. While 30 opencast mines in Lusatia were used for lignite mining during the socialist German Democratic Republic (GDR) era and employed 75,000 people, there are currently four active open pits left where people find work (Welzow-Süd, Jänschwalde, Nochten, Reichwalde). However, many workers are engaged in mining rehabilitation to give the region a new face and a future. Twelve open pits have been flooded in the Federal States of Brandenburg and Saxony since the political change in East Germany, forming a total water surface of about 8,000 ha. However, it should not be forgotten that since 1924, 80 villages were dismantled because of lignite mining (Förster, 1995); local memorials remind future generations of these modifications (Figure 7.13, left). In order to keep the mines dry, it was necessary to drain the groundwater down to depths of up to 150 m. As a result, streams and wetlands dried out, which now are partially irrigated. In addition, the soil structure changed, resulting in extensive land settlements (sometimes up to distances of 15 to 20 km in the Lusatian region). The use of land in the Lusatian lignite mining areas resulted in a total deficit of around 13 km^3 of groundwater in 1990. Today, the deficit amounts to around 6 km^3.

In Lusatia, an ambitious project has been developed with the Lusatian Lake District to give the region a new future after the remediation has been completed. One example is Lake Geierswald (Figure 7.13 right), a former lignite open pit that has been flooded since 1973, which now has a size of 653 ha and a depth of up to 34 m (Lausitzer Seenland, GmbH, 2019). However, the ability to link such visions for the future with plans for new mining ventures presuppose that during the active mining period, the mining operator will provide financial reserves to finance the refurbishment costs and to ensure that these are not transferred to the taxpayer. Most countries with active mining have such regulations in their mining laws to ensure this funding. In practice, however, these regulations are not always enforced.

The need to assess and mitigate the economic and social consequences of mine closure results in the conclusion that the involvement of the local service and material suppliers may be critical in terms of

Figure 7.13 Period after mining: memorial for a village that was removed for lignite mining in Lusatia, Germany (left), floating houses on a flooded open cast lignite mine for touristic purposes at Lake Geierswald, Germany. (Photographs: *Petra Schneider*)

obtaining the approval of the community and other local and national related authorities for carrying out the rehabilitation works. Consequently, the remediation concept from each site needs to be correlated as much as possible with these issues. The remediation strategy for a site also needs to consider feasible post-mining use scenarios that could be supported by remediation investments.

7.6 ROLE OF FINANCIAL GUARANTEES

Financial guarantee is defined as the obligation and responsibility of the natural or legal entities carrying out mining operations under an exploration license or permit, to supply the financial funds required for the environment rehabilitation. For the European member states, Directive 2006/21/EC of the European Commission (EC) on the management of waste from the extractive industries (EWD) (EU Parliament, 2006) provides 'measures, procedures and guidance to prevent or reduce as far as possible any adverse effects on the environment and any resultant risks on human health' from the management of extractive waste. This Directive obliges the competent authority to 'require a financial guarantee, prior to the commencement of any operations involving the accumulation or deposit of extractive waste in a waste facility'. The EWD requires the Member States to ensure that the operator draws up an extractive waste management plan (EWMP). This EWMP includes a 'proposed plan for closure, including rehabilitation, after-closure procedures and monitoring'. Beside the Directive 2006/21/EC, a main regulatory frame is given through the Directive 2004/35/EC of the European Parliament and of the Council of 21 April 2004 'on environmental liability with regard to the prevention and remedying of environmental damage' (EU Parliament, 2004). In general, 'a financial guarantee is a promise to assume responsibility for another entity's financial obligation if that entity is unable to meet its obligation' (Chaudhari, 2017). Directive 2006/21/EC recognises the 'importance of guaranteeing that the taxpayer is not left with the financial burden of environmental clean-up and rehabilitation of mining liabilities, as has often been the case in the past and thus, requires a financial guarantee to be lodged by the mine operator prior to the commencement of deposition operations in the waste facility' (MonTec, 2008).

7.7 CONCLUSIONS AND OUTLOOK

Mining activities cause large-scale and often irreversible interventions in nature and landscape, but are also required to ensure the provision of mineral resources. Water management plays a substantial role during exploration, development, mining and mining rehabilitation.

While the active mining phase involves hydrogeological site exploration, as well as planning and implementation of groundwater lowering, water supply and disposal, the construction of waste disposal facilities and flood protection in the rehabilitation phase with demolition of the mining facilities is in the focus of the activities. This involves the backfilling of cavities and flooding of the mines, the management of the groundwater recharge with the associated water drainage and treatment, as well as the sealing of waste rock dumps and tailings ponds including their rehabilitation.

Regardless of whether mining is in the active or decommissioning phase, the dimension and challenges of these tasks are usually enormous. Large quantities of rock and water have to be moved, and existing land uses have to be abandoned. This results in competing usage claims, and often, a high cost to residents when mining is officially approved. Often the water demands of other water users are subordinated to those of the mining industry. The priorities of residents and the long-term health of the local ecosystem should play a substantial role when mining projects are considered, licensed, and monitored.

ACKNOWLEDGMENTS

This chapter was prepared in the context of the project 'Water Management and Climate Change in the Focus of International Master Programs (WATERMAS)', funded by the Erasmus+ Programme of the European Union. As this manuscript only reflects the views of the authors, the European Union cannot be held responsible for these views or any future use of them.

All photos were taken by the authors during their work in the mining industry.

REFERENCES

Baker B. J. and Banfield J. F. (2003). Microbial communities in acid mine drainage. *FEMS Microbiology Ecology*, **44**(2), 139–152.

Blowes D. W., Ptacek C. J., Jambor J. L., Weisener C. G., Paktunc D., Gould W. D. and Johnson D. B. (2014). The Geochemistry of Acid Mine Drainage. In: Treatise on Geochemistry, H. D. Turekian and K. K. Holland (eds.), Elsevier, Oxford, pp. 131–190.

Bowles D. S., Giuliani F. L., Hartford D. N. D., Janssen J. P. F. M., McGrath S., Poupart M., Stewart D. and Zielinsk P. A. (2007). ICOLD bulletin on dam safety management. Proceedings NZSOLD-ANCOLD 2007 Conference – Dams – Securing Water for Our Future, Queenstown, 1–9.

Bungart R., Bens O. and Hüttl R. F. (2000). Production of bioenergy in post-mining landscapes in Lusatia: Perspectives and challenges for alternative landuse systems. *Ecological Engineering*, **16**, 5–16.

Chaudhari C. A. S. (2017). A Guide to Risk Based Internal Audit System in Banks. Notion Press, Chennai.

Dold B. (2008). Sustainability in metal mining: from exploration, over processing to mine waste management. *Reviews in Environmental Science and Bio/Technology Volume*, **7**(4), 275–285.

Edwards J. A., Llurdes I. and Coit J. C. (1996). Mines and quarries: Industrial heritage tourism. *Annals of Tourism Research*, **23**, 341–363.

Espana J. S., Pamo E. L., Santofimia E., Aduvire O., Reyes J. and Barettino D. (2005). Acid mine drainage in the Iberian Pyrite Belt (Odiel river watershed, Huelva, SW Spain): geochemistry, mineralogy and environmental implications. *Applied Geochemistry*, **20**(7), 1320–1356.

EU Parliament. (2004). Directive 2004/35/CE of the European Parliament and of the Council of 21 April 2004 on environmental liability with regard to the prevention and remedying of environmental damage.

EU Parliament. (2006). Directive 2006/21/EC of the European Parliament and of the Council of 15 March 2006 on the management of waste from extractive industries and amending Directive 2004/35/EC – Statement by the European Parliament, the Council and the Commission.

Evangelou V. P. and Zhang Y. L. (1995). A review: pyrite oxidation mechanisms and acid mine drainage prevention. *Critical Reviews in Environmental Science and Technology*, **25**(2), 141–199.

Fields S. (2003). The earth's open wounds: Abandoned and orphaned mines. *Environmental Health Perspectives*, **111**, A154–A161.

Förster F. (1995). Verschwundene Dörfer. Die Ortsabbrüche des Lausitzer Braunkohlenreviers bis 1993. *Schriften des Sorbischen Instituts*, Bautzen, 8.

Gazea B., Adam K. and Kontopoulos A. (1996). A review of passive systems for the treatment of acid mine drainage. *Minerals Engineering*, **9**(1), 23–42.

Grünewald U. (2001). Water resources management in river catchments influenced by lignite mining. *Ecological Engineering*, **17**(2–3), 143–152.

Heikkinen P. M., Noras P., Salminen R., Mroueh U. M., Vahanne P., Wahlström M., Kaartinen T., Juvankoski M., Vestola E., Mäkelä E., Leino T., Kosonen M., Hatakka T., Jarva J., Kauppila T., Leveinen J., Lintinen P., Suomela P., Pöyry H., Vallius P., Nevalainen J., Tolla, P. and Komppa, V. (2008) Mine Closure Handbook, Geological Survey of Finland, Espoo.

Johnson D. B. and Hallberg K. B. (2005) Acid mine drainage remediation options – a review. *Science of the Total Environment*, **338**, 3–14.

Kleinmann R. L. P., Crerar D. A. and Pacelli R. R. (1981). Biogeochemistry of acid mine drainage and a method to control acid formation. *Minerals Engineering*, **33**(3), 300–305.

Kondolf G. M. (1997). PROFILE: hungry water: effects of dams and gravel mining on river channels. *Environmental Management*, **21**(4), 533–551.

Lausitzer Seenland GmbH. (2019). Geierswalder See, https://www.lausitzerseenland.de/de/die-seen/artikel-geierswalder-see.html (accessed 15 September 2019).

Lintz G., Wirth P. and Harfst J. (2012). Regional structural change and resilience: from lignite mining to tourism in the Lusatian Lakeland. *Raumforschung und Raumordnung*, **70**, 363–375. doi: 10.1007/s13147-012-0175-x

Ludwig J. A., Hindley N. and Barnett G. (2003). Indicators for monitoring minesite rehabilitation: trends on waste-rock dumps, northern Australia. *Ecological Indicators*, **3**(3), 143–153.

Martínez N.M., Basallote M.D., Meyer A., Cánovas C.R., Macías F. and Schneider P. (2019). Life cycle assessment of a passive remediation system for acid mine drainage: Towards more sustainable mining activity, *Journal of Cleaner Production*, **211**, 1100–1111. doi: 10.1016/j.jclepro.2018.11.224

McHaina D. M. (2001). Environmental planning considerations for the decommissioning, closure and reclamation of a mine site. *International Journal of Surface Mining, Reclamation and Environment*, **15**, 163–176.

McKenna G. (2002). Sustainable Mine Reclamation and Landscape Engineering. Ph.D. Dissertation, University of Alberta, Edmonton, SK, Canada.

Melchers C., Westermann S. and Reker B. (2019). Evaluierung von Grubenwasseranstiegsprozessen. *Berichte zum Nachbergbau*, **1**, 1–130, Deutsches Bergbau-Museum Bochum, Bochum.

MonTec GmbH. (2008). Guidelines on Financial Guarantees and Inspections for Mining Waste Facilities – Final Report for European Commission – DG Environment. Report Nr. 2007/S 49-059670.

Neculita C. M., Zagury G. J. and Bussière B. (2007). Passive treatment of acid mine drainage in bioreactors using sulfate-reducing bacteria. *Journal of Environmental Quality*, **36**(1), 1–16.

Olías M., Nieto J.M., Sarmiento A.M., Cerón J.C. and Cánovas C.R. (2004). Seasonal water quality variations in a river affected by acid mine drainage: the Odiel River (South West Spain). *Science of the Total Environment*, **333**(1–3), 267–281.

Oswald K.-D., Schneider P. and Riedel W. (2018). Technical solutions for mining and processing of aggregates and the mining sites after-use: A Cleaner Production Guideline for Vietnam, Results of the project MAREX: Management of Mineral Resource Extraction in Hoa Binh Province – a Contribution to Sustainable Development in Vietnam (2015–2018); Issue 3, IÖR Eigenverlag Dresden.

Pepper M., Roche C. P. and Mudd G.M. (2014). Mining legacies – Understanding life-of-mine across time and space. Proceedings of the Life-of-Mine Conference 2014, Brisbane, Australia, 16–18 July 2014, 1449–1466.

Rapantova N., Grmela A., Vojtek D., Halir J. and Michalek B. (2007). Ground water flow modelling applications in mining hydrogeology. *Mine Water and the Environment*, **26**(4), 264–270.

Ritcey G. M. (1989). Tailings Management – Problems and Solutions in the Mining Industry. Elsevier, Amsterdam.

Sarmiento A. M., Nieto J. M., Olías M. and Cánovas C. R. (2009). Hydrochemical characteristics and seasonal influence on the pollution by acid mine drainage in the Odiel river Basin (SW Spain). *Applied Geochemistry*, **24**(4), 697–714.

Schneider P., Neitzel P. L., Osenbrück K., Noubacteb C., Merkel B. J. and Hurst S. (2001) In-situ treatment of radioactive mine water using reactive materials – Results of laboratory and field experiments in uranium ore mines in Germany. *Acta hydrochimica et hydrobiologica*, **29**(2–3), 129–138, doi: 10/ff64fb

Schneider P., Osenbrück K., Neitzel P. L. and Nindel K. (2002). In-situ mitigation of effluents from acid waste rock dumps using reactive surface barriers – a feasibility study. *Mine Water and the Environment*, **21**(1), 36–44.

Sheoran V., Sheoran A. S. and Poonia P. (2010). Soil reclamation of abandoned mine land by revegetation: A review. *International Journal of Soil, Sediment and Water*, **3**(2), Article 13.

Slingerland N. and Wilson G. W. (2015). End Land Use as a Guide for Integrated Mine Planning and Closure Design. Mine Closure, Vancouver, BC, Canada. Available online: https://www.researchgate.net/profile/Neeltje_Slingerland/publication/304626811_End_land_use_as_a_guide_for_integrated_mine_planning_and_closure_design/links/57c5b69f08ae0a6b0dc8d31f.pdf (accessed 15 September 2019).

Sweigard R., Burger J., Zipper C., Skousen J., Barton C. and Angel P. (2017). Low compaction grading to enhance reforestation success on coal surface mines. In: The Forestry Reclamation Approach: guide to successful

reforestation of mined lands, M. B. Adams, (eds.) Gen. Tech. Rep. NRS-169. US Department of Agriculture, Forest Service, Northern Research Station, Newtown Square, PA, 4-1–4-8.
Tiwary R. K. (2001). Environmental impact of coal mining on water regime and its management. *Water Air and Soil Pollution*, **132**(1–2), 185–199.
Wismut GmbH (2019). Stabilisation of tailings management areas (TMA). Available online: https://www.wismut.de/en/tailings_pond_remediation.php. (accessed 15 September 2019).
Wolkersdorfer C. (2008). Water Management at Abandoned Flooded Underground Mines – Fundamentals, Tracer Tests, Modelling, Water Treatment. Springer, Heidelberg.
Wolkersdorfer C., Nordstrom D. K., Beckie R., Cicerone D. S., Elliot T., Edraki M., Valente T. M., França S. C. A., Kumar P., Oyarzún Lucero R. A. and Soler A. I. G. (2020). Guidance for the integrated use of hydrological, geochemical, and isotopic tools in mining operations. *Mine Water and the Environment*, **39**(2), 204–228, doi: 10.1007/s10230-020-00666-x
Younger P. L. and Wolkersdorfer C. (2004). Mining impacts on the fresh water environment: technical and managerial guidelines for catchment scale management. *Mine Water and the Environment*, **23**, s2–s80, doi: 10.1007/s10230-004-0028-0

Chapter 8

Triple water reuse at Dow

Cornelis (Niels) Groot

Dow Benelux BV, Environmental Technology Center, Terneuzen, The Netherlands

Keywords: industrial reuse, water stress

8.1 INTRODUCTION

One of the bigger challenges industry faces today relates to water use, especially in areas where water-stress due to drought or sea level rise and competition for freshwater sources are most severe. The availability of freshwater in particular is of utmost importance for many industries, including the petrochemical industry where water is needed for heating and cooling. For the petrochemical industry to sustain its operations, it must secure reliable access to a freshwater supply to replenish the large amounts of water 'lost' due to evaporative cooling systems and the discharge of wastewater.

To ensure an adequate, reliable supply, industries must find ways to use water more efficiently. However, sound investments in water are difficult to justify on economic value only. Water pricing is often related to the social right of each individual to have access to healthful water. However, water is more 'valuable' than can be captured in its price alone. Hence, industries are challenged to develop a strategy for sustainable water use for long-term economically feasible operations that reduces its overall environmental impact and expresses its responsibility towards society.

For each location these strategies must be tuned to specific circumstances. In every case, though, highest on the ladder of options is avoiding or reducing water use where possible to lower the industry's intrinsic water footprint. The next level is reuse of water within the industrial processes. Reusing water also minimizes intermediate treatment (e.g. chemical addition) and may also reduce heat (energy) losses. An example frequently applied in the (petro) chemical industry is the reuse of steam condensates in adjacent processes. If designed well, opportunities exist to cascade water qualities in a number of applications within facilities to maximize its beneficial use while minimizing the interconnecting infrastructure.

DOI: 10.2166/9781789060676_0089

Finally, prior to discharge, wastewater may be further treated to return into the water cycle. Industrial water recycling emphasizes the importance of running a stable and efficient wastewater treatment plant, in which plant discharges are seen as 'a product' to be prepared for recycling. To this end, upstream segregation of storm water flows from process water often enables its cost-effective treatment and reuse.

Nevertheless, even after water use is reduced and all internal recycled water uses are maximized, there will remain a need for an external supply of freshwater for industry to make up for the permanent losses. In regions where industries must share limited water supplies with a variety of agricultural and domestic water users it is worthwhile to explore opportunities for collaboration with other stakeholders to seek synergy in sustainable solutions that serve everyone's mutual interests. This leads to the concept of a regional sustainable water management, as Dow demonstrated recently at its facility in Terneuzen, the Netherlands.

8.2 DOW'S INTEGRATED APPROACH TO SUSTAINABLE WATER USE

Dow's efforts to conserve water and preserve watersheds are embedded in its seven 2025 Sustainability Goals which seek to help lead the transition to a sustainable society Dow's water blueprint. Inspired by the goal 'World Leading Operations,' Dow identified six of its manufacturing sites as water-stressed, and is working to reduce freshwater intake at these sites by 20%. In pursuit of the goal 'Valuing Nature,' Dow is poised to deliver $1 billion of company value in business-driven projects that enhance nature, working with The Nature Conservancy to demonstrate how to value water and evaluate investments that sustain rivers and watersheds, including nature-based solutions like wetlands.

8.2.1 Analytical approach

To assess its water stressed locations, Dow applied both the Global Water Tool developed by the World Business Council for Sustainable Development (WBCSD – global water tool) and the Aqueduct tool developed by the Water Resources Institute (WRI) (Gassert *et al.* 2013). Table 8.1 shows how Dow weighted each of the assessment factors that the WRI tool uses to score overall water risk.

This analytical framework provided a prioritized list of six major production sites at separate locations vulnerable to and negatively affected by water stress. The six sites identified were the following:

- Tarragona (Spain)
- Bahia Blanca (Argentina)
- Freeport (USA)
- Seadrift (USA)
- Terneuzen (Netherlands)
- Boehlen (Germany).

As a result of this analysis, in 2011 Dow Tarragona began using treated municipal effluent as cooling tower make-up (https://corporate.dow.com/en-us/news/press-releases/dows-camp-de-tarragona-demoware-industrial-water-reuse-project.html), while in Bahia Blanca a large project was started to treat and purify municipal wastewater effluent for use in industry as an alternative for poor quality raw surface water. In the US, Dow's Freeport facility began recycling river water while their Seadrift facility directed the wastewater effluent after treatment to Dow owned and operated wetlands for a final polishing. At Dow Boehlen (Central Germany), plans are in place to install treatment wetlands for rain and process water reuse. By contrast, at Dow Terneuzen (as described in detail in the case study below) the optimal sustainable solution required close cooperation between a number of public and private sector partners.

Beyond these six, there is a list of almost 20 smaller locations which may suffer significantly from water scarcity. In order to include nature-based technologies in its toolset of sustainable solutions, Dow partnered

Table 8.1 WRI weight factors in evaluating water risk.

Dow Overall Water Risk	Weighting (%)*
Baseline Water Stress	24.2
Upstream Storage	12.1
Groundwater Stress	12.1
Access to Water	12.1
Inter-annual Variability	6.1
Flood Occurrence	6.1
Drought Severity	6.1
Return Flow Ratio	6.1
Media Coverage	6.1
Seasonal Variability	3.0
Upstream Protected Land	3.0
Threatened Amphibians	3.0

*Out of 100%.

with the Nature Conservancy to develop the 'ESII Screening Tool' which scores projects on their nature-related impacts.

8.2.2 Technical approach

Generally speaking, process water can be reused at several places within a site after sufficient treatment. Most large chemical complexes and industrial parks comprising multiple facilities with independent owners (I-Parcs) collect wastewater from all processes and operate a centralized wastewater treatment plant (WWTP) to remove contaminants before discharge. As indicated in Figure 8.1, depending on the complexity of the water to be treated, a WWTP may include two or more treatment unit operations, so if the wastewater cannot be reused directly, it may be suitable for reuse after additional treatment by one or more units at the WWTP. In some cases, it may be attractive to treat process water at its source, close to the production facility where it originates. This specifically applies to process waters with high concentrations of organics, where anaerobic treatment may be feasible for energy recovery and for treatment of streams with contaminants that are difficult to remove at the central wastewater treatment plant (e.g. metals, oil, and grease).

Typically, before such treatment technologies are incorporated into industrial processes, each approach must be tested in the laboratory and in pilot plants before progressing to a full-scale design. When Dow first began its reuse programme, there were few successful examples of full-scale industrial reuse available for Dow to reference. As a result, Dow undertook several research programmes over the years to enhance insight into available technologies, the water quality they produced and their impact on downstream users. Depending on the water source and its application, a mix of different treatment options has now been used. These include reverse osmosis (RO), nanofiltration, ultrafiltration, and electrodialysis reversal (for desalination); biological solids and nutrient removal and ultraviolet/ultrasound (for disinfection); ion exchange (for producing ultrapure water); natural and constructed wetlands (as tertiary treatment and raw

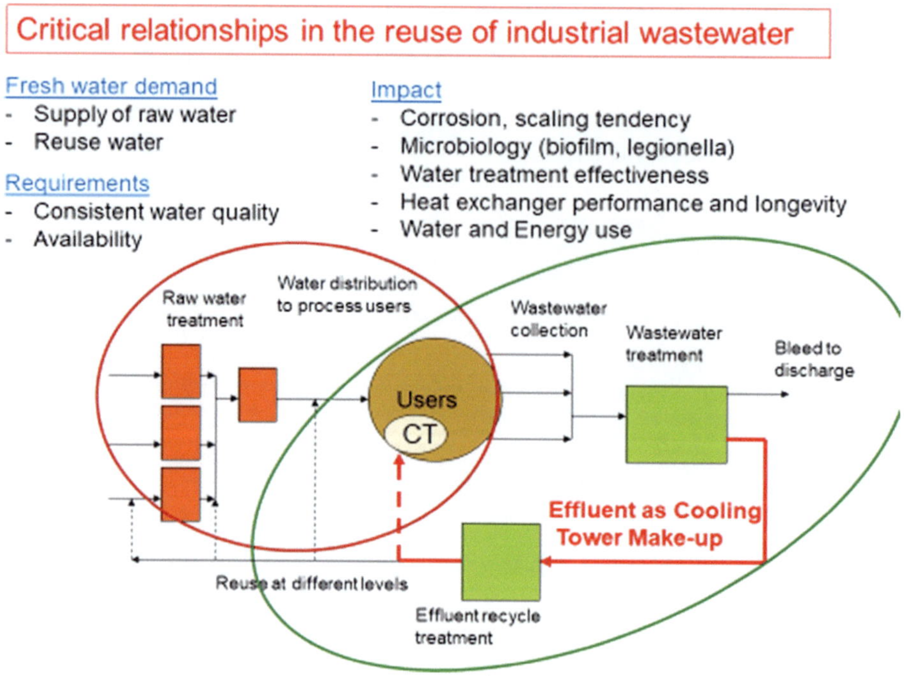

Figure 8.1 Industrial water reuse options and critical aspects for reuse as cooling tower make-up.

water stabilization); and Nalco Water's 3D TRASAR™ technology to reduce chemical addition and increase cycles of concentration in evaporative cooling towers.

Dow's interest in learning more about water treatment alternatives also extends to its commitment to non-competitive research with other parties on a regional, national, and international level. On a European level, Dow has participated in EU FP7 projects on desalination of brackish raw water streams (E4Water project involving Terneuzen E4Water (2016)) as well as to enhance water reuse solutions (through the DEMOWARE project at Tarragona). Also Dow Terneuzen participated in two regional EU projects (Interreg), i.e. IMPROVED (2019) (polishing contaminated steam condensates for direct reuse) and FRESH4Cs (identifying and demonstrating freshwater subsurface infiltration, storage, and abstraction). Nationally, Dow Terneuzen has participated in a number of collaborative programmes. These included Wetsus Research, NWO Water NEXUS by Wageningen Research Water NEXUS (2015), and the Eurydice and Steam Condensate Quality projects by the Institute for Sustainable Process Technology to recover salts from process water and improve the quality of steam condensate for reuse. On a regional level, in the Netherlands, Dow initiated collaboration with a broad range of partners to develop a robust water system for the region. Participants included: provincial and municipal government and water board representatives; private enterprises (Dow, Yara, ICL-IP, Evides, and North Seaport); nature and environmental groups; universities (HZ University of Applied Sciences); and agricultural representation (ZLTO). Similarly, in the United States, Dow facilities in Texas partnered with the Nature Conservancy to evaluate water options for the Freeport site and develop wetlands at the Seadrift facility.

In other instances, the production process itself may be reconfigured to use less water. This was the case when Dow and BASF collaborated to develop a new process for the production of propyleneoxide.

Compared to the previous formation reaction with chlorine, the direct oxidation of propylene with peroxide reduced freshwater use by 70%. The first new fully Dow-owned propyleneoxide plant was built in Thailand in 2011 and a second one in Saudi Arabia in 2017. Dow's desire to be good water stewards in their communities extends beyond its own sites. Dow also collaborates with the Great Lakes Blue Accounting platform to bring data on the Great Lakes together for decision makers. Dow employees also works with the Ocean Conservancy on worldwide coastal cleanups in watersheds where they live and work (#pullingourweight).

8.2.3 Educational approach

In addition to its research into water technology, Dow provides training to its employees who are involved in water and water treatment via on-line E-Learning courses. These courses range from basic to advanced learning modules including sustainable water use and associated practices. Specific courses have been developed for water risk management, especially focusing on water availability in water scarce areas. In specific cases class room trainings are provided.

8.3 SITE SPECIFIC APPROACH: TERNEUZEN, THE NETHERLANDS

Dow's manufacturing site in Terneuzen, The Netherlands, is in a watershed where freshwater is scarce. Growing populations and manufacturing needs are straining water sources, and rising sea levels threaten the freshwater supply. Additionally, the site withdraws fresh river water from near sensitive wetlands of international importance (Biesbosch area).

8.3.1 Reduce, reuse, recycle

To support the growth of the Terneuzen facility (Figure 8.2), Dow has invested in a rigorous water reuse scheme over the past two decades. After reducing both water use and pollution within its manufacturing plants, in 1995 Dow constructed a central wastewater treatment plant consisting of two parallel trains. This dual design was subsequently utilized five years later when the infrastructure was adapted to treat fresh and saline wastewater separately, enabling the reuse of the freshwater fraction after treatment.

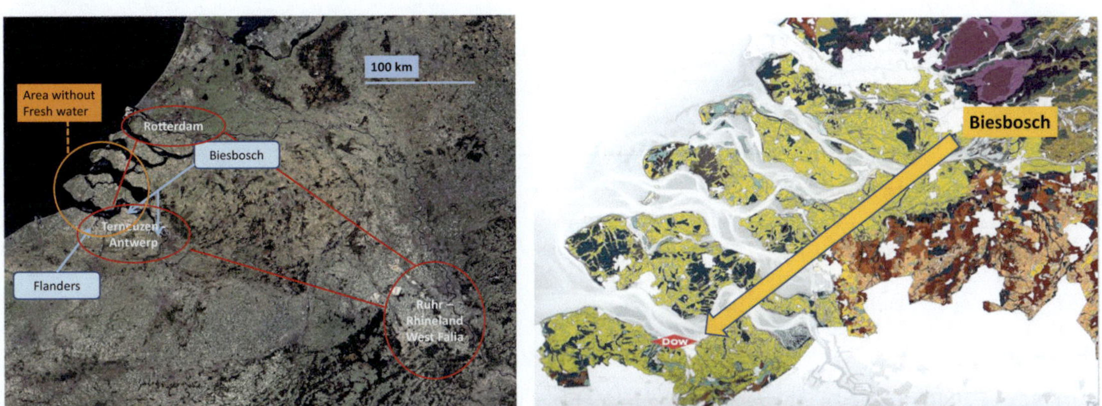

Figure 8.2 Dow Terneuzen, which is located in an economic triangle that is embedded in a brackish water region, imports freshwater from the Biesbosch wetland area.

Today about half of this fresh effluent is used as make-up for evaporative cooling, which comprises 20% of the site's total water demand. Internal reuse of steam condensates makes up another 25% of the total water demand. In addition, rainwater collected from roofs of warehouses and from green ditches is stored in a 70,000 m^3 basin for use in the site's firewater distribution system.

8.3.2 Collaboration is key

After developing a strategy for internal efficiency and water reuse, Dow looked beyond its fenceline for alternative reuse options. This resulted in a unique collaborative reuse solution that required close cooperation with the City of Terneuzen, Evides Industriewater, and the regional Water Board, which owns and operates the City's wastewater treatment plant. In 2007 the Terneuzen facility began taking effluent from the city's wastewater treatment plant. The effluent was pumped through a 13 km pipeline from the wastewater plant to the local water supplier (Evides Industriewater), which then conveyed it to the Dow facility for reuse. When it appeared that the secondary effluent caused excessive biological growth in the interconnecting pipeline, Dow worked with Evides to install a new membrane bioreactor (MBR) at the municipal wastewater treatment plant. In addition to improving effluent quality, the MBR unit also gave the City of Terneuzen the additional treatment capacity that the regional Water Board was looking for (Figure 8.3). About 15% of Dow's total water demand is now provided by this reuse of municipal effluent. The value of this collaboration was increased further when an amendment to Dutch law made reused municipal wastewater effluent exempt from additional discharge fees.

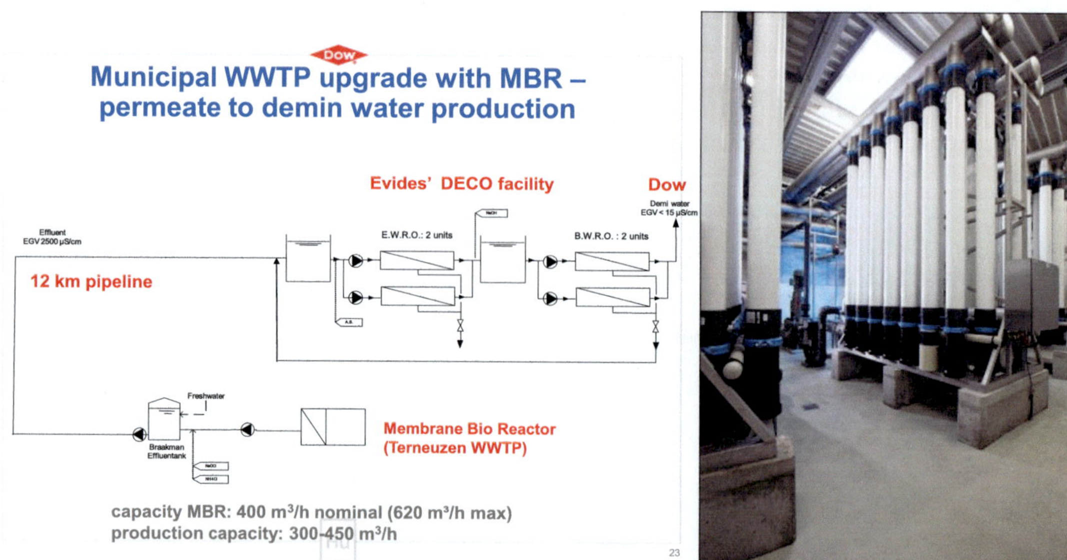

Figure 8.3 In 2010, the city's wastewater treatment plant was equipped with a membrane bioreactor (owned by Evides), which increased the treatment capacity of the municipal plant by 20% and dramatically improved the quality of the effluent (called permeate) sent to Evides' reverse osmosis treatment facility for the production of demin water. (*Credit: Evides Industriewater, Rotterdam, the Netherlands*)

8.3.3 Triple reuse and beyond

As a result of these efforts, as illustrated in Figure 8.4, water at the Terneuzen facility is used three times:

- when freshwater enters the plant and is used for the first time;
- when used water is reused directly or treated further for onsite reuse; and
- when treated effluent is returned from the municipal plant as source water for the Terneuzen facility.

Encouraged by the success of these efforts, Dow has set a new goal for the Terneuzen site to completely eliminate the use of remotely supplied potable water (currently 25% of plant demand) by 2025. Further alternatives are being explored within the region with both public and private partners, including brackish water desalination, use of natural systems ('green infrastructure' shown in Figure 8.5), and freshwater underground storage. Implementation of these projects is anticipated to take place by 2024 at the latest.

8.4 FUTURE OUTLOOK

In the face of climate change, with the threat of increasing salinization of freshwater due to sea level rise, Dow Terneuzen does not want to compete with other municipal water users for that resource. In Terneuzen collaboration with multiple partners created a 'robust water system', ensuring that water would be available for the region in the future for all purposes and under all conditions (including droughts and floods). This collaboration also extends further across the border with Belgian (Flemish) partners, for example in the FRESH4Cs project to demonstrate the sustainable use of underground creek ridges for storing excess rainfall during winter and subsequent abstraction in water stressed summer periods.

Future challenges to Dow's achievement of its 2025 Sustainability Goals remain in the areas of technology, economics, and social issues. Finding the right technologies to connect source water to application is not easy; however, by using data analysis, testing, piloting, and appropriate design, it can be accomplished. It remains difficult to build sound business cases to justify infrastructure-intensive water projects solely based on the price of water. Therefor also socio-economic factors should be taken into account. Social challenges include the cultural differences between collaborating partners (e.g. public, private, NGOs, academia). Each group has its own practices, criteria, and work processes, which can be a barrier to building consensus and reaching mutually acceptable decisions.

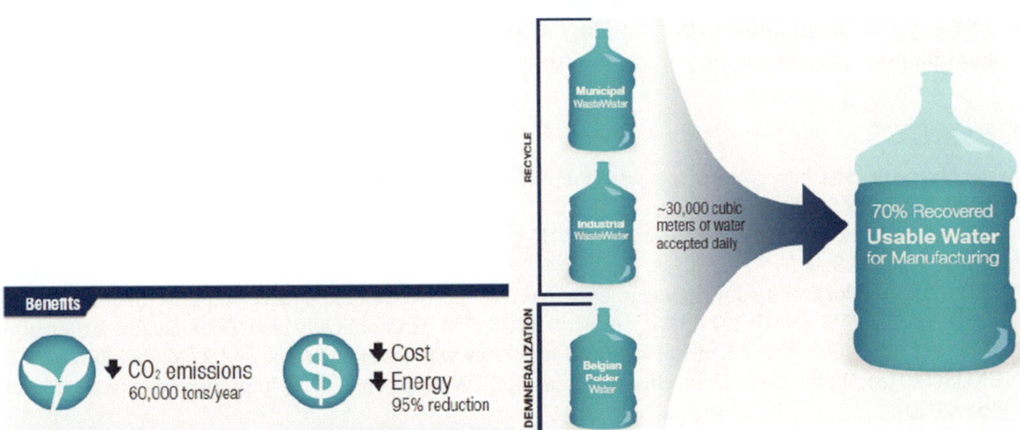

Figure 8.4 Triple reuse at Dow Terneuzen – water, associated energy, and chemical savings.

Figure 8.5 Dow, Evides, and Water Board partnering in research to use constructed wetlands for pretreatment of brackish water.

8.5 CONCLUSIONS

Growing populations and economies mean that there will continue to be a strain on the water resources available to industry. The chemical industry is increasingly aware of the impact current and future water stress has on its operations. It is important that the chemical industry works together with other industries and water users to develop strategies to enhance water recycling and promote sustainable water use as a whole.

We need to bring sustainable water management to the next level. Within industries water needs to be seen more as a valuable resource than just a cost of doing business that must be minimized. New process designs require close cooperation between process engineers and water technology specialists to decrease current high water footprints. More information on Dow's corporate sustainability projects and goals is available on its water blueprint page, www.dow.com/waterblueprint or in Dow's annual Sustainability Report.

REFERENCES

Dow's water blueprint page, https://corporate.dow.com/en-us/science-and-sustainability/2025-goals/blueprint/water (accessed 2 October 2020).

Gassert F., Landis M., Luck M., Reig P. and Shiao T. (2013). Aqueduct Metadata Document. World Resources Institute, Washington DC, U.S.A.

IMPROVED. (2019). Pilot facility development, https://www.improvedwater.eu (accessed 2 October 2020).

E4Water. (2016). EU FP7 (NMP.2011.3.4-1) project nr 280756 Economically and Ecologically Efficient Water Management in the European Chemical Industry, http://www.e4water.eu/ (accessed 2 October 2020).

Water NEXUS. (2015). Wageningen University & Research, STW project nr 14199, http://waternexus.nl/ (accessed 2 October 2020).

https://www.wbcsd.org/Programs/Food-and-Nature/Water/Resources/Global-Water-Tool-Flyer (accessed 2 October 2020).

Chapter 9

Severe water crises: Industry's role and response

Brent Haddad

University of California-Santa Cruz, Environmental Studies, Santa Cruz, USA

Keywords: drought, economics, industrial water use, sustainability

9.1 INTRODUCTION

Managing water supply during emergencies is a critical part of sustainable water management, and may be more common in the coming decades. Four times in the past decade cities on four continents came face to face with 'Day Zero' when water would run out: in the Murray-Darling Basin, Australia (Breyfogle, 2010); Big Spring, Texas (Martin, 2014); Cape Town, South Africa (Buder, 2019); and Chennai, India (Gupta, 2019). In each case they were saved by a combination of extreme conservation, new water supply technologies, and rain. There were also a number of urban water quality emergencies, including Newark, New Jersey and Flint, Michigan, where aging pipelines conveying incompatible supplies resulted in lead-poisoned water being delivered to thousands of homes. We only know about the events in open societies: it is possible that cities in nations with restricted press freedom experienced similar crises.

These events impacted whole cities and their integrated water systems. None were caused by industry, but industry was itself affected by the water shortages and quality breakdowns. This paper focuses on the role and treatment of industry during a water crisis, laying out general principles for how industry should be treated by government and then providing some examples of how the principles can be implemented.

9.2 RELATIONSHIP BETWEEN INDUSTRY AND WATER

Industry acquires, transforms, and combines materials into products, and influences the water sector in countless ways. These can be divided into (1) impact of industrial production on water supplies, including the use of water and discharge of wastewater; (2) impact of the use of industrial products on the water environment; and (3) industrial equipment manufactured for water supply and treatment.

DOI: 10.2166/9781789060676_0097

Industry is a water consumer, accounting for roughly 19% of water withdrawals world-wide (WWAP, 2019). Industrial uses of water include fabricating, processing, washing, diluting, cooling, transporting a product, incorporating water into a product, and sanitation needs within the manufacturing facility (USGS, 2019). Plant cooling is a major non-consumptive user of withdrawn water, while food processing, paper, and mining are major consumptive users. Industry is a predictable water consumer since industrial facilities operate at planned production rates and is capable of conservation, although many production methods have minimum water needs below which production will cease.

Industry also alters the chemistry of water returned to the environment. Water that leaves industrial facilities, especially chemical and pharmaceutical producers, often carries dangerous by-products that must be removed to prevent damage to downstream facilities. Cities normally require 'pretreatment' in the form of removal or neutralization of these chemicals before industrial effluent can be mixed with other urban wastewater and treated at a wastewater plant.

Water providers influence industry in other ways as well. Companies design products that use water on the basis that the user will have access to water of appropriate quantity and quality. Product design may be further constrained in markets where regulations limit water use and specify wastewater discharge quality. Industry also makes products for the water supply and wastewater treatment industry. The vast infrastructure of 'the urban water cycle' moves water to cities, treats it for use, and then treats the effluent for safe discharge or reuse. While natural conveyances (rivers) and storage systems (snowpack) are integrated into these engineered water systems, industry provides the multi-billion-dollar infrastructure of reservoirs, canals, tanks, pumps, pipes, treatment plants, laboratories, and computerized monitoring and operating systems.

9.3 CURRENT SOURCES OF WATER CRISIS

While societies have always managed their water use around expected deviations in flow and quality, five long-term trends have changed these historical averages, requiring that we update these plans. The first four factors that have influenced the quantity and quality of water available for human use are increases over the past century in the following:

- water infrastructure;
- population and economic output;
- scientific discovery and innovation; and
- greenhouse gas concentration in the atmosphere.

In addition, a growing awareness of the importance of water for 'non-human' (i.e. environmental) purposes constitutes a fifth factor that has further constrained our use of water. As discussed below, these trends have both separate and cumulative impacts, as each influences and is influenced by the other four.

The United States currently supplies the majority of its residents and businesses with water through about 51,000 individual (mostly) public and private water systems (Buckley *et al.*, 2016). Their collective water infrastructure provides year-round use of seasonal water supplies, safeguards public health, protects urban and agricultural regions from flooding, and even brings electric power to many regions. These water systems can also claim credit for reducing morbidity and extending life expectancy over the last century resulting in an explosion of human population. These systems are aging, however, as pipeline corrosion and deteriorating storage and conveyance systems threaten the reliability of the national water infrastructure. Over the next 25 years, US water agencies will be responsible for an estimated $1 trillion USD to restore their systems to meet future needs (AWWA, 2012).

While the availability of water has contributed to population growth, the lack of water is a potential constraint to the stability of population centers. According to the United Nations (2019), over two billion people live in countries with high water stress, while roughly four billion people experience severe water scarcity at least one month per year. Water conservation that de-links the connection between and water supply economic activity has become a standard part of the water supply portfolio, and suppliers continually search for new conservation opportunities as well as new water supplies, including recycled water and saline aquifers. Advanced water treatment technologies now allow impaired waters to be made suitable for any purpose. Urban wastewater, storm water, seawater, and brackish groundwater are all considered to be important new water supplies. Improved aquifer modelling supports better resource management, while advances in water quality testing enhance our ability to identify and correct potential problems.

These technological improvements take on even greater significance in the face of climate change, caused by the increased concentrations of greenhouse gases in the atmosphere. Pre-industrial carbon dioxide levels of roughly 280 ppm are now over 400 ppm, growing at a rate of 2 to 3 ppm per year. The resultant rise in average global temperatures has produced heat waves, extreme storms, severe droughts, and reduced snowpack and glacier volumes impacting freshwater flows, while warming water increases the concentration of microbial contaminants and pathogens. Even if the world suddenly discontinued the use of fossil fuels, decades of further warming would occur, increasing the uncertainty around water availability, challenging future plans, and putting at risk hundreds of billions of dollars already invested in water infrastructure.

Compounding the planning process is a growing awareness of the value of water in the environment. The combination of massive water diversions and reduced groundwater infiltration due to urbanization further threatens water-dependent natural systems that, if altered or impaired, could have unforeseen human impacts. Others factors include conversion of forests to farming and expanding use of pesticides. Ecologists describe our era as an 'extinction event.' While humans are not presently experiencing a drop in population, many species are declining and the loss in global biodiversity in general reinforces the importance of apportioning more water to open spaces. The era in which water could be withdrawn from the environment with little thought of the consequences is over.

Each of these trends impacts the others. Economic and population expansion increases demand for high quality water supplies that technological innovation enables through the reuse of wastewater and other non-traditional supplies. Global warming changes how freshwater supplies are naturally replenished, requiring new infrastructure to maintain economic productivity The loss of streams and wetlands and has a cumulative impact on food production and public health.

The cumulative effect of these combined trends is that water management is now more challenging and complicated than ever. Increasingly sophisticated tools are used to model and track water supplies, even as integrated regional systems attempt to use the same H_2O molecules over and over, inserting intermediate treatment, testing, storage, and transfers between uses. Much can go wrong when the quantity and quality of water available is so thoroughly managed. While a diverse water portfolio is more resilient than reliance on a single supply, the potential for failures in system coordination increases. These risks increase when financial capital is in short supply or governance systems are not up to the task, and they multiply still further when water managers stop paying attention or slow down their investment in water infrastructure. In Big Spring, Texas, for example, only immediate regional planning and significant investment in alternative technology were able to avert a water crisis through the implementation of a direct potable reuse program.

9.4 THE ROLE OF INDUSTRY DURING A WATER CRISIS

A water crisis is a sudden or emerging trend in water supply or quality that deviates from expected conditions and imposes actual or potential harm on society. When available water supplies drop to less than 50% of normal, severe stress results. Many cities do not plan past a 50% reduction in supply, considering it too improbable. For example, the City of Santa Cruz, California stops at 50% deficiency from normal peak season demand, calling it a Stage 5 drought condition that is 'an extraordinary crisis threatening health, safety, and security of the community' (2016, p. 8–9). Water quality crises present the same type of challenge, as water users are left without sufficient supplies at the expected quality and price.

While industry is affected by a water crisis, companies can in turn influence the severity of its impact on the community, either positively or negatively. For example, when industry exercises prior rights to water and continues to withdraw water it can exacerbate the effect of a water shortage on local residents. Industry also has the capacity to improve conditions by enhancing water quality, reducing its demand on local supplies or even creating new high-quality sources of water. And while individual facilities may be able to minimize the impacts of a regional water quality crisis, the most effective response involves a combination of actions by industry and government that preserves the benefits industry provides society while mitigating the effects of the crisis.

Governments face a choice on where to supply water when it is scarce. Direct human consumption is an obvious priority, but industrial, commercial, governmental, and agricultural users all make a claim on limited supplies. During a severe water crisis, regional or national governments will likely seek to control the distribution of available supplies. These efforts, however, are unlikely to be fully efficient, leaving the health of many at risk – especially poor families located in remote regions.

For this reason, one primary reason to supply water to industry during times of drought is due to the wages it pays to its employees. Well over 300 million people work in industry, which accounts for over 22% of employment in lower middle-income and richer nations (ILO, 2018). In times of water crisis, families with even a small amount of money have options to purchase water or water-saving appliances, or to buy water treatment devices or water substitutes that could help see them through a water crisis.

A related argument in favor of maintaining at least a minimal flow of water to industry during extreme water shortages is that industries that cease production could lose markets and market share as a result of a temporary cessation of water supply. On the reasonable expectation that wet years will return, and the continuing importance of regional economic activity, one should keep industry alive even in a water crisis as a matter of long-term planning for the region. This is a longer-term perspective than the immediate value of maintaining employment and wages, but equally important. Buder (2019) notes that about 30,000 mostly low-income jobs were lost during the Cape Town crisis.

Note that the argument for maintaining a minimum water supply to industry in a water crisis is not based on the value of the goods that industry provides. Local industry may occasionally demand water to provide essential products like medicine, just as farmers sometimes need water to keep crops from drying up or rotting in the field. Water provided to these uses can be justified. But regardless of the products they manufacture, the wages industry pays to local residents provide an important resource that facilitates a 'bottom-up' response to critical shortage.

9.5 POLICY APPROACHES TO INDUSTRIAL WATER SUPPLY BEFORE AND DURING A WATER CRISIS

While government bears the primary responsibility to prepare for a water crisis, industry has a larger capacity than any other sector to identify, acquire, and treat its own supplies. In doing so, industry can assist both in long-term planning as well as in emergency response consistent with a sustainability ethic that emphasizes industry's role and impacts on its region of activity.

9.5.1 Participate in planning

Successfully developing and implementing a resilient regional water supply program is a costly, long-term effort. Both government and industry operate in a budget-constrained environment where large investments to address future problems are frequently deferred for years – even decades. Industry can play an influential role in accelerating water infrastructure projects by helping to evaluate planning assumptions and advocating for timely implementation of projects that lower regional risk. From industry's perspective, investment in a secure water supply is an insurance policy for their operations as well as those of nearby suppliers and customers. They can use their resources and regional influence to support and promote such investments.

With respect to water quality, industry can also help by reviewing the adequacy of water quality testing systems and encouraging regulators to stay up to date on emergency response planning, including mutual aid and other networked responses. And industry can encourage and participate in table-top scenarios and other exercises that rehearse the regional response to an emergency.

9.5.2 Public–private partnerships

Beyond this advisory role, industry can also participate directly in a public–private partnership for water supply and treatment facilities. This open-ended role can include any aspect of project design, construction, ownership, and/or operation. With its financial and technical resources, industry can help build projects that would otherwise be out of reach to regional government. For example, one risk associated with the development of alternative water supplies is that during normal and wet years the new, more expensive, supply may not be needed. By helping to fund the facility, industry can reduce the burden of the 'stranded investment' ensuring that the cost is recouped despite its intermittent operation and revenue generation.

Another form of industrial public–private partnership, common in agriculture, is joint underwriting of water conservation projects. Industry can invest in long-term, large-scale projects like extensive replacement of leaking water mains in exchange for access to some portion of the water conserved.

9.5.3 Restrictions on industrial water use

During a water crisis, government has the primary responsibility to manage water supplies in a way that best protects society. However, as noted earlier, government action can be augmented by the behavior of other members of the community, including companies and their employees – provided that they continue to receive wages.

The ability to maintain a functioning industrial sector during water shortages should be integrated into water supply reduction strategies well in advance of a crisis. The State of California provides a framework for emergency reduction strategies that gives local jurisdictions the ability to set aside water for industry even during the most extreme events. After setting aside water for domestic use, sanitation, and fire protection,

> 'the regulations may establish priorities in the use of water for other purposes and provide for the allocation, distribution, and delivery of water for such other purposes….' (CA Water Code Section 354)

Cities and regions plan in advance as to how they will implement these powers. As an example, the City of Santa Cruz, California serves roughly 80,000 people in a system that is hydrologically independent from other regional and state-wide supply systems, and preferentially allocates water for industry in its drought emergency curtailment plan.

Table 9.1 Water supply allocation and customer reduction goals (Santa Cruz, CA)

Normal Peak Season Demand = 2,473 mil gal	No Deficiency		Stage 2 15% Deficiency		Stage 3 25% Deficiency		Stage 4 35% Deficiency		Stage 5 50% Deficiency	
	Delivery		Delivery		Delivery		Delivery		Delivery	
Customer Category:	%	Volume (mil gal)	%	Volume (mil gal)	%	Volume (mil gal)	%	Volume (mil gal)	%	Volume (mil gal)
Single Family Residential	100	1,031	84%	864	73%	753	62%	639	48%	495
Multiple Residential	100	524	87%	454	78%	411	69%	361	55%	287
Business	100	438	95%	416	92%	402	87%	381	70%	307
UC Santa Cruz	100	132	85%	113	76%	100	66%	87	52%	68
Other Industrial	100	23	95%	22	90%	21	85%	20	67%	15
Municipal	100	48	76%	36	57%	27	41%	20	28%	14
Irrigation	100	110	64%	70	34%	37	12%	13	0%	0
Golf Course Irrigation	100	106	73%	78	51%	54	34%	36	20%	21
Coast Agriculture	100	59	95%	56	90%	53	85%	50	67%	40
Other	100	2	95%	2	90%	2	50%	1	50%	1
Total	**100**	**2,473**	**85%**	**2,111**	**75%**	**1,861**	**65%**	**1,607**	**50%**	**1,247**
Demand Reduction %, Million gallons	**0**	**0**	**15%**	**-362**	**25%**	**-612**	**35%**	**-866**	**50%**	**-1,226**

Source: 2015 Santa Cruz Urban Water Management Plan (2016).

Santa Cruz allocates water to the various sectors through a process that is cognizant of water limits, urban planning goals, economic trends, state and regional water conservation rules, and rate structures. In Table 9.1, 'No Deficiency' indicates the full amount (100%) of water each sector utilizes during 'normal/wet' years when their use is constrained only by price, and 'Stage 5 50% Deficiency' indicates the percentage of normal water supply delivered to each sector during a Stage 5 Drought—the worst-case drought scenario in which the community as a whole has only half as much water available.

Deliveries are not cut 50% across the board. The sectors receiving the largest proportion of their shares are Business (commercial and industrial), Other Industrial (light industrial production facilities including the technology sector), and Coast Agriculture. Collectively, these sectors are required to reduce their water use by about one-third of normal usage. By contrast, Golf Course Irrigation (20%) and Municipal usage (28%) are reduced to only one fourth or one-fifth of their normal usage, while residential and commercial Irrigation (0%) is eliminated entirely. This plan is consistent with maintaining industrial production during a water crisis, which should enable companies to continue to operate and to pay their employees.

9.5.4 Voluntary reductions and quality improvement

The plan described above envisioned a maximum water reduction of 50%. Some cities have recently confronted even more severe shortages. Should supplies drop below half their normal amount, it is likely that Santa Cruz would need to revisit their allocation formulas and consider further mandatory cutbacks. In that case, industry can help by reducing water consumption voluntarily to a greater degree than required. Industry also has the capacity to improve the quality of effluent leaving its facilities, which can become a valuable emergency source of water through further treatment and reuse. This approach could be implemented as part of a larger process of public–private cooperation to manage the water crisis.

9.6 CONCLUSION

There is today an increasing possibility of extreme water shortages and water quality events that present a crisis to cities and their regions. While industry is not the sole cause of these water-related crises, it can play a positive role in preparing for and coping with water shortages. Industry can advocate for long-term public water investments, engage in public–private partnerships, and maintain a flow of wages to employees during a severe crisis. By thinking of its post-use water effluent as a possible source of usable water, it can improve effluent quality, thereby increasing the region's usable supply of water in a crisis.

REFERENCES

AWWA. (2012). Buried No Longer: Confronting America's Water Infrastructure Challenge. (February). American Water Works Association.

Breyfogle N. (2010). Dry days down under: Australia and the world water crisis. *Origins*, **3**(7), updated 1-26-11. http://origins.osu.edu/article/dry-days-down-under-australia-and-world-water-crisis (accessed 17 September 2019).

Buckely P., Gunnion L. and Sarni W. (2016). The aging of water infrastructure: Out of sight, out of mind? Deloitte Insights (March). https://www2.deloitte.com/us/en/insights/economy/issues-by-the-numbers/us-aging-water-infrastructure-investment-opportunities.html (accessed 17 September 2019).

Buder E. (2019). The Countdown to Day Zero. The Atlantic, March 5. https://www.theatlantic.com/video/index/584179/dry-city/ (accessed 17 September 2019).

City of Santa Cruz. (2016). 2015 Urban Water Management Plan. http://www.cityofsantacruz.com/government/city-departments/water/2015-urban-water-mgmt-plan (accessed 17 September 2019).

Gupta S. (2019). Indian water train arrives with desperately needed relief for Chennai. CNN World, July 12, 2019. https://www.cnn.com/2019/07/12/india/india-chennai-water-crisis-train-intl/index.html (accessed 17 September 2019).

International Labour Office. (2018). World Employment and Social Outlook: Trends 2018. International Labour Office, Geneva.

Martin L. (2014). 'Texas Leads the Way with First Direct Potable Reuse Facilities in the U.S.' Water Online, Sept. 16, 2014. https://www.wateronline.com/doc/texas-leads-the-way-with-first-direct-potable-reuse-facilities-in-u-s-0001 (accessed 17 September 2019).

USGS. (2019). Industrial Water Use. https://www.usgs.gov/mission-areas/water-resources/science/industrial-water-use?qt-science_center_objects=0#qt-science_center_objects (accessed 17 September 2019).

WWAP (UNESCO World Water Assessment Programme). (2019). The United Nations World Water Development Report 2019: Leaving No One Behind. UNESCO, Paris.

Chapter 10

The role of the consultant in supporting sustainable industrial water use

Arthur Umble

Stantec Consulting Services, Inc., Wastewater Treatment Sector, Denver, CO, USA

Keywords: barriers, community, consulting engineer, environmental awareness, regulatory sensitivity, stakeholder, water reuse

10.1 OUR GLOBAL WATER CHALLENGE

The global water, air and terrestrial environments, as we've known them, are being altered at unprecedented rates due to the on-going influences of our changing global climate. Utilities that manage water supplies and industries that rely on water to produce their goods are equally challenged to adapt to what appears to be a 'new normal' in water management. For public utilities, the uncertainties in the coming decades challenge decision-makers to provide sustainable, reliable water services to the public now and for the future. At the same time, industries that rely on water for their production must change their approach to water consumption and disposal in order to remain competitive.

Never before has the need been more critical for holistic water solutions that reach beyond technology and engage the economic, social and political sciences. And at no time in history has the role of environmental consultants been more crucial, or their obligation so great, to provide comprehensive guidance to decision-makers to navigate the complexities ahead. The focus of this chapter is the water consultant's role in helping to shape the future of the water environment.

10.2 THE ROLE OF THE WATER CONSULTANT

The role of the professional consultant with respect to both the private and public clients centers around four primary responsibilities:

(1) identify and deliver effective, sustainable technology solutions;
(2) minimize project environmental impact;

DOI: 10.2166/9781789060676_0105

(3) navigate regulatory complexities; and
(4) broker key stakeholder partnerships to maintain the entity's 'license to operate' in the community.

10.2.1 Technology selection

Water solutions today are more complex than ever. Historically, companies purchased their water from a local purveyor, and while the influent often needed further treatment before use in certain processes, after use the effluent was only treated to the extent mandated by the local sewer authority. Today, however, companies looking to increase their water reliability expect to treat industrial wastewater streams for reuse in plant processes, and for cooling, cleaning, firefighting, dust control and other non-potable purposes. Wastewater can even be purified now to the extent that, when it is blended with the utility water supply, no further treatment is needed for process use. This approach can lower water, energy and chemical costs, alleviate regulatory constraints and even reduce the impact on the local environment. In this case, the optimum technical approach must address both supply and disposal and the consultant's role is to map out the technology alternatives aligned with these desired outcomes. This may involve developing technical criteria for evaluating alternatives, researching technology performance in similar applications, conducting pilot studies and documenting regulatory hurdles. The consultant frequently needs to help the client generate a detailed business case including a 'Triple Bottom Line' analysis to demonstrate that the selected technology can be justified economically as well.

10.2.2 Environmental impact

Supplying, using and discharging used water each have the potential to impact the environment and must be carefully managed. Excessive industrial demand on aquifers can result in declining groundwater levels which in turn can lead to land subsidence, increased salinity or even seawater intrusion in coastal areas. Unless utilities properly schedule their withdrawals from surface water sources, reduction of seasonal water flows can disrupt the aquatic habitat and reduce ecological diversity. At the same time, wastewater discharge can pollute surface or groundwater water sources and erode natural waterways. Water reuse mitigates both types of impacts. The consultant should possess the technical expertise needed to provide the client complete information to make appropriate decisions.

10.2.3 Navigating the regulatory framework

Regulatory frameworks are designed to protect water quality and ensure its continued suitability for all designated uses. Complexity arises, however, due to divergent regulations from multiple jurisdictions (local, State and Federal), with different permit approval processes and even competing surface and groundwater rights. The consultant must be able to chart the regulatory roadmap and, more importantly, have experience navigating it successfully. Regulatory constraints represent significant costs to companies, and an uncertain permitting process makes it difficult for them to know when – or whether – to 'pull the trigger' on investment in water treatment and reuse. Both the private corporation and the public utility rely heavily on the consultant to provide guidance on such decisions.

10.2.4 Stakeholder partnerships

Finally, the consultant can broker crucial stakeholder relationships to help gain support for a company's water use and reuse. These stakeholders can include regulators, Chambers of Commerce and economic development agencies, community service clubs and social service agencies, non-governmental organizations, public health agencies, media, etc. Each of these stakeholders can either support or challenge the proposed water use, depending on whether they believe the company or utility is acting responsibly as it achieves its environmental and financial goals. Consultants can often bring these stakeholders together by demonstrating their ability to provide acceptable, holistic water solutions. Because the consultant represents the client, however, stakeholders may initially view them as biased. To overcome this suspicion, the consultant must demonstrate respect for stakeholder opinions throughout the life of the project, from planning and design through construction, startup, commissioning and continuing during operation.

10.3 CHARACTERISTICS OF THE RESPONSIBLE WATER CONSULTANCY

How well or inadequately a consultant fulfills these primary responsibilities can be gauged not only by how their projects perform, but also by the relationships that frequently outlast them. In each case, the responsible consultancy looks at the 'big picture' and determines how to meet the client's needs in a way that benefits the community in the long run.

10.3.1 Technical expertise

The professional water consultant is regarded as a technical expert whose knowledge and experience reach deep into the details of water resource management, as well as water and wastewater treatment engineering, science and technology. As a result, the responsible consultant has the expertise and practical experience to design and deliver a wide range of appropriate technical solutions, including conceptually pragmatic 'roadmaps' allowing water to be managed efficiently throughout the watershed – and sometimes between watersheds – with minimum environmental impact.

For example, the consultant must ensure that groundwater and surface water sources are protected from contamination by industrial activity. A sound wellhead protection plan eliminates the potential for groundwater contamination by defining aquifer zones to be kept free of industrial pollution, stormwater or agricultural runoff. Project designs should incorporate the use of green infrastructure where applicable such as grass filter strips along urban streams or specified tillage practices in agricultural areas to treat stormwater runoff.

10.3.2 Environmental awareness

Similarly, the consultant provides the client with solutions that concurrently maximize value while minimizing impact on the environment. Surface water, for instance, can be treated to meet potable water standards in a variety of ways, from sand filtration and chlorination at one end of the technological spectrum to advanced membrane treatment, ultraviolet disinfection and ozonation at the other. Whichever technology is selected, the consultant is obligated to consider impacts from disposal of treatment residuals whether backwash solids from filtration or reverse osmosis brines. This is one reason why, even when highly purified water is needed, a consultant may recommend an alternative to reverse osmosis like ozone with biologically activated carbon (O3/BAC) which can meet reuse standards with lower chemical and energy costs and no residual brines.

10.3.3 Regulatory sensitivity and relationship management

It goes without saying that industrial water projects must be designed to meet all regulatory compliance requirements and perform at a level that is acceptable to both the client and external stakeholders. In many instances, however, the consultant's role does not end there but continues through project optimization, technical training of company staff, and on-going involvement with the community. This is accomplished best by simply always being responsive in a timely manner. Unlike climate change, which is the cumulative result of diverse actions around the world, water use is local and its impact on the local economy and environment can be linked back directly to the company. For this reason, industrial water users rely on consultants to help them manage the many relationships they enter as responsible water users. Examples of these relationships and the consultant's role in maintaining them are described in the sections that follow.

10.3.3.1 Case study: copper mining in Arequipa, Peru

In today's world, the corporation must make water use decisions that take into account relevant socioeconomic realities of the communities in which they operate. This can require partnerships and investments which may in the short run negatively impact the company's 'bottom line'. In this case, the responsible consultant's job is to help the client build and maintain those relationships and develop projects that provide long-term value to the company. An example of this took place recently in the Arequipa region of southern Peru, an arid territory where water scarcity is a constant issue. A global copper mining company operating in the region – and a significant contributor to the region's economic growth – drew water from local rivers for its mining operations. Meanwhile, the regional water/wastewater utility was struggling to keep up with its growing urban population, and the rivers (which also provided water for local irrigation) were being polluted by untreated sewage discharges. The decline in river water quality increased the company's operating costs just as the company planned to extract even more water to expand its operations, and the both the utility and the mining company were losing public trust.

In this case, the consultant provided a cost-effective technical solution, and worked alongside both the mining company and the public utility to implement the project by gaining approvals from the regional regulatory agencies. The mining company partnered with the regional utility agency to construct a new centralized wastewater treatment plant at a site outside the urban area. The treated effluent improved the local river water quality which made the mining operations more efficient and even increased the safety of locally grown crops. The public began to regard the utility as a leader in improving their quality of life, and the mining company demonstrated its corporate responsibility by financing the much-needed wastewater treatment plant. (The consultant also advised the company on its public relations outreach to ensure that the message of the partnership was articulated effectively and understood.)

10.3.3.2 Case study: high tech water use in the USA

Another example of 'out-of-the-box' engineering took place when a consultant helped a large microelectronic manufacturing firm meet its water needs more sustainably in the rapidly growing Pacific Northwest region of the USA. The firm needed to expand its manufacturing process to maintain its competitive advantage in microchip technology, which required more water and resulted in more industrial wastewater discharged to the local wastewater utility. The role of the consultant in this case was to devise a technical solution to treat the industrial wastewater to a quality such that a large portion could be reused in the manufacturing process, while the rest could be discharged to the municipal wastewater plant in compliance with the municipality's pretreatment regulation.

Because of the uniqueness of the wastewater characteristics, this technical solution required extensive bench and pilot scale research, conducted by the consultant, to prove the treatment approach and provide design criteria for the full-scale system. The consultant also assisted the company in obtaining approval to discharge the treated wastewater by helping to develop an appropriate set of protective effluent discharge limits. This solution required a strong partnership between the company and the public utility, which was facilitated when the company's use of municipal potable water freed up supply for this rapidly growing suburban area.

10.4 THE BREADTH OF WATER CONSULTANT RELATIONSHIPS

The value a water consultant brings to the water industry depends a great deal on the breadth and depth of established, trusted advisory relationships with stakeholders who hold key interests in the community's sustainable water environment. Perhaps the three most important relationships a consultant has are with their clients, the regulatory authorities, and the academic community.

10.4.1 Client and community relationships

The consultant's relationship with the client centers around being a 'trusted advisor'. The client must be able to trust the consultant's judgments on technical matters that affect the policies and initiatives promoted by the client to best serve their customers. Consultants have the obligation to tell clients the truth about the hard realities they sometimes must face with respect to the reliability of water supplies, and the rising, and often-times hidden costs associated with water and wastewater treatment. Ideally, the consultant has appropriate knowledge about the environmental, economic and social impacts of climate change as it affects water quality and availability, able to provide science-based evidence to support decision-making around water use and reuse. It is the credibility, relevance and timeliness of this evidence that is foundational for the trusted advisor relationship.

Although the client in the following example is a municipal government rather than a private company, the Pure Water San Diego program illustrates how a consultant acting as a trusted advisor can advocate effectively for sustainable water management. For several decades, the city of San Diego, California (USA) sought to develop a potable water reuse project as a component of their holistic water management plan. In the late 1990s, initial efforts to launch a sustainable indirect potable reuse program stalled due to a lack of public support. Despite this setback, the city continued to work with key stakeholders and engaged a number of consultants to continue public outreach initiatives, refine technical solutions, and update *pro forma* financial analyses. Most important, they continued to build and operate demonstration projects to develop science-based evidence to verify project reliability. As a result, in 2014 when the city council launched Pure Water San Diego to provide potable recycled water to the community the public supported it. At this time, construction is underway, and the program is scheduled to be fully implemented by 2035, adding 83 MGD to the water supply through a combination of indirect and direct potable reuse.

10.4.2 Regulatory authorities

In addition to promoting public support for a water treatment project, a consultant can also help regulatory authorities develop sound, science-based regulations. This was demonstrated recently by a public utility seeking to permit an alternative wastewater disinfection technology, but the same task has been undertaken on behalf of corporate clients. A US utility using conventional hypochlorite for wastewater disinfection experienced periodic failures in complying with discharge regulations due to chronic maintenance challenges with the aged infrastructure. The utility evaluated a series of alternatives including chloramination, ultraviolet (UV), and peracetic acid (PAA) disinfection. Desktop evaluations indicated that PAA could meet objectives most effectively, but the regulator had no prior experience with the technology and needed to be certain that it would meet performance objectives without negatively impacting receiving water quality.

In order to demonstrate to the regulatory authority that PAA disinfection would protect receiving water quality and consistently meet permit limits, the utility crafted strategic partnerships with both the technology provider and the Water Research Foundation, through its 'Leaders Innovation Forum for Technology (LIFT)' initiative. A consultant was initially engaged only to review the technology developer's experimental plan and data collection program, but their duties evolved to include direct oversight of data analysis and an independent assessment of the results shared equally with the utility and the regulator. This role brought both transparency and credibility to the pilot program that gave the regulator the necessary confidence to approve the technology. Together the project team designed a three-year, full-scale pilot study of the PAA technology, which the regulator ultimately incorporated into the utility's NPDES (National Pollution Discharge Elimination System) permit as the approved method of disinfection.

10.4.3 Research institutions

A third type of relationship is maintained between the consultant and research institutions. This collaboration enhances the consultant's legitimacy because it demonstrates a commitment to ensuring sound, scientific-based decision-making. It also increases the likelihood that the consultant can recommend a new, improved technology for the industrial water user's application, boosting its competitiveness in the marketplace and enhancing its bottom line.

An example of an academic partnership is the Johns Hopkins Water Institute-Stantec Alliance (JHU-Stantec Alliance), established in 2013 '*to provide innovative, sustainable solutions to global environmental issues and natural resource limitations by conducting desktop-, bench-, pilot- and demonstration-scale investigations needed to support sustainability science*.' Stantec, a global water consultant, collaborates with the Johns-Hopkins University School of Public Health to study a host of global water issues that impact public health and the environment. The research areas include analytical and monitoring methods for contaminant identification in the water environment, mechanisms for contaminant degradation, alternative physical, chemical, and biological treatment schemes, and validation studies for emerging water technologies. The collaboration ranges from bench- and pilot-scale studies to full-scale demonstration projects. This alliance also provides an avenue for industrial clients to fund water research directly pertinent to their operations, and allows young academics (e.g., both doctoral candidates and post-docs) to explore career options in water research disciplines of interest to industry.

10.4.4 Business partners

A client is more likely to invest in a sustainable water project when the consultant can craft a business case with a reasonable payback period, which for private industry is typically 3 years or less. Since the 'market price' of water rarely yields such returns alone, the consultant must look at the full life-cycle cost that considers the value of the reusable water and materials embedded in wastewater. In addition to nutrients like nitrogen and phosphorous (for fertilizers) and organic chemicals (for fuels, bioplastics, pharmaceuticals, etc.), wastewater also contains power from embedded chemical and thermal energy. Each of these products potentially bolsters the economic value of a technical solution that recovers them and returns them to the economy to meet an array of 'fit-for-purpose' options.

The consultant can play a significant role here not only by designing treatment systems to recover these byproducts but also by helping the company either reuse them directly (e.g., energy) or by identifying markets and negotiating contracts with buyers. An example would be a consultant who assisted a utility with the sale of struvite (magnesium ammonium phosphate) generated by its wastewater treatment process to an agriculture fertilizer distributor. The distributor was able to utilize the struvite directly into a useful fertilizer product, while the long-term revenues helped the client stabilize its annual budget.

The consultant can also help the company's suppliers to enhance the sustainability of the entire supply chain. For example, agriculture is among the largest users of water globally, so a company's commitment to improvements in irrigation and fertilization practices among its agricultural suppliers can increase crop production while simultaneously reducing both demand and water pollution.

10.5 OBSTACLES TO SUSTAINABLE INDUSTRIAL WATER USE

If all consultants have the potential to provide industrial clients with holistic water solutions, why don't more companies use water sustainably? As suggested in the discussion above, industry's reluctance to invest in sustainable water use is not only due to the initial cost of the project but also to a fear of jeopardizing operations by changing water use technology and practice. This is why the consultant focuses on demonstrating the reliability of the recommended technology as well as making a business case for sustainable water use based on its long-term value to the company.

10.5.1 Financial barriers: shareholder vs. stakeholder

In the private sector, the shareholder generally has greater influence over bottom-line management decisions than the external stakeholder. On environmental issues, however, the stakeholder's perspective can override the shareholder's short-term concerns as companies recognize the importance of environmental protection to the communities – and the consumers – they serve. This connection is so unmistakable that in many cases shareholders have become the key stakeholders demanding environmental accountability from the businesses in which they are investing, even when such accountability requires investments that provide no direct monetary return (i.e. no immediate payback). Rather, the company benefits from the on-going endorsement by the community of its 'social license to operate'. This is particularly relevant for companies that require large quantities of water to support their process operations, and not just in regions of water scarcity.

Reducing greenhouse gas emissions, energy efficiency, water reuse, resource recovery – all these environmental initiatives have been implemented in response to demands from company customers and external stakeholders. Initial objections that these measures were too expensive, that they weren't required by regulation, that their payback period was too long, or their return on investment too low to be acceptable to the shareholder are eventually overcome when companies see how they have been implemented successfully by 'early adopters'. In this case the consultant plays a pivotal role by identifying the broad range of benefits associated with sustainable water use practices and including them in a more comprehensive cost-benefit analysis so that their value is reflected in the business case.

10.5.2 Fortitude barriers: making the first move

Just as the public utility is often the last to invest in a new technology, so many private companies resist changing processes, especially in areas like water treatment and reuse that are outside their 'core competency. Since water treatment equipment can be costly and no one wants to be left with expensive, stranded assets, it is understandable why public utilities and private companies both find themselves in a 'race for second place' when it comes to investing in new technology. While it may be prudent to wait to observe the results of someone else's trial before embarking down a similar path, one unintended consequence of this inertia is that it favors incumbent systems, further delaying advances in the industry. Here again the consultant can accelerate change by investigating advances in the field and keeping clients informed about best practices and methods, connecting them with other companies that are overcoming the barriers, sharing knowledge and technology.

10.6 OBSTACLES TO CONSULTANT ADVOCACY

Clients are not alone in their need to confront their concerns about the cost of adopting sustainable practices. Consultants, too, must face up to the challenge of advocating for sustainable solutions. Expertise in the advancing field of sustainable technology requires a significant investment in personnel and training. At the same time, giving a client the 'big picture' perspective when asked for a simpler, less appropriate solution might pose the risk of losing a client. Both these challenges require consultants to make a meaningful commitment to sustainable practice.

10.6.1 Financial barriers: the cost of competency

A consultant's expertise in sustainable, holistic technology water solutions for the water industry requires investment in staff and staff training. This generally means hiring PhD-level professionals to direct research and development initiatives – thought leadership – around sustainable technology. Consultants must collaborate with universities and research institutions to conduct further research and publish results to stay abreast so they can advise clients on best practices for managing their water resource. It also means discussing emerging technology and sharing the insights gained from sustainability research with regulators, professional associations, non-governmental organizations, elected officials and the general public. Publishing technical papers, presenting at conferences, preparing and circulating on-line videos, podcasts, blogs, webinars all take time and money, which must be taken from the consulting company's operating budget. On the other hand, to effectively communicate the value of sustainable practices, the consultant must change the historical view that wastewater is 'something to be gotten rid of' rather than reuse. This perception can only be overcome with proactive, timely and consistent communication of the benefits these recycled products provide to the public.

10.6.2 Fortitude barriers: keeping an eye on the future

While the consultant may clearly see the advantages of sustainable water use, the client may only be focused on its quarterly costs and revenues or its regulatory requirements. These immediate needs may eclipse the question of how the company might perform most profitably in the future. The consultant must listen carefully to the client's views and understand their current concerns in order to develop a strong relationship with the client. Based on this trust, the consultant can then provide constructive feedback to expand the client's perception of the value of sustainable water use.

To be blunt, encouraging clients to consider holistic solutions can sometimes mean stepping away from a potentially lucrative contract because the solution is simply 'the right thing to do' for the public and environmental good. While the value of integrating all aspects of the water cycle can be considerable, exercising this opportunity requires considerable planning. It often necessitates the development of a water management 'roadmap' that includes a triple bottom line (environmental, economic and social) analysis of benefits and costs to the company and key decision points in the implementation of water improvements. The goal is to ensure that each strategic decision is a 'no regrets' decision, meaning contingencies are in place to fall back on if the decision's outcomes differ from that which was intended or expected. Implementation of new technology can also call for the construction of pilot projects to prove the concept before advancing to full-scale application. This involves a risk that the pilot may fail to perform as intended, requiring an alternate solution to be found.

From this perspective one can see how a consultancy, eager to meet its own business targets, might be tempted to propose design and construction of a less sustainable solution which uses a conventional technology that is well-proven but consumes excessive amounts of energy and chemicals or produces more residuals. Nonetheless, it is incumbent on the responsible consultant to help the client look beyond short-term economics to appreciate the many benefits of a holistic solution.

Chapter 11

Government-industry partnership for sustainable water use: Insights from Pakistan

Arjumand Nizami[1]*, Jawad Ali*[1] *and Sophie Nguyen-Khoa*[2]

[1]*Helvetas Pakistan, Islamabad, Pakistan*
[2]*Helvetas Switzerland, Bern, Switzerland*

Keywords: appropriate technology, industry, Pakistan, role of government, water pricing, water use

11.1 INTRODUCTION

A recent study (Zia *et al*., 2019) in the Pakistan province of Khyber Pakhtunkhwa indicates that the industrial sector (including commercial enterprises) uses significant amounts of freshwater in competition with other users. This sector also generates employment for millions in the country and is important for the growth of the national economy. As a result, the industrial sector needs to be recognized as a stakeholder in the management of water resources and included in future plans to provide sufficient water to meet their needs. In Pakistan, as elsewhere, water management challenges are too complex and daunting to be solved by the government alone: water users from the industrial private sector must play an active role in reaching sustainability in the use of water resources. Notwithstanding their importance, however, industrial and commercial companies are seen merely as water consumers with few responsibilities. Until now, representatives of the industrial private sector have not been involved in the management and governance of water. A further constraint is that the majority of commercial and industrial companies are not registered, their use of water is not metered, and their wastewater discharge is not monitored.

The industrial private sector can help move in the direction of sustainable water use both voluntarily and in response to regulation. For example, several companies around the world have voluntarily invested in facilities that provide drinking water to communities in areas where they operate. These initiatives seek to improve water use efficiency in their production and processing, and provide technical and financial support to the government for sustainable water management. Preferably, the government would formally negotiate with national and multinational companies to determine and regulate their role in

DOI: 10.2166/9781789060676_0115

water management, and to assess their fair contribution to regional water and wastewater distribution and treatment. These companies would no longer act as economic 'free riders' transferring the cost of their water use to other stakeholders.

The industrial private sector and the government must establish a mutually supportive partnership according to a normative framework. Experience in the Khyber Pakhtunkhwa province in Pakistan suggests that, despite tensions, government authorities and industrial sector representatives can work together towards sustainable water management provided that they together address a range of key obstacles to their collaboration. These include insufficient information about commercial and industrial water use, treatment and discharge, and the lack of effective enforcement of existing environmental regulations. The analysis in this chapter reflects the benefit of several years of experiences and lessons acquired in the Water Productivity Project (WAPRO) in Pakistan, which is an effort of multiple partners to acquire effective water stewardship in commercial supply chains (rice and cotton). These include government of Punjab, Sustainable Rice Platform, international export companies (Mars Foods and Westmill) and local milling companies (Rice Partners' Limited and Galaxy Rice Mills).

11.1.1 Role of the government and the industry in water management

The role of the government in promoting sustainable water use by the industry includes responsibility for overall water management at both the national and provincial level. The industry can play an important role in reducing the volume of water and energy used in its operations, as well as in their suppliers' operations through management of supply chains. Working together, government and industry can make significant progress towards sustainable water use.

Governments around the world are responsible for providing their citizens with water for drinking and irrigation. They must establish appropriate policies and regulations and invest in and construct essential water and sanitation facilities. The UN Sustainable Development Goal (SDG) #6 calls for ensuring water availability and sustainable management of water and sanitation for all, water use efficiency, and integrated water resources management. This remains a daunting task for many countries; the UN 2017 on Global Analysis and Assessment of Sanitation and Drinking Water (WHO, 2017) reported that even though national budgets for Water Sanitation and Hygiene are increasing at an annual rate of 4.9%, 80% of the surveyed countries claimed they did not have enough financial resources to meet the Water Sanitation and Hygiene targets.

A recent study (Newborn & Mason, 2012) observed that, in addition to formulating and overseeing water management regulations and national water resource policies, governments are also frequently responsible for allocating water for competing uses. To do this effectively, and to support private sector efforts towards sustainable water use, governments must first understand how the industry functions to formulate polices and legislation encouraging efficiency and transparency in industrial use of water. These policies aim at a triple-win in Public–Private Partnership: they need to be socially desirable, ecologically sustainable and economically viable for both the government and the industry (Figure 11.1).

11.1.2 The importance of data to the role of government

While the role of the government is to provide water for all and protect the water resources environment, it remains the responsibility of the industry is to reduce its water use and to decrease its impact on the environment. According to several current studies and reviews of the literature, private companies have developed a range of instruments for water management and water use efficiency, and they increasingly seek certification through water stewardship standards (GoP, 2018; Morrison & Gleick, 2002; Newborn & Mason, 2012; Water Futures Partnership, 2012). For example, a coalition of multinational

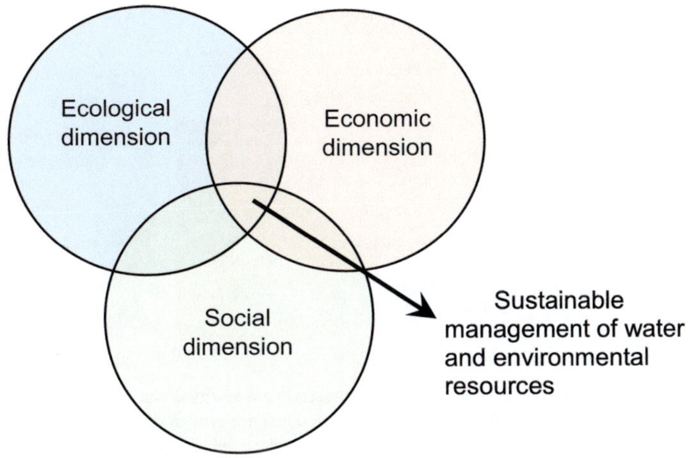

Figure 11.1 Triple-win Situation for Public-Private Partnership.

companies, aid agencies, international conservation agencies and partner governments in several countries together formed the 2030 Water Resource Group (2009), a 'public-private-expert-civil society platform' organized to facilitate sustainable use of water. This group intends to play a role in the management of water at basin scale – a role traditionally played by the public sector alone – by encouraging the development and sharing of data on: water supply and demand in catchments, water infrastructure assets and their condition, costs and performance, estimated trends in demand for services, etc.

The focus of the 2030 Water Resource Group on data reflects their belief that 'many countries struggle to shape implementable, fact-based water policies, and water resources face inefficient allocation and poor investment patterns because investors lack a consistent basis for economically rational decision-making.' This perspective was echoed by a recent UN World Health Organization report (WHO, 2017). As illustrated in Figure 11.2, below, out of 65 countries surveyed, almost three-quarters stated that they analyzed and used available information in making at least some decisions about water or sanitation investments, while the remaining 25–30% reported limited availability or incomplete information. In the opinion of both the 2030 Water Resources Group and WHO, such limitations are an obstacle to effective planning and policy formulation. As discussed further below, lack of useful information about the nature and extent of the use of water by the private sector can also hinder the ability of the government to establish working partnerships with industrial water users.

11.1.3 Summary

In summary, government and industry would have the following responsibilities in the management of water resources:

Government:

- Collect and manage data on water use and wastewater discharge to support decision-making;
- Formulate policy and develop regulations;
- Oversee implementation of policies and ensure implementation of rules and regulations;
- Identify and address problems related to access to water and sanitation services;
- Equitably distribute water resources; and
- Fulfill international obligations.

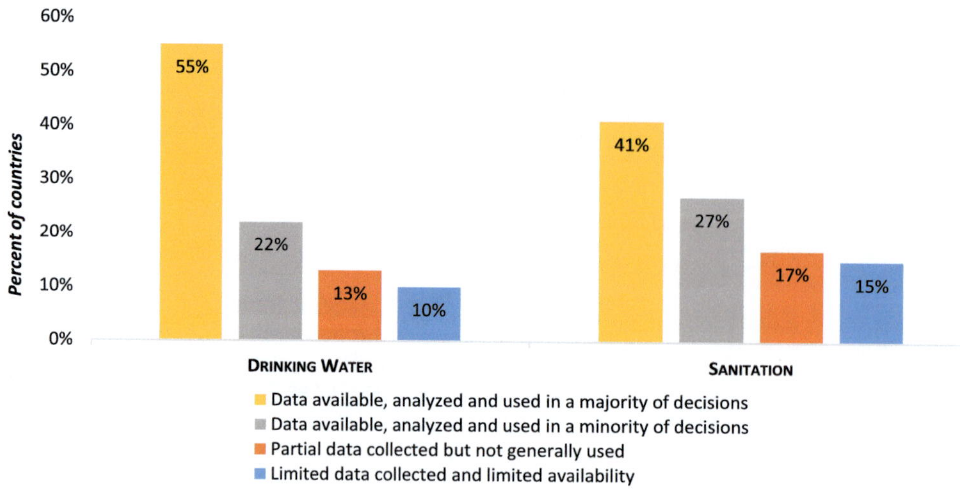

Figure 11.2 Are data collected and used to inform decisions on resource allocations? (n = 65). (*Source: GLAAS 2016/2017 country survey in WHO, 2017*).

Industry:

- Reduce water use in own operations and increase water efficiency in operations of suppliers and growers in supply chains;
- Treat wastewater prior to discharge to reduce environmental impact;
- Assist the government and others in collecting the data needed for decision-making on water management;
- Finance water research and invest in water infrastructure as an example of corporate social responsibility towards neighbors and affected people;
- Supplement government's investments in water infrastructure;
- Advocate for sustainable water resources management and policy-making.

11.2 WATER MANAGEMENT IN KHYBER PAKHTUNKHWA PROVINCE

In the Khyber Pakhtunkhwa province, characterizing the roles of the government and industry in water management underscores the importance of data collection and management. To support the industrial sector specifically, the government first needs to know and understand how this sector uses water: the size of the enterprise, its water-related needs (quality and quantity) and the resources it can deploy to increase public water resources and capacities. This knowledge helps government officials to recognize the industry as a partner in water use, rather than perceive the industry as merely a water user and polluter. Unfortunately, this information base is generally thin, especially in developing countries. Recent experience in the Khyber Pakhtunkhwa province illustrates the impact of information gaps on the development of partnerships for sustainable water use.

11.2.1 Private sector water use

Complete, reliable data on the size of the private sector in Pakistan (including the Khyber Pakhtunkhwa province) is not available. According to the Bureau of Statistics (GoKP, 2016), the province has an extensive agriculture-based industry that produces various products including tea, tobacco, match boxes, vegetable ghee and sugar. However, out of an estimated total of 12,000 industrial units referenced in the Industrial Policy Report, only 2,299 are registered with the Directorate of Industries, of which approximately 1,821 are functional. While some commercial entities are regulated, cottage industries are neither registered nor regulated. Fish farming, poultry farming, and commercial dairying are all important commercial activities for which the government has no reliable data, and there is no data at all for smaller commercial entities like shopping centers, individual shops, hotels and the many car wash stations now mushrooming in the area.

11.2.2 Source of water for industrial and commercial companies

Like in the whole country, the use of water by industrial and commercial companies in Khyber Pakhtunkhwa province is not well regulated, and complete information on the sources of water exploited by the industrial sector is not available. As reported in the 2002 National Water Sector Strategy (GoP, 2002), larger industries (e.g., manufacturers) make their own arrangements for water supply, generally by digging boreholes at their own cost to access groundwater. By contrast, smaller commercial entities (e.g., shops and small hotels situated within the cities and towns) may use water provided by various government departments, for which the source is also groundwater. Hence the main source feeding industrial and commercial companies is groundwater, either from owned boreholes or from piped water supplied by the government.

11.2.3 Quantity of water used

Although complete data on the quantity of water used by the private sector in Pakistan are not maintained by any institution, as illustrated in Figure 11.3, some references suggest that the large manufacturing units in the country use 2.5 percent of the total available freshwater (UNDP, 2016). While this is significantly lower than the global average for industrial water use estimated at around 20 percent of freshwater (FAO, 2016), available data from some industries suggest that the amount of water used by the industrial sector in Pakistan in certain areas is substantial enough to result in competition with municipal and agricultural

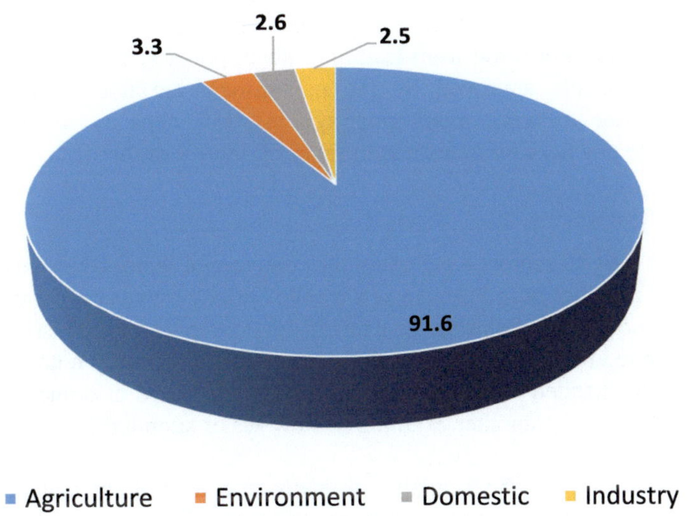

Figure 11.3 Percentage of water use by sub-sectors. (*UNDP, 2016*).

water users. For example, while a single car service station needs on average 16 m^3/day of water a day (Zahir *et al.*, 2017), the water requirement of an industrial park in Karachi was estimated to be about 681,374 m^3/year (MG, 2018). Since approximately 3 to 4 liters of freshwater are required to produce a single liter of soft drink (Haroon *et al.*, 2012), a medium-sized bottled drink company may use 20,000–30,000 m^3/year and a large one may demand upwards of 250,000 m^3/year. From the available secondary data on industrial freshwater production and consumption, it is estimated that the six beverage factories in the province may use on average over 1,500,000 m^3/year (Bhatti & Nasu, 2010) – enough for the daily needs of around 500,000 persons (based on 30 liter/per capita/day). The gravity of this issue was first highlighted in the media when a committee constituted by Supreme Court of Pakistan for a judicial inquiry on bottled companies found that none of the 44 water and beverage plants had flow meters, and the overexploitation of aquifers was reported to have resulted in an increase in arsenic content in groundwater. Similarly, seven cement factories (Sharma *et al.*, 2013) would use water volumes equivalent to the daily needs of nearly 1 million persons. This not only underscores the significant needs of the industry for freshwater, but also the increasing pressure on competing uses in the wake of water scarcity in the province.

11.2.4 Water contamination

According to available reports, only 1% of industrial wastewater in Pakistan is treated before being discharged into rivers and drains (Khan, 2015). As a result, about 3 billion liters per year (or 8.2 million liters per day) of freshwater is annually converted into wastewater. This is about the amount which could be used by 274,000 people for drinking purposes (Dawn News, 2005, 2019). Due to poor monitoring of environmental compliance by the government, companies do not acquire necessary environmental approvals for industrial water use. Hence companies neither have facilities to treat wastewater prior to discharge nor the ability to monitor the quality of water discharged to waterbodies. In Khyber Pakhtunkhwa, 80,000 m^3 of industrial effluents containing high level of pollutants are discharged every day (Khan, 2015). Pollutants released by the commercial sector include detergents, dyes, acids, sodas, salts, and heavy metals (Figure 11.4). Contaminated runoff from agricultural fields is also likely leaching down to the groundwater, and residues are left in wastewater flowing to the gorges and streams, and in the air. The quality and quantity of agrochemicals applied in agriculture are neither estimated nor monitored. This situation does not result from lack of relevant policies, but because the relevant National Environment Quality Standards approved by the government are not properly enforced. A weak enforcement of laws is due to weak monitoring systems, low capacity of enforcement agents and unethical or unsustainable practices by industries in favor of short-term benefits.

11.2.5 Water tariffs

As stated earlier, water use by all sectors – including the commercial sector – has never been well-regulated in Khyber Pakhtunkhwa province or in Pakistan as a whole. So long as water is provided by the state, water is largely considered 'a free good' in Pakistan. As a result, the private sector has historically used 'free' water or subsidized water supplied by various government departments. Our study notes that some medium-size industries even use water intended for residential customers and billed at domestic supply rate for their commercial processes. This is yet another example of how water supplied by the state is misused due to poor monitoring.

'Free' water, however, will not be available anymore due to scarcity and competing uses. Water management authorities in Pakistan and elsewhere are under increasing pressure to implement or increase water tariffs in a differentiated manner for different users in order to ensure the long-term

Figure 11.4 This Neelam Valley gorge of the Kunhar River (Balakot Pakistan) is threated by industrial water pollution. (*Credit: Tahir Saleem, 2019*).

availability of water supply and water resources. In addition, in Pakistan the role of the private sector in depleting and contaminating water resources is being criticized. For example, the Supreme Court of Pakistan recently reminded the bottled water industry that groundwater is a property of the government. The court order included levying price from bottled water industry for using groundwater for commercial purposes. Even before the Supreme Court passed down this directive, various government agencies in the Khyber Pakhtunkhwa province had revised water tariffs for piped drinking water so that commercial water users paid more than domestic customers. For example, the agency supplying piped drinking water in parts of the capital city charges 424 Pakistani Rupees per month (Rs/mo) or $2.74/mo for domestic uses and 4,830 Rs/kgal ($31.16/mo) for service stations and small hotels (exchange rate 1 USD to 154,75 Pak Rupees in 2019). Another agency charges 532 Rs/mo ($3.43/mo) for domestic compared to 5,844 Rs/mo ($37.70/mo) for gas stations. In both cases, commercial tariffs are more than ten times higher than the domestic rate. However, in another city where water is piped for free for domestic uses, some private companies also take advantage of these facilities without paying. Likewise, where private companies have their own pumps and facilities to access groundwater, they frequently fail to recognize that water still belongs to the state and needs to be paid for and used judiciously.

As noted in the above example, tariffs are flat rates per connection and not based on volume used. Interestingly, the discussion related to water tariffs for industry has focused on increasing the flat rate and not on introducing volumetric tariffs systems. While the industry seems agreeable to pay a flat rate per connection, in the long-term a volumetric system will be more transparent and will benefit both the government and the industry. Recognizing the benefit of a transparent volumetric billing system can contribute to developing Public–Private Partnership dialogue between the government and industry.

11.2.6 The 'polluter pays' policy

In Pakistan, discussions on water tariffs have focused on water use, rather than pollution control. To date, a 'Polluter Pays' policy has yet to see the light, and very little has been done to make polluters pay the cost of extracting pollutants from their discharged wastewater. This is not due to a lack of legislation: there is no dearth of policies and legal provisions requiring the control of pollution in general and of water pollution in particular. In fact, a number of policy and regulatory documents have sections which directly and indirectly pertain to controlling water pollution by the private sector, but they do not specify any mechanisms for controlling industrial use of water. For example, Section 15.2 of the National Water Policy 2018 states that *'industry shall be required to carry out in-house treatment of their wastewater before transfer to municipal sewer as per National Environmental Quality Standards and the 'Polluter Pays' principle shall be strictly enforced'* (GoP, 2018). On the other hand, there are few effective directives on how to enforce these laws. For instance, the Khyber Pakhtunkhwa Drinking Water Policy (2015) identifies overlapping roles for water sector stakeholders, resulting in uncoordinated and inefficient use of resources (GoKP, 2015). The policy also states that there is no direct accountability or link between cost recovery and service provision in drinking water.

Figure 11.5 Failure to implement pollution control regulations jeopardizes the water quality in Shandur Lake seen here at Shandur Pass, Chitral District, Khyber Pakhtunkhwa. (*Credit: Tahir Saleem, 2018*).

In Khyber Pakhtunkhwa province (Figure 11.5), the most significant barrier to effective water pollution control is simply that the responsible institutions have failed to implement regulations. This is partly due to a lack of actionable information: a tariff might have been levied on companies releasing untreated contaminated water, but successful implementation of such a tariff requires accurate data about the size and number of companies in the province, the volume of water they use, the extent to which their wastewater is contaminated, and proper monitoring of wastewater to ensure compliance. Most often these data are not available on time, so polluters get away without paying.

11.3 LEGISLATION AND REGULATION OF WATER USE AND POLLUTION

Existing environmental protection laws give authorities the ability to regulate water use and to control water pollution, and all these laws are applicable to the private sector. For example, Pakistan's National Water Policy endorses the regulation of groundwater withdrawals to curb over-abstraction and prescribes metering all industrial users to promote financial sustainability and to facilitate enforcement of water pollution regulations by municipal entities. Similarly, the Khyber Pakhtunkhwa Environmental Protection Act 2014 gives enormous power to the provincial Environmental Protection Agency to control pollution by any user. However, despite the authority granted by policy, law and regulation, water use is not well managed and water pollution has not been controlled as desired.

Progress is being made, however. In some cities of the Khyber Pakhtunkhwa province separate tariffs have been introduced to price water based on volume of water used, in order to incentivize the use of water-efficient technologies. Measurement of water used and discharged is also a first step to regulate wastewater and reduce pollution. In Peshawar city, one modern car service station operates as an example of water-efficient technology, using less water and fewer chemicals to wash cars, and treating wastewater prior to discharge (Figure 11.6). Such examples should be encouraged either by directly charging or by raising water tariffs for those who fail to adopt sustainable practices. Either way, the regulation must result in a better business model for sustainability, as previous experience confirms that attempts to encourage appropriate technology based solely on environmental values will not succeed. A wiser pathway would be to find a 'win-win' system of investment and incentives for the industry and the private sector.

Another work in progress is the Integrated Water Resource Management (IWRM) strategy the Government of Khyber Pakhtunkhwa has drafted to address multiple issues. The strategy reflects a clear intention of the government to achieve the strategy goals by regulating private sector and industry, but also by promoting public–private partnership in achieving efficient use of water. One out of four priority areas of the strategy deal with effective participation of private sector.

11.4 CONCLUSIONS

Based on experiences in Pakistan and in the province of Khyber Pakhtunkhwa in particular, the following recommendations and conclusions may be drawn with respect to the challenges faced by government agencies in promoting sustainable industrial water use:

11.4.1 Challenges

- Lack of information on the industrial sector (including commercial enterprises) regarding:
  - the size of the sector
  - its needs and demand for water
  - its water use (piped water supplied by the authorities, groundwater and canal water)

(a)

(b)

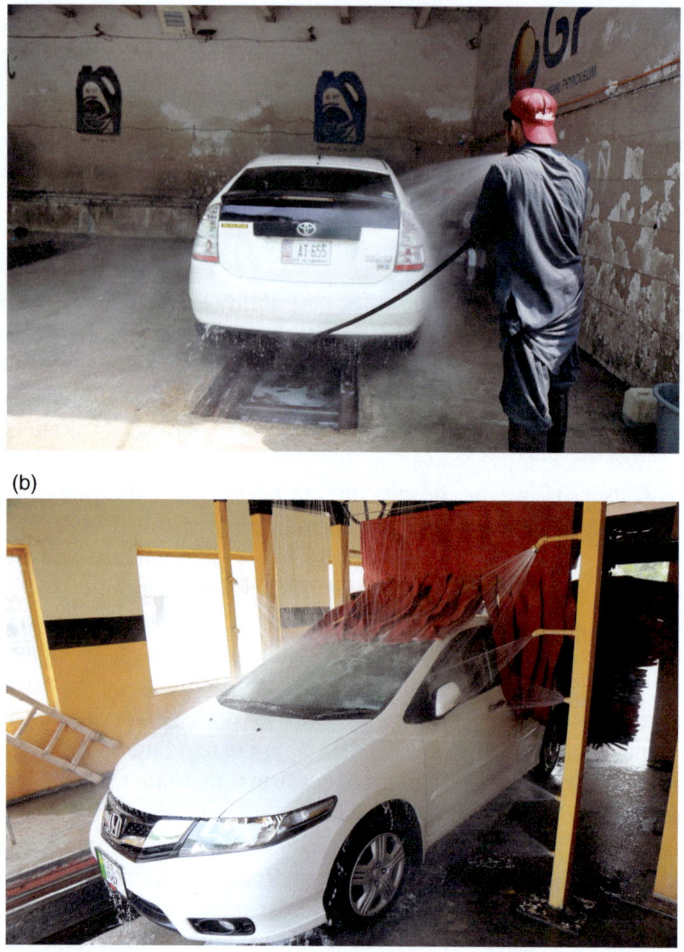

Figure 11.6 (a) A conventional car wash without water recycling and (b) modern car wash with altered nozzle and water recycling. Peshawar, KP Pakistan. (*Credit: Tahir Saleem, 2019*).

- Ambiguities in the roles and responsibilities of various water management authorities;
- Inefficiencies in collection of payment due to political influences;
- Low water rates and use of non-commercial rates by industries.

11.4.2 Recommendations

- Water metering for all groundwater and piped water used;
- Collection of data on number and size of commercial and industrial water users, volume and rate of water use, and volume and quality of wastewater discharged;
- Tariffs based on volumetric water use instead of fixed charges or flat rates;
- Legislation for regulating water use by the private sector through licenses and water quality and quantity controls;

- A thorough study on the use of water by the largest private companies to set the baseline and best practice and for taking well-informed decisions on legislation and water tariffs;
- Strong enforcement of available laws in collaboration with the private sector;
- The industrial sector should not be seen as a culprit but as an actor contributing to devising solutions and strategies with the government;
- Promulgation of a single authority to keep record of all private activities (including commercial) to avoid missing records and lack of overview on the private sector;
- Encouraging use of appropriate technologies through subsidized connection and monthly fees (differentiated fees for private entities using improved and appropriate technology compared to those using conventional technology);
- Investment in prototype manufacturing of water efficient domestic and industrial equipment, e.g., manufacturing of water efficient water taps which are already in use internationally;
- Promoting the adoption of water stewardship standards, especially in the private sector;
- Encouraging Public–Private Partnerships for water conservation. Ideas include recycling, reuse of wastewater (mandatory treatment, especially in all industries and other commercial entities requiring large quantities of water); mandatory rainwater harvesting in all new construction, especially government buildings and private sector societies.

REFERENCES

2030 Water Resources Group. (2009). Charting Our Water Future: Economic Frameworks to Inform Decision-Making. International Finance Corporation, p. 4.

Bhatti A. M. and Nasu S. (2010). Domestic Water Demand Forecasting and Management Under Changing Socio-Economic Scenario. 2010. Kochi University of Technology, Kochi, Japan. Society for Social Management Systems SSMS-2010, pp. 1–8.

Dawn News. (2005). The Pakistan Council of Scientific and Industrial Research (PCSIR) has proposed wastewater treatment plants to be set up at vehicle service stations. Bureau report. https://www.dawn.com/news/160161 (accessed 15 August 2019).

Dawn News. (2019). How Pakistan wastes its water. Article by Dr. Syed Muhammad Abubakar. https://www.dawn.com/news/1428966 (accessed 30 September 2019).

FAO. (2016). AQUASTAT website. Food and Agriculture Organization of the United Nations (FAO). http://www.fao.org/nr/water/aquastat/didyouknow/index2.stm (accessed 24 September 2019).

GoKP. (2015). The Khyber Pakhtunkhwa Drinking water Policy 2015. Department of Public Health Engineering, Government of Khyber Pakhtunkhwa (GoKP), Peshawar.

GoKP. (2016). The Khyber Pakhtunkhwa Industrial Policy 2016. Department of Industries, Commerce and Technical Education, Government of Khyber Pakhtunkhwa (GoKP), Peshawar.

GoP. (2002). National Water Sector Strategy Volume **2**. Ministry of Water and Power, Government of Pakistan (GoP), Islamabad.

GoP. (2018). National Water Policy 2018. Ministry of Water Resources, Government of Pakistan (GoP), Islamabad.

Haroon H., Waseem A. and Mahmood Q. (2012). Treatment and reuse of wastewater from beverage industry. *Journal of the Chemical Society of Pakistan*, **35**(1), 5–10.

Khan N. H. (2015). On site waste management and industrial symbiosis of Hayatabad industrial estate, Peshawar. http://prr.hec.gov.pk/jspui/handle/123456789/9217 (accessed 22 November 2018).

MG. (2018). Industries in SITE Super Highway area face imminent threat of closure owing to serious water crisis. https://mettisglobal.news/industries-in-site-super-highway-area-face-imminent-threat-of-closure-owing-to-serious-water-crisis/ (accessed 18 November 2018).

Morrison J. and Gleick P. (2002). Freshwater Resources: Managing the Risks Facing the Private Sector. Pacific Institute Oakland, California, USA. https://www.pacinst.org/reports/business_risks_of_water/business_risks_of_water.pdf. (accessed 18 November 2018).

Newborn P. and Mason N. (2012). The private sector's contribution to water management: Re-examining corporate purposes and company roles. *Water Alternatives*, **5**(3), 603–618.

Sharma K., Jain U. and Singhal A. (2013). Treatment of waste generated from cement industry and their treatment – a review. https://pdfs.semanticscholar.org/d603/bc728d3de038edad48bf40aa5d8a24c6ab9f.pdf (accessed 20 September 2019).

UNDP. (2016). Water security in Pakistan: issues and challenges. *Development Advocate Pakistan*, **4**(3). https://www.pk.undp.org/content/pakistan/en/home/library/development_ policy/development-advocate-pakistan--volume-3--issue-4.html (accessed 3 December 2018).

Water Futures Partnership. (2012). Water futures: Beyond 2012. Woking & Godalming: UK; Eschborn, Germany: SABMiller, WWF: UK & GIZ.

WHO. (2017). Financing universal water, sanitation and hygiene under the sustainable development goals. UN-Water Global Analysis and Assessment of Sanitation and Drinking-Water. GLAAS 2017 Report. World Health Organization (WHO).

Zia Z., Ali J. and Nizami A. (2019). Use of Water by Private Sector. Provincial Status Report. Government of Khyber Pakhtunkhwa, Pakistan.

Zahir Q. *et al.* (2017). Appraisement, source apportionment and health risk of polycyclic aromatic hydrocarbons (PAHs) in vehicle-wash wastewater, Pakistan. *Science of the Total Environment*, **605–606**, 106–111.

Chapter 12

Sustainable solutions to the impact of industrial water pollution on the environment and community health

Mutshiene Deogratias Ekwanzala[1,2], Phumudzo Budeli[1] and Maggy Momba[1]

[1]Tshwane University of Technology, Department of Environmental, Water and Earth Sciences, Pretoria, South Africa
[2]Kyung Hee University, Environmental Engineering, Department of Civil Engineering, Yongin-si, Gyeonggi-do, Republic of Korea

Keywords: bacteriological impacts, communities, environmental impacts, health impacts, industrial effluent, industrial wastewater, metals, pollutants, reuse, sustainable solutions

12.1 INTRODUCTION

For many centuries, water has been considered as a renewable and unlimited resource. In recent decades, however, it has become evident that freshwater is in fact limited, and industrial water pollution has become a global cause for concern. This is because these sparingly available water resources are negatively impacted by anthropogenic activities, among other factors (Kamika & Momba, 2011). Over the years, manufacturing industry has been identified as one of the main contributors to environmental pollution. Industrial wastewater effluent is the second largest contributor of pollution to freshwater, only after sanitary sewage. Poorly treated industrial wastes damage the environment and pose serious health hazards to communities worldwide. To some extent, this is due to the fact that governments worldwide are often challenged to manage water resources effectively, and industrial effluents do not always comply with discharge standards. In many developing world countries large quantities of untreated industrial effluents are discharged into receiving water bodies used for irrigation (Sherli & Kani, 2018). The use of this water for irrigation leads to the transfer of toxic chemical and biological pollutants from the water to the soil to the plant tissues, where they have adverse effects on both the plants themselves and the people who consume them (Eijsackers *et al.*, 2019). The following sections provide a detailed

DOI: 10.2166/9781789060676_0127

summary of the variety and sources of industrial pollutants in water, their impact on the environment and human health, and current approaches to treating and removing them.

12.2 INDUSTRIAL POLLUTANTS – SOURCES AND TYPES

Due to rapid urbanization and the continuous use of metals and organic substances in industries, health risks posed by industrial effluents have assumed severe dimensions. Although effects of industrial effluents are well documented in the literature, there is a dire need for additional studies aimed at understanding the problem of industrial effluents in its entirety. Industries that contribute most significantly to water pollution include paper mills and metal platers; the food industry (sugar mills, sago factories, distilleries, plus the agricultural fertilizer and pesticide manufacturers); clothing manufacturers (textiles, tanneries); thermal power plants and oil refineries (Kyzas, 2015); and mining (Keshri, *et al.*, 2015). These industries and others, along with the chief pollutants they discharge, are summarized in Table 12.1. As discussed further in the sections below, the discharge of inadequately treated effluents leads to physical, chemical, and biological contamination of the receiving water bodies, threatening both the environment and human health (Dadi *et al.*, 2018; Mattsson *et al.*, 2018).

12.3 ENVIRONMENTAL IMPACTS

Numerous studies conducted all over the world indicate industrial effluent negatively impacts the physical, chemical, and biological state of the receiving aquatic environment. According to a recent study, anthropogenic activities resulted in the presence of metals (e.g. Pb, Cr, Zn) in one Indian river where amphibians and reptiles were found to have unacceptable levels of mercury from industrial waste (Venkatramanan *et al.*, 2018). In South Africa, where metal contamination of water sources by mines is a widespread problem, the discharge of untreated mine wastewater continues unabated with serious consequences for human and animal health as well as significant environmental problems (Keshri *et al.* 2015). A similar study in Nigeria showed that continuous discharge of untreated industrial effluents into a river in Ilorin (Kwara State) resulted in unacceptable levels of contamination, to the extent that water from a well dug in the vicinity of the river had high turbidity and a thin layer of oil at the surface. The water was also found to have high conductivity, suggesting that the dissolved solids are mostly mineral salts, making it perilous for human consumption (Eunice *et al.*, 2017).

12.3.1 Temperature and Biotoxicity

Since every biotic component in the hydrosphere live within a specific temperature range, a sudden change in temperature caused by the discharge of industrial effluent (e.g. power stations and heat-exchange towers) has resulted in their decreased reproduction and growth, and even death. In addition, discharge of excessive amounts of nutrients (P, nitrites, and nitrates) in industrial wastewater can result in eutrophication of freshwater ecosystems, creating environmental conditions that favour the growth of cyanobacteria, which produce toxins hazardous to aquatic organisms and humans alike (Walakira & Okot-Okumu, 2011).

12.3.2 Depletion of dissolved oxygen

Since many aquatic organisms use oxygen in respiration, the concentration of dissolved oxygen (DO) is an indication of the health of aquatic ecosystems. While discharge of large volumes of industrial effluent with low DO can dilute the oxygen concentration in receiving streams, a more pervasive problem is the discharge of high concentrations of organic contaminants, which stimulate the growth of aerobic (oxygen-consuming) microorganisms. These microorganisms compete for oxygen, depleting the DO concentration in freshwater

Table 12.1 Major industrial polluters and their pollutants.

Industries	Examples of Pollutants
Battery manufacturing	Metals (Cd, Cr, Co, Cu, CN, Fe, Pb, Mn, Hg, Ni, Ag, Zn); oil & grease
Cement mills	Heavy metals; Sulfur and nitrogen oxides; Waste soil, by-product gypsum, coal ash
Distilleries	Organics (glucose, polysaccharides, ethanol, glycerol, amino acids, proteins, caramels) and organic matter; High concentration of salts and sulphates
Dye and dye intermediates/textiles	Acids, salts, and heavy metals; Pigments and dyes; Organochlorine-based pesticides; Polycyclic aromatic hydrocarbons (PAHs)
Electric power plants	Metals (As, Cd, Cr, Hg, Pb, Se); Nitrogen compounds (nitrates and nitrites)
Fertilizers	Organics, ammonia, nitrate, phosphorus, and fluoride; Cd and other heavy metals; Suspended solids
Food industry	Organic matter and suspended solids; Biochemical oxygen demands (BOD)
Integrated iron and steel	Ammonia, cyanide, benzene, naphthalene, anthracene, phenol, cresol Heavy metals
Mining industry	Heavy metals (Cu, Pb, Zn, Hg) and metal oxides (CdO, CaO, Na_2O, BaO, Cu_2H_2O, ZnO, LiO, MnO, MgO, SiO2) Sulphates and chlorine oxides; gypsum; Hydroxides, carbonates, and sulphates; cyanide and S.
Nuclear industry	Radioactive waste
Organic chemicals manufacturing	Benzene, chloroform, naphthalene, phenols, toluene vinyl chloride and BOD Metals (Cr, Cu, Pb, Ni, Zn)
Pesticides	Volatile aromatics, phthalates, and PAHs; Halomethanes, haloethanes, and and haloethers, Heavy metals
Petroleum/Petrochemicals	Oil, acid, soda sludge, H_2S, Pb sludge, hydrocarbons, spent filter clay, ethylene glycol, 1,4-dioxane
Pharmaceuticals	PAHs, As_2O_3, heavy metals, halogenated and non-halogenated solvents, organic chemicals, residual pharmaceutical chemicals (chlorambucil, epinephrine, cyclophosphamide, nicotine, daunomycin, nitroglycerin, melphalan, physostigmine, mitomycin C, physostigmine salicylate, streptozotocin, warfarin over 0.3%, uracil mustard), sludge and tars, test animal remains
Pulp and high-quality paper	High concentrations of NaOH, Na_2CO_3, Na_2S, Cl_2, ClO_2, CaO, HCl, organic halides, toxic pollutants, lime mud, wood processing residuals, traces of heavy metals, and pathogens
Sugar	Floor washing waste, sugar cane juice, molasses
Tanneries	Organics, heavy metals such as Cr, ammoniacal nitrogen, acids, salts, sulphides, suspended solids, dyes, fats, oil
Thermal power plants	Fly ash and heavy metals; Coal and oil; Suspended solids
Wood and wool processing	As, Cu, Cr; Abnormally high or low pH; Phenols; oil and grease, COD and suspended solids.

Modified from Ranade and Bhandari (2014).

ecosystems, leading to the creation of anoxic or hypoxic conditions (dead zones). The degree to which organic contaminants can deplete oxygen in freshwater is measured as biochemical or chemical oxygen demand (BOD, COD). To sustain life in a freshwater ecosystem, the BOD and COD values of treated industrial wastewaters should not exceed 10–30 mg/L and 250 mg/L, respectively (Jouanneau *et al.*, 2014).

For example, a study by Idu (2015) revealed that oil companies in Nigeria are responsible for the discharge of inadequately treated effluents containing higher BOD and COD in freshwater streams. Excessive nutrient loading (e.g. N and P) can also leads to the proliferation of microorganisms that rapidly deplete the DO in freshwater ecosystems, impacting the growth and reproduction of freshwater biota, especially fish (Momba *et al.*, 2006).

12.3.3 Metals

Heavy metals are one of the most persistent groups of contaminants in industrial wastewater. Unlike organic pollutants, heavy metals resist biodegradation, accumulating throughout the ecosystem and threatening the health of aquatic organisms. Elevated concentrations of metals have been shown to impede the microbial growth in the environment (Monteiro *et al.*, 2009; Fu & Wang, 2011). As indicated in Table 12.1, the metals most commonly discharged through industrial wastewater are: As, Cd, Ca, Cr, Co, Cu, Fe, Pb, Mg, Mn, Hg, Mo, Ni, K, Se, Na, V, and Zn. Among these, cadmium, lead, and mercury are known as 'non-essential' metals as they have no known role in biological function and often have the most adverse environmental effects (Sharma & Agrawal, 2005). Cadmium has been identified as the chief anthropogenic pollutant released into the environment through industrial effluent discharges (Fu & Wang, 2011). The toxicological effects of cadmium on microbial species are well documented, and various mechanisms of cadmium toxicity on microbial populations have been proposed (Tchounwou *et al.*, 2012; Wasi *et al.*, 2013). Plants absorb cadmium deposited in contaminated soils which accumulates in their roots, stems, and leaves rendering edible crops unfit for human consumption (Ansari & Malik, 2009; Liu *et al.*, 2011). Soils and sediments are also important sinks for lead, retaining 70% of the lead accumulated in the ecosystem and serving as the main reservoir for lead in the microbial community (Rinklebe *et al.*, 2016). Elevated levels of lead in surface soil horizons negatively affect the soil organic fraction, redox, and pH. In addition to its effects on soil, lead toxicity has been reported in aquatic plants such as *Elodea canadensis* (waterweed) and *Hydrilla verticillata* (waterthymes) (Dogan *et al.*, 2009; Singh *et al.*, 2013). In both aquatic plants, the toxic effects resulted in general reduced plant growth and especially chlorophyll function. The widespread presence of mercury in the soil is also well documented in the literature, and its accumulation is hazardous to both the soil structure and edible crop production (Hu *et al.*, 2016).

12.3.4 Constituents of Emerging Concern (CECs)

Contaminants of emerging concern (CECs) are chemicals and other substances 'that have no regulatory standard, have been recently 'discovered' in natural streams and potentially cause deleterious effects in aquatic life at environmentally relevant concentrations' (USEPA, 2008). CECs include but are not limited to agrochemicals (e.g. pesticides and fertilizers); pharmaceuticals and personal care products or PPCPs (e.g. antibiotics and sunscreen agents); industrial chemicals (e.g. brominated flame retardants); and nanomaterials (e.g. carbon nanotubes (CNTs)). Pesticides and fertilizers enter the freshwater ecosystem through the release of polluted water from industries that use them. Many of these pesticides, such as chlorinated hydrocarbons, were an improvement over old methods of controlling pests with toxic arsenic, cyanide, creosote, lead, mercury, and tars (Chau, 2018). However, these newly introduced compounds were often persistent and bio-accumulative and both acutely and chronically toxic to freshwater biota, especially mammals and birds. Some have also been found to be both carcinogenic and have endocrine-disrupting effects (Ortiz-Santaliestra *et al.*, 2018). Although their full effects in water bodies are not yet known, PPCPs have also been found to alter development of freshwater biota. For example, oestradiol from birth control pills has been implicated in the feminization or masculinization of fish populations (Ortiz-Santaliestra *et al.*, 2018). The presence of antibiotics and antibiotic resistance genes in the environment is particularly alarming, and antibiotic resistance is portrayed as one of the most significant issues of the 21st century alongside climate change. Antibiotic overuse helps speed up selection for resistant bacteria that cannot be treated with the currently available antibiotics (Fiorentino *et al.*, 2019).

Nanoparticles are of particular scientific interest due to their application in a wide variety of sectors (electronics, cosmetics, pharmaceutical, information technology, and even water treatment). Their increased use, however, has resulted in high discharges of nanoparticles in industrial effluent and detected presence in aquatic biota (Meli *et al.*, 2016). Metal oxide nanoparticles, such as titanium dioxide, aluminium oxide, silicon dioxide, and zinc oxide in particular, have received attention due to their widespread industrial, medical, and military applications. Nanoparticles could have detrimental effects on ecosystems through their interactions with existing environmental contaminants (Wiesner *et al.*, 2009).

A study has shown that emerging contaminants leave industries through industrial effluent discharge to conventional wastewater plants (Gadipelly *et al.*, 2014). One review of municipal wastewater influent revealed cumulative concentration of CECs between 0.007 and 56.63 μg/L. The removal rates for some compounds were significantly lower than conventional pollutants; analgesics, anti-inflammatory drugs, and beta-blockers in particular showed removal rates of only 30–40% (Qiu *et al.*, 2016). Since these emerging pollutants, by definition, are not included in industrial discharge regulations, we must improve our knowledge of their impacts and adopt efficient and environmentally friendly approaches to preventing their discharge into environmentally sensitive environments.

12.4 HUMAN HEALTH IMPACTS

Just as industrial wastewaters affect the freshwater ecosystem, they also affect human health with both acute (direct) and chronic (long-term) impacts. Figure 12.1 illustrates how the industrial pollutants listed in Table 12.2 alter the function of various vital human organs.

12.4.1 Metals

As reflected in Table 12.2, industrial effluents primarily release pollutants of a chemical nature, which reach humans through inhalation or direct ingestion of polluted water, or from eating contaminated food (fish, plants, animals). Metal pollutants in industrial wastewater – chiefly mercury, lead and arsenic – have negatively impacted the health of human communities. For example, studies have shown respiratory effects in humans and animals resulting from inhalation of nickel released to the atmosphere by electroplating, nickel-cadmium battery production, fossil fuel combustion, and nickel-refining operations. Nickel exposure has also been associated with cancer, hypophosphatemia, neurological disorders, liver damage, and cardiac disorders. Since metal pollutants are bioaccumulated, their effects are magnified as they move up the food chain; as the final consumers, humans would be at the highest risk of health impact from prolonged exposure or ingestion (Castro-González & Méndez-Armenta, 2008).

12.4.2 Bacteriological Impacts

The immediate impact of microbial pollutants is the worldwide incidence of gastrointestinal infection, which causes diarrhoea. While most industrial effluents do not carry microbial pollutants, the chemicals they do

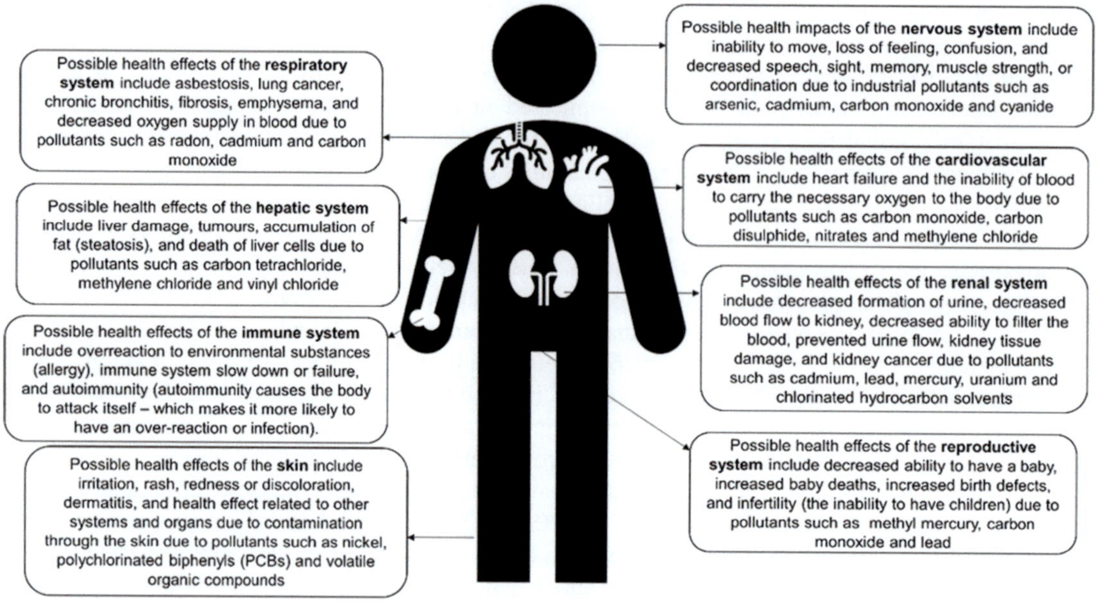

Figure 12.1 Health effects of industrial pollutants on human body systems. (Adapted from US Centers for Disease Control Agency for Toxic Substances and Disease Registry, 2011)

contain (e.g. metals, pharmaceuticals, antibiotics) can render pathogenic microorganisms untreatable. It has been shown that pollutants such as nanomaterials, metals, pharmaceuticals, pesticides, and fertilizers all induce antibiotic resistance in microorganisms (Verlicchi *et al.*, 2010). Diarrhoeal episodes brought about by such microorganisms are hard to treat, and some are untreatable.

On the other hand, some wastewater treatment plants can themselves be a source of bacteriological pollution. Analyses of the maturation ponds of the Nigerian Bottling Company (Owerri, Nigeria), revealed a variety of pathogenic microorganisms including *Staphylococcus, Bacillus, Lactobacillus, Streptococcus, Klebsiella, Escherichia, Proteus*, and *Serratia*. Species of *Lactobacillus* and *Proteus* were also isolated from the final discharge point (Kanu & Achi, 2011), implying that soft drink processing effluent is a source of bacterial contamination of water.

12.5 SUSTAINABLE SOLUTIONS

To tackle industrial wastewater pollution, industries should use sustainable wastewater treatment technologies. For many industrial processes, technologies are now available that cost less, use less energy, require less equipment, and generate no solid waste. Sustainable solutions such as these are the best way to alleviate the environmental and health impacts of industrial pollution for a healthy earth. Measures implemented at the industry level should be able to reduce the pollution entering the ecosystem. The recovery of valuable materials from industrial effluents, and the reuse of waste heat and water from industrial processes themselves will all help to ensure environmental sustainability.

Recent applications have shown that industrial effluents contain valuable materials that can be recovered (Puyol *et al.*, 2017), in addition to industrial wastewater itself that can be recycled for reuse in water-scarce regions (Sato *et al.*, 2013). Future investigations should focus on reducing the cost of recovering chemicals

Table 12.2 Possible human health effects of industrial wastewater pollutants.

Industrial Wastewater Pollutants	Possible Hazards
Pb	Complications in the nervous system and red blood cells Reduction in cognitive development and intellectual performance Death among children
Cd	Renal tubular dysfunction, associated with a high risk of lung and breast cancer Osteomalacia and osteoporosis
Cr	Ulcer, perforation of nasal septum, respiratory cancer
Cu	Gastroenteritis with symptoms such as nausea, vomiting, and abdominal pain, liver and kidney disorders
As	Associated with dermal, respiratory, nervous, mutagenic, and carcinogenic effects
N	Associated with dermatotoxicity, lower body weight, and fetotoxicity among pregnant women
Mn	Central and peripheral neuropathies
Hg	Linked to cardiovascular, reproductive, and developmental toxicity, neurotoxicity, nephrotoxicity, immunotoxicity, and carcinogenicity
Sn	Central nervous system disorders, visual defects, and EEG changes, pneumoconiosis
Antimicrobials	Impaired intestinal flora (e.g. tetracyclines) Drug-resistant pathogens (e.g. quinolones) Hypersensitivity and anaphylactic shock (e.g. macrolides) Kidney damage and nephropathy (e.g. sulphonamides)
Benzo[a]pyrene	Mutagenicity and carcinogenicity DNA damage and oxidative stress Impaired male fertility Respiratory diseases Cognitive dysfunction among children
Chlorpyrifos	Neurological symptoms
DDT	Neurological symptoms Endocrine disruption
DDT and other OCPs	Infertility and foetal malformation
Dioxins and PCBs	Delays in language development Disturbances in mental and motor development
PCBs	Neurological disorders
EDCs	The feminization of male and masculinization of female
Nanomaterials	Disrupt the Immunological system and inflammation of the lungs, or signs of asthma, interstitial fibrosis, genotoxicity and human carcinogen

EEG: electroencephalogram; OCP: organochlorine pesticide; PCB: Polychlorinated biphenyl; EDC: endocrine disrupting chemicals. (Modified from Thompson and Darwish, 2019) such as phosphorus, phosphates, salts, metals, and other substances from industrial effluents. This process could be facilitated using treatment technologies involving immobilized enzymes, immobilized resins, membranes, adsorbents, and other enhanced transport mechanisms.

Many water-intensive industries now treat their generated effluents for reuse in their own processes (Ranade & Bhandari, 2014). For example, water from a heat-exchange process can be allowed to cool before recycling into the process again, avoiding the direct release of heated water which causes adversely affects the biota (Klemeš, 2012). When industrial wastewater is recycled within a closed system, companies achieve savings by using less clean water, in competition with household use. Industrial water reuse also reduces wastewater discharge and can lower the overall carbon footprint of the facility. Using cutting-edge technologies, industrial wastewater can also be treated to the point that it is satisfactory for human consumption (Singh *et al.*, 2019). Treated industrial effluent can also be used for agricultural irrigation where its nutrients can act as fertilizer ('fertigation'), as when dairy wastewater is used to irrigate cropland. This also reduces the reliance of the agricultural sector on potable water, making more drinking water available to the community (Libutti *et al.*, 2018; Qadir *et al.*, 2010).

While many developed world countries reduce environmental pollution through stringent national and state industrial effluent control programs, this is not the case in many developing countries. Therefore, local regulation of industrial wastewater should be given priority to ensure that industries do not discharge effluent without adequate treatment. This would encourage industries to adopt efficacious industrial wastewater treatment methods such as adsorption, electrochemical oxidation, chemical advanced oxidation, membrane filtration, and photocatalysis. Since the term 'green chemistry' was coined in 2002, a plethora of 'green' chemical treatments as an alternative to chemicals that are hazardous with toxic by-products. Among the advanced oxidation process technologies photochemical methods use light to break down molecules in wastewater (e.g. disinfection with ultraviolet light), while non-photochemical techniques make use of relatively non-toxic chemicals that produce hydroxyl radicals without light energy. These include ozone, Fenton's reagent (hydrogen peroxide and iron), wet air oxidation (dissolved oxygen), and electrochemical oxidation (Bokare & Choi, 2014).

Another treatment alternative is membrane bioreactor (MBR) technology, which produces high-quality effluent by substituting membranes for gravity settling and filtration in biological treatment systems. Although use of membrane technology is more expensive than traditional wastewater treatment methods, there are numerous environmental and health benefits associated with this technology. The advantages of MBR treatment include (1) higher effluent quality (2) smaller plant footprint (3) less generated sludge, and (4) shorter hydraulic retention times. Recent research also shows a drastic reduction of oestrogenicity effects of wastewater effluents with membrane technology (Iorhemen *et al.*, 2016).

Where these treatment methods are ineffective in removing trace pollutants, enzyme-based bioremediation techniques may succeed in taking xenobiotics out of wastewater (Gadipelly *et al.* (2014). Due to recent advances in enzyme immobilization and DNA recombinant engineering, enzymes can be produced in larger quantities and immobilized on a solid to thrive under extreme environments, making enzymatic treatment relevant as an effectual method for industrial wastewater xenobiotic removal. For example, Falade *et al.* (2018) demonstrated the use of an enzyme-based tertiary treatment unit for removal of CECs and other micropollutants, further protecting sensitive aquatic environments.

12.6 CONCLUSION

In conclusion, untreated industrial wastewater has detrimental effects on the receiving environments and poses serious health hazards to communities worldwide. The negative health and environmental effects of these discharges have been illustrated in this chapter, as well as alternative approaches to treatment and use of industrial water that could protect both communities and the environment. Technologies for protection of nature and human communities are available, and regulations in many developed countries require their use. However, the health and ecosystems of developing countries will continue to be at risk either until the industries in these countries apply new treatment and reuse technologies, voluntarily or when forced to do so by external regulation.

REFERENCES

Ansari M. I. and Malik A. (2009). Genotoxicity of agricultural soils in the vicinity of industrial area. *Mutation Research/ Genetic Toxicology and Environmental Mutagenesis*, **673**(2), 124–132. https://doi.org/10.1016/j.mrgentox.2008.12.006

Bokare A. D. and Choi W. (2014). Review of iron-free Fenton-like systems for activating H2O2 in advanced oxidation processes. *Journal of Hazardous Materials*, **275**, 121–135. https://doi.org/10.1016/j.jhazmat.2014.04.054

Castro-González M. I. and Méndez-Armenta M. (2008). Heavy metals: implications associated to fish consumption. *Environmental Toxicology and Pharmacology*, **26**(3), 263–271. https://doi.org/10.1016/j.etap.2008.06.001

Chau A. S. Y. (2018). Analysis of Pesticides in Water: Volume I: Significance, Principles, Techniques, and Chemistry of Pesticides. CRC press.

Dadi D., Mengistie E., Terefe G., Getahun T., Haddis A., Birke W., … Van der Bruggen B. (2018). Assessment of the effluent quality of wet coffee processing wastewater and its influence on downstream water quality. *Ecohydrology and Hydrobiology*, **18**(2), 201–211. https://doi.org/10.1016/j.ecohyd.2017.10.007

Dogan M., Saygideger S. D. and Colak U. (2009). Effect of lead toxicity on aquatic macrophyte *Elodea canadensis* Michx. *Bulletin of Environmental Contamination and Toxicology*, **83**(2), 249–254.

Eijsackers H., Reinecke A., Reinecke S. and Maboeta M. (2019). Heavy metal threats to plants and soil life in Southern Africa: present knowledge and consequences for ecological risk assessment. *Reviews of Environmental Contamination and Toxicology*, **249**, 29–70. https://doi.org/10.1007/398_2019_23

Eunice O. E., Frank O., Voke U. and Godwin A. (2017). Assessment of the impacts of refinery effluent on the physico-chemical properties of Ubeji Creek, Delta State, Nigeria. *Journal of Environmental & Analytical Toxicology*, **7**(01), 1000428. https://doi.org/10.4172/2161-0525.1000428

Falade A. O., Mabinya L. V., Okoh A. I. and Nwodo U. U. (2018). Ligninolytic enzymes: versatile biocatalysts for the elimination of endocrine-disrupting chemicals in wastewater. *MicrobiologyOpen*, **7**(6), e00722. https://doi.org/10.1002/mbo3.722

Fiorentino A., Di Cesare A., Eckert E. M., Rizzo L., Fontaneto D., Yang Y. and Corno G. (2019). Impact of industrial wastewater on the dynamics of antibiotic resistance genes in a full-scale urban wastewater treatment plant. *Science of The Total Environment*, **646**, 1204–1210. https://doi.org/10.1016/j.scitotenv.2018.07.370

Fu F. and Wang Q. (2011). Removal of heavy metal ions from wastewaters: a review. *Journal of Environmental Management*, **92**(3), 407–418. https://doi.org/10.1016/j.jenvman.2010.11.011

Gadipelly C., Pérez-González A., Yadav G. D., Ortiz I., Ibáñez R., Rathod V. K. and Marathe K. V. (2014). Pharmaceutical industry wastewater: review of the technologies for water treatment and reuse. *Industrial & Engineering Chemistry Research*, **53**(29), 11571–11592. https://doi.org/10.1021/ie501210j

Hu Y., Cheng H. and Tao S. (2016). The challenges and solutions for cadmium-contaminated rice in China: a critical review. *Environment International*, 92–93, 515–532. https://doi.org/10.1016/j.envint.2016.04.042

Idu A. M. (2015). Threats to water resources development in Nigeria. *Journal of Geology and Geophysics*, **4**, 1000205.

Iorhemen O. T., Hamza R. A. and Tay J. H. (2016). Membrane bioreactor (MBR) technology for wastewater treatment and reclamation: membrane fouling. *Membranes*, **6**(2), 33. https://doi.org/10.3390/membranes6020033

Jouanneau S., Recoules L., Durand M. J., Boukabache A., Picot V., Primault Y., Lakeld A., Sengelin M., Barillon B. and Thouand G. (2014). Methods for assessing biochemical oxygen demand (BOD): a review. *Water Research*, **49**, 62–82. https://doi.org/10.1016/j.watres.2013.10.066

Kamika I. and Momba M. N. B. (2011). Comparing the tolerance limits of selected bacterial and protozoan species to nickel in wastewater systems. *Science of The Total Environment*, **410–411**, 172–181. https://doi.org/10.1016/j.scitotenv.2011.09.060

Kanu I. and Achi O. K. (2011). Industrial effluents and their impact on water quality of receiving rivers in Nigeria. *Journal of Applied Technology in Environmental Sanitation*, **1**(1), 75–86.

Keshri J., Mankazana B. B. J. and Momba M. N. B. (2015). Profile of bacterial communities in South African mine-water samples using Illumina next-generation sequencing platform. *Applied Microbiology and Biotechnology*, **99**(7), 3233–3242. https://doi.org/10.1007/s00253-014-6213-6

Klemeš J. J. (2012). Industrial water recycle/reuse. *Current Opinion in Chemical Engineering*, **1**(3), 238–245. https://doi.org/10.1016/j.coche.2012.03.010

Kyzas G. Z. (2015). Green Adsorbents. Bentham Science Publishers, Sharjah, United Arab Emirates.

Libutti A., Gatta G., Gagliardi A., Vergine P., Pollice A., Beneduce L., Disciglioa G. and Tarantino E. (2018). Agro-industrial wastewater reuse for irrigation of a vegetable crop succession under Mediterranean conditions. *Agricultural Water Management*, **196**, 1–14. https://doi.org/10.1016/j.agwat.2017.10.015

Liu J., Zhang X.-H., Tran H., Wang D.-Q. and Zhu Y.-N. (2011). Heavy metal contamination and risk assessment in water, paddy soil, and rice around an electroplating plant. *Environmental Science and Pollution Research*, **18**(9), 1623–1632. https://doi.org/10.1007/s11356-011-0523-3

Mattsson K., Jocic S., Doverbratt I. and Hansson L.-A. (2018). Nanoplastics in the aquatic environment. In *Microplastic Contamination in Aquatic Environments* (pp. 379–399). Elsevier. https://doi.org/10.1016/B978-0-12-813747-5.00013-8

Meli K., Kamika I., Keshri J. and Momba M. N. B. (2016). The impact of zinc oxide nanoparticles on the bacterial microbiome of activated sludge systems. *Scientific Reports*, **6**(1), 39176. https://doi.org/10.1038/srep39176

Momba M. N. B., Osode A. N. and Sibewu M. (2006). The impact of inadequate wastewater treatment on the receiving water bodies–Case study: Buffalo City and Nkokonbe Municipalities of the Eastern Cape Province. *Water SA*, **32**(5).

Monteiro D. R., Gorup L. F., Takamiya A. S., Ruvollo-Filho A. C., de Camargo E. R. and Barbosa D. B. (2009). The growing importance of materials that prevent microbial adhesion: antimicrobial effect of medical devices containing silver. *International Journal of Antimicrobial Agents*, **34**(2), 103–110. https://doi.org/10.1016/j.ijantimicag.2009.01.017

Ortiz-Santaliestra M. E., Maia J. P., Egea-Serrano A. and Lopes I. (2018). Validity of fish, birds and mammals as surrogates for amphibians and reptiles in pesticide toxicity assessment. *Ecotoxicology*, **27**(7), 819–833. https://doi.org/10.1007/s10646-018-1911-y

Puyol D., Batstone D. J., Hülsen T., Astals S., Peces M. and Krömer J. O. (2017). Resource recovery from wastewater by biological technologies: opportunities, challenges, and prospects. *Frontiers in Microbiology*, **7**. https://doi.org/10.3389/fmicb.2016.02106

Qadir M., Wichelns D., Raschid-Sally L., McCornick P. G., Drechsel P., Bahri A. and Minhas P. S. (2010). The challenges of wastewater irrigation in developing countries. *Agricultural Water Management*, **97**(4), 561–568. https://doi.org/10.1016/j.agwat.2008.11.004

Qiu L., Dong Z., Sun H., Li H. and Chang C. C. (2016). Emerging pollutants–Part I: occurrence, fate and transport. *Water Environment Research*, **88**(10), 1855–1875.

Ranade V. V. and Bhandari V. M. (2014). Industrial Wastewater Treatment, Recycling and Reuse. Butterworth-Heinemann, Germany.

Rinklebe J., Shaheen S. M., Schröter F. and Rennert T. (2016). Exploiting biogeochemical and spectroscopic techniques to assess the geochemical distribution and release dynamics of chromium and lead in a contaminated floodplain soil. *Chemosphere*, **150**, 390–397. https://doi.org/10.1016/j.chemosphere.2016.02.021

Sato T., Qadir M., Yamamoto S., Endo T. and Zahoor A. (2013). Global, regional, and country level need for data on wastewater generation, treatment, and use. *Agricultural Water Management*, **130**, 1–13. https://doi.org/10.1016/j.agwat.2013.08.007

Sharma R. K. and Agrawal M. (2005). Biological effects of heavy metals: an overview. *Journal of Environmental Biology*, **26**(2), 301–313.

Sherli M. C. and Kani M. (2018). Need of life cycle thinking for effective utilisation of resources based on developed and developing countries: a scenario for future generation. *Environmental Science and Pollution Research*, **25**(8), 7280–7286. https://doi.org/10.1007/s11356-018-1375-x

Singh A., Kumar C. S. and Agarwal A. (2013). Effect of lead and cadmium on aquatic plant *Hydrilla verticillata*. *Journal of Environmental Biology*, **34**(6), 1027.

Singh R. P., Kolok A. S. and Bartelt-Hunt S. L. (2019). Water Conservation, Recycling and Reuse: Issues and Challenges. Springer, Singapore.

Tchounwou P. B., Yedjou C. G., Patlolla A. K. and Sutton D. J. (2012). Heavy metal toxicity and the environment. In *Molecular, Clinical and Environmental Toxicology* (pp. 133–164). https://doi.org/10.1007/978-3-7643-8340-4_6

Thompson L. A. and Darwish W. S. (2019). Environmental chemical contaminants in food: review of a global problem. *Journal of Toxicology*, **2019**, 2345283. https://doi.org/10.1155/2019/2345283

US Centers for Disease Control Agency for Toxic Substances and Disease Registry. (2011). https://www.atsdr.cdc.gov/emes/public/docs/Health%20Effects%20of%20Chemical%20Exposure%20FS.pdf. (accessed 2 October 2020).

USEPA. (2008). 'Aquatic Life Criteria for Contaminants of Emerging Concern: Part 1, General Challenges and Recommendations.' Prepared by the OW/ORD Emerging Contaminants Workgroup (June 03, 2008) https://www.epa.gov/sites/production/files/2015-08/documents/white_paper_aquatic_life_criteria_for_contaminants_of_emerging_concern_part_i_general_challenges_and_recommendations_1.pdf. (accessed 2 October 2020).

Venkatramanan S., Chung S. Y., Ramkumar T. and Selvam S. (2018). Ecological risk assessment of selected heavy metals in the surface sediments of three estuaries in the southeastern coast of India. *Environmental Earth Sciences*, **77**(4), 116. https://doi.org/10.1007/s12665-018-7294-9

Verlicchi P., Galletti A., Petrovic M. and Barceló D. (2010). Hospital effluents as a source of emerging pollutants: an overview of micropollutants and sustainable treatment options. *Journal of Hydrology*, **389**(3–4), 416–428. https://doi.org/10.1016/j.jhydrol.2010.06.005

Walakira P. and Okot-Okumu J. (2011). Impact of industrial effluents on water quality of streams in Nakawa-Ntinda, Uganda. *Journal of Applied Sciences and Environmental Management*, **15**(2). DOI: 10.4314/jasem.v15i2.68512

Wasi S., Tabrez S. and Ahmad M. (2013). Toxicological effects of major environmental pollutants: an overview. *Environmental Monitoring and Assessment*, **185**(3), 2585–2593. https://doi.org/10.1007/s10661-012-2732-8

Wiesner M. R., Lowry G. V., Jones K. L., Hochella M. F., Jr, Di Giulio R. T., Casman E. and Bernhardt E. S. (2009). Decreasing uncertainties in assessing environmental exposure, risk, and ecological implications of nanomaterials. *Environmental Science and Technology*, 6458–6462. https://doi.org/10.1021/es803621k

Chapter 13

Fit-for-purpose water reuse in the food processing industry

Susanne Knøchel

University of Copenhagen, Department of Food Science, Copenhagen, Denmark

Keywords: food processing, HACCP, regulations, risk based approach, water reuse

13.1 INTRODUCTION

Food processing industries are major users of potable water and often large producers of waste water. The accessibility and price of water have made water use an issue of increasing importance, as has the cost and impact of discharging polluted wastewater. In some countries (e.g. Denmark) the discharge of water is heavily regulated and the cost of discharge far exceeds the cost of potable water. In addition, the possibility of recovering products, compounds, and energy from processing streams may influence the economics of pollutant removal and is receiving more attention. Many companies have found that considerable water savings can be achieved, and more and more food industries are now looking with interest at treating and reusing process water.

However, the introduction of water sources other than drinking water in the food industry creates new challenges for producers, technology providers, and – not least – for regulatory authorities. Over the years, a major constraint for water reuse in food processing has been uncertainty regarding safety issues, regulations, and customer requirements (Campden BRI, 2015; Casani *et al.*, 2005; Palumbo *et al.*, 1997). This chapter explains some recent regulatory developments and describes current progress towards a risk based approach that will support the safe reuse of 'fit-for-purpose' water in the food processing industries.

13.2 WATER USE IN THE FOOD PROCESSING INDUSTRIES

13.2.1 Amount of water used

Water consumption within the food industry varies depending on the different sectors, the products, the size and efficiency of each operation, the country of production, and the seasonal demands for products. Even

DOI: 10.2166/9781789060676_0139

within the same sector, as water consumption data from the dairy industry indicate (Rad & Lewis, 2014; Vourch, 2008), enormous differences have been observed with volume of water consumed per unit of milk produced ranging from less than 1 to more than 20 L/kg. Several other examples of the wide range of water consumption and discharge levels reported within comparable productions can be found in a recent comprehensive EU report (Santonja *et al.*, 2019), underlining that there is a great potential for savings. The document also contains information on current best available techniques for limiting water consumption in the food, drink, and milk industries and reports on the 'indicative environmental performance levels' obtained. Further reductions in benchmark values will likely follow with new developments in water saving practises and methods.

It should be emphasised that very substantial savings have in many cases been obtained by simple **reduction** at the source which involves staff training, monitoring of consumption, feedback mechanisms, removal of redundant lines, and focus on avoiding spillage and excessive use of water etc. These savings require attention and effort by management and staff in **rethinking** habits and processes in order to make them more water efficient. They are mostly low cost improvements which do not present specific problems from a food safety or regulatory viewpoint since they still operate with the use of potable water. Once these more obvious savings have been implemented, however, further savings may require water **reuse** and in-house water treatment, which may be more challenging both economically, technically, and from a regulatory viewpoint (Casani *et al.*, 2005).

13.2.2 Types of water use

Food and beverage manufacturers use water for many different purposes. In addition to heating, cooling, and cleaning, water in the food processing industry can be used to wash and transport food, and may also be mixed directly with the product for consumption. Figure 13.1 shows an example of how water is being used in a dairy setting.

In order to make a 'fit-for-purpose' evaluation of the water, it is necessary to differentiate between the various intended reuses of the water. This can be done as follows:

Water with no intentional contact

- E.g. water used in cooling towers, closed loop recirculation, toilet flushing, vehicle washing, or other external uses. Provided the physical barrier is intact, this water does not need to be part of the food processing safety management.

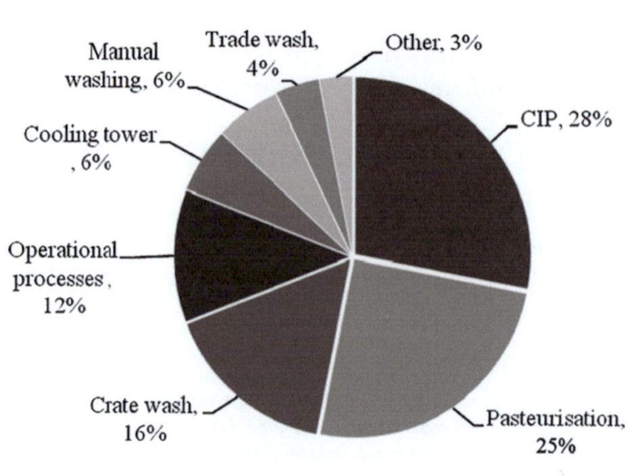

Figure 13.1 Water consumption at a market milk processor (Rad & Lewis, 2014 cf Prasad *et al.* (2004)).

  - There may, however, still be other relevant specifications not related to food safety but related to occupational safety (e.g. growth of *Legionella*), animal or plant safety (e.g. infectious agents being transferred by vehicles), scale, and corrosion-associated parameters etc.

Water with intentional or potential contact

- Water used as an **ingredient** in food;
- Water which has intentional **direct contact** with food such as water used for washing, fluming (transporting) soaking, or scalding food, as well as water used for directly heating and cooling food products;
- Water which has intentional **indirect contact,** including water used for cleaning and/or disinfection of surfaces that may come into contact with food;
- Water, not intended for contact but used for purposes such as floor or wall cleaning in production rooms **where contact cannot be completely excluded** at all times.

All of the above contact uses present some risk of contamination, and the potential for any waterborne hazardous microbiological or chemical contamination should be managed as a part of the company's food safety assurance programme.

13.2.3 Management of water flow in a food processing facility

Before water use and wastewater discharge in a food processing facility can be minimised through reuse, managers must have a good understanding and control of the infrastructure through which it flows and the manner in which it is utilised. For instance, all the hygienic design recommendations made for drinking water distribution and storage are also important for water used at a food processing facility (EHEDG, 2018). Flaws in design and materials may affect cleaning efficacy, time and temperature conditions, contamination routes, etc., and may lead to local growth of bacteria and/or biofilm formation, increases in turbidity, build-up of organic matter, deposits of iron or other compounds, increased nitrite levels (if nitrate-rich streams are recirculated) unwanted sensory changes, and more.

From a food production perspective, the factory flow will be important, especially for 'ready-to-eat products' since water and air streams should be cleanest where the finished product is exposed. In a high risk zone (e.g. where slicing of pre-cooked 'delicatessen-style' meat is performed), the demands for the washing of the conveyor belt or floor will be higher than in a receiving/pre-cooking area from a microbiological perspective. Water streams other than drinking water should be in separate and clearly marked pipes without possibility of backflow.

Preparing and maintaining an accurate, clearly marked flow diagram indicating the current source, treatments, transportation, and use of water throughout the food processing facility is of vital importance and will provide the framework of a hazard analysis necessary to ensure that water used in the facility can be reused in a manner that is protective of public health and in compliance with all relevant national, state, and local regulations.

13.3 WATER REUSE IN FOOD PROCESSING

The definitions of 'water reuse' refer to a wide range of applications, including augmentation of raw water supplies and irrigation with treated municipal wastewater. This mixture of terminologies may constitute a problem for the perception of water reuse applied in the food industry, both in terms of customer acceptance and in terms of the hazard identification. For the sake of consistency, we will use the terms for reuse within the food industry discussed by the Codex Committee on Food Hygiene (CAC, 1999), a UN supported activity on internationally adopted food hygiene guidelines. According to this, 'reuse water' in a food processing context is any water that has been recovered from a processing step – including from the food components – and that after reconditioning treatment(s) is intended to be reused in the same, prior, or subsequent food processing operation. In that sense, reuse water includes reclaimed, recycled, recirculated, and reconditioned water defined as follows:

First use water – potable water used in any food manufacturing process.

Reclaimed water – water that was originally a constituent of food and has been removed from it in a processing operation (e.g. evaporation). Note that this 'water' in some contexts is also defined as food (e.g. permeate from whey water is sometimes termed cow or milk water).

Recycled water – water, other than first use or reclaimed water, which has been obtained from a food processing operation (e.g. counterflow vegetable washing).

Recirculated water – water reused in a closed loop for the same processing operation (e.g. a cooling system in which the water circulates).

Reconditioning – refers to the treatment of water intended for reuse by means designed to reduce or eliminate microbiological, chemical, and physical contaminants according to its intended use.

Note that any of the defined water types can be 'reconditioned for reuse,' and the risk from all types of reuse can be quantified in the same manner.

Types of water use and corresponding water reduction strategies are illustrated in Figure 13.2, below. As noted earlier, optimising current water use or redesigning unit operations to use less water are often the least costly ways to conserve water. Once those savings have been realised, however, it may be possible to further cut water use and significantly reduce wastewater discharge by reusing water before it leaves the facility.

Whichever terminology is employed, reuse of water in food processing facilities is subject to both national regulations and international standards and guidelines designed to protect public health by reducing the risk of food contamination.

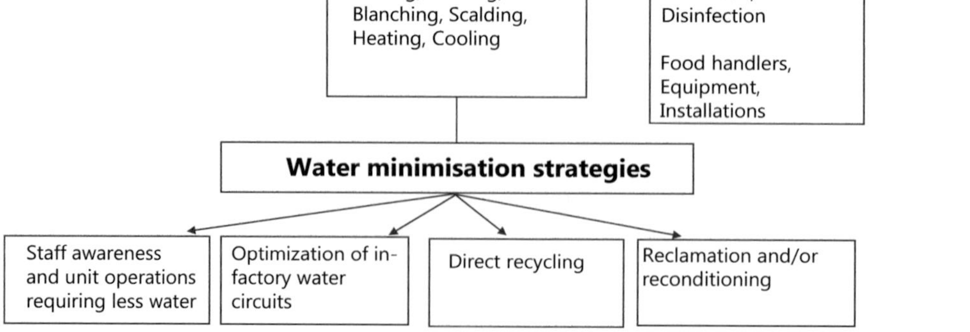

Figure 13.2 Water minimisation strategies in food processing (modified from Casani, 2004).

13.4 WATER REUSE RISKS AND REGULATIONS

It is generally required of food producers that they should have access to 'an adequate supply of potable water.' Once the water has passed the entrance tap and is being used in food production, the responsibility for the water lies with the company. The Codex Alimentarius, the United States Food Code, the European Council Regulations on the hygiene of foodstuffs, the Australia New Zealand Food Standards Code and many other national regulations refer to potable water (as defined by the World Health Organization Guidelines on Drinking Water Quality, the United States Environmental Protection Agency National Primary Drinking Water Regulations, the European Council Drinking Water Directive, the Australian Drinking Water Guidelines, etc.), as the default water quality in food processing but they also operate with the possibility of using other water qualities provided the competent authorities are satisfied that the wholesomeness of the foodstuff is not affected. It is also generally stated that sanitary water cannot be used as source water. Some key provisions of these regulations are shown in Table 13.1.

Of particular note is that both Canadian and Australian rules specifically require the use of a formal risk assessment approach to ensure that every aspect of water reuse is appropriately designed and inspected, and that all personnel involved in planning and operations, sampling, and analysis should have adequate training and competences. The consensus of regulations suggests that water can be safely reused in food production if adequate care is taken to assess the risk of contamination, determine the required quality to avoid that risk, and evaluate the reliability of the treatment process to ensure that the quality is maintained (e.g. 'multiple barrier' approach). The Canadian regulations (CFIA, 2020) describe this approach thus:

> Given the many different ways water is used, and the many ways food can be exposed to or come into contact with water, the level of risk and the level of control needed may vary, depending on the scenario… Relying on the categorisation of water as either potable or non-potable may not always be sufficient depending on the intended use… You should **classify each source of water you use based on the potential hazards it presents** and intended use [and] you should **identify all potential hazards from water that could be a risk of contamination to a food** regardless of whether or not you have direct control over the hazard or its source. [Emphasis added.]

A number of more general water hygiene guidelines from different industry associations have been elaborated (Campden BRI, 2012; EHEDG, 2018; ILSI, 2008, 2013) and in the recent FAO/WHO (2019) report on 'Safety and Quality of Water Used in Food Production and Processing' a general decision tree is presented to help a risk assessment based decision on the options for reusing water, i.e. for no contact or potential contact purposes. If the water is used in direct or indirect food contact, the food producing companies should apply the Hazard Assessment and Critical Control Point (HACCP) principles, a familiar concept in the food industry, as suggested by Casani and Knøchel (2002) and recommended by Kirby *et al.* (2003), ILSI, and others. The development of an HACCP plan has traditionally been divided into seven principles, as follows:

(1) Conduct a hazard analysis
(2) Determine critical control points (CCPs)
(3) Establish critical limits
(4) Establish monitoring procedures
(5) Establish corrective actions
(6) Establish verification procedures
(7) Establish record-keeping and documentation procedures

Table 13.1 Selected text examples from various national and international guidelines and regulations on use of water in food processing applications (note that these may change).

Country – Guideline or Regulation	Recommendation for Water Reuse in Food Processing
United Nations – Codex Alimentarius[1]	• Reuse water shall be safe for its intended use and shall not jeopardise the safety of the product through the introduction of chemical, microbiological or physical contaminants in amounts that represent a health risk to the consumer • Reuse water shall be subjected to ongoing monitoring and testing to ensure its safety and quality. • Treatment of water must be undertaken with knowledge of the types of contaminants the water may have acquired from its previous use.
European Union – Regulation (EC) 852/2004	• Recycled water used in processing or as an ingredient is not to present a risk of contamination. It is to be of the same standard as potable water, unless the competent authority is satisfied that the quality of the water cannot affect the wholesomeness of the foodstuff in its finished form • Where heat treatment is applied to foodstuffs in hermetically sealed containers it is to be ensured that water used to cool the containers after heat treatment is not a source of contamination for the foodstuff
United States – 21 CFR 110[2]	• Each plant shall be equipped with adequate sanitary facilities and accommodations, including a water supply that is sufficient for is intended use, and water that is safe and of adequate sanitary quality. • Water may be reused for washing, rinsing and conveying food providing it does not increase the level of the contamination of the food.
United States – 9 CFR 416.2 (g)[3]	• A supply of running water that complies with the National Primary Drinking Water regulations (40 CFR part 141), at a suitable temperature and under pressure as needed, must be provided in all areas where required (for processing product, for cleaning rooms and equipment, utensils, and packaging materials, for employee sanitary facilities, etc. • Reconditioned water that has never contained human waste and that has been treated by an onsite advanced wastewater treatment facility may be used on raw product, except in product formulation… Product, facilities, equipment, and utensils coming in contact with this water must undergo a separate final rinse with non-reconditioned water that meets the criteria prescribed in paragraph (g)(1) of this section.
Canada – Safe Food for Canadians Regulations[4]	• Any water that might come into contact with a food must be potable, unless it does not present a risk of contamination of the food, and must be protected against contamination. • Any system that supplies water that meets the requirements of subsection (1) must not be cross-connected with any

(Continued)

Table 13.1 Selected text examples from various national and international guidelines and regulations on use of water in food processing applications (note that these may change) (*Continued*).

Country – Guideline or Regulation	Recommendation for Water Reuse in Food Processing
	other system, unless measures are taken to eliminate any risk of contamination of a food as a result of the cross-connection.
Canada – CFIA Guidance[5]	• Operators should have, in their establishments, a supply of water that is safe, appropriate for the intended use and of a quantity and pressure sufficient for the operational needs. You should consider the state of the water you use for processing, as potable or non-potable, and its suitability for the intended use. The suitability of water is determined by doing a thorough assessment of the hazards it presents and the risk of contamination to a food • Reusing and reclaiming water helps conserve water and is a more efficient use of the available water resources
Australia – The New South Wales Food Authority (2008)[6]	• In addition to not adversely affecting the safety of the food, the reuse of water should not adversely affect the suitability or quality attributes (flavour, colour, texture) of the food product. • All reused water must be safe for its intended use and must not jeopardise the safety of the product through the introduction of chemical, microbiological or physical contaminants in amounts that represent a risk to the consumer. • Any food business intending to reuse water must apply the principles of Hazard Analysis Critical Control Point (HACCP) and risk assessment to implement appropriate control measures to address the identified hazards. • A food business must have adequate safeguards in place to ensure that non-potable water cannot be reused (unless previously demonstrated that it is safe to do so), and must have adequate verification measures in place to ensure appropriate reconditioning of wastewater. There should be no physical connection between the potable and non-potable water supply. • Where a food business reconditions industrial wastewater to a potable standard through the application of appropriate technology for direct reuse, the business must use a multiple barrier approach (ie utilise more than one treatment process to ensure that if one step fails at least one other treatment step will control the identified hazard). The business must provide evidence of adequate validation of treatment methods in accordance with the validation guidelines specified in this document.

(*Continued*)

Table 13.1 Selected text examples from various national and international guidelines and regulations on use of water in food processing applications (note that these may change) (*Continued*).

Country – Guideline or Regulation	Recommendation for Water Reuse in Food Processing
	• Where reuse of water is for areas of direct contact with food, the record keeping and monitoring systems for the water treatment system must be incorporated into the business's Food Safety Program. The food business must ensure that regular independent audits are made on the suitable operation of the wastewater treatment system.

[1]Codex Alimentarius.
[2]US 21 CFR 110 Good Manufacturing Practice In Manufacturing, Packing or Holding Human Food.
[3]9 CFR 416.2. Animals and Animal Products – Establishment grounds and facilities.
[4]Safe Food for Canadians Regulations Part 4 – Preventive Controls; Division 4 – Maintenance and Operation of Establishment; Subdivision B, 'Conditions Respecting Establishments'; 70.1 'Water – Contact with Food;' 70.3 'Water – Cross-Connections'.
[5]Canadian Food Inspection Agency (CFIA) (2019) https://inspection.gc.ca/preventive-controls/preventive-control-plans/water-for-use-in-the-preparation-of-food/eng/1511377944601/1511377945080
[6]New South Wales Food Authority (2008) Water reuse guideline NSW/FA/FI023/0805 (May 12, 2008).

Detailed instructions for implementation of an HACCP program are beyond the scope of this article, but it is a necessary part of setting up a water reuse operation in a food processing facility. As each food processing sector (e.g. meat processing, dairy, beverage production) and each sub-sector (e.g. poultry processing, cheese-making, breweries) have distinct systems, equipment, and protocols, input from industry leaders in these areas will be critical to establishing national sectoral guidelines based on best available knowledge. The final responsibility for the food products will always be the individual food producers and there will always be site and production specific issues but sector codes aligned with regulatory requirements will greatly facilitate the process for the individual production site.

13.5 REGULATORY BARRIERS AND KNOWLEDGE GAPS

13.5.1 Regulatory barriers

Drinking water and food production regulations are normally dealt with separately and in different parts of the administration. Water experts have usually concentrated on drinking water or urban/industrial wastewater where there is strong emphasis on fecal contamination risks and chemical residues while food safety experts have regarded potable water as a water source of reliable and safe quality. Typically, the microbiological and chemical issues in the food production will be quite different. The requirement for a risk-based approach to water reuse and need to evaluate on a 'case-by-case' basis is not only a challenge for the food producers but also for the regulators and individual inspectors who, as the competent authorities, will have to approve the companies' own safety programs when the companies are reusing water on their premises. There is a risk that lack of familiarity with the area and the potential problems associated with it may lead to overly cautious rejections to be on the safe side or, on the other hand, that obvious flaws in the safety assurance program and its implementation will be overlooked. Solutions may also depend on local conditions, e.g. some countries may rely more on chlorination while others do not support chlorination of water in contact with food. There are by now many full-scale operations running with reuse of reconditioned water within a range of industries and it will be important to learn from both successful and less successful operations.

13.5.2 Knowledge gaps

There are many knowledge gaps to be filled from consumer perception, business incentives, and sustainability issues with water *versus* energy consumption, etc., but in terms of microbiological challenges, there is definitely a need for better understanding of removal/inactivation efficiencies of different treatment methods. While much is already known, as indicated in Table 13.2, the variety of equipment available to treat water sources of disparate quality creates wide variations in removal or inactivation efficiencies which make it difficult to foresee the efficacy in a specific scenario. Rapid monitoring techniques with feedback in order to pick up hazardous events or treatment failures are also needed.

While safety is the main issue for regulators, producers must also avoid problems with product quality, such as the growth of spoilage organisms in membranes, pipes, and tanks, corrosion or precipitations, and more. Additional research in this area is particularly useful, as many of these problems occur over time or under certain circumstances (e.g. warm weather). It is therefore vital that regulators, industry, consultants, and researchers collaborate in order to streamline some of the guidelines and build up experience and competence.

Finally, industry executives, facility managers, line staff, regulators, and inspectors would all benefit from collaborating in order to facilitate the process. Improved access to education and training, which could include workshops and e-learning material, would be of great help.

Table 13.2 Range of inactivation of microorganisms by various treatment methods (EPA, 2012).

	Indicator microorganisms			Pathogenic microorganisms				
Type of Microorganism	*Escherichia coli* (indicator bacteria)	*Clostridium perfringens*	Phage (indicator virus)	Enteric bacteria (e.g., *Campylobacter*)	Enteric viruses	*Giardia lamblia*	*Cryptosporidium parvum*	Helminths
Bacteria	X	X		X				
Protozoa and helminths						X	X	X
Viruses			X		X			
Indicative Log Reductions in Various Stages of Wastewater Treatment[1]								
Secondary treatment	1 - 3	0.5 - 1	0.5 - 2.5	1 - 3	0.5 - 2	0.5 - 1.5	0.5 - 1	0 - 2
Dual media filtration[2]	0 - 1	0 - 1	1 - 4	0 - 1	0.5 - 3	1 - 3	1.5 - 2.5	2 - 3
Membrane filtration (UF, NF, and RO)[3]	4 - >6	>6	2 - >6	>6	2 - >6	>6	4 - >6	>6
Reservoir storage	1 - 5	N/A	1 - 4	1 - 5	1 - 4	3 - 4	1 - 3.5	1.5 - >3
Ozonation	2 - 6	0 - 0.5	2 - 6	2 - 6	3 - 6	2 - 4	1 - 2	N/A
UV disinfection	2 - >6	N/A	3 - >6	2 - >6	1 ->6	3 - >6	3 - >6	N/A
Advanced oxidation	>6	N/A	>6	>6	>6	>6	>6	N/A
Chlorination	2 - >6	1 - 2	0 - 2.5	2 - >6	1 - 3	0.5 - 1.5	0 - 0.5	0 – 1

(Sources: Bitton, 1999; EPHC, 2008; Mara and Horan, 2003; NRC, 1998; NRC, 2012; Rose et al., 1996; Rose, et al., 2001; EPA, 1999, 2003, 2004; WHO, 1989)

[1]Reduction rates depend on specific operating conditions, such as retention times, contact times and concentrations of chemicals used, pore size, filter depths, pretreatment, and other factors. Ranges given should not be used as design or regulatory bases—they are meant to show relative comparisons only.
[2]Including coagulation
[3]Removal rates vary dramatically depending on the installation and maintenance of the membranes.
N/A = not available

13.6 CONCLUSIONS

There is a potential for large savings of freshwater in the food processing industry. Much can be obtained by simple savings while further minimisation requires reconditioning and reuse. While use of water other than potable water in the food processing industry has previously been controversial for safety reasons, there are now strong drivers for reuse from many viewpoints: water accessibility as well as pollution due to discharges are important sustainability issues; economically, the price of water and its discharge is rising and valorisation of compounds in the water streams is feasible in some industries; in terms of regulation, there is now increasing openness for using 'fit-for-purpose' concept and risk based approaches; technically, reconditioning to different quality levels is possible, and from a societal/customer point of view, there is a more positive attitude towards environmentally friendly solutions. There are still also many barriers and knowledge gaps. Uncertainty of regulations and their implementation is a major barrier for many food processers. This will hopefully diminish as both producers, technology providers, researchers, and the regulatory system gain and share experience.

REFERENCES

CAC (1999). Codex Alimentarius Commission: Codex Committee on Food Hygiene. Discussion Paper on Proposed Draft Guidelines for the Hygienic Reuse of Processing Water in Food Plants. Joint FAO/WHO Food Standards Programme, Thirty-second Session, Washington, D.C., United States, November 29-December 4.

Casani S. and Knøchel S. (2002). Application of HACCP to water reuse in the food industry. *Food Control,* **13**, 315–327.

Casani S. (2004). Microbiological quality of reused process water in the food industry. Ph.D thesis. Royal Veterinary and Agricultural University. Copenhagen (Now part of University of Copenhagen).

Casani S., Rouhany M. and Knøchel S. (2005). A discussion paper on challenges and limitations to water reuse and hygiene in the food industry. *Water Research*, **39**(6), 1134–1146.

Campden BRI (2012). Guidelines on the reuse of potable water for food processing operations. J. Holah (ed.). Guideline No. 70, Chipping Campden.

Campden BRI (2015). Water from alternative sources: uses and treatments. E. Maguire, A. Alldrick and P. Voysey (eds). R&D report No. 384, Campden BRI. Chipping Campden.

Canadian Food Inspection Agency (CFIA). (2019). 'Water for Use in the Preparation of Food' https://inspection.gc.ca/preventive-controls/preventive-control-plans/water-for-use-in-the-preparation-of-food/eng/1511377944601/1511377945080 (accessed November 29 2020).

EHEDG (European Hygienic Engineering Design Group). (2018). Doc. 28 Safe and Hygienic Treatment, Storage and Distribution of Water in Food and Beverage Factories.

EPA. (2012). Guidelines for water reuse EPA/600/R-12/618. https://www.epa.gov/sites/production/files/2019-08/documents/2012-guidelines-water-reuse.pdf (accessed 12 October 2020).

FAO and WHO. (2019). Safety and Quality of Water Used in Food Production and Processing. Meeting Report. Microbiological Risk Assessment Series no **33**, Rome.

ILSI (2008). Considering water quality for use in the food industry, Brussels. Available at http://ilsi.org/publication/considering-water-quality-for-use-in-the-food-industry/ (accessed 20 March 2020).

ILSI (2013). Water recovery and reuse: Guideline for safe application of water conservation methods in beverage production and food processing. Available at http://ilsi.org/wp-content/uploads/2016/05/Guideline-for-Water-ReUse-in-Beverage-Production-and-Food-Processing.pdf (accessed 20 March 2020).

Kirby R., Bartram J. and Carr R. (2003). Water in food production and processing: Quantity and quality concerns. *Food Control*, **14**, 283–299.

Santonja G. G., Karlis P., Stubdrup K. R., Brinkmann T. and Roudier S. (2019). Best Available Techniques (BAT) Reference Document for the Food, Drink and Milk Industries; EUR 29978 EN; doi: 10.2760/243911

Palumbo S.A., Rajkowski K.T. and Miller A.J. (1997). Current approaches for reconditioning process water and its use in food manufacturing operations. *Trends in Food Science and Technology*, **8**(3), 69–74.
Prasad P., Pagan R., Kauter M. and Price N. (2004). Eco-efficiency for the Dairy Processing Industry. (Southbank Victoria: Dairy Australia).
Rad S.J. and Lewis M.J. (2014). Water utilisation, energy utilisation and waste water management in the dairy industry: A review. *International Journal of Dairy Technology*, **67**, 1–20.

Chapter 14

Musings of a former regulator, or 'how can we do better?'

Felicia Marcus

Environmental Policy and Organizational Management Consultant; Member, Water Policy Group (Former Administrator, EPA Region IX and Chair, California State Water Resources Control Board), Emeryville, USA

Keywords: alternative compliance, Common Sense Initiative, environmental regulation, innovative regulation, National Performance Review, Project XL, regulatory reinvention, results-based Regulation

14.1 INTRODUCTION

The United States has made undeniable progress over the past 50 years regulating industrial water pollution. Current industrial regulation is based on a bottom-line set of rules centered around pollutant-by-pollutant or technology-based limits. Over a dozen statutes and attendant regulations, including the Clean Water Act and the Safe Drinking Water Act, regulate a variety of constituents across all media, down to the part per billion. These rules represent the best level of protection that legislatures and regulatory agencies can achieve through an inherently political process over years of advocacy, litigation, and conflict. Although the rules are adjusted periodically in response to new information and shifts in policy or political winds, this regulatory framework is relatively static, and it is certainly not optimal.

In theory, this adversarial process will ultimately protect public health and safety and preserve the environment. But we need more than bottom-line regulation to respond to the challenge of climate change and reach the goal of sustainability: we need to harness the power of industry to create innovative solutions that span traditional silos and geographic divides, and provide greater benefits at lower cost and at a more rapid pace. To do that, we must be able to inspire and incentivize companies to do more than meet the minimum standards required by law. In an arena fraught with mistrust and

DOI: 10.2166/9781789060676_0151

frustration, the process of regulating industry looks more like combat than the kind of strategic collaboration capable of producing the innovative solutions we desperately need.

For decades, regulators in the United States have experimented with a regulatory system strict enough to prevent pollution but agile enough to encourage multiple benefit, cross-media solutions. Indeed, some of the most creative multi-benefit solutions have resulted from enforcement cases. In Philadelphia, Washington DC, and other places, the USEPA has settled enforcement cases against sewage spills from combined sewer overflows of sewer and stormwater systems by allowing cities to use urban greening and stormwater capture to help solve their problems, instead of forcing them to build multimillion or even multibillion dollar treatment facilities to achieve the same water quality goal. This solution achieves that goal, but also creates other value for scarce local dollars.

Having served many years at all levels of government – local, state, and federal – I can share some reflections about the regulatory road we have travelled, as well as some ideas about where we need to go from here. To be clear, a strong regulatory framework is still the backbone of an effective regulatory scheme. Strict, enforceable, and intelligible environmental standards protect public health and the environment, and they create a level playing field for business competitors. Strong enforcement also supports pollution prevention programs and similar innovations. However, regulations need not be limited to 'one-size-fits-all' maximum contaminant or technology-based limits. When properly adapted to specific sectors or companies and embraced by willing participants, they can incentivize innovation and lead to low-cost, multi-benefit solutions that lead to greater environmental and public health protection in parallel with a strong regulatory bottom-line or backstop. I'm not suggesting we do away with traditional, enforceable rules in the hope that industry will magically wake up one morning and decide to protect the environment, but that we make a space in our regulations for creative agreements with companies or other regulated entities that actually want to achieve better environmental and public health results.

With the benefit of the US experience, countries just starting out on the road to regulation may find a shorter path to developing a system that can both deter polluters and motivate and reward superior performers. From my perspective, and as I describe in more detail below, here are the key elements of a rugged yet adaptive environmental management program:

(1) A clear set of *bottom line regulations* that are protective of public health and safety.
(2) Strong *enforcement* of those rules, including the option for public enforcement through the courts to bolster under-resourced enforcement agencies.
(3) Room for *better alternate projects* that achieve multiple environmental and public health benefits at less economic cost and provide greater flexibility and quicker decision-making.
(4) *Transparent data and accountability systems*, so that incentives for going beyond the minimum can be appropriately awarded and poor performance identified and appropriately dealt with. It is key that progress, or lack thereof, can be assessed and transparent to all.

14.2 A LOOK BACK AT US INDUSTRIAL REGULATION

The environmental movement in America is often traced to public outcry against industrial pollution during the 1960s and 1970s. On the East Coast, an ongoing fight to prevent Consolidated Edison from building a hydropower plant on scenic Storm King Mountain confirmed the right of US citizens to sue to preserve the environment (Schuyler & Gallay, 2018). Perhaps equally important, the Hudson River Fishermen's Association became the first client of the nascent Natural Resources Defense Council (NRDC), which went on to become one of the most effective environmental advocacy groups in the country. In the

Midwest, the Cuyahoga River – choked with petrochemicals and slowed by debris – was ignited by sparks from a train near Cleveland, Ohio (Hartig, 2019). Photos of the river on fire graphically showed the results of failing to regulate industry, while in nearby Illinois a vigilante (known only as 'The Fox') poured cement into industrial outfalls and sent bottles of waste and boxes of dead animals to industry executives (Gathman, 2015). On the West Coast, an oil platform six miles off the coast of Santa Barbara, California experienced a blowout on January 28, 1969, releasing 100,000 barrels of oil into the ocean and blackening beaches with tar and the carcasses of thousands of seabirds (Hamilton, 2019). These stories and others like them precipitated the first Earth Day (April 22, 1970) and put a spotlight on the need to regulate industry to protect both public health and the environment. In the burst of legislation that followed, the US Congress passed many laws (including the Clean Water Act) and gave the newly formed Environmental Protection Agency sweeping power to regulate industrial discharges (Copeland, 2016; EPA, 1972; EPA, 2020a).

14.2.1 The seventies: building the basics

Although the Clean Water Act initially covered a wide array of pollutants, the extent of regulation and regulatory enforcement developed in waves, each one hard-fought. For several reasons, USEPA and state agencies first focused on reducing pollution by business. Public utilities already had been regulated at some level to protect the public health from pathogens in wastewater, while runoff from 'normal' farming activities was generally exempted from federal law and left to the states because it tended to be diffuse ('non-point'). (Sympathy for farmers also was no doubt a part of the political calculus.) By contrast, readily identified industrial pollution was viewed unsympathetically as an 'externality' that companies passed on to taxpayers, making the public unfairly pay to clean up their mess or pay through harm to our shared environment and public health. The Clean Water Act and other acts were therefore designed to force companies to more seriously internalize these costs, giving rise to the logical notion of 'polluter pays' which figured in many early environmental policies and regulations and continues to do so to this day.

From industry's point of view, though not frequently voiced in public, it also made sense to create a 'level playing field' by regulating all companies equally. Even within the business community, there was support for some type of consistent regulatory framework so that companies that properly managed their wastes were not put at a competitive disadvantage compared to those that didn't. It was common knowledge within regulatory agencies that many enforcement tips about polluting companies came from law-abiding competitors in their neighborhood or their business line that did not want these scofflaw companies to escape the cost of compliance. This is not to say that business supported unbridled regulation; they did not, but over time industries began to lobby for less costly rules for their sectors or product lines and greater regulation for others. As an example, an industry/labor group, the California Council for Environmental and Economic Balance, advocated for increased inspection of pollution control devices on individual cars and trucks ('mobile sources') to avoid stricter requirements limiting emissions from industrial smokestacks ('stationary sources'), which would have been far more expensive to implement per ton of pollution reduction.

14.2.2 1980–2000: consolidating command and control and branching out

EPA often prioritizes certain types of enforcement for a period of time in waves known as 'national compliance initiatives.' The decades between 1980 and 2000 saw the extension of environmental enforcement from industry into several arenas, including regulation of municipal utilities, federal agencies, and animal feedlots. These regulations took the form of limits on both concentration and mass of pollutants, and in some cases included the requirement to use specific technologies. This period also heralded the rise of criminal prosecution of environmental crimes.

During the 1980s Federal enforcement took aim at large municipal utilities (e.g., Boston, Los Angeles, and San Francisco) where inadequate sewage treatment fouled beaches and grabbed headlines. In 1993 EPA also started to bring enforcement against other federal agencies under a variety of federal statutes. Although Section 313 of the 1972 Clean Water Act explicitly made federal agencies subject to the act – a fact which was reaffirmed in 1978 by Presidential Executive Order – many government agencies previously enjoyed immunity from prosecution, under the ancient privilege of sovereigns who were held to be above the law. Eventually, in the 1990s, EPA began to bring enforcement actions against these agencies.

EPA also took a second look at the agricultural sector, regulating what are known as concentrated animal feeding operations (CAFOs). These are large animal production sites created when economic and other pressures led to consolidation of smaller facilities into larger more densely populated ones. While 'agriculture' is generally exempt under the Clean Water Act, these concentrated operations had produced myriad environmental and public health threats like bacterial contamination and nutrient pollution of eastern waterways and nitrate pollution of groundwater drinking water sources that put them well beyond the concept of 'normal' farming operations (EPA, 2020f).

The overall regulatory system that emerged after passage of the Clean Water Act can best be described as 'command and control.' Its top tool was setting discharge limits pollutant by pollutant, and applying technology-based limits by sector. Discharge concentrations of metals, chemicals, pathogens, and other pollutants were limited down to the parts per billion level. In some cases EPA would specify technology-based limits (e.g., 'best practicable control technology' for industries, full secondary or tertiary treatment of wastewater for municipal agencies), and would also set limits for particular contaminants, while in other instances they would also require the discharger to immerse 'indicator species' in effluent samples to enforce 'toxicity limits' to make sure that the regulations hadn't let something through that was causing harm. In those instances, the discharger would measure the ability of the selected species to live, grow, and reproduce as an indication of the potential toxicity of the effluent, potentially due to the presence of constituents yet to be regulated. If found to be 'toxic,' the discharger needed to take remedial action to reduce the harmful effects.

In addition to pollutant-by-pollutant regulation and technology-based limits, in the 1990s, and frequently as the result of litigation, EPA also began exercising its authority to set limits for dischargers based on the character of individual waterbodies. Waterbodies characterized as 'impaired' by Section 303(d) of the Clean Water Act could be protected by constituent limits that took into account the extent to which the waterbody was already polluted. States were required to identify which of their rivers, lakes, and streams were impaired with respect to any of their beneficial uses (e.g., drinking, recreation, irrigation, fisheries, etc.) and to develop limits based on how much of a pollutant the waterbody might absorb (its 'assimilative capacity') and still be suitable for that use. States then allocated each discharger a maximum amount of pollutant (by weight) it could discharge per day – the 'Total Maximum Daily Load,' or TMDL (EPA, 2020g). Those limits found their way into individual discharge permits.

During the1990s the EPA also began the challenging task of assigning responsibility for preventing 'non-point' discharges from diffuse urban sources like industrial site drainage and runoff from city streets that found their way into water bodies. In theory, this more holistic assessment should have led to the 'fishable, swimmable' waters promise of the Clean Water Act through a more customized assessment of responsibility for pollution reduction. For instance, it could allow a discharger to help pay to clean up non-point source pollution if that option was less expensive than removing a point-source pollutant at its facility. It also allowed trading of pollution allowances ('effluent trading'), whereby a company could sell pollution reductions beyond its discharge reduction targets or where a third party could reduce pollution that was unregulated and then sell the 'credits.' This practice is more common in the air quality regulatory arena, where it has been refined as a market for selling pollution allowances, but has been applied in the context of water quality regulation, primarily for nutrients and sediments, as well as for effluent temperature limitations (ACWA, 2016; EPA, 1999a; NNWQT, 2015). Implementation here has been slow going and with some exceptions, has become a process-heavy exercise to assure accountability.

14.2.3 Other tools in the regulatory toolbox

In addition to the limit-based regulations described above, federal law made available several other strategies to help ensure that companies took care to reduce their impact on the environment. The 'citizen suit' provision allowed private parties to bring legal action against regulated companies to augment government enforcement. The 1980 Comprehensive Environmental Response, Compensation and Liability Act (CERCLA) –also known as 'Superfund' – held companies responsible for restoring polluted industrial sites, while the Toxics Release Inventory (TRI) forced them to reveal the nature and extent of the hazardous chemicals they used. Hefty fines, the broad liability that follows companies even after they've left town, and the added transparency of a toxic chemical list all served to drive down pollution.

Several of the federal statutes previously described have what are known as 'citizen suit' provisions. In the absence of federal action, these provisions allow non-governmental actors (i.e. private citizens) to file suit in place of the federal government. This tool has been particularly powerful in raising awareness in the private sector of the need to comply with federal water laws, where government resources are too scarce to inspect every outfall of every company. While federal inspectors can't be everywhere at all times (an ever-present 'cop on the beat'), concerned citizens can crop up anywhere, and successful enforcement of a citizen's suit can lead to the award of substantial penalties paid to the United States and attorney fees to the plaintiffs in addition to remedial action.

Under CERCLA, the federal government can clean up badly polluted ('Superfund') sites and assess treble damages against polluting companies as compensation. This provides a powerful incentive for Potentially Responsible Parties (PRPs) to step up to manage Superfund clean-ups themselves. It also applies an important legal application known as 'joint and several liability,' in which any one of the PRPs can be forced to pay the entire cost of a clean-up. (They can bring in other PRPs to help share the cost, but the burden is on them to do so.)

The Toxics Release Inventory (TRI) was established in response to Union Carbide's deadly release of methyl isocyanate in Bhopal, India. Authorized under the 1986 Emergency Planning and Community Right-to-Know Act (EPCRA), it required companies to publish a list of all hazardous chemicals they store and use (EPA, 2020b). It was expanded dramatically between 1992 and 2000 to include all emissions to air, land, and water; the USEPA required submission of the amounts of pollutants emitted by individual companies each year, then publicized the results. The very act of having quantities of toxic emissions linked to corporate names resulted in dramatic reductions in those emissions. In this scheme, the regulation is in favor of transparency of data to the public, and that transparency leads to business decisions by companies to reduce their emissions to not look bad on the list to their customers, communities, and their own employees.

14.2.4 Reactions to regulation

Although some executives understood the need for rules and appreciated the advantage of uniform regulations for all, in general industry protested loud and long the emergence of serious environmental regulations. Even as some in the environmental community claimed the new rules didn't go far enough to protect the environment, the regulated business community by and large reacted with a sense of outrage. Companies claimed that they would go out of business, that environmental protection measures would deprive society of oil and gas, automobiles, cell phones, computer chips, food, and spell the end of civilization. Heavy industry (e.g., oil refining, rail transport) whose expensive equipment lasts many years may have been more resistant than those industries like electronics whose manufacturing equipment is frequently refreshed, refined, or replaced, but as a whole industry roundly rejected regulation. Their arguments varied from disputing the authority of the government to regulate commerce to quarrelling over the degree of harm caused by the prohibited pollutants, but their protests only attested to the necessity of strong regulations to protect public health and the environment, which did not lead to the parade of economic and other horrors they predicted.

14.3 SOME INTERESTING ATTEMPTS AT ADAPTIVE REGULATION

Once the regulatory framework was firmly established, however, rather than simply continuing to push back against the new legal limits, some industries suggested alternate means of achieving the same objectives. Similarly, as time passed, regulators realized that there might be more than one way to achieve environmental goals. By the early 1990s, it became apparent that while many industries continued to resist regulation, some were sincerely trying to comply with the new rules, while still others were doing better than required. Agency officials were cheered by the enthusiasm of these early adopters; they hoped they might presage the arrival of a 'tipping point,' after which regulatory compliance would become commonplace and many companies would do even better than just meeting standards (Fox, 2011). Sometimes these compliant companies were motivated by a new generation of senior leadership (or their employees) who genuinely cared about the environment. Others saw a marketing advantage in producing goods in a more environmentally friendly manner. In any case, it was apparent that different industries had different capacities to reduce pollution, and could do more to protect the environment than they were required to. Some in the regulatory community began to realize that even if they enforced every one of the laws and regulations on the books (an impossible feat) they could produce even better environmental results by also leveraging the innovation of industry and the market. More cynically, regulators admitted that they would never have sufficient resources to enforce existing limits, even though these standards were themselves often the product of political compromise and not as strong as the agency experts felt they should be.

Over time, companies open to innovative compliance strategies could be distinguished from the industrial 'laggards,' prompting regulators to look for alternative approaches to regulation. Many of these experiments began in the late 1980s under USEPA Administrator William K. Reilly, and fell under the rubric of 'pollution prevention' (EPA, 2020d). Administrator Reilly referred to these as programs to 'reward good behavior rather than simply punishing bad behavior.' One initiative found that the cost of revamping treatment to meet benzene emission regulations at the Amoco Yorktown Refinery was five times higher than the cost of preventing even more pollution from unregulated tankers taking gasoline from the refinery (Schneider, 1993). Amoco asked to lower the pollution from those tankers rather than cutting benzene emissions from the refinery at greater cost, and while this particular project was not ultimately a success it illustrated the potential to gain greater environmental gains outside the existing regulatory system. In 1994 this concept became part of an entire phalanx of 'reinvention' experiments led by Vice-President Al Gore (NPRI, 2001) and USEPA Administrator Carol Browner, who outlined the EPA part of the effort in the agency's 'New Generation of Environmental Protection' (Browner, 1998; EPA, 1994). This strategy laid out guidance for a variety of innovative incentives, and 'Faster, cheaper, smarter' regulation became the driving theme of a variety of experiments. To take just two examples of the ensuing projects, Project XL developed alternative regulatory schemes for pilot companies after a competitive selection process, while the Common Sense Initiative sought to adapt regulations to match differing industry characteristics rather than a one-size-fits-all approach. The hope of these experiments was to demonstrate that the traditional regulatory scheme could be supplemented, and improved upon, on a parallel track that might someday lead to different forms of more customized regulation.

14.3.1 Project XL

In response to industry's complaint that the regulatory process itself ('red tape') added too much time and effort to compliance, USEPA launched Project XL (1995–2002) to prove that greater environmental results could be produced in exchange for greater regulatory flexibility (EPA, 2016a). The eight projects initially selected for this pilot project, as well as the subsequent 17 that were added, spanned the range of industrial sectors (EPA, 2016b). Between 1995 and 2002, 25 pilot projects were implemented (2000a). In one of the more successful examples, Intel Corporation proposed to construct a chip manufacturing facility in Chandler, Arizona, that would reduce air emissions below the 'major source' threshold and meet a variety of 'performance goals,' recycle its process water, reduce and recycle solid waste, and more, in exchange for faster permitting in the future. Many of the environmentally positive projects, for example, recycling process water, were not things that regulations required. Of special note and interest, it would publish online monthly all its water use and emissions data – well beyond any requirements – so that there would be transparency to the regulators and community on an ongoing basis. This approach was very different than the prevailing system where the permittee would submit piles of paper periodically on their compliance at designated intervals with a limited set of information. For the public to obtain access to those records required a specific request to the agency. It was very labor intensive and limited in utility. In the present day, such records are more readily available online but still meet only designated criteria set in permits.

The projects were not without controversy from both within and outside the agency. For instance, although the Intel project had gone through a stakeholder review process, at least one national environmental group critiqued it in the national media (Bloomberg, 1996). Other problems also emerged at the outset, as the administration chose to announce the selection of the 10 pilot projects at a White House ceremony as an example of 'reinventing government' before all of the details were worked out. Unfortunately, this ceremony gave the companies the sense that the details of their pilots had also been accepted, which reduced the agency's leverage in reaching agreement on permits that met the proposals' promise, which added enormously to the negotiating time. Meanwhile, some USEPA staff felt support for these innovative efforts signaled a disinterest in enforcement, which lagged when permit writers and lawyers took time to work on them instead of their traditional enforcement cases. In fact, the administration supported *both* enforcement and innovation, but that message was not understood at first by a skeptical staff. Alternative strategies like pollution prevention and other innovative programs actually require a strong enforcement backstop, both to incentivize the innovation and to maintain a level playing field. These challenges do not negate the importance of the experiment, however, and project XL yielded a great many lessons learned for how to do it better. Unfortunately, with a new administration this meaty experiment lapsed when it could have been built upon.

14.3.2 Common Sense Initiative

The Common Sense Initiative applied the principles of flexibility and adaptability reflected in Project XL, and sought to apply them on an industry-wide basis (EPA, 1994). It was based on the notion that different industries would have different regulatory issues and attitudes, and that developing targeted regulations for each of them would be more efficient and effective than the traditional 'one-size-fits-all' approach. As with Project XL, the goal was to achieve greater environmental performance more rapidly and at less cost – a win/win for both the environment and for the economy. The industries chosen were:

- Automobile manufacturing
- Computers and electronics
- Iron and steel
- Metal finishing
- Petroleum refining, and
- Printing.

Stakeholder groups were developed for each sector that, in addition to EPA staff, included state regulators and representatives from industry, environmental groups, and labor (perhaps not surprisingly, some of the greatest resistance from industry was to the inclusion of labor participation).

These industries varied greatly with respect to their attitudes towards regulation. The computer and electronics sector was motivated by a desire for speedy permitting since speed was a competitive advantage in this rapidly changing technological sector. By contrast, the petroleum refining and automotive sectors tended to believe that when it comes to regulatory change, 'slower is better' since they require significant investment to retool their heavy infrastructure. While the theory behind the Common Sense Initiative (CSI) was sound, disputes among stakeholders and limited personnel to help resolve them meant that EPA could not deliver as many new regulations or as much systemic change as hoped. That said, some significant projects did emerge through this process that highlighted the benefits of an industry-specific approach to regulations. For example, EPA learned that its existing lead limits prevented recycling a lot of computer glass. Revising these rules allowed cathode ray terminal (CRT) glass to be remanufactured into new computer screens (EPA, 1998a). The metal finishing, printing sectors, and computer and electronics sectors all achieved significant agreements (EPA, 1999b). Again, there is much that could be built on here and many lessons learned on stakeholder engagement, potential for tailored regulation, and performance-based regulation.

14.3.3 Other results-based regulatory initiatives

The EPA continued to look for ways beyond the conventional 'hide and seek' approach to enforcement by which they could encourage industry to reduce its environmental impact. For example, the 1995 Audit Policy invited industries to perform detailed 'enforcement style' inspections of their own facilities to find and fix violations. It was designed *'to enhance protection of human health and the environment by encouraging regulated entities to voluntarily discover, promptly disclose and expeditiously correct violations,'* in exchange for which EPA would eliminate or reduce civil penalties and even forego criminal prosecution (EPA, 2000b). This was appealing to the agency because its enforcement resources were always limited, so self-disclosure had the potential to correct more violations than piecemeal enforcement measures. In addition, the Office of Enforcement and Compliance Assurance updated the policy for obtaining 'Supplemental Environmental Projects' (SEPs) as part of enforcement agreements in lieu of a portion fines levied again violators and paid to the federal treasury. In this way, they allowed violating industries to make up for their bad behavior by funding SEPs in the communities harmed by their prior violations (EPA, 1998a). This allowed the community that had been harmed to receive a benefit in addition to the cessation of the harm, and gave the violators a chance for public atonement.

The self-audit measures were reinforced by the growth of sustainability reporting demanded by consumers and employees and promoted by CERES, ISO9000 and 14000, and other public reporting initiatives at the national and international level. Corporate accountability projects were growing and shareholders were also demanding greater environmental performance. This was also the era when companies began hiring environmental compliance or sustainability officers, and they were eager to get credit for fixing problems proactively. At the same time, many governments around the world were investigating creative approaches to achieving environmental, ecological, and societal results on a holistic basis. Called 'Green Plans' in the United States, these approaches were developed and tested in the Netherlands, New Zealand, Singapore, and elsewhere and allowed governments to contract with industry to invest in a range of integrated environmental and ecological programs that would add up to nationally set goals in lieu of a command and control type of approach alone (RRI, 2020). For understandable reasons, this may be easier to do in a smaller setting than the United States or when fewer large industries are responsible for the lion's share of pollution, but the option of tailoring regulation to results achieved at less cost, and by an easily enforceable agreement, is an appealing one if you can get to it.

In another 'alternative strategy,' the EPA created the Brownfields Initiative (EPA, 1998b) that included grants to local governments to test clean-up sites that might be suitable for redevelopment, and more broadly took other measures and created a movement to redevelop previously contaminated sites. These sites didn't rise to the level of 'Superfund' sites requiring heavy clean-up but Superfund cast a long shadow. This initiative was developed in response to EPA's realization that the powerful 'joint and several liability' provision of its Superfund enforcement law had the unexpected consequence of scaring people from touching a contaminated site for fear of Superfund liability. Moreover, the agency saw many clean-up projects tied up in court for years. While the provision of joint and several liability allowed the agency to sue the largest companies responsible for pollution without having to chase down every individual contributor, it also allowed those parties to sue other contributors which could drag in thousands of other parties in cascades of litigation whose actual contribution to the environmental damage was remote. As a result, land redevelopment stalled as potential developers, or redevelopers, feared buying into an endless chain of litigation in addition to the property itself. In essence, fear of 'Superfund' liability or litigation halted productive and beneficial local projects and economic activity. One tool the agency used to remedy this unintended consequence was to issue both local governments and landowners 'comfort' letters from the federal government assuring prospective developers that their site would not be subject to future Superfund enforcement. In addition to supporting redevelopment of impaired sites, the Brownfields Initiative officially released thousands of small 'partially responsible parties' from groundwater enforcement sites as a way to both let them get on with their businesses and lives and to eliminate a source of great delay in reaching resolution of enforcement actions. This initiative continues.

At the state level, there have been a variety of experiments and alternative compliance approaches that show promise but are not in any way ubiquitous. The nutrient and sediment trading schemes, mentioned above, have gotten the greatest traction (ETN, 2020). These are most frequently seen in the Midwest, Northwest, and eastern parts of the US, with programs in the Chesapeake Bay being the most well-known. In California, efforts by stakeholders through the 'Brake Pad Partnership' led to legislation to eliminate copper from brake pads, which was found to the greatest uncontrolled source of copper to urban water bodies, and was not essential for braking safety. Stakeholders from the wastewater treatment, stormwater management, and environmental worlds joined together in that effort, recognizing a mutual interest in reducing copper at least cost and greatest impact (SusCon, 2020). In Southern California, a permit scheme is currently underway that gives cities more time to meet stormwater quality standards if they capture it to achieve multiple benefits like urban greening and groundwater infiltration in addition to water quality. The goal is to achieve multiple objectives rather than simply building plants to treat stormwater and discharge it to the ocean, which would meet regulatory requirements. This promising scheme is supported by a local funding measure, but only time will tell if they can achieve the water quality benefits along with all the other pressing water issues they are also trying to solve (TPR, 2019).

14.4 THE WAY FORWARD

Results of these experiments and others like them are hard to gauge. Having worked on these efforts at all levels of government, I know that on a case-by-case basis a well-crafted alternative to enforcement can create greater benefits for business and the environment. But I also know how challenging it can be to propagate these 'bespoke' programs throughout an industry and inspire regulatory innovation, only to have the program rolled back by the next political administration, especially when a different political party takes control. Even as an experiment, these customized permits take up a tremendous amount of time and personal energy – resources that would otherwise go to traditional enforcement. In addition, one of the greatest challenges we face in the United States is that the shelf-life of regulatory innovations is frequently limited to the four or eight-year term of each administration, with succeeding governments alternately rolling back regulations ramping up regulations, rolling back enforcement, and stepping up enforcement. Even administrations that support environmental protection frequently oppose innovative programs, as frustrated activists and policy-makers roar back into office keen to restore all the rules weakened by previous governments. This see-saw drama eats up lots of energy that could be better spent strategically focused on results, if it were possible to build on what has come before that had promise.

The better approach is to build in *both* a strong regulatory bottom-line and a robust alternative track for those who can demonstrate commitment to achieving better, multi-benefit results that can be verified. This can avoid the false dichotomy between creativity and regulation that prevents consistent progress using a variety of tools and creates an 'either/or' discourse rather than a strategic 'both/and' approach to achieving environmental results. It is clear that innovative regulation of industrial water use requires a commitment to strong enforcement to protect public health and the environment. Only a strong bottom line of protection can create credible incentives for innovation, and, it is only with commitment to that strong bottom line that innovation can and should be supported by activists and agency staff. At the same time, enforcement efforts can and should be coupled with the opportunity to reward those who go beyond the minimum standards to create a more sustainable, equitable, efficient, and less polluting means of production.

On a global level, we have the best opportunity ever to create custom-tailored approaches to achieving environmental results. Industrial leadership roles are increasingly being filled by younger people for whom environmental protection is an important value. Younger employees and consumers increasingly demand greener facilities and products. The sustainability, resilience, corporate accountability, circular economy, and related movements have gained traction and a gravity and momentum of their own in the corporate world since the 1990s. Technological innovations in sensors, real-time monitoring, and decentralized treatment systems make possible greater accountability that the public will demand in exchange for customized options. Web-based reporting (coupled with audits) allows the public to see results, and this greater transparency gives the public a fighting chance to 'trust but verify' that better environmental results have in fact been delivered.

Developing a modernized system that acknowledges and works with these more recent drivers can accelerate the environmental progress that we desperately need. One framing of the more modern setting talks about four quadrants of environmental drivers, of which 'law-based systems' (Public governance) is only one. The other three include 'risk-brand management' (Private environmental governance), 'Greener' technologies (Closed Loop systems for footprint reduction, also referred to as 'circular economy' in other fora), and 'Big data' (online sharing platforms) (ELI, 2019).

Creating a stronger incentive to leverage all facets of corporate accountability (e.g., financial, reputation, competitive marketing advantage, employee morale and recruiting, demographic shifts) is a 'both/and' rather than an 'either/or' consideration.

The need for a strong regulatory backstop will never disappear. Strong rules and strong enforcement are necessary to achieve the objective of public health and safety and to provide a uniform economic environment ('level playing field') for industry. Clear standards are also essential to determine whether a given innovation merits the time and energy it takes to evaluate and monitor its promise and progress. Similarly, that baseline, coupled with robust monitoring and reporting, is essential for innovative companies to demonstrate to outside stakeholders and agency staff alike that their customized programs are not just excuses to pollute. In short, there is a need for what Australian water official Anthony Slatyer (2019) refers to as 'policy scaffolding' against which an innovative scheme can be measured. It can either form a strong alternative backstop to give political will to achieve results, or it can provide structure to ensure that results have something to be measured by. The presence of such a framework assures that well-intended time and effort will not be wasted on weak schemes at a time when we need innovation badly to meet the challenges of the future.

In summary, the following are the key requirements of any innovative regulatory framework capable of producing environmental results over time superior to minimum regulatory standards:

(1) Promise of superior environmental results substantially and significantly better than the minimum legal requirement, to justify all the time that it takes to develop and monitor and review customized permits/projects.
(2) Incentives that include streamlined permitting, reasonably longer timeframes for meeting certain deadlines, and customized requirements.
(3) Transparency in monitoring, reporting, and review by the agency and the public using the most advanced tools possible.
(4) Audits, regularly scheduled reporting, public engagement, and other demonstrations of accountability.
(5) Clear consequences for failure to live up to the terms of the customized permit/agreement (e.g., sufficient penalty or fine, revocation of the permit/agreement, immediate enforceability of existing standards).

Defining the superior performance itself is critical to the integrity of the effort. Examples could include:

- reducing emissions significantly and substantially below regulatory thresholds (most important);
- prioritizing important community needs not always susceptible to regulatory improvement (e.g., provision of alternative sources of safe drinking water);
- developing projects that achieve multiple community benefits (e.g., a combination of flood control, water supply, water quality, and urban greening benefits);
- participating in multiple-benefit regional projects (e.g., combining smaller permit requirements into contributions to regional projects that address stormwater management issues by creating stormwater projects at greater scale).
- projects that prioritize high priority pollutant reductions more rapidly than traditional permitting (e.g., reducing pollutants from unregulated non-point sources to achieve better objectives more quickly in a given community).

In summary, while the United States' water quality regulatory history contains many great accomplishments, the nation has not yet achieved its promise of fishable, swimmable safe waters across the country, nor has it restored and maintained the chemical, physical, and biological integrity of the nation's waters as envisioned would be done by the mid 1980s. We can build on what has come before by strategically coupling a strong and well-enforced regulatory baseline with incentives to do even better. Other countries looking to the United States as a model should be advised not to wait half a century or even a decade to develop programs that incentivize faster, cheaper, and smarter results. They can devise programs suited to their unique scale, dynamics, geographies, and industries to provide cleaner safer waters more rapidly, and should integrate modern technologies that provide data, monitoring, and the transparency that is essential to focusing on results and building trust. Underpinning all of that, they should include a backstop of regulatory authority with integrity that can both incentivize innovation and guarantee minimum environmental and public health protection.

REFERENCES

Airlie House Workshop Report: Re-Imagining Environmental and Natural Resources Law. (November 18–19, 2019). (Environmental Law Institute, George Washington School of Law, Meridien Institute) (ELI) at 6.

Association of Clean Water Administrators (ACWA) and Willamette Partnership. (2016). The Water Quality Trading Toolkit. Version 1.0, August, 2016.

Bloomberg Business News. (1996). 'EPA, Intel sign 'Alternative' Environmental Strategies Pact.' Los Angeles Times, 11/20/96 https://www.latimes.com/archives/la-xpm-1996-11-20-fi-949-story.html. (accessed May 11 2020).

Browner C. (1998). Remarks Prepared for Delivery to The Brookings Institution,10/13/98. https://archive.epa.gov/epapages/newsroom_archive/speeches/a63f9ffc3923ae198525701a0052e428.html, (accessed May 11 2020).

Copeland C. (2016). 'Clean Water Act: A Summary of the Law.' Congressional Research Service. Available at. fas.org/sgb/crs/misc/RL30030.pdf

Environmental Trading Network (ETN). (2020). State Trading Programs Website. http://www.envtn.org/water-quality-trading/state-programs. (accessed May 24 2020).

EPA. (1972). EPA History: Water - The Challenge of the Environment: A Primer on EPA's Statutory Authority https://archive.epa.gov/epa/aboutepa/epa-history-water-challenge-environment-primer-epas-statutory-authority.html. (accessed May 11 2020).

EPA. (1994). The New Generation of Environmental Protection (EPA 200-B-94-002) Summary available online https://nepis.epa.gov/EPA/html/DLwait.htm?url=/Exe/ZyPDF.cgi/400009NN.PDF?Dockey=400009NN.PDF. (accessed May 12 2020).

EPA. (1998a). 'Common Sense Initiative (CSI) Council Recommendation on Cathode Ray Tube (CRT) Glass-to-Glass Recycling.' 6/4/98. Official website, https://www.epa.gov/sites/production/files/2016-01/documents/csi-crt.pdf. (accessed May 12 2020).

EPA. (1998b). 'EPA Delivers on Superfund Reform Progress Making 1997 a Standout Year for Faster Cleanups at Lower Costs.' EPA Press Release 2/3/98. Official website, https://archive.epa.gov/epapages/newsroom_archive/newsreleases/88695df97a0f1fc5852565a0005a8fa0.html. (accessed May 22 2020).

EPA. (1999a). A Summary of US Effluent Trading and Offset Projects. Report prepared by Environomics (Bethesda, MD) for Dr. Mahesh Podar, USEPA Office of Water, November, 1999. http://www.environomics.com/Effluent-Trading-Summaries_Environomics.pdf (accessed May 22 2020).

EPA. (1999b). 'Analysis and Evaluation of the EPA: Common Sense Initiative.' Report prepared by Kerr, Greiner, Andersen, and April, Inc. July, 1999. Available from National Service Center for Environmental Publications https://www.epa.gov/nscep. (accessed May 24, 2020).

EPA. (2000a). Project XL: Summary of Current Pilot Projects EPA 100-F-00-018 Official website, https://nepis.epa.gov. Accessed October 2, 2020.

EPA. (2000b). Incentives for Self-Policing: Discovery, Disclosure, Correction and Prevention of Violations.' 4/11/2000. 65 FR 19617 Official website. https://www.federalregister.gov/documents/2000/04/11/00-8954/incentives-for-self-policing-discovery-disclosure-correction-and-prevention-of-violations. (accessed May 12 2020). Original policy referenced 60 Fed. Reg. 66706 (Dec. 22, 1995).

EPA. (2016a). 'What is Project XL?' Official website, https://archive.epa.gov/projectxl/web/html/file2.html (accessed 12 May 2020). (accessed May 12 2020).

EPA. (2016b). 'Project XL Projects' Official website, https://archive.epa.gov/projectxl/web/html/projects.html. (accessed May 12 2020).

EPA. (2020a). Clean Water Act (as amended). 33 U.S.C. §1251 et seq. (1972). Official website https://www.epa.gov/sites/production/files/2017-08/documents/federal-water-pollution-control-act-508full.pdf. (accessed May 11 2020).

EPA. (2020b) 'History of the Toxics Release Inventory (TRI) Program (List)' Official website. https://www.epa.gov/toxics-release-inventory-tri-program/history-toxics-release-inventory-tri-program-list. (accessed May 11 2020).

EPA. (2020d). 'William K. Reilly: Oral History Interview.' Official website, https://archive.epa.gov/epa/aboutepa/william-k-reilly-oral-history-interview.html. (accessed May 11 2020). 'Reilly also encouraged EPA to address regional pollution problems which forced the Agency to strive to design cross-media regulatory strategies. These strategies had been discussed but deemed too unmanageable and complex to devote vast amounts of energy in the face of more pressing needs by previous Administrations.'

EPA. (2020f). Former National Compliance Initiative: Preventing Animal Waste from Contaminating Surface and Ground Water. Official website, https://www.epa.gov/enforcement/former-national-compliance-initiative-preventing-animal-waste-contaminating-surface-and. (accessed May 22 2020).

EPA. (2020g). Overview of Identifying and Restoring Impaired Waters under Section 303(d) of the CWA. Official website, https://www.epa.gov/tmdl/overview-identifying-and-restoring-impaired-waters-under-section-303d-cwa. (accessed May 22 2020).

Fox C. (2011). 'The ERM tipping point: ERM has reached critical mass – time to get on board'. *Risk Management*, **58**: 9, Nov. 2011, 22 ff.

Gathman D. (2015) 'Real-life anti-pollution hero the Fox may be headed to the big screen'. *Chicago Tribune*, **7/3/15**. https://www.chicagotribune.com/suburbs/elgin-courier-news/ct-ecn-fox-movie-st-0706-20150703-story.html. (accessed May 11, 2020).

Hamilton J. (2019) 'How California's Worst Oil Spill Turned Beaches Black And The Nation Green' National Public Radio (NPR), Broadcast January 28, 2019 https://www.npr.org/2019/01/28/688219307/how-californias-worst-oil-spill-turned-beaches-black-and-the-nation-green. (accessed May 11 2020).

Hartig J. (2019) 'Great Lakes Moment: The Event that Transformed Cleveland's Flats and Influenced National Policy' GreatLakesNow.org (July 1, 2019) https://www.greatlakesnow.org/2019/07/a-great-lakes-moment-from-john-hartig-13/. (accessed May 11 2020).

National Network on Water Quality Trading. (2015). Building a Water Quality Trading Program: Options and Considerations. NNWQT website, http://nnwqt.org/building-a-water-quality-trading-program/ p. 8. (accessed May 22 2020).

National Partnership for Reinventing Government. (2001). 'A Brief History of Vice President Al Gore's National Partnership for Reinventing Government During the Administration of President Bill Clinton. https://govinfo.library.unt.edu/npr/whoweare/historyofnpr.html. (accessed May 22 2020).

Planning Report. (September 3, 2019). (TPR). 'Katy Young-Yaroslavsky Unpacks Passage of Measure W' https://www.planningreport.com/2019/09/03/katy-young-yaroslavsky-unpacks-measure-w-implementation-la-s-safe-clean-water-program. (accessed May 24 2020).

Resource Renewal Institute. (2020). 'Green Plans' Website https://www.rri.org/green-plans. (accessed May 5 2020).

Schneider K. (1993). 'Unbending Regulations Incite Move to Alter Pollution Law,' New York Times 11/29/93. Available online https://www.nytimes.com/1993/11/29/us/unbending-regulations-incite-move-to-alter-pollution-laws.html (accessed May 11 2020).

Schuyler D. and Gallay P. (2018). 'The Battle for Storm King: In celebration of the mountain and river that helped launch the modern environmental movement'. *Scientific American Weblog*, **8/30/18** https://blogs.scientificamerican.com/observations/the-battle-for-storm-king/. Riverkeeper Mission Statement at https://www.riverkeeper.org/. (accessed May 11 2020).

Slatyer A. (2019). 'Getting on Track with SDG6 – A Role for Multilateral Processes.' Speech delivered at Towards a Sustainable Water Future Conference, Bengaluru, India September 2019. http://waterpolicygroup.com/wp-content/uploads/2019/09/Water-Future-conference-Slatyer-session-speech-final.pdf. (accessed May 13 2020).

Sustainable Conservation (SusCon). (2020). Brake Pad Partnership. Website https://suscon.org/project/brake-pad-partnership/. (accessed May 24 2020).

Chapter 15

Recovering from disaster: Holding industry accountable for restoration

Karla Raettig and David Muth

National Wildlife Federation, Reston, USA

Keywords: Deepwater Horizon, environmental restoration, Gulf Environmental Benefit Fund, Gulf Renewal Project, National Wildlife Federation, Natural Resource Damage Assessment, Oil Pollution Act, RESTORE Act

15.1 INTRODUCTION

As the largest non-profit education and conservation organization in the United States, the mission of the National Wildlife Federation (NWF) is 'uniting all Americans to ensure wildlife thrive in a rapidly changing world'. To do this, NWF educates and advocates for science-based policy to protect and restore wildlife populations and habitat. In the Gulf of Mexico region, NWF is engaged in several projects to restore the Mississippi River Delta, the Gulf, and the Everglades as well as Texas rivers and streams. That ongoing work took a dramatic turn in 2010 when the largest marine oil spill in history occurred in the Gulf of Mexico. Despite regulations requiring oil companies to operate safely with legal penalties for pollution, federal law and policy alone were ill equipped to mitigate a disaster of this magnitude.

As this chapter relates, in addition to its disastrous impact on the Gulf Coast ecosystem, the Deepwater Horizon oil spill also presented a challenge to national environmental protection and habitat preservation organizations and partners in the Gulf region. Their collective task was to make sure that the federal government vigorously enforced all regulations and applied all available remedies towards significant, permanent restoration of damaged ecosystems. NWF met this crisis with a combination of political advocacy and scientific fieldwork to ensure that those responsible would pay to mitigate the damage, and that funds collected would be allocated where they would do the most good.

DOI: 10.2166/9781789060676_0165

15.2 SETTING THE STAGE: DEEPWATER HORIZON DISASTER AND ITS ENVIRONMENTAL IMPACTS

On April 20, 2010, an explosion occurred during the drilling of an exploratory oil well 41 miles off the coast of Louisiana. The explosion of the Deepwater Horizon tragically killed 11 people and started spilling oil into the Gulf of Mexico until it was finally capped on July 15, 2010. During those 87 days, an estimated 130 million gallons of oil spilled into the Gulf of Mexico, making the Deepwater Horizon disaster the largest oil spill in recent United States history.

It took years to assess the full environmental impact of the spill and those impacts continue to this day. Because of the biological interconnectedness of the marine ecosystems, the spill impacted the entire food chain throughout the Gulf (Figure 15.1). Five years after the spill, National Wildlife Federation released a report (NWF, 2015) documenting the extent of impacts:

- In 2014, dolphins on the Louisiana coast were found dead in numbers four times greater than historic rates, and there is increasing evidence that these ongoing dolphin deaths are connected to the 2010 oil disaster.
- Between 27,000 and 65,000 Kemp's ridley sea turtles are estimated to have died in 2010, and the annual numbers of Kemp's ridley nests have declined in the years since the spill.
- Twelve percent of the brown pelicans and 32 percent of the laughing gulls in the northern Gulf are estimated to have died as a result of the BP oil spill.
- Oil and dispersant compounds have been found in the eggs of white pelicans nesting in three states – Minnesota, Iowa, and Illinois.
- Exposure to oil has been shown to cause abnormal development in many species of fish, including mahi-mahi, Gulf killifish, and bluefin and yellowfin tuna.
- Spotted seatrout, also known as speckled trout, spawned less frequently in 2011 in both Louisiana and Mississippi than in previous years.

Figure 15.1 Oil from the Deepwater Horizon disaster seeps past barriers into vulnerable coastal Louisiana marshes. (*Credit: National Wildlife Federation, Amanda Moore*)

- Both 2010 and 2011 had the lowest numbers of juvenile red snapper seen in the eastern Gulf fishery since 1994.
- Coral colonies in five separate locations in the Gulf – three in the deep sea and two in shallower waters – are showing significant oil damage.
- Sperm whales are spending less time foraging in the area around the wellhead.
- Oil has been found in sediments deep in the Gulf of Mexico, in a 1,200-square-mile area surrounding the wellhead.

April 2020 will mark the 10-year Anniversary of the disaster. As detailed in the NWF follow-up report, '*10 Species, 10 Years Later: A Look at Gulf Restoration after the Deepwater Horizon Disaster*' and other studies the oil and its impacts are still being felt throughout the Louisiana marsh and Gulf ecosystem (Fleeger *et al.*, 2018; NWF, 2020). To mitigate those impacts, a massive restoration effort throughout the region is now underway, funded by money from Deepwater Horizon owner BP (formerly British Petroleum) and other parties responsible for the oil spill. But while the money was obtained through enforcement actions related to existing regulations (and associated settlement agreements), fines and penalties alone could not help restore the Gulf coastal environment. What was needed was a plan sweeping enough to mitigate the damage caused by the spill with adequate funding to put it into motion. Thanks to the efforts of its Congressional champions and the broader environmental community (including NWF), just such a plan was supported by ground-breaking legislation, which, was signed into law in 2012.

15.3 FROM REPORTING TO RESTORATION: THE ROLE OF NWF IN DEEPWATER HORIZON RECOVERY

Regulation and enforcement have long been recognized as an integral part of environmental management, especially with respect to industrial water pollution. Hauling companies into court for violating environmental laws, however, will not in itself undo the harm caused by those violations, or even prevent further misconduct in many cases. What is needed is a process that fully accounts for the extent of the environmental damage, assesses proportional penalties, and ensures that the money collected is invested in long-term habitat enhancement. This is illustrated in the response of NWF and other non-governmental organizations (NGOs) in the aftermath of the Deepwater Horizon disaster. The immediate post-spill response followed by the careful creation and implementation of what came to be called the RESTORE Act and the restoration funds provides a lesson in the role played by environmental advocates in holding industry accountable. In the case of the Deepwater Horizon disaster, NWF responded in three phases: emergency assessment, policy advocacy, and penalty allocation.

15.3.1 Environmental response

As oil continued to spill into the Gulf, NWF worked to focus the national spotlight on the on-going devastation. Dozens of NWF staffers travelled to the state during the summer of 2010, taking up residence in Venice, Louisiana, where the press was documenting the ongoing disaster. Local charter boat captains – sitting idle as vacationers cancelled trips and emergency work cut them off from traditional fishing spots – were hired by NWF to take journalists and scientists to areas feeling the worst impacts of the spill. As a result, the press was able to see the damage 'up close and personal', preventing those responsible for the disaster from sweeping it 'under the rug' (Figure 15.2).

Figure 15.2 NWF staff taking press out to document oil spill impacts. (*Credit: National Wildlife Federation, Amanda Moore*)

In addition to these on-the-ground efforts, NWF worked with national, state, and local NGOs to press for congressional and agency-level oversight. Most important, the NGOs combined to pressure Congress to ensure that any penalties collected from responsible companies would be used to mitigate the disaster. NWF was in a unique position to mobilize these efforts as it had been working with several other NGOs to address land loss in the Mississippi River Delta (RMRD, 2020). Supported by the Walton Family Foundation, the coalition included NWF, Environmental Defense Fund, Audubon, Coalition to Restore Coastal Louisiana, and the Lake Pontchartrain Basin Foundation. These organizations brought federal and state policy, advocacy, scientific, and outreach expertise to restore the Delta. As the damage continued throughout the summer, the Delta coalition joined forces with other NGOs working throughout the Gulf to support a congressional proposal to direct the money that the corporate responsible parties paid would be directed to restoration, instead of the general treasury funds.

15.3.2 Policy response

Twenty years before Deepwater Horizon, the largest release of crude oil into US waters occurred when the ship *Exxon Valdez* ran aground in Prince William Sound, Alaska, on March 24, 1989. The volume of oil spilled into the Sound – nearly 11 million gallons – was a drop in the bucket compared to the gusher in the Gulf, but in terms of political response, the contrast between the two disasters was striking. Spurred by the *Exxon Valdez* debacle, the US Congress responded with landmark legislation, drafting and passing the Oil Pollution Act (OPA) in less than 18 months. Among other provisions, that transformative law set up the Natural Resource Damage Assessment (NRDA), a mechanism that evaluates and restores wildlife, habitats, and human resources impacted by oil spills and other human-made disasters.

Like most environmental law, OPA was criticized by advocates on both sides as being either too harsh or too lenient. Nevertheless it passed the Senate unanimously and was approved by the House the next day by a bipartisan vote of 360-0, with 70 Democratic and Republican members from energy-producing states abstaining, before being signed into law by Republican President George H. W. Bush. It is worth noting that even some representatives from energy-producing Gulf States were moved to vote for the bill (e.g. Texas, Louisiana), while not a single member of either chamber was willing to go on record against it.

Twenty years later, following a spill more than ten times larger, it was a different story. Despite the reform recommendations of a bipartisan blue ribbon Oil Spill Commission appointed by President Barack Obama, Congress passed no new law. A reform bill passed in the waning days of the 111th Congress by the Democratic-controlled House with virtually no Republican support died in the Senate just before the historic midterm elections of 2010 (Howell, 2011). With that election, the balance of power in Congress shifted dramatically and become far more partisan on the issue of regulating industry. What had been a matter of division within parties, had become division between parties.

Some policy changes followed the Deepwater Horizon disaster, however. President Obama implemented most of the reforms that did not require Congressional action that were recommended by the bipartisan Oil Spill Commission. Co-chaired by former US Environmental Protection Agency (EPA) Administrator William Reilly (a Republican) and former Florida governor and senator Bob Graham, (a Democrat), the Commission issued broad recommendations for reform of Outer Continental Shelf practices. These included a complete re-organization of the agency within the Department of Interior that managed and regulated oil drilling in offshore waters (National Commission, 2011) including the formation of a new the Bureau of Ocean Energy Management (BOEM) (BOEM, 2020).

Under President Obama, the new BOEM issued a series of new regulations over the objections of many in industry and their supporters in Congress. Most germane to the Deepwater Horizon disaster were new rules to increase the safety and reliability of blow-out preventers, since a failed blow-out preventer was the last line of defense in the cascading series of failures that preceded the disaster. Industry particularly chafed at these stringent new blowout preventer rules which they claimed were expensive and unnecessary, arguing that – notwithstanding the evidence of Deepwater – the industry could be relied upon to police itself. A majority in Congress apparently agreed with industry and refused to ratify any of the Obama reforms (OSCA, 2013). Moreover, the new Republican majority attempted to move legislation intended to expand opportunities for offshore energy development, causing a stalemate between Congress and the President until the end of the Obama Administration.

After the inauguration of Donald Trump in January 2017, BOEM's leadership shifted in a decidedly pro-industry direction, and began dismantling some of the key reforms it had just implemented. Symbolically, on May 2, 2019, the Bureau of Safety and Environmental Enforcement finalized 'improved' blowout preventer and well control regulations that relaxed many of the post-Deepwater Horizon reforms (US DOI, 2019).

15.4 DEEPWATER HORIZON PENALTY RESPONSE

15.4.1 Overview of penalties for environmental disasters

From the outset, NWF was clear about its priorities with respect to the economic impact of the disaster (NWF, 2015):

- The Department of Justice must hold the parties responsible for the Deepwater Horizon spill fully accountable for their violation of federal environmental laws, including the Clean Water Act and the Oil Pollution Act.
- Federal, state, and local officials must direct the fines and penalties paid by the parties responsible for the Deepwater Horizon oil spill to ecological restoration efforts that will make the Gulf healthier and more resilient for people and wildlife.

- Any settlement of claims must include a reopener clause to hold the responsible parties accountable for damages that may be identified in the future.

Following the Deepwater disaster, NWF and its NGO partners were primarily concerned with the magnitude of the penalties assessed and the amount of funds that would ultimately become available for environmental restoration. At the time of the explosion, existing US law provided several different forms of liability for oil spills. Two of the main laws are the Clean Water Act (CWA), the landmark water pollution legislation passed after the first Earth Day, and the above- mentioned Oil Pollution Act (OPA), which followed the Exxon Valdez spill into Prince William Sound. The CWA prohibits releasing oil and hazardous materials into the waters of the United States and doing so can result in both civil and criminal penalties (US Code, 1972). Under OPA, 'responsible parties' are liable for the costs of clean up and damages, including for damages to natural resources (US Code, 1990a). All these fines, penalties, and costs are managed under separate rules, by different entities, and are held in separate accounts.

There were good reasons to be concerned. Both civil and criminal penalties under the CWA would be enforced by the federal EPA and assessed by the courts, based in part on the extent to which EPA could prove that the spill resulted from gross negligence or wilful misconduct. Civil penalties assessed under OPA, on the other hand, would be focused on remedying the damages from the oil spill. OPA's NRDA process requires federal and state trustees to carry out the assessment of damages and create a plan to restore the affected resources (US Code, 1990b). In this case, the Trustee Council included the federal Department of Commerce, Department of Interior, EPA, Department of Agriculture, and representatives from the state of Alabama, Florida, Louisiana, Mississippi, and Texas.

15.4.2 RESTORE Gulf Coast Restoration Trust Fund for civil penalties

While the natural disaster was still unfolding in the Gulf, advocates met with members of Congress to discuss an ambitious plan for long-term restoration funding, over and above what might be assessed through CWA fines and OPA restoration costs. Congressional champions and NGO advocates realized that a change in the law could direct the CWA civil penalties to restoration instead of the general treasury. Gulf ecosystem needs went well beyond the direct damages from the spill, and billions of dollars were at stake. On July 11, 2011, Senators Mary Landrieu (D-LA) and Richard Shelby (R-AL) introduced the RESTORE Act, which dedicated 80% of the penalties to a newly created Gulf Coast Restoration Trust Fund and 20% to the Oil Spill Liability Trust Fund. It also created a process for spending the money that would help ensure the bulk of the money would be dedicated to environmental restoration.

To get the bill passed, NWF and its NGO partners formed a collaboration that came to be informally named the Gulf Renewal Project (GRP). Funding for this effort came from the Walton Family Foundation (WFF) as well as individual donations to organizations helping wildlife recover from the spill. In addition to NWF, the Gulf Renewal Project included the Environmental Defense Fund and Audubon (our national partners in Louisiana) along with The Nature Conservancy and The Ocean Conservancy. The GRP, focused nationally working with WFF-funded OxFam America, also coordinated efforts with WFF supported NWF instate affiliates including the Florida Wildlife Federation, Mississippi Wildlife Federation, and Louisiana Wildlife Federation; Audubon chapters and state offices, as well as environmental and community NGOs in each of the five Gulf states that joined in the effort. President Barack Obama signed the RESTORE Act on July 6, 2012 as an amendment to a broader transportation bill (a generally popular and bipartisan piece of legislation which satisfied a broad array of interests).

15.4.3 NFWF Gulf Environmental Benefit Fund for criminal penalties

The RESTORE Act was drafted to direct the disposition of the civil penalties under CWA. While it worked its way through Congress, the Department of Justice (DOJ) negotiated plea agreements under the criminal provisions of the CWA with BP and Transocean, which, in addition to facing civil penalties, had been charged with crimes. On November 15, 2012, DOJ and BP announced a settlement that would direct $2.394 billion, to be paid over five years, to a public-private intermediary, the National Fish and Wildlife Foundation (NFWF). On January 4, 2013, DOJ negotiated a similar agreement with Transocean for $150 million to be paid over a two-year period.

NFWF, chartered by Congress in 1984, had played this role before, serving as a conduit for funnelling settlement dollars from environmental legal settlements to targeted restoration projects through Impact-Directed Environmental Accounts. At $2.544 billion, this would be, by far, the largest such Account in history. To manage a program of this size, NFWF created the Gulf Environment Benefit Fund (NFWF, 2020). The plea agreement language laid out some specifics for how the money was to be used by state and project type, but NFWF, working with each of the Gulf States has broad latitude in project selection.

15.4.4 Keeping the big picture in focus

With the passage of the RESTORE Act, the pieces were in place to begin the largest restoration undertaking in US history. However, while it was clear that unprecedented funding would eventually be available, no one knew what the final total would be. The court cases for civil charges under CWA and NRDA damages under OPA were still being litigated in the Federal District Court in New Orleans. Restoration work had already begun, however, since BP had agreed to make a $1 billion 'down payment' on its NRDA responsibilities, and NFWF had begun to select projects to fund with criminal penalties paid into the Gulf Environment Benefit Fund. The process of choosing and constructing restoration projects made headway, even as the trial of BP and the other defendants continued.

NWF and its Gulf Renewal Project work centered on two main efforts: (1) mobilizing public support to see that any trial verdict or settlement would provide meaningful funding; and (2) advocating for a science-based analysis of the problem, a coordinated response by the parties responsible for distributing the funds, and investment in projects of significant scale. This second effort was complicated by the fact that restoring the Gulf's already stressed ecosystems to 'pre-spill' health was not enough. Restoring a seabird nesting beach on an island that could disappear in 20 years would be a waste of money. Aside from this, over the past several decades, the rivers that nourish the Gulf of Mexico have been blocked behind dams, forced into artificial channels, polluted by urban and agricultural uses, and diverted for use in distant cities. Successful restoration projects would need to overcome past degradation and survive climate change; they would need to be largescale and comprehensive, based on cutting-edge scientific analysis (NWF, 2014).

Fortunately, even before the Deepwater disaster, NWF and its Gulf Renewal Project partners had been working with the State of Louisiana's Coastal Protection and Restoration Authority to develop what by 2012 had become Louisiana's Coastal Master Plan, 'Comprehensive Master Plan for a Sustainable Coast' (LCPRA, 2017). Born from the earlier tragedies of 2005's hurricanes Katrina and Rita which devastated much of the Gulf Coast from northwest Florida to southeast Texas, the Coastal Master Plan utilized science, engineering, socio-economic analysis, and modelling of present and possible future conditions to select the most robust set of projects that could be sustained over the coming decades, even against future climate and sea level predictions. The Coastal Master Plan process thereby served as a model to help guide project selection with Deepwater Horizon generated funds.

Figure 15.3 At the one year anniversary of Deepwater Horizon, NWF staff and volunteers planting grasses in marshes impacted by the oil spill. (*Credit: National Wildlife Federation, Amanda Moore*)

As a case in point, prior to the Coastal Master Plan, state and federal agencies had spent tens of millions on coastal restoration projects designed to address specific individual problems. It had become apparent, however, that while these projects might be addressing symptoms, they were not addressing the underlying disease (National Research Council, 2006). The Gulf Renewal Project did not want to see potentially transformational funding atomized by individual state and federal agency priorities instead of being effectively applied in a coordinated effort to address big picture challenges. To prevent that, they mounted a grassroots effort to pressure decision-makers through public comments at meetings, letter writing, media stories and interviews, social media, and outreach coordination with everyone from neighborhood organizations and church groups to local elected officials (Figure 15.3). It also meant building scientific, policy, outreach, and communications capability within our organizations in order to make a credible case for our recommended priorities and policies. It meant forging relationships with political leaders and their staffs in the Gulf States and in Washington DC; with agency heads and with agency staff in DC and in five state capitals; and with opinion leaders among scientists, in the news media and in the business community.

15.4.5 Putting the money to work

On April 4, 2016, Federal District Judge Carl Barbier approved the settlement of outstanding civil charges against BP and the other responsible parties under the terms of the CWA and OPA. The settlement required BP to pay in excess of $13 billion in civil penalties, over and above the $5 billion in criminal fines it was responsible for. (Transocean, the owner of the drill rig, had settled its civil CWA and OPA claims for $1.4 billion in January 2013.) The BP settlement included $5.5 billion in funding from CWA civil penalties to be distributed under the RESTORE Act and up to $8.8 billion for restoration to address natural resource injuries. The CWA civil settlement of $5.5 billion, included $4.4 billion available for restoration (with other eligible uses in some cases). That money was divided as shown in Table 15.1.

The bulk of the restoration funding, however, came from the NRDA settlement of $8.8 billion, including $7.1 billion in new restoration funding, the $1 billion 'down payment' BP made towards the NRDA prior to

Table 15.1 Distribution of RESTORE portion of Clean Water Act penalties ($4.4 billion).

$B (%)	Distribution	Beneficiaries	Notes
$1.54 (35%)	Gulf States	5 Gulf States	TX, LA, MS, AL, FL (@7%)
$1.32 (30%)	RESTORE Council	Restoration	Funding for comprehensive restoration
$1.32 (30%)	RESTORE Council	Spill Impact	Distributed to states based on impacts (by formula)
$0.11 (2.5%)	Centers of Excellence	5 Gulf States	Research centers for environmental, health, economic, and energy issues in each Gulf state
$0.11 (2.5%)	NOAA Science Program	NOAA	Publicly accessible Gulf-wide environmental data management system

judgement, and an additional $700 million for NRDA damages undetected at the time of the agreement and for adaptive management. Importantly, the settlement allowed for a payout schedule over 15-plus years, which began in April 2017 and will continue until at least 2032. At the same time, BP's criminal penalty payments to Gulf Environmental Benefits Fund stretched from 2013 to 2018. Unlike other restoration projects with long time horizons which depended on uncertain annual federal and state appropriations (e.g., Great Lakes, Chesapeake Bay, the Everglades), these court-ordered payments were secured over 20 years. As a result, trustees could ensure that NGOs would continue to work on gulf restoration for the duration of the 15-year payment schedule.

15.5 CONCLUSION

The Deepwater Horizon disaster polluted miles of sensitive ecosystems with 130 million gallons of crude oil, for which the companies responsible forfeited nearly $15 billion in civil fines, criminal penalties, and legal settlements. To the extent that these funds are now available to produce significant, long-term improvements in Gulf Coast habitat, the National Wildlife Federation and its partner NGOs can take some measure of satisfaction. To achieve the settlement, NWF and its partners took a variety of actions, including:

- Ensuring that the media covered the ongoing disaster and its impacts on wildlife and the community;
- Launching an advocacy and media campaign to ensure that the Agencies and Courts assessed penalties to cover restoration costs as well as deter future violations;
- Working closely with congressional and state elected officials to direct the penalties to restoration efforts instead of the general treasury.

There is a long road ahead, but given the scale of the project, the investment in time, money and energy will be well worth it. At this point, our remaining task is to guard against allowing states and federal agencies, in pursuit of their worthy but parochial interests, to lose sight of the need and opportunity for truly comprehensive and lasting restoration in the Gulf of Mexico.

REFERENCES

BOEM (Bureau of Ocean Energy Management) (2020). The Reorganization of the Former MMS. Website https://www.boem.gov/about-boem/reorganization/reorganization-former-mms (accessed 4 June 2020).

Fleeger J., Riggio M., Mendelssohn I., Lin Q., Deis D., Johnson D., Carman K., Graham S., Zengel S. and Hou A. (2018). What promotes the recovery of salt marsh infauna after oil spills? *Estuaries and Coasts*. doi: 10.1007/s12237-018-0443-2 cited in Malmquist, D. (2019) Study shows continuing impacts of Deepwater Horizon oil spill. Phys.Org, https://phys.org/news/2019-04-impacts-deepwater-horizon-oil.html (accessed 5 May 2020).

Howell K. (2011). A Year After BP's Oil Spill, Congress Sits Idly By. Greenwire in *New York Times*, 4/15/11 https://archive.nytimes.com/www.nytimes.com/gwire/2011/04/15/15greenwire-a-year-after-bps-oil-spill-congress-sits-idly-29261.html?emc=rss&partner=rss (accessed 5 April 2020).

Louisiana Coastal Protection and Restoration Authority (2017). Committed to Our Coast: Louisiana's Comprehensive Master Plan for a Sustainable Coast. http://coastal.la.gov/our-plan/2017-coastal-master-plan (accessed 6 April 2020).

National Commission on the BP Deepwater Horizon Oil Spill and Offshore Drilling (2011) Deep Water: The Gulf Oil Disaster and the Future of Offshore Drilling. Report to the President, January, 2011. https://www.govinfo.gov/content/pkg/GPO-OILCOMMISSION/pdf/GPO-OILCOMMISSION.pdf (accessed 5 April 2020).

National Fish and Wildlife Foundation (2020). Gulf Environmental Benefit Fund (GEBF). https://www.nfwf.org/gulf/Pages/home.aspx (accessed 6 April 2020).

National Research Council (2006) Drawing Louisiana's New Map: A Review of Restoration Plans in Coastal Louisiana. https://www.nap.edu/download/11476 (accessed 6 April 2020).

NWF (National Wildlife Federation) (2014) Restoring the Gulf of Mexico for People and Wildlife: Recommended Projects and Priorities.

NWF (National Wildlife Federation) (2015) Five Years and Counting: Gulf Wildlife in the Aftermath of the Deepwater Horizon. National Wildlife Federation. p. 3. https://www.nwf.org/~/media/PDFs/water/2015/Gulf-Wildlife-In-the-Aftermath-of-the-Deepwater-Horizon-Disaster_Five-Years-and-Counting.pdf (accessed 6 April 2020).

NWF (National Wildlife Federation) (2020) 10 Species, 10 Years Later: A Look at Gulf Restoration after the Deepwater Horizon Disaster. https://restorethegulf.nwf.org/wp-content/uploads/2020/04/nwf_gulfreport2020_web.pdf (accessed 8 April 2020).

OSCA (Oil Spill Commission Action) (2013). Three years after BP spill, report card notes improvements, but more to do. http://oscaction.org/press-release-osca-assessment-april-2013 (accessed 6 April 2020).

Restore the Mississippi River Delta (2020) Land Loss. http://mississippiriverdelta.org/our-coastal-crisis/land-loss/ (accessed 5 April 2020).

US Code (1972) Federal Water Pollution Control Act 33 U.S.C. §1251 et seq. Accessed via EPA Website Summary of Clean Water Act. https://www.epa.gov/laws-regulations/summary-clean-water-act (accessed 6 April 2020).

US Code (1990a) Federal Oil Pollution Act 33 U.S.C 2702(a). Accessed via EPA Website 'Summary of Oil Pollution Act'. https://www.epa.gov/laws-regulations/summary-oil-pollution-act and Legal Information Institute https://www.law.cornell.edu/uscode/text/33/2701

US Code (1990b) Federal Oil Pollution Act 33 U.S.C 2706(b). Accessed via EPA Website 'Summary of Oil Pollution Act'. https://www.epa.gov/laws-regulations/summary-oil-pollution-act and Legal Information Institute https://www.law.cornell.edu/uscode/text/33/2701

US Department of Interior (2019) BSEE Finalizes Improved Blowout Preventer and Well Control Regulations. 5/2/19 https://www.doi.gov/pressreleases/bsee-finalizes-improved-blowout-preventer-and-well-control-regulations (accessed 6 April 2020).

Part 2

Incentives and Barriers

The chapters in *Incentives and Barriers* reflect insights on the incentives, supports, barriers, disincentives, and penalties that currently shape water use. The authors also provide recommendations on how individuals and organizations can work together to design and implement incentives and disincentives that will effectively modify the way that water is used by industry.

While some companies have implemented strategies to reduce their long-term exposure to water-related risk, their success should not obscure the difficult path confronting companies still struggling to adopt sustainable water use practices. Early adopters often have sufficient resources to invest in process change, and strong motivation to protect their corporate brand, reducing their 'reputational risk'. Later adopters, on the other hand, are often challenged to take on technical tasks like re-engineering their core processes, and non-technical tasks requiring increased collaboration with suppliers, policy-makers, regulators, water utilities, environmental associations, and others. Their progress, however slow, contrasts markedly with companies unaware of the need to use water more sustainably as well as those content to use and dispose of water as cheaply as possible, regardless of the social or environmental costs.

This section examines the motivations that drive industries towards sustainable water use, and the incentives, penalties, and obstacles that can accelerate or impede their progress.

16. **Davis** *Sustainable water use by industry: What can we do?*
    Taking the broadest view, **Davis** suggests that addressing sustainable use of water by industry requires not only an understanding of technical problems but also an understanding of human motivation, and the factors likely to change behavior. She notes that, even as individuals, we don't reliably act in our own best interests, observing that 'we all arrive in the world as narcissists', inclined to choose short-term pleasure over long-term gain and put ourselves ahead of others. She points out a connection between these tendencies and unsustainable use of water that damages communities and the environment. She also suggests that in designing incentives and penalties to influence industrial water use, we should consider how different types of

DOI: 10.2166/9781789060676_0175

companies might respond to different strategies. She describes four categories of companies – Early Adopters, Potential Adopters, Profiteers, and the Resource Limited – and suggests incentives and disincentives that may be most effective in changing the behavior of companies in different groups. Drawing on results from the survey circulated among the authors who contributed to this volume, Davis also provides a list of distinct actions that can be taken to promote sustainable water use by individuals regardless of the sector where they work or volunteer their time. This includes our options as consumers who can change our buying behavior—a change that 'requires no legislation or advanced technology'. Davis also calls for more effective communication on the part of sustainability advocates, which could combine language and symbols that have emotional appeal with sophisticated new communication tools, and discusses possible long-term implications of the coronavirus pandemic on global water issues.

17. **Chesnutt** *The economics of sustainable industrial water use, reuse, and the value of water*
Chesnutt focuses on the economic dimension of human relations – specifically, how the value of water is expressed to industry economically, and how that impacts the decisions businesses make about water use. Industrial water is different from most other goods, he claims, in that it is most often sold by a public monopoly at a price based not on market value of water but on the cost of transporting it to the customer. This sends the wrong signal to industries, as these artificially low prices discourage conservation and mask the true cost of adequately maintaining water systems. The gap between value and price is further widened when utilities base their water rates on their historical costs which don't take into account either today's demands or tomorrow's challenges. Chesnutt warns, 'There are dangers to driving through life while staring in the rearview mirror', and he encourages that utilities use instead forward-looking cost accounting methods. Two strategies he recommends to help companies appreciate the value of water conservation and reuse are: (1) calculating the value of avoiding future water shortages; and (2) considering the social and environmental benefits of reuse in terms of increased market share. This latter benefit is the result of preferences expressed by customers for sustainably produced goods, and he notes that the growing desire on the part of retail customers to 'buy green' has inspired some companies to use water more sustainably in order to ensure their future profitability. At the same time, awareness of water risks has even encouraged some companies to invest in recycled wastewater to become independent of their local utility, an economic response Chesnutt describes as 'system bypass'.

18. **Maennicke and Hoenerhoff** *Development finance: Encouraging sustainable water use by industry*
Where wholesale industrial customers make decisions based on price alone, or when a company can't afford to invest in water improvements, financial institutions can play a key role in incentivizing sustainable water use. This is especially true for small- to medium-sized enterprises (SMEs) in developing countries. These enterprises may have a hard time accessing capital to invest in sustainable practices. In these cases, development finance serves as a catalyst to attract and mobilize other sources of private capital. According to **Maennicke and Hoenerhoff**, Development Finance Institutions (DFIs) such as the German Development Finance Institution DEG can lower the burden and risk associated with businesses and attract private investors. DEG provides tailored solutions to private enterprises operating in developing countries with emerging markets, including long-term financing, promotional programs, and individual advice. One special focus is SMEs, which they support with loans as well as with training in technical and business practices like due diligence assessment. As a further incentive, DEG helps introduce qualifying companies to European markets. For example, companies like EDEKA (Germany's largest supermarket chain)

require suppliers to certify their sustainable water practices especially in areas designated as 'high risk' by the Water Risk Filter Assessment. By financing water management investments through DEG, companies can gain the expertise to certify their sustainable practices and sell products globally.

19. **Crespi Reghizzi** *Reducing pollution from industrial wastewater in developing and emerging countries*
Investment banks can also help by encouraging governments in the countries where they operate to improve water management. The World Water Assessment program estimates that 70% of industrial wastewater is discharged without treatment. According to Crespi Reghizzi, companies pollute because the market allows them to exclude the cost of their impact on the environment from the prices they charge consumers. These externalities can only be addressed by regulation, by creation of economic incentives, or by setting up a system where pollution is recognized as an impairment of property rights. Within France, levies and incentives promote sustainable water management at the river basin scale. This success has encouraged the French Development Agency (AFD) to provide financing for both public and private projects in developing and emerging countries that contribute to achievement of UN Sustainable Development Goals (SDG). Since reducing industrial water pollution furthers UN SDGs related to water, infrastructure, biodiversity, and others, AFD funds upgrades to industrial facilities so that they use fewer chemicals and less water. ADF also invests in appropriate municipal wastewater treatment technologies; for decades, ADF has financed municipal water and wastewater plant upgrades in Morocco. Today ADF also works directly with industries to reduce water consumption and improve treatment prior to discharge. In Senegal, where 80% of industry is located near Dakar and discharges waste into polluted Hann Bay, ADF is funding municipal wastewater treatment plant upgrades and working with the government to set up a structure to bill industry for water and sanitation services and implement a sustainable policy.

20. **Vicencio** *The banking sector as an intermediary in supporting sustainable use of water by industry*
Briefly stated, the banking industry can play a key role in supporting sustainable industrial water use, especially in developing countries. As **Vicencio** explains, banks have extensive knowledge about the behavior of different industries, markets, and business models that they can share with motivated companies. They can do this by providing technical, financial and legal expertise; validation and certification; connections to markets; and favorable terms for investment in sustainable practices. In Ecuador, Podubanco (a subsidiary of Promerica Financial Corporation) has established a successful 'Green Lines' program to connect businesses to legal, academic, and other resources to enhance their sustainability. The program helped on one milk processing facility in the Ecuadorian Highlands to reduce water consumption by over 33% through replacement of machinery and automation of a clean-in-place (CIP) process. In another example, Produbanco's investment allowed an Ecuadorian rose grower to install drip irrigation and other improvements that reduced water use by more than 70%, increasing the company's profits as well.

21. **Mariluz and Antunez de Mayolo** *The Certificado Azul: Peru's innovation for encouraging sustainable use of water by industry*
National governments can also implement programs to encourage more sustainable use of water by industry. As explained by **Mariluz and Antunez de Mayolo**, the Peruvian government knows the country is particularly susceptible to the impact of climate change as much of its water supply comes from shrinking glaciers. To stimulate companies to reduce their water use, the National

Water Authority of Peru has implemented the Certificado Azul Program, which supports watershed-level analysis of water use and recognizes documented achievements by companies that implement more sustainable practices. In order to qualify for recognition, companies are required to calculate their water footprint using an approved measurement tool (the Alliance for Water Stewardship, the Water Footprint, or ISO14106) and develop a plan to address water use within at least one of its facilities. Companies must also submit a plan to contribute to improved watershed management in the watershed where the facility is located. As an example, the company may make an audited investment that will reduce water loss in a channel that brings water to farmers who grow produce that is part of the company's supply chain. Although, recognition as a 'Certificado Azul' company does not currently carry a significant financial benefit, the National Water Authority is considering a wide range of economic and non-economic incentives for the future.

22. **Spencer** *The culture of water needs to change*
While incentives can encourage sustainable water use, **Spencer** points out that, too often, progress towards sustainability is hindered by not including industry in developing future water solutions. He argues that the 'culture of water' needs to change, so that utilities bring water customers, industry, government, and other stakeholders into the process. Spencer suggests that the dominance of water professionals in identifying water problems leads to the definition of technical and engineering solutions even in areas where 'communication and collaboration skills would be more relevant.' He concludes that the water industry must shift from its exclusive focus on technical solutions to one that acknowledges the need for behavioral change. This will require financial incentives and strong collaboration among government, customers, civil society, and industries willing to share their practical experience modifying facilities to use water more sustainably.

23. **Caldwell** *Designer water: One utility's unique approach to industrial sustainability*
The potential benefits of utility/industry collaboration are illustrated by West Basin Municipal Water District (WBMWD) in southern California (USA), which is often cited as an example of a successful utility–industry partnership. **Caldwell** traces the history of WBMWD's well-known program, explaining how manufacturing 'designer water' allows a utility to provide water particularly suited to industrial use. WBMWD currently offers no fewer than five types of recycled water tailored to the specific requirements of its large industrial water customers; the types of water produced are: (1) irrigation water; (2) cooling tower water; (3) seawater barrier and groundwater replenishment water; (4) low-pressure boiler feed water; and (5) high-pressure boiler feed water. To be successful in this sort of arrangement, utilities must maintain regular communication with industrial customers.

24. **Rosenblum** *Incentivizing sustainability: How utilities can support industrial water conservation and reuse*
Notwithstanding the positive example set by WBMWD, **Rosenblum** maintains that utilities often drive industry away from the table. He claims that, from the perspective of business, the prospect of collaborating with government can be daunting, and highlights several approaches a utility can take to encourage industries to use municipally treated effluent (i.e., recycled water). He specifically recommends that utilities should strive to understand the concerns of business and adapt their outreach methods to address those concerns. They should acknowledge the risks and challenges associated with the use of recycled water, and candidly discuss the fiscal impact of implementing water reuse, providing financial assistance in the form of grants or loans to help cover the costs. Most important, utilities should take pains to respond promptly to industry

inquiries, and provide support as soon as it is requested. Rosenblum also references a chartering tool used to facilitate joint industry–utility participation in water reuse projects.

25. **Kehoe and Chang** *Onsite water reuse: A collaborative strategy to manage water*
Another example of successful collaboration between utilities and industry is the program launched by the San Francisco Public Utilities Commission (SFPUC), a municipally owned water/wastewater utility, to promote onsite wastewater treatment and reuse. As **Kehoe and Chang** relate, in 2015 SFPUC adopted an ordinance requiring all new commercial, mixed-use and multi-family development projects greater than 250,000 square feet (approximately 23,000 m^2) to capture, treat, and reuse wastewater. With the technical assistance and guidance provided by the City, it has proven feasible to apply this approach to multiple sites, ranging from office buildings and transit facilities to breweries.

26. **Groot** *Challenges in regional collaboration*
Just as utiltiies can provide incentives to promote sustainable water management practices, following Spencer's recommendation, companies themselves can encourage local governments to include them in developing water solutions. **Groot** provides two examples – one from Spain and one from the Netherlands – where Dow Chemical collaborated with local utilities to develop an integrated wastewater treatment and reuse strategy, doing together more than either could accomplish alone. In Catalunya, in Spain, the government headed up a consortium of public and private sector participants to install a 19,000 m^3/day (5 mgd) water reclamation plant to treat water from the Tarragona and Vilaseca wastewater treatment plants, blend it with Ebro river water, and supply it as make-up cooling water for the Tarragona Petrochemical Complex plants. The water reclamation plant facility was built by Veolia, owned by ACA (the Catalan Water Agency), and operated by Veolia and AITASA. The process in Catalunya was top-down, driven by clear mandates for individual entities to participate in the collaboration process. As a result, there was a common incentive at a sufficiently high level for all parties to cooperate which ensured there was widespread awareness and support. By contrast, in the Netherlands, Dow Terneuzen funded construction of a mebrane bioreactor (MBR) system to purify treated effluent from the local utility so they could reuse the water in their facility. Looking back, Groot observes that it was more difficult for a private sector company to lead the effort, as the utility producing the wastewater and the industry taking the treated water have different motivations and different notions of what is fair. Nevertheless, based on Dow's experience at various water-stressed locations in Europe, he maintains that by cooperating with local utiltiies, industries can reduce their water footprint more than they can by staying within their own fence line.

CONTRIBUTOR SURVEY RESULTS: INCENTIVES AND BARRIERS

As part of the process of preparing this anthology, the editors asked each author to complete a survey reflecting their views on sustainable use of water by industry. The survey gave authors the opportunity to suggest incentives, supports, barriers, disincentives, and penalties that they believed would encourage more sustainable use of water by industry.

Contributors cited both the low cost of water and the high price of technology as barriers to improved water use, and identified economic incentives and penalties as the most effective stimuli to motivate sustainable use water by industry. Incentives in the form of tax breaks for investments in treatment were mentioned, as were higher fees for water and wastewater discharge and increased fines for permit violations. Even some industry authors welcomed tougher discharge limits and more aggressive enforcement, because they would help support their business case for investing in advanced wastewater treatment and reuse schemes.

As highlighted by **Chesnutt**, another incentive identified by industry and advocacy groups alike was the importance of customer preference. One industry respondent admitted that progress towards sustainable use of water in the textile industry has been driven primarily by the apparel companies' need to court consumers who buy 'green' products. Competitive companies today cannot afford to associate their brands with water pollution and waste. Environmental advocacy groups reported that companies also respond to their pressure for more transparency in corporate accounting for water use.

While these factors may work for larger companies, they may not be sufficient to motivate small- and medium-sized enterprises in developing countries. According to contributors from the financial sector, these companies (referred to by **Davis** as 'resource limited') first need to find the right solutions, then they need to get the funds to pay for them. Echoing **Vicencio**, they identified favorable loans and technical support from experienced bankers as key to helping companies improve their production processes. Development banks should dispel the notion that investing in water technology is a form of 'philanthropy' by giving sustainable companies access to markets that require environmental certification, while investors can recognize the importance of water by giving them higher valuations.

Not all barriers can be overcome by financial incentives or market inducements, however. Academic, consultant, government, and industry respondents all noted the absence of formal watershed- or catchment-based planning as an obstacle to sustainable use of water by industry. This problem was also noted by **Groot** as he described the relative ease with which industry contributed to a government-sponsored regional solution compared to one in which his company was forced to take the lead. In addition, utility contributors pointed out the need to train operations and maintenance staff in new treatment technologies. To round it out, virtually all survey respondants noted that the public as a whole is too often disengaged, knowing and caring little about their local water supplies until the tap runs dry. As one author put it, 'One key difference between the oil crisis and the current water crisis is that the oil crisis simultaneously affected the entire world, whereas water issues tend to be local and regional. The oil crisis catalyzed a paradigm shift in how the world perceived oil, but this paradigm shift has not yet occurred for water.'

The chapters in this section provide numerous insights on the incentives and barriers that impact current use of water by industry, and recommendations for changing those incentives and barriers in a way that will create a paradigm shift in how water is viewed and used.

Chapter 16

Sustainable water use by industry: What can we do?

Cheryl Davis

CKD Consulting, Piedmont, USA

Keywords: collaboration, communication, communities, environment, ethics, incentives, motivation, regulations, social media

16.1 INTRODUCTION

We know from history and personal observation that much human behavior is both self-interested and short-sighted. Alain de Botton has suggested that most of the damage we do is not intentional but 'a by-product of the constant collision of blind competing egos in a world of scarce resources' (de Botton, 2020). In relation to water, the impact of this collision has become more damaging, as technological progress has increased our capacity to wreak havoc on nature and fellow humans, eliminating many species and endangering our own. As with our personal lives, no amount of planning or prudent action can guarantee a good outcome, but some choices – in this case, combining current practices with a growing global population and climate change – can be relied on to ensure a bad one.

We have access to extensive documentation on the effects of unsustainable water use, as well as a plethora of tools to help us do better. The mystery is how to motivate decision-making and resource allocation that reflect both rationality and compassion. This might be considered an ethical or even a religious issue, as many religious traditions have valued water and suggested that the welfare of others should be taken into account. However, even religious communities have sometimes claimed absolute dominion of humans over the earth, as reflected in the notes of a town meeting held in Milford, Connecticut in 1640: 'Voted, that the earth is the Lord's and the fullness thereof; voted, that the earth is given to the Saints; voted, that we are the Saints (Offill, 2020).

In order to protect water and everything that depends on it, we need to design clear-eyed strategies that reflect a realistic view of how humans operate – our values, our motivations, and our limitations. To the

DOI: 10.2166/9781789060676_0181

extent we are successful in this analysis, we will be better prepared to construct and implement an effective portfolio of incentives, supports, barriers, and penalties that promote sustainable use of water by industry.

16.2 WHAT DO WE KNOW ABOUT WHO WE ARE?

Humans respond to multiple influences in decision-making, and respond to the same factors differently at different times. This means that we need to be creative, resourceful, and vigorous in developing multiple varieties and layers of incentives and penalties in relation to water use. It's good, but not sufficient, to define business cases for right action based on long-term consequences. Even as individuals, we don't reliably act in our own long-term best interests, even when fully informed of possible consequences. (Readers who have ever eaten or drunk too much may silently raise their hands.) Our proclivity to pick short-term satisfaction over long-term benefits is supplemented by the fact that, as one friend advised me (in a different context but just as applicable to denial of climate change, water degradation, and water scarcity), '*People know what they want to know.*'

Sometimes our resistance to absorbing the implications of observable facts reflects our helplessness in the face of a threat we are powerless to avert. In 1665, Samuel Pepys noted that over 6000 Londoners were reported to have died that week from the plague, but that he believed the true number was closer to 10,000, taking into account deaths of the poor 'that cannot be taken notice of through the greatness of the number' (Latham, 1985). But in the case of unsustainable water use, often what is missing is the will to apply what we know. A panel of water experts in a session on Incentives and Barriers to Sustainable Use of Water by Industry at the 2019 Innovation Conference in Guayaquil, Ecuador (co-sponsored by the International Water Association (IWA) and the Inter-American Development Bank (IDB)) identified the issue of how to motivate changes in human behavior relating to water as an area where we most needed additional research (Maennicke, 2019).

Since we all arrive in the world as narcissists, it is worth considering the cultural and social factors that encourage us to develop a broader perspective. Adoption of sustainable practices in some measure depends on moral imagination – the ability to visualize and value human beings beyond our social group, our culture, and our time, as well as other species and entities in the natural world. During a plenary session at the same Innovation Conference, Dave Archambault, a Standing Rock Sioux tribal leader, described how indigenous cultures see human beings as progressing from an infant's narcissism to a mature concern for all – human and non-human, animate and inanimate, and living or not-yet-born (Archambault, 2019). In that context, he contrasted indigenous peoples' understanding of water as a spiritual entity with its more common definition by modern society as a means to economic gain.

Within the United States, the conflict between respect for water as an intrinsic good and the definition of water as a unit of production is reflected in the controversy surrounding construction of the Dakota Access Pipeline. As described in more detail in Archambault's contribution to this volume, the economic resources of the oil industry were pitted against the environmental and cultural values of a tribe that owned land that the oil pipeline would traverse. A similar conflict is also being played out in Brazil, where mining and cattle interests endanger the remaining Amazon rainforest. In his campaign for President, Jair Bolsonaro promised to support economic development of the Amazon with the slogan, '*Where there is Indigenous land, there is wealth underneath it.*' Following his election, he announced that he had 'put an end [to] astronomical fines' against companies that violated environmental laws. Reduced enforcement encouraged additional land grabs by loggers, miners, and cattle ranchers; according to Brazil's National Institute for Space, in the one year between August 2018 and July 2019 over 1600 square miles of forest cover was slashed – a 74% increase over the same period the year before. In February, 2020, he presented a bill that would authorize oil and gas exploration and the construction of hydropower plants and legalize outlaw mining ventures that have polluted rivers and torn down large swaths of the Amazonian forest (Londoño & Casdo, 2020).

Our perspectives on such conflicts are affected not only by our culture, but our times. Within the United States, concern for the protection of the environment and communities has been inconsistent during my lifetime. I entered young adulthood in the United States during the 1960s, a time of changing views on the environment, civil rights, gender and racial equality, poverty, and war. Many of us who participated in marches and demonstrations for these causes were raised by parents who had prospered, relative to their own parents, in the US economic boom following World War II. We expected our generation to do even better. We saw no downside to sharing, no need for sacrifice; there seemed to be plenty for everyone. The increased environmental awareness that generated the first Earth Day in 1970 led to tighter protection of water bodies, investment in municipal wastewater treatment, and increased regulation of industrial effluents. Over time, however, resistance to national regulations emerged and commitment to environmental protection has sometimes languished even in prosperous times (Marcus, 2021). At the time of writing, it is unclear how the economic impacts of the coronavirus pandemic will affect our will to be generous to others, the environment, and future generations.

In some parts of the globe, both knowledge and resources to implement sustainable use of water are insufficient; in prosperous countries where both knowledge and resources are abundant, motivation may be lacking or inconsistent. Recognizing these hurdles, we can expect that sustainable use of water by industry will require substantial modifications to the organizational ecosystems in which companies operate. Rather than hoping all businesses will recognize and act on the long-term consequences of unsustainable water use, we need to implement programs that have more immediate impacts.

16.3 WHICH INCENTIVES AND PENALTIES WILL BE MOST EFFECTIVE?

A wide range of potential incentives and disincentives have been implemented over time to modify corporate use (and misuse) of water. Incentives and supports include:

- programs that educate the public, government, educational institutions, NGOs, and other stakeholders about damage caused by unsustainable water use, opportunities for improvement, and positive effects of industry initiatives;
- provision of social and financial rewards, including financing, for sustainability initiatives;
- funding and technical assistance to support investments in sustainable practices and technologies;
- water charges based on metered consumption, with rates high enough to encourage conservation;
- limitations on water consumption, and requirements for water reuse;
- requirements relating to wastewater discharges and other environmental impacts;
- capacity-building programs for industry employees and suppliers;
- training and adequate pay for regulators, to encourage competent and reliable enforcement of regulations;
- pressure on corporate boards to consider water sustainability as part of the company's bottom line;
- ability to attract and retain high-quality employees whose values include protection of communities and nature;
- increased ability to successfully market goods to customers who value sustainable practices; and
- public/private partnerships that benefit the community.

Barriers and penalties include:

- restrictions on consumption of potable water;
- creation and enforcement of regulations on water reuse and wastewater discharges;

- loss of stock value due to reduced investor confidence, resulting from financial liability associated with damage to people and nature;
- reputational loss that impacts sales to customers; and
- criminal prosecution.

Many of these rewards and penalties can be implemented only by government, which in some areas may itself be more focused on short-term economic gains than long-term environmental and social benefits. However, stakeholders can have an effect on both laws and government actions, and many tools are in the hands of non-governmental entities. A significant shift in the allocation of power in the world occurred with the widespread availability of digital phones that can photograph and distribute instant documentation of the effects of unsustainable water use, making international shaming an achievable option for individuals who may be relatively powerless at the local level.

16.3.1 Industrial taxonomy

Various taxonomies have been developed for categorizing sectors of industry by the products they produce, the revenue they generate, or their size. However, our ability to strategize about the range and type of incentives and penalties needed will improve if we can gain a better understanding of the values, constraints, and motivators of different types of companies. Otherwise, our strategic thinking may be blurred by optimism generated by the examples of well-resourced international leaders in the area of sustainable water use, or outrage against companies whose use of water is flagrantly damaging. The primitive taxonomy offered here is intended to encourage further thinking about the actions, laws, regulations, and programs required to have an effective impact on companies at different levels of motivation and functioning. The categories suggested here are Early Adopters; Potential Adopters; Profiteers; and the Resource-Limited, with the last being further subdivided into 'Knowledge limited' and 'Finance limited' and 'Subsistence Business'. Each has a different disposition towards sustainable water use, and can be expected to respond to different incentives, and disincentives.

16.3.2 Early adopters

Early Adopters of sustainable water use practices are often multi-national businesses with significant financial and technical resources. People speak their brand-names in many languages, and they are sensitive to reputational risk. These companies can afford to take a long-term perspective, have the internal capacity to fund, test, and use new technologies, and employ staff with the skills needed to work effectively with other stakeholders, such as governments, consultants, educators, and community groups. They set standards for themselves and their suppliers, and often provide sustainability-related training. Their contributions may extend to water supply and water quality improvements and investments on watersheds where they operate, in addition to their own facilities. Similar practices are sometimes adopted by small business entities with a strong ethical commitment to social and environmental values.

Support for Early Adopters usually involves incentives, such as customer appreciation for sustainable water use. However, some companies now providing leadership in relation to sustainable water use were at some point in their corporate history subject to international revulsion generated by environmental or social damage associated with their operations, permanently sensitizing them to the desirability of green branding. Early Adopters benefit from diligent government application of barriers and penalties for unsustainable water use, as they are otherwise at a competitive disadvantage relative to companies that have not been required to invest in sustainable water practices and technologies.

16.3.3 Potential adopters

Potential Adopters have the financial and technical resources to adopt more sustainable practices and technologies, but lack sufficient internal motivation or external drivers to do so. The full range of incentives, barriers, and penalties could potentially impact companies in this category. If supplied with visible models for improved practice and external pressures to conform, they have the capacity to comply. There may be individuals within the company who value sustainability and are proponents of change, who are unable to make a successful business case for investments in water sustainability because the cost of water is low and there is no visible risk to the supply or quality of the water that will be available to the company in the future.

16.3.4 Profiteers

Profiteers are companies that display, at the corporate level, the characteristics of narcissism, addiction to short-term benefits, and denial of the negative impacts of their choices. Despite access to considerable financial and technical resources, they persist in pollution and over-consumption. They do not invest in infrastructure, best practices, or technologies that would protect their employees, communities, or nature. These flourish in social ecosystems where water is cheap, governance is weak, regulators are inadequately trained and/or susceptible to corruption, and communities have insufficient influence to protect themselves and the environment. Without firm barriers and penalties, profiteering companies can exploit resources without restraint. Incentives that might be effective with companies in other categories are unlikely to work with them, as the financial rewards other entities can offer are unlikely to compete with the immediate financial pay-offs associated with unfettered operations. Where governments are unable to monitor, regulate, control, and apply penalties, behavior change may have to be driven by customers and/or investors.

One difference between Potential Adopters and Profiteers is that the former may be encouraged by the promise of substantial long-term benefits, while the latter may only be moved by severe short-term consequences. A recent example – again from Brazil – illustrates how appropriate incentives – and disincentives in the form of escalating public, governmental, and financial penalties can impact corporate decisions.

In November of 2015, the Fundáo tailings dam that collected water and sediment effluent from an iron ore mine jointly operated by Vale and GHP Billiton near the city of Mariana in the Minas Gerais region in Brazil, failed, releasing about 40 million liters of water and sediment containing arsenic, lead, and mercury into the Doce River. Five thousand streams and 10,000 hectares of land were affected; Figure 16.1 reflects the extent of the pollution. Two years later, when I visited one of the farming communities that had been destroyed, the once-clear river was still a murky orange-brown, fish were gone, churches and homes that had drowned in sediment lay vacant, and the crops cultivated previously couldn't grow in the toxic soil. A young man then employed by Vale to work with displaced inhabitants said that the company had spent considerable money to rehouse families, but had never acknowledged culpability for the event, or (to his knowledge) made changes to prevent a recurrence.

In January of 2019, a tailings dam owned by Vale failed in the Minas Gerais region, this time near Brumadinho (Silva de Sousa *et al.*, 2019). Over 300 people (predominantly Vale employees) died; Vale stocks fell 24% the day following the disaster; and three Vale employees and two consultants who had recently attested to the reliability of the dam were arrested. The magnitude of the penalties associated with this disaster had an industry-level impact. In February, Freeport-McMoRan Inc. Chief Executive Richard Adkerson sent a memo to his 29,000 employees telling them to immediately report any safety concerns about the scores of dams the company operates. The company was already spending several

Figure 16.1 Comparison of Minas Gerais water quality in a river unaffected by mine discharge (left) and Doce River downstream of tailings dam failure (right). (*Credit: Joao Guedes*)

hundred millions of dollars per year on tailings dam upkeep and had not had a tailings dam failure since acquiring Phelps Dodge in 2007. However, the Vale failure underscored the importance of caution. Adkerson and 26 other CEOs, including leaders from BHP and Vale, agreed to form a panel to set international design and maintenance standards for dams and study ways to reduce the volume of water stored behind the dams in waste rock (Scheyer, 2019).

16.3.5 Resource-limited

Many companies are **Resource-Limited**, lacking the understanding of water sustainability issues, technical expertise, and financial resources required to implement sustainable water use practices and technologies. In this situation, it is often appropriate to focus on incentives and support. At a workshop on Sustainable Use of Water by Industry at the Water and Development Congress held by the International Water Association in Buenos Aires in 2017, panelists from South America pointed out that many companies in their region were unaware of the concept of water sustainability, and the tools and technologies available to improve industrial water use. This observation led to convening the 2019 IWA/IDB Innovation Conference in Guayaquil, which included a seven-session track on sustainable use of water by industry, including a session on financing support for sustainable water use (Industrial Track, Innovation Conference, 2019). For many small businesses, investment in new technologies and practices is not financially feasible without support from external sources such as development banks. Workforce capacity can also be an issue. Uneven educational systems in low-income communities can impact the technologies that both industries and local utilities can reliably operate and maintain. Staff training may be required in order for employees and suppliers to change practices and use new technologies effectively.

Early Adopters sometimes provide significant leadership to resource-limited small businesses in developing and emerging countries where knowledge of sustainability practices and technology is limited and governance is weak. They can, for example, make small and individual business owners aware of environmental criteria relating to wastewater effluent, require suppliers to collect data on the constituents of their wastewater effluent, analyze, compare, and publish information on the relative performance of different suppliers on critical sustainability performance issues, and encourage suppliers to invest in wastewater treatment or reuse that will help them meet international standards.

Some of the most grievous examples of unsustainable industrial practice are not the work of big companies, but individuals and micro-businesses who scratch out a living at a subsistence level. Examples include sand miners in India and individual mineral miners in Ecuador. Unlike the Profiteers, who may cause damage without adverse impacts to their own daily lives, these individuals often bear the brunt of the damage their actions cause. Where the only way for parents to feed themselves and their children is to damage water quality or the environment, resources from external entities are required. Regulatory penalties that have been enacted but are not reliably enforced are not a realistic deterrent in areas where governance is weak. Resources to support more sustainable water use can be provided by government, NGOs, financing institutions, academic institutions, or private businesses operating in the same watershed (Greig & Rathjen, 2021).

Although the taxonomy proposed here for categorization of the relative openness of different companies to modifying their practices and technologies to support sustainable use of water are conceptual rather than data-based, it may be useful for those considering the potential impacts of different possible incentives and penalties to think through the potential responses of businesses with different attributes. Illustrative examples are provided in Table 16.1.

16.4 WHERE ARE WE IN THIS PICTURE?

Because many of us play multiple roles in our work and social lives, any one individual might show up in the ecosystem of industrial water use in numerous locations. In research related to development of this anthology, the co-editors surveyed the views of authors from all sectors on the type of contributions that could be made by individuals in different sectors. Survey respondents were assured that their responses would be reported, without attribution, so that they could speak freely on their views on other sectors as well as their own. A sample of their responses reflects the reality that individuals can affect how industry uses water from numerous vantage points.

Industry – In order to implement sustainable practices, companies need a combination of awareness, commitment, and capability. As an example, Early Adopters with suppliers in developing and emerging countries can set goals relating to sustainable water use, collect data, and compare the performance of suppliers, providing both leadership and an incentive for suppliers to invest in improved technologies.

Government policy-makers – Policy-makers can increase water rates, limit potable water consumption, require reuse, and set standards for effluent discharged from industrial processes, and define penalties for non-compliance (including definition of processes for directing fines collected to environmental restoration).

Regulatory officials – Regulators should not only enforce regulations, but understand the processes and challenges of industry, so that standards can be customized for different industries and situations, with a focus on maximizing overall benefits rather than meeting minimum standards.

Industry associations – Industry associations can help connect businesses with other sectors (e.g., academia, research, government policy-makers, and water utilities and associations) to promote best practices (e.g., through conferences and workshop).

Professional water associations – Need to realize that water is not just an engineering problem, and create partnerships with industry and other stakeholders so that they can work together to develop solutions at the watershed level.

Table 16.1 Examples of differential impacts.

Differential Impacts	Early Adopters	Potential Adopters	Profiteers	Resource Limited
Incentives and Supports				
Awareness-raising programs on the damage caused by unsustainable water use, opportunities for improvement, and positive effects of industry initiatives	Positive information on sustainability efforts supports the company's branding	May be an incentive for change	May not outweigh short-term financial benefits of unsustainable use	May motivate change but does not provide resources needed for implementation
Funding and technical assistance for sustainability initiatives	Early Adopters often have sufficient resources without external support	Availability of funding support or technical assistance could make a difference	Resources offered might not outweigh immediate financial benefits of unsustainable use	Key to incentivize and support sustainability efforts by resource-limited businesses
Water charges based on metered consumption, with rates high enough to encourage conservation	Avoided costs will be used to highlight benefits of investments in sustainable use	A cost increase might impact decision-making	Rate increase unlikely to be high enough to affect decision-making as long as sufficient water is available for production	Could have an impact on water use
Creation and vigorous enforcement of wastewater effluent regulations	Early Adopters benefit when competitors are required to match their investments	Will improve business case for new practices, technologies	Will only invest if penalties unavoidable from vigorous monitoring, enforcements	Will increase motivation but not automatically create capacity

Ability to successfully market goods to customers who value sustainable practices	An important consideration for Early Adopters	May be a motivation for change	May inspire green-washing	Affects suppliers that sell their products to Early Adopters
Barriers and Penalties				
Financial penalties issued by governmental agencies for violation of regulations	Supports their business case for investments in sustainability, when others are penalized for misuse	Will increase motivation for change	Might motivate change if substantial	May be unable to either pay penalties or change without assistance
Financial liability associated with damage to people and nature; and loss of stock value due to reduced investor confidence	May benefit if competitors lose stock value due to misuse	Would radically affect business case for sustainability	Can force some level of change	May be unable to change without assistance
Reputational loss that impacts sales to customers;	Place a high value on avoiding this outcome	Would increase motivation for change	Likely to result in at least the appearance of change	May be unable to change without assistance
Criminal prosecution	Place a high value on avoiding unflattering publicity	Would probably ensure some level of change	Would force at least the appearance of change	May be unable to change without assistance

Environmental/sustainability associations – Environmental/sustainability associations need to work with industry rather than look at industry as an enemy, taking into account that economic and production processes are also important for developing countries, and use social media to increase engagement.

Social/community associations – Social/community associations should make sure that their voices are heard by government officials so that action is taken to reduce the impact of industry on their health and their communities.

Financial institutions – Financial institutions such as development banks should supply small and micro-businesses in developing countries with both the funding and technical assistance needed to produce products more sustainably.

Water/wastewater utilities – Utilities should work with their industrial customers to support more sustainable use of water. For example, wastewater utilities could provide customized varieties of wastewater to service to meet the process needs of different customers, to provide training for operators and maintenance workers in industrial facilities, to introduce them to newer technologies.

Universities and research organizations – Universities and researchers should not limit themselves to conceptual research, but also support pilot projects and full-scale experiments, to help industry investigate and work through the challenges of implementing changes at the facility level.

Press – Media helps form public opinion. They can create or support discourse that favors sustainable use of water by industry by disseminating information on sustainable practices and technology. They can also encourage responsible water use by the public. They can create a demand for new research findings by providing information that will open the eyes of policy-makers to new options that are available.

Public – The public can have a major influence as a pressure group on industry. One positive contribution they can make is to adopt the principle of 'fit for use,' rather than expect potable water to be used to meet all needs.

Many readers of this chapter will locate themselves in several of the categories above. However, it is also worthwhile to think about our role as consumers in a global ecosystem that enables irresponsible use of water. For example, most of us own several items of clothing, implicating us in how water is used to in the textile and fashion industries. The lead article in the February, 2020 Awareness Issue of *Vogue Italia* cited seven key apparel life cycles which needed to be revolutionized (fiber production, yarn preparation, fabric making, dyeing and finishing, manufacturing and assembly, and end of the life cycle) and described five risk areas that urgently needed to be taken into account: water usage, waste and chemical flow, climate change stress, biodiversity loss, and social welfare. The article included references to both the UN Sustainable Development Goals and the Global Footprint Network, noting the 8 billion people or more who would need clothes by 2030, and one unifying reason to take action: life (Ward, 2020). One fashion designer/producer featured in the same issue recommended that customers buy upcycled, reused, or completely biodegradable, organic clothing; she said her company had been using deadstock fabric that would otherwise be discarded, but wanted to move toward 'use of materials that could return to the ground as food for the earth and not as poison.' She acknowledged that sustainability could be used as a marketing tool without regulations, monitoring, and enforcement: 'There must be regulations and laws put in place, and there needs to be someone who comes in and checks the facts' (Taymor, 2020). Another designer, noting that 'The world is bursting with stuff,' recommended that in a globalized world with limited resources, we needed to redefine luxury, placing value on heritage, tradition, craft, and quality (Goutos, 2020).

A recent article in *Harper's Bazaar* acknowledged that Americans throw away an estimated 16.9 million tons of textile waste each year; buy five times as much clothing as they did 40 years ago, and the percentage of a family's budget spent on clothing has declined from approximately 12 percent to about 3%: 'we buy more, but not better' (Omerod, 2020). There is increasing recognition that radical changes in the textile and fashion industries will require the support of consumers, whose purchases currently support a take-make-dispose model, consuming finite resources to produce textiles that are unused or used for a short time, then sent to a landfill or incinerated.

The take-make-dispose model also applies more broadly to our relentless pursuit of new/better everything. A realistic strategy for changing the consumer behavior that drives much corporate production would require some insight into our collective lust for physical possessions. It has been suggested that compulsive acquisition and rapid wasting are stimulated by advertising that convinces consumers that every level of Maslow's Hierarchy of Needs (physiological, love/belonging, esteem, and self-actualization) can be achieved through purchase of material goods (de Botton, 2020). Peter Wenz has proposed, as an alternate to Maslow's Hierarchy of Needs, Wenz's Lowerarchy of Worry: 'When one source of worry is put to rest by an appropriate purchase, some matter less inherently or obviously worrisome takes its place as the focus of concern. … Toxic wastes are produced in this context' (Wenz, 2004).

Historically, we have relied on consumption of goods to generate the value flow associated with capitalism. Services, however socially valuable, are sometimes treated as a lower level of financial transaction. The salaries of grade school teachers in most countries would be one illustration of this point. One possibility would be to assign more value to production of services directed toward goals (such as education and healthy communities) that more directly address our higher-level needs (de Botton, 2020).

Another possibility is that consumers could reuse and/or redistribute the goods they purchase, and purchase fewer, more valuable products with a longer use-span. Individual behavior change requires no legislation or advanced technology (as witnessed by the saying passed down from the Great Depression: 'Use up, wear out, make do, or do without.')

It is also possible for us to directly engage with industrial firms in our communities, to help them understand how their goals relate to those of the community. Individuals working within companies can point to how community-engagement can serve the company's best interests; businesses that support and enable community volunteer work in the communities where they are located get the benefit of less rigid community barriers to their own plans (Chesnutt, 2020). In summary, any place where we appear in the picture could be a location for positive action.

16.5 HOW CAN WE COMMUNICATE MORE EFFECTIVELY?

Regardless of our location, our efforts will bear more fruit if we are able to communicate effectively with people whose perspectives differ from our own. One challenge is to convince a target audience that your assessment is correct; the second is to persuade them to care. Advocates for better practices often rely on statistics and science to make their case for change, using numbers, acronyms, charts, and graphs to communicate with stakeholders who would prefer a trip to the dentist: Sometimes we provide information that is incomprehensible to its audience, fails to address valid concerns (e.g., immediate economic needs of low-resource families and businesses), and displays inadequate listening skills, questionable timing (asking for input when it is too late for input to have an effect), and unwillingness to acknowledge or express values.

Although we've learned to think in numbers, human beings have a much deeper history of being taught and moved to action by images, emotions, aesthetics, and cultural values. This is particularly true in relation to water, which has inspired art, architecture, poetry, music, and dance since prehistoric times. As Priscolli and Llamas have observed, 'Water is one of our enduring human symbols of life, regeneration, purity, and hope. It is one of our potent links with the sacred, with nature, and with our cultural inheritance' (Priscolli & Llamas, 2001). An example is the condomblé religion, built on Yoruba religious traditions brought from

Figure 16.2 Images of comdomblé deities: Left–Yemanjá, orixá of the ocean; Center-—Oxun, orixá of freshwater (center), and Right–Shangó, orixá of rain and thunder. (*Credits: Left and Center images courtesy of Brasarte and Lee Davis; Right image courtesy of Lee Davis*)

West Africa to Brazil (Bastide, 1978). As explained in The African Religions of Brazil, 'the blacks, herded into the slave ships, could bring nothing with them except their cultural values.' The condomblé religion includes multiple orixás (deities) associated with water. As illustrated in Figure 16.2, these include an orixá of the sea (Yemanjá); an orixá of fresh water (Oxun); and Shangó, an orixá associated with rain, lightning, and thunder. These entities are still honored in Brazil with song, dance, and gifts.

The Hindu tradition considers water a medium of purification and a source of energy, while the Islamic Shariah reflects the principle that water is proof of God's existence, proof of God's care, and proof of resurrection, as water restores life every day' (Priscolli & Llamas, 2001). Many religious traditions have used baptism and other forms of cleansing to symbolize redemption and renewal using words such as these from a Christian hymn book: 'Now wash me, and I shall be whiter than snow' (Nicholson, 1938).

In the Basilica Cistern constructed in Istanbul in the sixth Century by Emperor Justinian, the 336 columns with Doric and Corinthian capitals supporting the reservoir were so stunning that the Turkish name for the reservoir was Yerebatan Saray or 'Underground Palace.' The Mae d'Agua das Amoreiras Reservoir, completed in Lisbon in 1834, was constructed with a 'waterfall' where water cascaded into the storage basin. The building's layout suggests the design plan of a church, and EPAL, the utility that owns the reservoir and the associated Water Museum, describes this as consistent with 'the holy nature of the space' (EPAL, 2020). EPAL uses the site for art exhibitions, concerts, ballets, and theatre performance that express love for water. In 1910, San Franciscans constructed the Sunol Water Temple (Figure 16.3) to celebrate a significant point in their water transmission system.

Unfortunately, this rich cultural inheritance has been largely ignored in our efforts to speak effectively about water. We have been more inclined to rely on the language of scientists, economists, academicians, and engineers, languages not designed to connect with the emotional component of human decision-making. As an expert on water-related communications has observed, 'Public relations practitioners know that in order to drive change, you have to reach people on an emotional level, whether you are communicating with public officials or the general public. The key is to craft a narrative that identifies long-term solutions and appeals to their self-interest' (Lauppe-Rhodes, 2020).

Figure 16.3 Sunol Water Temple, constructed by Spring Valley Water Company in 1910 and restored by the San Francisco Public Utilities Commission in 2002. (*Photo courtesy of the San Francisco Public Utilities Commission*)

If we want stakeholders from many walks of life to feel moved to protect water, we need to expand our communication skills – learning to use tools that are relatively new (e.g., social media) as well as time-tested forms of communication with which we are unfamiliar. The capacity for social media (e.g., Twitter, Snapchat, Facebook, Reddit, Twitch, Nextdoor, TikTok) to change attitudes was recently demonstrated with respect to race relations in the United States. Statistics on US racial inequity have been widely circulated for decades, but on May 25, 2020 a Minneapolis teenager posted a video of a black man, George Floyd, dying while a policeman knelt on his neck, for over eight minutes. This video provoked a reaction among both whites and people of color even more visceral than the outcry at news photos 70 years earlier showing civil rights protestors attacked with dogs and firehoses by southern sheriffs. National polls now reflect a significant increase in the percentage of Americans who view racial inequality as a serious problem, as marches and demonstrations have erupted in cities across the country. The message has even spread into my predominantly white neighborhood where I now see slogans like 'Resist Racism' and 'Black Lives Matter' (often including the symbol of a fist) printed on hand-made posters attached to palatial homes and artfully scrawled on sidewalks in pastel chalk. Portraits of victims of racism have been set up in several well-tended yards; I saw one neighbor precariously perched on a chair carefully hanging paper tags on the limbs of a tree in her front yard, each tag bearing the name of a black person wrongly killed by police.

While the focus of this learning has been racial injustice rather than damage to the environment, it illustrates the power of social media to change minds by widely and rapidly distributing clear visual evidence of a true, terrible story. While few images have the emotional impact of a live murder, everyone with a mobile phone can now bear witness by photographing the serious damage inflicted on water, people, and wildlife by irresponsible water use. Full utilization of this tool will require transmission not only of images, but also of messages that convey the global and long-term implications of local water misuse, but there is no doubt that transmitting that information on social media platforms has the potential to inspire unknown others elsewhere on the globe to empathize with and respond to problems that will ultimately affect us all.

The speed with which information can be disseminated through social media contrasts with the historically slow dissemination of technical knowledge into the business world. In the race against population growth and climate change, it seems unlikely that we will change current practices in time if we limit our communication to professional conferences and academic journals. The conventional communication methods of the water industry have merit, but we need to reinforce them with older traditional modes (stories, visuals, images, and symbols) as well as new online tools.

16.6 WHAT WILL WE LEARN FROM THE CORONAVIRUS PANDEMIC?

Since it can seem easier to be considerate of others when we feel comfortable with our own situations, there is a risk that the economic losses sustained during the pandemic will have an adverse impact on sustainability efforts. I do not expect, in my lifetime, to see a reflowering of the optimism about easy answers that flourished in my youth. The excess resources that allow generosity without sacrifice will almost certainly be reduced for some time.

On the other hand, the sudden, jarring shifts we have experienced in the world as we knew it will allow us to reimagine things that once seemed fixed. For example, the shelter-in-place orders that required many office workers to work remotely could have a long-term impact on our assumptions about work-space, with consequent effects on traffic flows, public transit use, and air pollution.

The incredible reach and speed of an international pandemic has illustrated our vulnerability to the afflictions of others. As Albert Camus (1948) noted in *The Plague*: 'This had a lessons for us all… there is no island of escape in time of plague.' While clean water and sanitation goals associated with the United Nations Sustainable Development Goals may have seemed a matter of altruism to citizens of affluent countries, the pandemic has demonstrated our own vulnerability to the suffering of others. In the San Francisco Bay Area, long characterized by a combination of affluence, high housing costs, and significant homeless populations, one pandemic response supported by the State of California was temporarily housing residents of homeless encampments in hotels at government expense. Unsanitary conditions endured by the homeless suddenly posed a frightening risk not only to them, but also their better-housed neighbors. Similarly, lack of sufficient water for basic hygiene and sanitation could become a focus of more international investment because of its implications for international health. Realization of our interconnectedness is a lesson that is valuable but not new. Nearly 2000 years ago, the Roman emperor and Stoic philosopher Marcus Aurelius advised: 'Always think of the universe as one living organism, with a single substance and a single soul…Remark the intricacy of the skein, the complexity of the web' (Staniforth, 1964).

Our experience of the reality that some of our basic assumptions about how the world works can shift almost overnight could affect how we invest our resources in the future. From a clothing designer and manufacturer, 'It would be a real disservice if we didn't come out of this moment with a deeper level of thought as to what we do and why we make things and how we make them' (McCartney, 2020).

We tend to learn what we are ready to know, but we have shared an opportunity to learn about the human capacity for service. From Roberta Terzi, 66, Physician in the Infectious Disease Unit, Luigi Sacco Hospital, Milan: 'I went for a walk in the hospital's garden and took in the smell of it, and I said to myself, 'Do you really want to go into retirement to be with the grandchildren? On a day like this, where would you rather be?' And my answer to myself is: 'I want to be here, at the side of these patients. I want to heal these patients'' (Horowitz, 2020).

The lessons we learn will be influenced by what we hear, creating a responsibility for advocates of sustainable use of water to frame and communicate messages that are responsive to this moment and our hopes for the future.

16.7 CONCLUSION

So many different things happen simultaneously that we can find credible data to support either hope or despair. During a trip to Southern Ethiopia in 2014, I visited, in a single day, a government-supported research center (with non-functioning sanitation facilities) where academicians and testers did research on the coffee and grain species best suited to the local soil and climate; saw walking on the side of the road five stooped, blank-faced women in clothing worn to rags who were carrying heavy loads of brush for burning, contributing to the further deforestation of the land (Figure 16.4); and a Japanese-owned coffee plantation, located on a deeply rutted road, where the primary operator was proud to show me a conscientiously constructed facility for treating the effluent from coffee processing before it was released into the river.

Positive changes are occurring, but not necessarily at the level and speed required to counter-balance population growth and climate change. The options of hope and despair are also supplemented by

Figure 16.4 Statue of Ethiopian woman carrying brush on her back. (*Image courtesy of Ivan Dimas*)

Figure 16.5 'My Precious Water, I Kiss You' by Parkpoom Poompana, age 15 Ft. Myers, FL. USA. (*Credit: River of Words, 1996*)

outrage, which can distract us from our own culpability while serving the always-pleasing function of helping us feel virtuous.

The first thought that came into my mind when I saw the bent-over women in rags carrying brush on the road in Southern Ethiopia was that I should give away most of the clothes I owned and never buy more; this idea did not translate into action. Luckily, those of us who are unable or unwilling to do the very best thing can still do something. Doing what we can where we can, is rarely heroic or wholly successful, but the staggering number of serious problems in the world – one of the most dire being unsustainable use of water – means that there is never a shortage of opportunities to do useful work. Comparing what the world needs and what we have to offer, we can find tasks consistent with our interests, talents, and limitations. Being realistic about our own capabilities and motivations, as well as those of industry, we can, from wherever we are, help shape a world that protects water (Figure 16.5).

REFERENCES

Archambault D. (2019). IWA/IDB Innovation Conference, Quito, Ecuador, 2019. https://www.globalsustainablewater.org/uploads/8/2/0/3/82035866/03_archambaultdave.pdf (accessed October 5 2020).

Bastide R. (1978). The African Religions of Brazil, trans. Helen Sebba. The John Hopkins University Press, Baltimore, p. 36. English translation published in 1978.

Camus A. (1948). The Plague trans. Stuart Gilbert, Random House, New York, p. 228.

Chesnutt T. (2020). E-mail correspondence.

de Botton A. (2020). The School of Life: An Emotional Education. The School of Life Press, London, pp. 293, 244–249.

EPAL. (2020). https://www.epal.pt/EPAL/en/menu/water-museum/permanent-collection-and-associated-heritage/m%C3%A3e-d'%C3%A1gua-das-amoreiras-reservoir (accessed 1 June 2020).

Goutos C. (2020). quoted in Bellini, E. *et al.* 'The New Avant-Garde,' in *Vogue Italia* 'Talents' special issue, p. 15.

Greig and Rathjen. (2021). Promoting Sustainable Industrial Water Use: Scotland's Hydro Nation at Home and Abroad. In: Sustainable Industrial Water Use: Perspectives, Incentives, and Tools, C. Davis and E. Rosenblum (eds.), IWA Publishing, London.

Horowitz J. (2020). The Life and Death Shift (*New York Times*, April 12, 2020), p. 39.

Industrial Track, Innovation Conference. (2019). IWA/IDB Innovation Conference on Sustainable Use of Water: Cities, Industry, and Agriculture. https://www.globalsustainablewater.org/ (accessed 5 October 2020).
Latham R. (1985). The Shorter Pepys. Penguin Classics, London, p. 519.
Lauppe-Rhodes B. (2019). Email correspondence.
Londoño E. and Casado L. (2020). 'Bolsonaro Covets Wealth of the Indigenous Land,' (*New York Times*, April 19, 2020).
McCartney S. (2020). quoted in *Vogue* (US) 'Our Common Thread' special issue, p. 79.
Maennicke O. (2019). Personal notes. IWA/IDB Innovation Conference, Quito, Ecuador.
Marcus F. (2021). Musings of a Former Regulator: We Can Do Better in Davis and Rosenblum. In: Sustainable Industrial Water Use: Perspectives, Incentives, and Tools, C. Davis and E. Rosenblum (eds.), IWA Publishing, London.
Nicholson J. (1938). Whiter than snow. In: The Cokesbury Worship Hymnal, Whitmore and Smith (ed.), Abingdon-Cokesbury Press, New York, p. 146.
Offill J. (2020). Weather. Knopf, New York. Prologue.
Omerod K. (2020). How to Edit Your Closet (*Harper's Bazaar*, February, 2020), p. 56.
Priscolli J. and Llamas M. (2001). International Perspectives on Ethical Dilemmas in the Water Industry in Navigating Rough Waters: Ethical Issues in the Water Industry. American Water Works Association, Denver, pp. 58–60.
Scheyer E. (2019). Rattled by Vale Disaster, mining CEO's move to change industry (Reuters UK, February 26, 2019).
Silva de Sousa M., Kaiser A. and Pregaman P. (2019). 40 dead, many feared buried in mud after Brazil dam collapse. Associated Press (Online) January 26, 2019. https://apnews.com/article/963aab329c2b42aba9052852d0086831 (accessed 27 November 2020).
Staniforth M. (Translator). (1964). Meditations by Marcus Aurelius. Barnes and Noble Press, New York, p. 73.
Taymor H. (2020). quoted in Bellini, E. *et al.* 'The New Avant-Garde,' in *Vogue Italia* 'Talents' special issue, p. 14.
Ward M. (2020). 'Ten Reasons for Radical Change' Vogue Italia, 'Talents' special issue, p. 4.
Wenz P. (2004). Just garbage: environmental injustice. In: Environmental Ethics: Divergence and Convergence, 3rd edn, S. Armstrong and R. G. Botzler (eds.), McGraw-Hill, New York, p. 109.

Chapter 17

The economics of sustainable industrial water use, reuse, and the value of water

Thomas Chesnutt

A&N Technical Services, Encinitas, CA, USA

Keywords: best practices, business case, economics, enterprise value, externalities, industrial water use, sustainability, subsidies, triple bottom line, value of water

17.1 INTRODUCTION

This article describes some of the economic aspects of sustainable industrial water use often overlooked by water professionals and the general public alike. Most industrial plant managers understand the importance of water as an input to their industrial processes. Most utility managers – and elected officials – understand the importance of industrial plants to the local economy – tax receipts, jobs, and incomes. What does an understanding of economics add to this picture?

Economics has been described as the science of scarcity. In the ideal market (the one that only exists between the covers of an economics textbook) the price of a good mediates the relationship between the seller and the buyer; this is enshrined in the popular concept of 'supply and demand.' Going back a step further, the cost of production acts as a floor, determining the lowest price the supplier can charge for goods, while the value to the buyer dictates the highest price they will pay. Between these two poles, the market will set the actual price, which determines how many units are offered for sale and how many are bought and consumed (Renzetti, 2002).

But industrial water as a product is different from most other goods, including other commodities. The difference has a significant impact on how companies make decisions about water, and how they understand the value of sustainable industrial water to their enterprise. This article will explore the nature of this difference, and the economic forces that even now are inspiring some companies to use water more sustainably to ensure their future profitability.

DOI: 10.2166/9781789060676_0199

17.2 A UNIQUE MARKET

As a result of monopoly, subsidies, and externalities, industrial water is bought and sold quite differently than the goods in the ideal market. Where textbook transactions are mediated by the rational voice of price, in reality the industrial water use market is more likely to be a shouting match between the holders of 19th century water rights and the voters in the last election. Each of these factions alters the function of the market; together they make buying and selling water a unique departure from the classical pattern.

17.2.1 Monopoly

Water is the prototype of 'the naturally-occurring monopoly.' The reason for this is simple: It does not make sense for competing water utilities to build and maintain parallel pipelines in every street. Without competition, though, what is to prevent the monopolistic supplier from charging exorbitant fees for its product – especially a product as vital to human survival as water? As the economics textbook will further explain, monopolistic suppliers must be constrained by regulation, either formal or informal, to prevent predatory pricing (i.e., 'price-gouging'). So instead of allowing the market to determine prices, prices are set through negotiation in the public arena between the utility monopoly and the government authority responsible for approving rates. This regulatory process is often imperfect which results in water bills that don't reflect the actual cost of water or its value to the customer. As we'll discuss in more detail below, this causes a multitude of problems, especially when future water supplies are threatened by droughts or unknown viruses, water pollution, and over-allocation of resources. In the face of such future uncertainties, and without accurate price signals to guide us, it is no longer enough to merely do as we have always done before.

17.2.2 Social subsidies

The rates a private utility charges for water are commonly regulated by an independent board, while a public utility's rates may be set by an elected board or city council. In both cases, the regulatory board and governing body are subject to political pressures, either directly or indirectly. The utility may lobby the board to raise rates to cover operating and maintenance expenses and capital costs (and in the case of private utilities, to turn a profit). Ratepayers on the other hand (i.e., the public) may reward elected officials who keep rates low. One way to avoid rate increases – at least in the short run – is to put off maintenance and defer equipment replacement and other infrastructure investments. While customers may appreciate the lower rates, they don't see the corresponding reduction in water quality and reliability resulting from these deferred investments in water infrastructure. Another way to keep rates low is to attempt to find state or federal grants to pay for water improvement projects. These 'social subsidies' are common, complicating the determination of the full costs of water. The discrepancy between the price of water and its true economic value can be even further widened when utilities use historical cost accounting to determine water rate since future costs can differ from the past. Forward-looking cost accounting provides better cost signals for customers about the cost consequences of their consumption decisions. There are dangers to driving through life while staring in the rearview mirror (AWE, 2014).

17.2.3 Externalities

The market price for water rarely includes all its environmental and social costs. In economic terms, an 'externality' occurs whenever, during the normal course of its manufacture, distribution, and use, a product impacts a third party who was not part of the transaction between buyer and seller (Papandreou, 1998). Water pollution is the classic example cited in economics textbooks to illustrate the meaning of externality: when one town's water consumption results in wastewater that pollutes the water supply of downstream towns. The need to prevent such impacts (or pay to mitigate them) is one of the reasons economists point to in order to justify government regulation of the market. 'Life cycle' cost accounting is the attempt to account for all costs throughout a product's lifecycle – costs of production, consumption, and disposal.

17.3 EFFECT OF AN IMPERFECT WATER MARKET ON INDUSTRY

In short, the water market is imperfectly regulated. The utilities that supply water to industrial customers do so within the context of a variety of outside pressures – economic, political, and social – that determine how much they can charge for water, limiting their ability to control the quality and reliability of the product they sell. In the past industry was not seriously inconvenienced by these limitations, since in general water was plentiful and the waste discharge requirements were easy to meet. Over time, however, as the mandatory environmental flows were added to growing municipal and agricultural water demand and wastewater regulations ratcheted downward, water and wastewater services began to struggle to meet industry requirements, and companies sometimes found themselves needing higher quality and more reliable water than utilities could provide.

At that point, it is often economically advantageous for a company to invest in the treatment and reuse of its own wastewater. This is the overall context in which a business case for onsite industrial water reuse can be created. Economically speaking, the forces that drive a consumer to become a self-producer of the needed service breaks the presumed market monopoly of the networked utility. Using jargon from the utility perspective, this is termed 'system-bypass' and is viewed negatively by utility planners desiring a simpler planning problem. From the customer or market perspective, self-provision can be a rational lower cost approach to achieving desired service levels.

17.4 MAKING A BUSINESS CASE FOR SUSTAINABLE WATER USE

Company managers looking to improve the quality and reliability of their industrial water supply must typically develop a business case to invest in water reuse. A business case compares the investment costs with expected benefits, and projects are favored when benefits exceed costs. By creating a business case, facility managers can increase their understanding of the challenges, costs, and advantages of projects. It creates a transparent, common ground from which the company can reach a 'go/no go' decision by placing a value on benefits that are hard to quantify, including avoided cost of service interruptions, reputational risk, internal capacity building, and other factors.

Construction of a business case for industrial water reuse can be hampered, however, by the fact that the 'true cost of water' – that is inclusive of all lifecycle costs and exclusive of subsidies and externalities – is rarely reflected in its price. As discussed above, the price of water is depressed by embedded subsidies and deferred maintenance, and the 'market value' of water reuse does not include the external benefit of reducing wastewater discharge. Many managers mistakenly assume that non-monetary factors should not enter into the decision ('If you cannot count it, it does not count'). As a result, most business cases tend to weigh financial impacts more heavily than equally important considerations such as influence on the local economy, society, and environment; public health effects; shareholder value; guaranteeing worker safety; reputational risk; and supply chain reliability.

Two strategies that often help tip the decision in favor of reuse are 1) to include the value of avoiding a water supply interruption and 2) consideration of the social and environmental benefits of reuse by internalizing or monetizing these externalities.

17.4.1 Best practices for making a business case

Since many of the benefits related to water reuse are diffuse (spread out), the best practice is to start with identification of costs and benefits. All costs and benefits can be named and qualitatively discussed, even before any attempt is made to assign a dollar value to them (Hein *et al.*, 2020). Since business decision-makers, in general, are very averse to any threat to existing production processes, some important benefits are related to reducing the risk posed by unpredictable, hard-to-control costs. For example, interesting, successful business cases have been made for reuse projects on the basis of reduced risk of supply disruption.

It has also been noted that recently water and sewer rates in the United States have been increasing at more than twice the general rate of inflation (customer expenditure data from the Bureau of Labor Statistics, 2019). To reflect the full benefit of water reuse, then, a business plan can also include the avoided cost of future rate hikes, as well as energy costs, staff time, and service provider fees (e.g., cooling towers operation) resulting from reduced water use. It can also address the additional avoided costs of damage to existing assets, reputational degradation, and the benefit of improved operational insights.

17.4.2 Sustainable profits

Another major development in economic practice is to include in the business case factors related to reputation degradation and operational insights. We note that evolutionary economics research explicitly ties sustainability and survivability to organizational culture and leadership (Sonntag, 2018). While most competitive industries are appropriately 'cost-phobic,' even within a purist's understanding of profit as 'the only valid objective for an ethical business decision-maker' (Friedman, 1970), an economic case can still be made for industrial reuse based on the current trend among customers demanding 'green' products. Sustainability initiatives can be an important 'signal,' to use a term from information economics (Spence, 2002), to both investors and buyers. Recognizing this, many public agencies have developed programs to recognize sustainability achievements by companies in their service areas to provide further incentive to invest in water and energy conservation projects (Figure 17.1). The market success of 'sustainability' exchange traded funds demonstrates this point, as investment giant Blackrock wrote recently to its investors: 'Our investment conviction is that sustainability- and climate-integrated portfolios can provide better risk-adjusted returns to investors' (Gilbert, 2020). Within this understanding, implementation of industrial reuse can be both rational and ethical for profit-maximizers.

Furthermore, a recent survey by the Environmental Defense Fund (EDF) shows that emerging technologies now allow companies to implement on-site treatment and reuse to raise the bar on sustainability performance, and reports that in the last five years 72% of business leaders see a greater alignment between business and environmental goals (EDF, 2019). In short, competition now embraces sustainability, and agile business leaders are leading the way.

17.5 THE INDUSTRIAL DECISION-MAKERS' PERSPECTIVE

Decision-makers for industry today frequently embrace the 'triple bottom line approach' to accounting, which includes a metric for social and environmental benefits. Introduced in 1994 by John Elkington, triple bottom line accounting was applied to the issue of measuring three types of factors: financial, societal, and environmental (Elkington, 2018). Standards for the triple bottom lines have been developed by the Global Reporting Initiative (GRI) and can be found at their website: www.globalreporting.org (GRI, 2019). A comparable approach expands the non-economic elements into Environmental, Social, and Governance factors (ESG), where 'governance' relates to corporate ethics, board diversity and

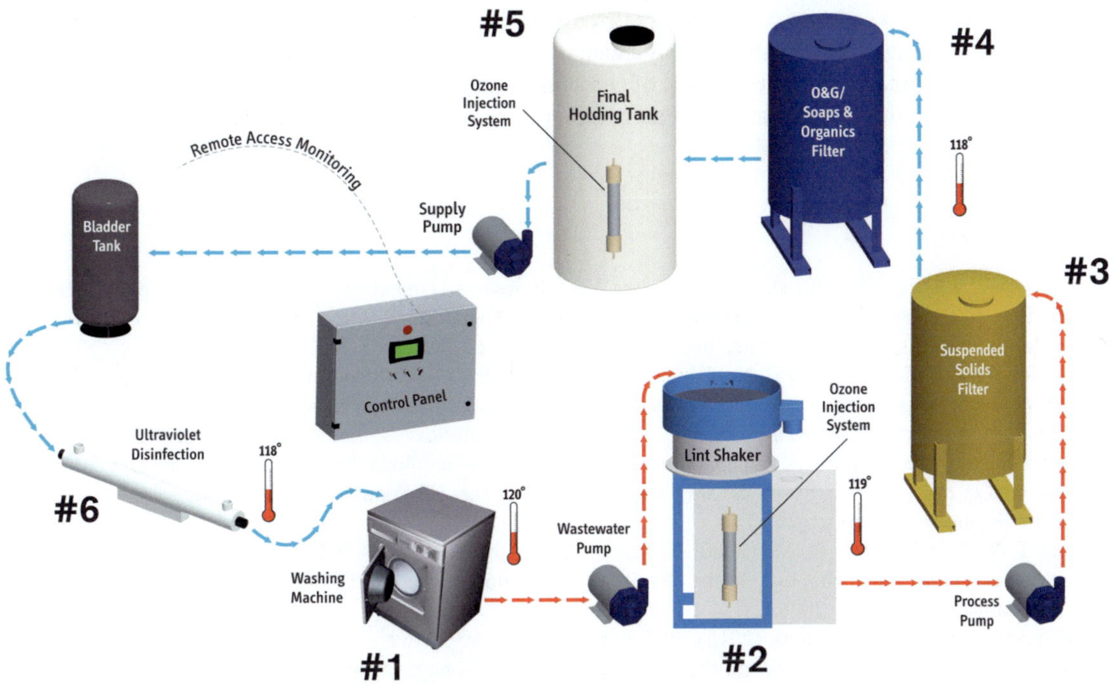

Figure 17.1 A treatment system to reuse laundry water at the 1000-room Westin Bonaventure Hotel & Suites was part of an energy and water recovery project recognized as 'Water Efficiency Project of the Year' by the Los Angeles 'Better Buildings Challenge,' co-sponsored by the LA Department of Water and Power. (*Source: AquaRecycle*)

composition, shareholder rights, supply chain engagement, and transparency. For public companies, there are sustainability disclosures that feed the Dow Jones Sustainability Indices (Robeco, 2020) and a recent application by the US WateReuse Foundation supports 'triple bottom line' analysis of alternative water supply projects (Stanford, 2016).

Company-generated social benefits can include community employment/income, effects on human health, labor relations, and drought resilience (USEPA, 2020). Company generated environmental benefits can include reductions in the pollution of ecosystems (measured by reduced eutrophication potential or ecotoxicity potential) or to the carbon footprint (avoided greenhouse gas emissions), and preservation of a natural resource (USEPA, 2014). Meanwhile, industrial firms that ignore sustainability can make themselves less competitive by missing strategic opportunities and overlooking risks material to their company's ongoing performance and enterprise value (AWE, 2015).

17.6 CONCLUSION

Effective business cases need to qualitatively identify effects of a water reuse project across the triple bottom lines (the company, the community/society, and the environment). The heart of the issue for many potential industrial reuse projects will be that the project investment costs are direct, upfront, and obvious. As noted, the reuse project's benefits may be diffuse, difficult to quantify, and spread over time (Chesnutt and

Pekelney, 2006). Value-adding firms have realized that their firm's competencies provide a competitive advantage in creating a sustainable supply-chain that also adds value to their community and their (operating) environment. Recent events of unpredictable uncertainties, such as pandemic-induced disruptions, point to the values that firms can add by means of sustainable industrial water use.

REFERENCES

Alliance for Water Efficiency. (2014). Building Better Water Rates in an Uncertain World, A Water Rates Handbook for AWE as part of the Financing Sustainable Water project, August 2014. http://www.financingsustainablewater.org/tools/building-better-water-rates-uncertain-world (accessed October 5 2020).

Alliance for Water Efficiency. (2015). Net Blue: Water Offset Policies for Water-Neutral Community Growth. https://www.allianceforwaterefficiency.org/resources/topic/net-blue-supporting-water-neutral-growth.One innovation being tested in multiple water-scarce North American communities is implementation of 'Water Neutral' developments where industrial expansion in water-scarce areas is allowed if the new development 'offsets' its incremental demand load by purchasing water savings in the community or off-site water reuse (including load-reducing stormwater capture)

Chesnutt T. W. and Pekelney D. M. (2006). A Review of Planning Methods and Tools Potentially Applicable for Advanced Treatment Technology in the Net New Water Supply Study (NEWAS), a report for the U.S. Bureau of Reclamation, February 2006.

EDF (Environmental Defense Fund). (2019). Business and the Fourth Wave of Environmentalism. https://www.edf.org/sites/default/files/Business-and-the-Fourth-Wave-of-Environmentalism_2019.pdf (accessed October 5 2020).

Elkington J. (2018). '25 Years Ago I Coined the Phrase 'Triple Bottom Line.' Here's Why It's Time to Rethink It.' Harvard Business Review, June 25, 2018. https://hbr.org/2018/06/25-years-ago-i-coined-the-phrase-triple-bottom-line-heres-why-im-giving-up-on-it (accessed October 5 2020).
Most recently, Elkington asserts that TBL has been turned into an accounting exercise that narrows the economic bottom line to a financial analysis and omits the rethinking necessary for a healthier capitalist economy. 'But the TBL wasn't designed to be just an accounting tool. It was supposed to provoke deeper thinking about capitalism and its future …'

Friedman M. (1970). 'The Social Responsibility of Business is to Increase its Profits', op-editorial in *The New York Times*, Sept. 13, 1970.

Gilbert M. (2020). 'Blackrock just became a champion of green investing, but its activism has limits.' Bloomberg News, reported *in Los Angeles Times*, 1/14/2020. https://www.latimes.com/business/story/2020-01-14/blackrock-climate-activism (accessed March 10 2020).

Global Reporting Initiative. (2020). 'GRI (Global Reporting Initiative) is an independent, international organization that helps businesses and other organizations take responsibility for their impacts, by providing them with the global common language to communicate those impacts. GRI provides the world's most widely used standards for sustainability reporting – the GRI Standards. https://www.globalreporting.org.

Hein L, Bagstad K. J., Obst C., Edens B., Schenau S., Castillo G. *et al.* (2020). 'Progress in natural capital accounting for ecosystems'. *Science*, **367**(6477), 514–515. doi: 10.1126/science.aaz8901

Papandreou A. A. (1998). Externalities and Institutions. Clarendon (Clarendon paperbacks), Oxford. Papandreou makes the forceful point that the entire concept of externalities has no meaning without an understanding of the institutions involved.

Renzetti S. (2002). The Economics of Water Demands. Kluwer Academic Publishers, Norwell MA. For an intermediate-level introduction to the economics of industrial water demand, see chapter 4; and for a discussion of the demand for sewer disposal and waste assimilation, see chapter 6.

Robeco, Inc. (2020). 'Dow Jones Sustainability Indices.' https://www.spglobal.com/esg/csa/indices (accessed October 5, 2020).

Spence M. (2002). 'Signaling in retrospect and the informational structure of markets.' *American Economic Review*, **92** (3), 434–459. doi: 10.1257/00028280260136200

Sonntag M. (2018). The biological foundation of an evolutionary economy and its implications for organizational culture and leadership: A new framework for strategic decision-making. In Positive Impact Investing. A Sustainable Bridge Between Strategy, Innovation, Change and Learning, K. Wendt (ed.), Cham, Switzerland: Springer Nature: Springer (Sustainable finance), pp. 231–255. https://doi.org/10.1007/978-3-319-10118-7_12

Stanford B. (2016). 'Methodology for a Comprehensive Analysis (Triple Bottom Line) of Alternative Water Supply Projects Compared to Direct Potable Reuse' WateReuse Research Foundation Report No 14-03 (Alexandria: 2016)

USEPA. (2014). An Introduction to Environmental Accounting as a Business Management Tool: Key Concepts and Terms. https://www.epa.gov/sites/production/files/2014-01/documents/busmgt.pdf (accessed October 5 2020).

USEPA. (2020). Full Cost Accounting Website https://archive.epa.gov/wastes/conserve/tools/fca/web/html/questions.html (accessed February 18 2020).

Chapter 18

Development finance: Encouraging sustainable water use by industry

Oliver Maennicke[1] *and Jens Honerhoff*[2]

[1]*Independent Consultant, Vienna, Austria*
[2]*German Development Finance Institution, Cologne, Germany*

Keywords: agriculture, Deutsche Investitions- und Entwicklungsgesellschaft mbH (DEG), ESG, pulp & paper, responsible investment, small and medium-sized enterprises (SMEs), stakeholder engagement, sustainable finance, water risk, water stewardship

18.1 THE ROLE OF DEVELOPMENT FINANCE

Development finance serves as a catalyst to attract and mobilize other sources of private capital, in areas where a higher level of risk is inhibiting private investment. Development Finance Institutions (DFIs) such as the German Development Finance Institution (DEG) lower the burden and risk associated with businesses, in order to attract private investors. DEG provides private enterprises operating in developing and emerging markets countries with tailored solutions, including long-term financing, promotional programmes, and individual advice. One special focus is small and medium-sized enterprises (SMEs) in those countries.

SMEs are valuable drivers of innovation, employment, and growth, but their financial needs are often too large for traditional lenders, such as local banks. Large banks tend to bypass the SME market due to the amount of administrative work required in order to work effectively with them, the limited information available from SMEs to allow thorough due diligence on the risk associated with the investment, and uncertainty regarding their credit risk. In many developing countries, long-term bond markets and other alternatives for innovation funding are not available as potential sources of funding.

DFIs such as DEG provide options to bridge these gaps, drawing on low risk capital from their own financial base. In doing so, they have a great responsibility towards their stakeholders as well as an opportunity to act based on best practices for sustainable use of water. The compliance with international

DOI: 10.2166/9781789060676_0207

environmental and social standards and supporting the clients in achieving sustainable business practices is part of the cooperation with the companies. DEG has to ensure that investments are consistent with sustainable development and DEG's Environmental and Social Standards that are based on internationally agreed on principles and standards.

DEG aspires to not only support but also act upon latest principles and practices for sustainable investment – 'walk the talk' so to speak, as an insurance and credibility mechanism for their clients and with stakeholders respectively.

18.2 SUSTAINABILITY IS RISK MITIGATION

Given their characteristics (see box), DEG is well equipped to attract higher risk and long-term investments with businesses that are prepared to change and willing to practice innovation and improvement, envisioning the opportunity to strengthen and building resilience for their businesses and become more attractive to private investors in their own region in the medium and long term. DEG can strengthen companies that are crucial in the supply chain of German businesses. Customer success stories show that DEG is 'walking the talk' on sustainable investment, producing results that can be multiplied and scaled.

KFW DEG

DEG (https://www.deginvest.de/International-financing/DEG/) is one of the major development finance institutions for private companies. DEG is a wholly owned subsidiary of KfW Group, the German development bank for reconstruction. What today is called DEG was founded as the 'German Association for Economic Cooperation' in 1962. It provides companies in developing and emerging markets with long-term investment capital in form of loans or equity and advice and works with customers to create investments and companies professionally, efficiently, and sustainably. Further, DEG provides funding to financial institutions and funds that provide small and medium-sized enterprises with reliable access to debt and equity financing.

DEG, through KfW, operates as a signatory to the United Nations Principles of Responsible Investment (UN PRI), with the Investment Exclusion criteria of the Exclusion list, integration of ESG-criteria (Environmental, Social, Governance) and promotes their programmes with companies in the developing and emerging-market countries. Due to these institutional commitments, the sustainable use of water by client companies is a key factor of success that encourages development of successful models that can be replicated by others (Further information on the United Nations Principles of Responsible Investment is available at program is available at https://www.unpri.org/pri/an-introduction-to-responsible-investment/what-are-the-principles-for-responsible-investment.).

In addition to two investment cases presented in the main text, DEG has provided Business Support Services (BBS) funding and continues to provide expertise to co-develop a water risk assessment tool for businesses, the WWF-DEG Water Risk Filter launched in 2012 (described in Section III of this book). This tool has subsequently been applied to operations and asset and production portfolios by DEG, their clients, and many other businesses and organizations. In addition, the tool is used directly with clients as a business support, water risk mitigation tool, and decision-support-system.

DEG is engaged specifically on water-related topics with various clients around the world, providing financing solutions for improvement measures, business support services, and staff training.

DEG in particular and DFIs more broadly have various tools at hand that can be put into action in order to achieve impact, which include the following:

Lending: the activity of lending money to people and organizations which they pay back with interest (Cambridge Dictionary, 2020);

Promotion: activity of promoting best practices and state-of-the-art thinking on environmental, social, and governance issues;

Training support: activity in providing financial, content advice, establishing networks, or direct training to (prospect) clients and client stakeholder groups in sustainable practices;

Due diligence practices: the detailed examination of a company and its financial records, done before becoming involved in a business arrangement with it (Cambridge Dictionary, 2020) – in particular including sustainability elements on environmental, social, and governance issues; and

Technical assistance/business support services. For example, DEG offers non-repayable co-financing and conceptual assistance for eligible projects that help companies (prospective and existing customers) to improve their performance, growth, and developmental impact. DEG connects companies with external experts to realize tailor-made advisory solutions and help customers to design coherent development projects. Measures that support investments contribute to advancing developmental broad-based and structural effects and help further professionalization.

The above mentioned tools are all in the scope to achieve investments in/ with renewable energy, climate resilient agriculture, low-emission manufacturing and green lending at financial institutions and can also be derived from and/or founded on the Commitment and Principles for Responsible Investment (United Nations Principles of Responsible Investment), a set of aspirational investment principles that incorporate **e**nvironmental, **s**ocial, and **g**overnance (ESG) issues into investment practice.

18.2.1 How development finance helps encourage sustainable practices

This type of financing support is key to setting precedents in developing and emerging countries, helping to build confidence among SMEs that it is possible to build thriving businesses while using sustainable business practices, bringing opportunities and positive financial impacts to local economies in developing nations. DEG provides business support to its clients to co-finance improvement measures and sector initiatives to improve cooperation between the private and public sector.

Through DEG's long-term private sector investment with companies that share DEG's take on sustainability practices, the long-term capacity of companies is enhanced (e.g., through education, training, technology transfer, and know-how). This expands the value chain in the supply of goods in the community, contributing to the creation of decent local employment, government revenue, and foreign exchange earnings that can be invested into the community's infrastructure, education, and health – contributing to sustainable development.

For DEG, investing in local business in developing countries which are suppliers for German businesses has a ripple effect, ultimately strengthening German companies and German's economy as a whole, and improving their resilience and lowering economic risks.

18.2.2 Why water matters – a company's perspective

Water, lack thereof or over-abundance (e.g., flooding), is an important input as well as external factor. Lack of local infrastructure for water supply, distribution, and treatment is often a key issue in developing and emerging countries, as are water-related ecological issues.

Investment decisions are driven by prioritizations of sustainability issues to be addressed in each investment case. Having water that is sufficient in terms of both quantity and quality is key for production. There might be competing demands with other water users, paired with weak governance and/or enforcement. Even when water is quantitatively abundant, it may be insufficient due to quality problems, where it is not feasible or economically viable to treat the water to the quality parameters required. In many emerging and developing countries, more than 80% of water is returned untreated, and therefore polluted, into natural water bodies, causing water scarcity due to pollution.

If a company can prove economic impact in relation to their water management in their basin, it might influence stakeholders, in particular upstream, to change their habits on managing surface- and groundwater and environmental contaminants, such as chemicals. But how can the habits of upstream users be changed? A company needs to set an example by committing and investing in their own improvement. However, providing funding for such ventures can be difficult and a relatively high risk, particularly if funding does not come along with ongoing education and training so that the employees will have the motivation, knowledge, and skills to maintain new approaches, infrastructure, and technologies.

Some of the companies that DEG invests in and works with, produce goods for the import market of economically stronger countries. Consumer and customers in these countries, and therefore the retailers in the developed countries that import supplies, are becoming more conscious of sustainability issues, and increasingly likely to demand action based on sustainable principles at the origin of the product – SMEs in emerging and developing countries.

These exporting businesses feel the customer pressure and are aware of issues on the ground. With the right mindset, recognizing the value of more sustainable practices, they can benefit from development finance both financially as well as through knowledge building – growing their expertise and becoming more resilient. Success stories showcase the validity of the business case of long-term development to grow strength and resilience, both in the local market and in similar markets globally. Through these mechanisms DEG can encourage and ensure impact on industrial water use and water management practices.

18.3 CASE STUDIES

18.3.1 Agriculture in Latin America

Peruvian agriculture is active in a desert-like environment on the Pacific coastline; agriculture here depends on water from the Andes for irrigation. DEG supported the evaluation of the costs of water risks for Peruvian agricultural companies. These analyses highlighted the need for long-term investments in (1) preventive measures to reduce the impacts of extreme weather events (e.g., reservoirs as an emergency backup water supply); (2) collaborative action by all companies in the irrigation area to improve the catchment (e.g., reforestation); and (3) control of groundwater levels (e.g., more efficient drip irrigation techniques). Through more large-scale measures in the basin, new markets could be developed.

European importers/supermarket chains look into the water management of their suppliers. Some (e.g., EDEKA, the largest German grocery chain) strongly encourage or demand certification for sustainable water management (Alliance for Water Stewardship, AWS) from those suppliers located in areas which are 'red -flagged' as high risk in a Water Risk Filter Assessment. DEG is a financing partner for companies who look for a sustainable business model, including sustainable water management where investments are supported by business support services and knowledge transfer. For example, DEG supports conducting necessary training, in addition to financing water management measures for the client.

18.3.2 Two investment examples in Peru (AWS, 2017)

Danper Trujillo S.A.C. produces asparagus, artichokes, peppers, mango, and other fruits. DEG provides training through using certified training providers for employees on sustainable water management and water stewardship based on the principles of the Alliance for Water Stewardship. In 2017, Danper used this knowledge to become the first company in Peru that obtained the 'Alliance for Water Stewardship' Certification. This certification validated their sustainable water use practices, which supported their goal to export more of their products to Europe.

Virú Group (KFW, 2018) produces and buys fruits and vegetables from smallholders and processes the products at three processing plants. Through DEG's Business Support Services DEG co-financed and assisted Virú in implementing a water efficiency management system that included planning, construction, and operation of a state-of-the-art wastewater treatment plant; this plant enables the reuse of wastewater, reducing the water footprint of the company.

18.3.3 Pulp and paper (P&P) in India (IFC, 2017)

India has a strong and growing pulp and paper market. A variety of different technological development levels exist within the sector in terms of water use and demand, chemicals used in processing and water treatment levels before using effluent as irrigation water (a common re-use of effluent from pulp and paper factories). Some plants in India are reported to use more than 200 cubic meters per metric ton of produced pulp and/ or paper, while global industrial best practice is in the range of 50 cubic meters per metric ton of produced pulp and/or paper (Figures 18.1 and 18.2).

Figure 18.1 Asparagus field in Peru where DEG-sponsored training in irrigation methods helped this enterprise acquire '*Alliance for Water Stewardship*' Certification. (*Credit: DEG*)

Figure 18.2 Drip irrigation system for asparagus. (*Credit: DEG*)

DEG, in connection with the International Finance Corporation (IFC), a member of the World Bank Group, and other lenders, invested in pulp and paper production in India (e.g., JK Paper Limited), where sustainable water management practices are part of the improvement measures. JK Paper is a leading Indian producer of office papers, packaging boards, printing and writing papers, and specialty papers. They have a sustainable, Forest Stewardship Council (FSC) controlled wood management supply chain of more than 1000 km^2 of forest from independent farmers to meet their virgin pulp demand. This means that forest operators need to avoid negative impacts on water quality quantity, and act when problems occur (FSC, 2020). Farmers are supported with the company-run Farm Forestry Program. In their plants, JK Paper has established high environmental standards and systems, producing within global best practices exemplified by a 50 m^3 water usage for the production of one metric ton of pulp or paper. Further, they are treating all their effluent water (i.e., waste water) to quality standards, so it can be used for agricultural irrigation or can be safely returned as treated effluent into the local river.

DEG has invested in capacity and efficiency improvements, which have resulted in implementation of better technology, process innovation, recycling, reuse, and minimization of wastewater discharges. These measures have helped significantly to reduce freshwater consumption and effluent generation per metric ton of paper in recent years, which also led to operational cost savings. Treated sewage water is used as cooling water, and a significant portion of the treated effluent is supplied to nearby villages for agricultural irrigation, which improves harvests, the local groundwater balance, and the relationship with communities (Figures 18.3 and 18.4).

18.4 KEY CHALLENGES

While the focus of this chapter is on improving water management activities, it is important to understand that the water issues addressed in the case studies above were identified in a prioritization process for the presented investment case. To evaluate lending to customers, DEG cannot afford to only look at a single issue such as water management, but requires thorough due diligence on all sustainability topics. It then becomes a challenge to prioritize the sustainability issues that have been identified (e.g., biodiversity in the supply chain versus the sustainable management of water resources).

Figure 18.3 Produced paper rolls at JK Paper plant in India where DEG in cooperation with IFC and others invested in Indian pulp and paper production. (*Credit: DEG*)

Often businesses, and particularly small and medium-sized enterprises, do not have sufficient information to identify, quantify, or compare the urgency of different sustainability issues. That makes it necessary to commission relevant intelligence – gathering activities to gain a full picture of

Figure 18.4 Environmental parameters of treated effluent are displayed publicly at the entrance gate. (*Credit: DEG*)

circumstances and build a base to make investment decisions that reflect sound data and can withstand careful scrutiny. While water issues are often present, in some cases there may be even more pressing issues that should be addressed first.

Another challenge is to not only initiate changes by providing financing, but to ensure progress is made on an ongoing basis, working sequentially on the list of identified priorities, or even changing priorities when needed, based on new knowledge or events.

18.5 THE ROAD AHEAD

While work with companies is important to strengthen businesses and make them more attractive to private investors and improve their water management, scaling impact is imperative to ensure broader impact on economies and the environment. DEG recognizes that cooperation between the private and public sector will be required to achieve this scaling impact effect on investments.

Larger investments and funding are at stake when the public sector is not adequately involved. Public sector policy is the basis of private sector action. Both the private and the public sector can benefit from cooperation, building confidence in each other through partnership that will support brining higher funding capacity to the economy.

Such public–private mechanisms are often used for infrastructure provisioning, such as water and sewerage systems. Clean water and sanitation services are often lacking in developing and emerging countries. We see examples of waste water being treated by a local company, fulfilling a service usually held by a local government entity.

Integrating better water specific approaches for sustainable business will show the business case for improvements to water balance; water quality; water governance; important water-related areas; and safe water, sanitation, and hygiene for all. These are all aspects of water stewardship, explained further in the chapter on the Alliance for Water Stewardship in Section III of this book. This water stewardship approach also shows the necessary upscaling from a facility to a river basin scale in order to spread the benefit of sustainable water use by industry to the wider community and the environment.

Water management improvement has to be integrated with other sustainability challenges and standards, such as climate change and land use. Since stakeholder-inclusive processes are at the base of many sustainability approaches, structures originally created to support more sustainable water management can be nurtured, further developed, and broadened to address additional United Nations Sustainable Development Goals, such as health and well-being, education, gender equality, and sustainable communities.

REFERENCES

Alliance for Water Stewardship. (2017). Alliance for Water Stewardship Assessment Report Prepared for DANPER TRUJILLO S.A.C. http://a4ws.org/wp-content/uploads/2017/10/DANPER-AWS-Report-Public-Report-2017.pdf (accessed 11 April 2020).

FSC (Forest Stewardship Council). (2020). International Generic Indicators Standard (STD) V (2-0) FSC-STD-60-004. https://fsc.org/en/document-centre/documents/resource/262 (accessed 11 April 2020).

IFC (International Finance Corporation). (2017). IFC PROJECT INFORMATION & DATA PORTAL: JK Paper III. https://disclosures.ifc.org/#/projectDetail/ESRS/39821 (accessed 11 April 2020).

KFW. (2018). Virú contributes to Peru's sustainable development (February 2018). https://www.deginvest.de/DEG-Documents-in-English/About-us/What-is-our-impact/EN_Case-Study_Viru_2018_final.pdf (accessed 11 April 2020).

Chapter 19

Reducing pollution from industrial wastewater in developing and emerging countries

Olivier Crespi Reghizzi[1]

Agence Française de Développement, Paris, France

Keywords: AFD, Agence Française de Developpement, biodiversity, climate change, France, Morocco, pollution, Sénégal, SDG, Sustainable Development Goals, wastewater, regulation

19.1 INTRODUCTION

Reducing pollution and contamination from industrial wastewater is a key step to achieving many of the UN Sustainable Development Goals (SDGs), including water-related and infrastructure goals (SDG 6 and SDG 9 Target 9.4) as well as targets focused on the preservation of the natural environment, biodiversity, and reduction of marine pollution (SDG Target 14.1 and SDG 15) (UN, 2020). Such a goal can be reached by reducing the quantity of chemicals used in the production process, by upgrading and modernizing the production process to use less water, and by investing in appropriate wastewater treatment technologies.

In many areas of the world, the projected impacts of climate change include a decrease in the quantity of water resources available and an enhanced water stress. In these cases, reducing the quantity of untreated wastewater discharged is also a key factor of adaptation to climate change as it will contribute to preserve the quantity of good quality water resources available.

While the industrial sector does not consume as much water as agriculture or domestic uses, industrial wastewater has a high concentration of contaminants and must be treated prior to discharge. Unfortunately, the World Water Assessment Programme estimates that 70% of industrial wastewater in developing countries is discharged without an appropriate treatment (UN, 2009). This includes both untreated industrial wastewater discharged directly to a receiving stream and industrial wastewater put into an urban sewer system not designed to treat it before discharge. In some countries, there are no laws

[1]The author wishes to thank Margot Moreau, Frederic Maurel, and Hassane Belguenani for their valuable revisions and inputs.

DOI: 10.2166/9781789060676_0215

or other regulatory tools to limit these industrial discharges; in others regulations exist but are not enforced. In both cases, the result is similar: industries do not receive sufficient incentives to invest in technology to improve the quality and reduce the quantity of wastewater discharged. This leads to severe negative impacts on water resources, on the biodiversity and on the natural environment.

The French Development Agency (*Agence Française de Developpement* or AFD) has provided economic and technical assistance in several countries to help address problems associated with industrial water pollution. The following chapter offers analysis of the role that AFD has played in Morocco and Sénégal with respect to reducing pollution from industrial water use.

19.2 AGENCE FRANÇAISE DE DÉVELOPPEMENT MISSION AND STRATEGY

The French Development Agency group (*Agence Française de Developpement* or AFD) is a bilateral development bank owned by the French state. AFD's mission consists in helping 'construct a world in common that preserves and protects five important common goods: people, planet, prosperity, peace and partnerships' (AFD, 2020a). To fulfill its mission the AFD group operates from its headquarters in Paris and from a network of more than 85 local offices throughout the world (AFD, 2020b). The AFD group (AFD and Proparco) (Proparco, 2020) finances projects and provides technical assistance in developing and emerging countries through grants and loans awarded to states, local authorities, and other government institutions, and state-owned companies and utilities, as well as private companies, banks, and financial institutions. Projects and activities financed by the AFD contribute to reach Sustainable Development Goals, and must be 100% compatible with the Paris Agreement, thus contributing either to the mitigation or adaptation to climate change.

19.3 THE ECONOMIC PERSPECTIVE: THE NEED FOR GOVERNMENT INTERVENTION AND INCENTIVE

The case of water pollution caused by an industry is analysed in environmental economics as an example of 'market failure due to an externality' (Hindriks and Myles, 2013). An externality is present in a transaction 'whenever one economic agent's welfare (utility or profit) is directly affected by the action of another agent' who is not a party (i.e., 'external') to the transaction. When an externality takes place, the perfect competitive market model breaks down, because the full costs of the transaction are not captured in the market, and the competitive equilibrium is inefficient. In economic language, a non-efficient outcome fails to maximize social surplus; in lay language, bad decisions are made that fail to consider 'external' costs like toxic streams, decimated fisheries, and polluted water supplies.

According to public finance literature, Rosen and Gayer (2010) a failure due to externalities like industrial water pollution can only be solved if government intervention induces all parties to include external effects in their decisions. Such an intervention can be based on three different tools: environmental regulations that prescribe the required behaviour; economic incentives (e.g., Pigouvian taxation); or encouragement of bargaining based on enforced property rights (Coase theorem). These three forms of policy inducement are explained in more detail below.

19.3.1 Prescriptive regulation

Many environmental regulations are prescriptive in nature ('command-and-control'), prescribing how much and what kind of pollution an individual source or plant is allowed to emit and/or what types of control equipment it must use to meet such requirements (EPA, 2014). Despite the introduction of potentially more cost-effective methods for regulating emissions, this type of regulation is still commonly used and is almost always available as a 'backstop' if other approaches do not achieve desired pollution limits.

19.3.2 Economic incentives and Pigouvian taxation

When the market inefficiency is due to externalities that result in a divergence between social and private benefits, taxes can force the company causing the impact to pay an amount equal to the marginal damage caused. Named for the British economist Arthur Pigou, Pigouvian taxation provides the theoretical background to justify the 'polluters pays principle,' according to which 'polluting entities should bear the costs of their pollution' (Mamlyuk, 2010). In theory, Pigouvian taxation would imply personalized prices on the offending party based on the impact they caused. While this is not often possible to implement in practice, some approaches approximate a variable taxation.

19.3.3 Coase theorem and bargaining

According to the Coase theorem (named for economist Ronald Harry Coase), in theory the allocation of resources will be efficient and invariant with respect to legal rules of entitlement. In other words, in a competitive economy with complete information and zero transaction costs, economic agents may resolve the problem of externalities themselves, without government intervention, provided that property rights are clearly defined. In that case, Coase argued, those affected by an externality – that is, those whose property rights are being violated – will find it in their interest to reach private agreements with those causing the problem to eliminate such a problem (market failure). In Coase's theory, the equilibrium obtained is invariant to how the property rights are assigned; if the polluters have the right to pollute, the 'pollutees' will pay them not to pollute, or if the pollutee have the right not to be polluted the polluters might pay them to accept their pollution. From a normative point of view, assigning rights to the pollutee makes Coase's approach compatible with the polluter pays principle.

19.4 FRANCE: A HYBRID SYSTEM FOR MANAGING INDUSTRIAL WASTEWATER

In France, the regulatory approach on industrial wastewater discharge combines prescriptive regulation (Legifrance, 2020a, 2020b) and ear-marked levies and incentives managed at the river basin scale. Industries must obtain an authorization to discharge from the local representative of the national ministry for environment. The administrative process is more or less complex depending on the classification of the industry and its impact on the environment (classification ICPE). Authorities set pollutant discharge limits level based on a number of factors, including the characteristics of the receiving water; the water quality objectives set by the river basin agency master plan (SDAGE); the best available treatment techniques; and the economic viability of treatment for the facility. Once authorized, the industrial plant must implement an autonomous monitoring and control system and allow inspections at regular intervals (SCE, 2015).

In addition, industrial entities discharging wastewater are subject to the fee mechanisms of the French river basin agencies. In 1964, France split jurisdiction of its waterways into six river basin areas, each with a Comité de basin (CB) representing water users and an executive body, the Agence de l'eau. Every five years the CB votes a water protection investment programme funded by levies on water abstractions and pollution discharges for all users category (farmers, industries, and domestic water users). Direct discharging industries pay a pollution levy based on the pollution discharged and on the water quality of the receiving body. For indirect discharge, industries pay for investments in wastewater treatment plus a levy for the modernization of the sewer system and a sanitation fee. On the other hand industries may receive some financial aids from the river basin agency both to finance feasibility studies and part of the investment costs.

From an operational point of view, with national legislative limits and discharge fees administered at the river basin level the French regulatory framework can be considered a hybrid approach. In fact, French industries are subject to a unique variety of institutional tools, combining prescriptive regulations, 'polluters pays' levies (the cornerstone of Pigouvian taxation), and fees agreed by all users at the river basin scale (suggestive of the Coase bargaining approach) (Barraqué, 2009).

19.5 MOROCCO: REGULATION OF WASTEWATER DISCHARGE BY INDUSTRY

In Morocco, more than 50% of the country's 8000 industrial dischargers are in the Casablanca-Kenitra area near the Atlantic coast, where most are located close to a sewerage system. Four industrial sectors with potential for water pollution include the food industry (organic pollution), chemical industry (organic and toxic pollution), textile and leather industry (toxic pollution mainly), and the mechanical, metal, and energy sectors (ASPA, 2016).

In 1995, Morocco made major changes to the water sector regulation through the creation of river basin agencies (*Agences de Bassin Hydraulique* or ABHs) responsible for collecting levies for water withdrawals and discharges to the inland water environment. Loosely modelled on the French system, wastewater dischargers in Morocco are subject both to a discharge authorization process and to a levy payment system. The various ABHs are responsible only for monitoring and charging industries that discharge directly into the inland water environment, while indirect discharging industries should pay a sanitation fee to the utility managing the sewerage system which is then responsible of paying a pollution levy to the ABH. The utility is also responsible for monitoring discharged wastewater quality, enforcing pretreatment requirements.

The effective implementation of this framework is still in progress. Indeed some of the administrative aspects of the 1995 Water Act were only published in 2006; limits on industrial discharges were approved as recently as 2013 for implementation in 2018, and limits for some specific industry sectors are still in progress (Département de l'eau, 2014). Sanitation utilities also face a number of challenges. For example, some existing sewerage system agreements are not sufficiently detailed for industrial customers who do not considered them binding and water utilities lack 'police power' to force compliance. In addition, some industrial customers withdraw water directly from groundwater resources without buying it from the water utility and evade paying an appropriate sanitation fee.

Despite these challenges, an even more ambitious water act was approved in 2016 which explicitly authorized ABH to give some financial assistance for industrial wastewater treatment investments. It also clarified that authorization of indirect discharge into sewers is subject to discharge limits (still to be approved by decree), and established fines for non-compliance with discharge limit values. The 2016 Water Act also gave police power to enforce industrial discharge compliance to employees in public institutions outside the ABH concerned, although the list of these institutions and of the procedure for awarding police power has yet to be established.

In addition to these regulatory tools, an industrial depollution fund (FODEP) was created in 1998, financed by German development institutions and in 2011 a second fund was underwritten by the European Union (MVDIH). In both cases these mechanisms allowed authorities to finance 40% of investment designed to reduce water pollution by industry. In all, nearly 130 projects have been financed through these mechanisms with a total project value of roughly 70M euros, 25M euros of which was received in the form of grants.

In Morocco in the last 20 years AFD has awarded more than 250M euros in loans to Moroccan water and sanitation utilities, including sewerage system and wastewater treatment plants (e.g., Nador, Agadir.) Since 2010, together with the German Development Bank (KfW), the European Investment Bank (EIB), and the European Union (EU), AFD has been the leading European organization contributing to Morocco's national utility (ONEE) for the National Sanitation Programme. In 2016, an EU grant managed by AFD has been awarded to support the ministry of internal affairs in improving the sustainability of the sanitation sector including concerning the regulation of industrial wastewater discharge in sewers.

In 2017, AFD and the European Investment Bank (EIB) approved two low-interest loans of 10M euros each to the Moroccan Bank of Africa (formerly BMCE) to finance a 'Blue Credit line' for investments in adaptation to climate change in the water sector. Among the eligible investments were investments to reduce industrial water consumption and treat wastewater before discharge. AFD and EIB are also financing technical assistance to help the bank analyse water-related projects, and to help industries prepare feasibility studies. The logic behind this operation is to incentive industries to reduce their water pollution thanks both to the technical assistance and to the low interest rate. It is too early to have results of this operation but we notice that step-by-step, some industries are willing to borrow through this facility in order to invest in technologies allowing them to reduce their wastewater pollution.

19.6 SENEGAL: REGULATING INDUSTRIAL WASTEWATER DISCHARGE IN THE HANN BAY AREA

In Sénégal, 80% of the industries are located around Hann Bay, a body of water roughly 20 km long located just outside the capital city Dakar. In the past, the Hann bay was well known as a resort, but nowadays it is one of the most polluted areas of the country. Untreated wastewater discharged by industries is one of the main causes of this environmental degradation, together with untreated domestic wastewater and stormwater drainage from this fast-growing urban area (Figure 19.1).

To improve the environmental situation in the bay area, the government of Sénégal and its state owned sanitation utility ONAS (Office National d'Assainissement du Sénégal) have launched the Bay area depollution project consisting in the construction of a 58-km sewer system (13-km principal and 45-km secondary sewer), a wastewater treatment plant, and a 3-km long outfall pipe. The project is jointly financed by the government of Sénégal, the Dutch cooperation, the European Union, and AFD. The Hann wastewater treatment plant is designed to treat both domestic wastewater and industrial wastewater after pretreatment.

Industrial wastewater discharge is regulated by the 1983 (revised 2001), Environment Act (*Code de l'environnement),* the 2009 Sanitation Act (*Code de l'assainissement)*, the Senegalese technical standard NS 05-061 setting the discharge limits values, and various administrative decrees. According to this environmental regulation, a pollution levy has to be paid by industries discharging directly polluted wastewater in the environment. The levy is presently set at a relatively low level (180 FCFA/kg pollution or roughly 0, 27 Euros/kg). A special department of the ministry of environment (DEEC) is in charge of controlling the polluting industries, computing and collecting the levy. However, the collection rate in 2013 was quite low as DEEC has limited means to control polluters and no fines or other sanctions were present in case of non-payment.

In order to discharge into a public sewer, the 2009 sanitation act requires an industry to sign a discharge agreement with the utility (ONAS). This allows ONAS to ensure that the wastewater discharged into the sewer system complies with specific limits. If not, the industry must install wastewater pretreatment

Figure 19.1 Untreated wastewater (a) from animal carcasses and (b) from industries flows in open air through the beach before reaching the sea in the Hann Bay. (*Credit: Margot Moreau and Laurent Raspaud, AFD*)

facilities before connecting to the sewer system. When the agreement is signed, the industry pays an industrial sanitation fee to cover ONAS's costs induced by wastewater treatment.

In 2013, a study executed by the French consultancy ARTELIA and financed through an AFD grant, made the following recommendations to the Senegalese government:

- Significantly increase the pollution levy.
- Set the industrial sanitation fee for industries that discharge their wastewater into the sewer system as a function of the quantity of pollution discharged up to the maximum acceptable discharge limits. This obligates high polluting industries to pretreat their wastewater in order to be connected to sewer system.
- Set a specific, higher fee for companies that exceed the maximum discharge limits.
- Define a practical way of billing and collecting the levy and the industrial sanitation fee such that:
    - the analysis and sampling costs will be paid by the industries
    - billing and collection of the pollution levy will be done by the DEEC
    - billing and collection of the industrial sanitation fee will be the responsibility of the sanitation utility ONAS.

The Senegalese government is willing to move forward on this regulation in parallel with the construction of the Hann Bay area sewer and wastewater treatment system.

Presently, a study financed by AFD and performed by SEURECA – MAZARS is assessing the technical and financial support required by the industries located in the Hann Bay Area to install pre-treatment facilities. This initial support to industries is key for this 'polluter-pays' policy to be sustainable.

19.7 AFD IN INDUSTRIAL WASTEWATER

AFD's strategy in the water and sanitation sector is defined by the Water and Sanitation Sectoral intervention Framework (AFD, 2017) which defines various operational objectives including:

- to support the definition of clear, effective and inclusive sectoral framework;
- to provide effective and sustainable access to services for all; and
- to preserve water resources in the context of climate change.

Reducing pollution from industry is one of AFD's key challenges, both in terms of preservation of water resources as an adaptation to climate change, and in terms of helping national and local authorities define an appropriate regulatory framework. In recent years, AFD has financed studies and technical assistance activities to various national authorities (not only in Morocco and Sénégal) in order to help them defining a regulatory environment concerning industrial wastewater and industrial pollution. AFD also regularly finances technical assistance to various sanitation utilities to define their strategy and tools concerning the way of managing, monitoring, and control the interaction between urban sewers and industrial wastewater treatment and discharge. In Morocco again, the Agadir water and sanitation utility RAMSA received a technical assistance grant in 2006 with the creation of the Great Agadir sewer system. In Tunisia, in the framework of the DEPOLMED project co-financed with EIB and KfW, a 1M euro EU grant managed by AFD allows : (1) provision of technical assistance to the sanitation utility ONAS in order to upgrade its industrial industry database, install new monitoring and sampling equipment, and implement training sessions; and (2) provision of technical assistance to industries in order to manage their wastewater pollution.

As with the 'Blue credit line' in the case of Morocco, the AFD group has also financed environmental upgrade of industries in various countries of the world through credit lines (Sunref, 2020) signed with various banks and financial institutions. Using these funds, the local banks can give loans to private industries in order to allow them to make investment on various environmental issues such as energy efficiency and air or water pollution reduction. These credit lines have attractive interest rates which makes them attractive to borrowers. Most often technical assistance is also provided with the loan to guide the bank through the water/environment/energy sector, to revise the design of the planned investment and to ensure correct implementation. In some cases, technical support is also provided directly to industries.

19.8 CONCLUSION

We introduced this chapter by noting that, while appropriate industrial wastewater treatment is essential to reach the UN SDGs and to adapt to climate change, industries in developing countries are not investing in technologies to reduce their pollution discharges. We observed that economic theory explained this as an example of 'market failure' due to externalities, which calls for government intervention. In all, we can see that industries in developing and emerging countries are not investing in the reduction of water pollution for a number of reasons:

- They do not receive appropriate incentives to invest.
- They do not feel the risk of sanctions or taxes sufficiently to change the status quo.
- They lack the knowledge to plan and design such investments.
- They lack awareness of the importance of improving environmental standards to keep their businesses sustainable in the long term.
- They lack access to appropriate financing or credit to build needed systems
- Their business return might not be sufficiently robust to add additional costs.

On a positive note, we also note significant progress in many countries based on the following observations:

(1) National authorities, sanitation utilities, and other institutions are improving their regulatory framework and increasing effective enforcement. This is a huge and uneasy challenge and it is important that development institutions such as AFD continue to give support in this field.
(2) Industrial associations are providing knowledge and technical support and sensitizing their members to the importance of environmental enhancement and the reduction of industrial wastewater pollution.
(3) Local banks are working with international development financing institutions (such as AFD) to finance industrial environmental upgrades, including wastewater pollution reduction technologies. This kind of initiative contributes to giving incentives to private industries to invest in this field.
(4) Worldwide, public opinion reflects more awareness of environmental issues, including the importance of preserving water resources, which can have a positive impact both on the political agenda and on the private sector priorities. This is an important field where the civil society has a key role to play.
(5) In some contexts local industries are investing in treatment systems to reduce wastewater pollution to comply with environmental requirements established by international client companies. This attitude of international brands seems to us a promising way of encouraging an environmental and water management upgrade of local industries in developing and emerging countries.

REFERENCES

Agence Francaise de Developpment (AFD). (2017). Cadre-intervention: L'eau ET l'assainissement https://www.afd.fr/sites/afd/files/2017-07/cadre-intervention-eau-assainissement.pdf; https://www.partenariat-francais-eau.fr/?ressource=water-sanitation-sectoral-intervention-framework-2014-2018. (accessed January 2020).

Agence Francaise de Developpment (AFD). (2020a). https://www.afd.fr/en/le-reseau-des-agences. (accessed 11 February 2020). 'AFD is present on five continents where it finances and supports projects that improve living conditions for populations, It has an international network of 85 offices, including in the French Overseas Communities and in Brussels.'

Agence Francaise de Developpment (AFD). (2020b). https://www.afd.fr/en/le-reseau-des-agences. (accessed 11 February 2020).

ASPA. (2016). Feasibility study for AFD for the Blue Credit line in Morocco.

Barraqué B. (2009). Valuing water: the Economics of an 'Impure Public Good'. Water as a Catalyst for Change: Making the Passage through the XXIst Century - FLAD-Lisboa, 2005, Fundação Luso-Americana para o Desenvolvimento, pp. 191–232, 2009. ⟨hal-00783359⟩

Département de l'eau. (2014). '(Valeurs Limites de Rejet à respecter par les déversements (Normes de pollution))' http://water.gov.ma/wp-content/uploads/2016/01/4.3.3.Valeurs-Limites-de-Rejet.pdf. (accessed 11 February 2020).

EPA. (2014). Guidelines for Preparing Economic Analyses (Washington DC: USEPA Office of Policy, National Center for Environmental Economics) May 2014. Chapter 4: Regulatory and Non-Regulatory Approaches to Pollution Control. https://www.epa.gov/sites/production/files/2017-09/documents/ee-0568-04.pdf. Accessed January 13, 2021.

Hindriks J. and Myles G. D. (2013). *Intermediate Public Economics*. MIT Press, Cambridge, MA.

Legifrance. (2020a). Arrêté du 2 février 1998 and by the Arrêté du 22 juin 2007. https://www.legifrance.gouv.fr/affichTexte.do?cidTexte=JORFTEXT000000276647. (accessed 11 February 2020).

Legifrance. (2020b). Code de la Santé Publique L.1331-10 and L.1337-2), code Général des Collectivités Territoriales (L.2224-12-2 et L.2224-19-1 à 11); Code de l'Environnement et particulièrement l'article R.214-5). https://www.

legifrance.gouv.fr/affichCodeArticle.do?idArticle=LEGIARTI000031928286&cidTexte=LEGITEXT000006072665&dateTexte=20160128. (accessed 11 February 2020).

Mamlyuk B. N. (2010). Analyzing the polluter pays principle through law and economics. *Southeastern Environmental Law Journal*, **18**, 43. https://ssrn.com/abstract=1679245. (accessed 11 February 2020).

Proparco. (2020). https://www.proparco.fr/en. (accessed 11 February 2020). 'Proparco, the private sector financing arm of Agence Française de Développement (AFD), has been working to support development in Southern countries for 40 years.'

Rosen H. S. and Gayer T. (2010). *Public Finance*. McGraw-Hill/Higher Education, Boston, MA.

SCE. (2015). Study on industrial sanitation in various countries for AFD. Unpublished report.

SDG 6 https://sustainabledevelopment.un.org/sdg6 'By 2030, improve water quality by reducing pollution, eliminating dumping and minimizing release of hazardous chemicals and materials, halving the proportion of untreated wastewater and substantially increasing recycling and safe reuse globally' (https://www.un.org/sustainabledevelopment/infrastructure-industrialization/)

SDG 9.4 https://www.un.org/sustainabledevelopment/infrastructure-industrialization/ 'By 2030, upgrade infrastructure and retrofit industries to make them sustainable, with increased resource-use efficiency and greater adoption of clean and environmentally sound technologies and industrial processes, with all countries taking action in accordance with their respective capabilities.

SDG 14.1 https://sustainabledevelopment.un.org/sdg14 By 2025, prevent and significantly reduce marine pollution of all kinds, in particular from land-based activities, including marine debris and nutrient pollution.

SDG 15 https://sustainabledevelopment.un.org/sdg15 'Protect, restore and promote sustainable use of terrestrial ecosystems, sustainably manage forests, combat desertification, and halt and reverse land degradation and halt biodiversity loss.'

Sunref. (2020). Website. https://www.sunref.org/en/ (accessed 11 February 2020).

United Nations. (2009). United Nations World Water Assessment Programme. The World Water Development Report 3: 'Water in a changing world.' 2009.; quoted by Mallory Rousseau, Guide de Bonnes Pratiques, Assainissement Industriel, internal document AFD 2016.

United Nations. (2020). Sustainable Development Goals. UN Knowledge Platform Website https://sustainabledevelopment.un.org (accessed 11 February 2020).

Chapter 20

The banking sector as an intermediary in supporting sustainable use of water by industry

Santiago David Vicencio, Chico

Produbanco, Grupo Promérica Quito, Ecuador

Keywords: corporate strategy, efficient use of water environmental and social impacts, Environmental and Social Management Systems, ESMS, sustainability

20.1 SUMMARY

Throughout history the banking sector has played a leading role in the creation and development of productive sectors of the economy. Today's complex derivatives can trace their origins as far back as ancient Mesopotamia, where ancient lenders provided seeds to farmers in return for a share of their crops. We can see the influence of banking on civilization in the development of currencies and national economies, and the evolution of economic models in parallel with political systems, as feudalism and monarchy grew into capitalism and modern democracy.

Since its inception, the financial sector has acted as an organized and educated intermediary, allocating scarce resources to optimize its benefit to society and maximize the growth of the economy as a whole. We now see that the role of the financial sector can progress even further, from being an intermediary to becoming an advisor and strategic partner, helping the economy develop not only from a quantitative perspective (i.e., increased profits), but also from the perspective of social equity and environmental sustainability.

In recent years, the banking sector has developed new skills and tools to improve decision-making (e.g., customized scores, digital services and solutions, financial applications), reducing the gaps between supply and demand at all industrial levels, taking into account the requirements of regulators and others, and creating more focused, efficient, and sustainable solutions for all. This chapter discusses the role of the banking sector in supporting sustainable use of water by industry, as exemplified by the activities and initiatives of Produbanco (Ecuador) and other branches of Promerica Financial Corporation

DOI: 10.2166/9781789060676_0225

operating in Central America. It illustrates how financing can serve as a tool that promotes economic, social, and environmental benefits, while highlighting the particular challenges in emerging markets.

20.2 THE ROLE OF BANKING IN SUPPORTING SUSTAINABILITY

Since the introduction of the word 'sustainability' into the business lexicon, a number of environmentally friendly and socially responsible practices have been developed and implemented by the private sector. For example, as part of action plans within corporate strategic planning, many corporations have changed their business models (at least in part) from production – that is, manufacturing and selling products, parts, and components – to services, in which customers pay for use rather than the products themselves. As an example, in the print and photographic sector, instead of selling the print machines companies now charge for the copies or prints customers create using the machines: this practice allows an exhaustive cost control related to the use of paper and prints. Reducing use of paper (one of the most use-intensive sector in terms of water) generates large environmental impacts considering deforestation, soil erosion, use of water, etc.

Individual efforts have provided useful models for the inclusion of environmental and social considerations within the strategic planning of corporations. Examples include corporations that have worked with suppliers in their supply chain to implement more sustainable practices and obtain certifications that verify use of more sustainable processes. For most companies, however, the concept of sustainability does not yet extend to planning and managing the impacts of all activities. Broader application of sustainability principles would include consideration of the following factors:

- Knowledge of operations and industrial processes including: location, resources, materials, life cycle issues, and associated technologies;
- Assessment of the effects of industrial water use on the communities where economic activity takes place;
- The role of other stakeholders, such as government and regulators, universities and research institutions, unions and associations in the sector, customers, suppliers, and distributors.

Integration of sustainability into business management means more than 'green branding.' A sustainable approach to business requires cultural change, even modification of the company's DNA – its interaction with stakeholders and its perspective on the future of the organization. Full implementation of sustainable practices requires commitment from each level within the company, from senior management to entry-level employees, supported by a strong communication plan and a clear organizational strategy.

Banking institutions, which maintain interaction with all industries, have a significant role to play in disseminating information on what sustainability is and what it entails. Through its wide range of financial support services (e.g., money management, investments, and loans) a banker is aware of all the challenges faced by its client, throughout its supply chain. As a result, banks have collectively gained extensive knowledge about the behavior of different industries, markets, and business models. Their expertise includes understanding the intricacies of each industry: its key risks, a range of average performance ratios, efficient and successful production processes, trends and business opportunities, and increase possibilities for growth. This expertise positions banks to play a strong and proactive role in the dissemination and implementation of concepts related to the sustainable use of water by industry. A key capacity of banks in this regard is the ability to modify the credit rating of a company based on a holistic view of its business model, processes, and capabilities, and including sustainability affairs.

Having the ability to communicate directly and quickly to the industrial sector carries significant responsibility, especially when the message comes from institutions that manage the resources of thousands of people and organizations. It is important for banks not only to encourage sustainable practices by others, but also to put them into practice themselves. The internal processes of banks should reflect all of the following:

(i) Efficient, environmentally friendly, inclusive, and socially responsible practices;
(ii) Commercial policies that require adherence to environmental and social standards by its customers and their suppliers (Environmental and Social Management Systems, Sustainable Products, etc.);
(iii) Transparency (annual reports, official publications, research, etc.); and
(iv) Continuous improvement and ongoing training for its employees.

20.2.1 Environmental and social management system, a control system created for customer support

A tool now available to the banking sector to support its work in this area is the Environmental and Social Management System (ESMS). ESMS can be used to help a company analyze and manage its social and environmental risks within a specific geographical area and throughout its supply chain. Successful use of the ESMS system can allow banks to quickly identify risks and opportunities for its clients, avoiding potential exposures that could harm the performance of the client (and its credit rating). Use of the system can help banks to identify good practices and highlight success stories that can be shared with other companies in the sector to exemplify how to implement sustainable practices.

At a minimum, ESMS must analyze the following variables: (1) regulatory requirements (e.g., water and wastewater treatment plant limits, health standards); (2) environmental regulations and licenses (e.g., those required for the management of hazardous wastes); and (3) regulations related to social responsibilities (e.g., those related to worker safety and health).

Each of the dimensions of ESMS, in addition to alerting clients to potential risks of non-compliance with relevant regulations, help clients generate action plans to close gaps. Implementation of these plans is integrated into the conditions for extension of credit to the client, which reduces its risk of non-compliance and encourages use of sustainable technologies.

With respect to water, ESMS can alert clients of the minimum local requirements as well as 'best practices' and international standards of infrastructure, equipment, machinery, and processes associated with specific operations. For example, wastewater treatment plants are required to process chemicals and hazardous materials within water before being discharged safely to the environment. Efficient reuse of treated wastewater to irrigation of crops can qualify agricultural clients for sustainability certifications, and ensuring efficient use of water for both industrial cultivation and food processing is particularly important in arid areas.

For example, considering daily operations in a dairy producer: these require large quantities of water and are subject to strict sanitary and environmental standards due to the potential exposure of the public to high risk of zoonotic disease. More than 30% of water use by the dairy industry is linked to the washing and disinfection of raw materials, equipment, floors, etc. The wastewater generated by these washing and disinfection processes needs to be treated prior to discharge to the environment.

A project in the Ecuadorian Highlands reduced water consumption by one milk processing facility by over 33% through technological upgrades that included the replacement of machinery (the new machinery uses re-circulation systems), automation of the pressing process eliminating the use and washing of molds, and modifying the cleaning process for milk storage tanks through a clean-in-place (CIP) process. As a result, the quantity of water needed to produce a single liter of milk has been reduced from 1.9 to only 1.2 liters.

20.3 LINKING AND INTERACTION OF INDUSTRIAL ACTORS

Creating a sustainable model, then, requires an analysis of all the economic and social interaction between actors. Within all industries, we find six predominant actors that interact in each productive ecosystem. As illustrated in Figure 20.1, these actors are:

- Government
- Companies
- Customers
- Academies
- Support Organizations
- Environment.

The banking sector supports every industry, not only from the perspective of generation and delivery of financial services and products, but also in terms of collaboration about rhythms of expansion and use of technology. Large enterprises often have sufficient resources to bridge the gaps identified by the ESMS system, including their own internal research programs and laboratories. They also have the financial capacity to contract for the services of consultants, research centers, and universities. In addition, they have the ability to engage with government and regulatory entities, and initiate new initiatives and processes that will support their market growth.

For small and medium enterprises, microenterprises, and entrepreneurships, however, the banking sector can play a critical role in their survival and growth by providing support in the areas of allocation of resources, financial advisory services, and money administration. Banking institutions can also help smaller companies build relationships with research and development institutions, producers of efficient machinery and equipment, technological and technical support organizations, educational organizations,

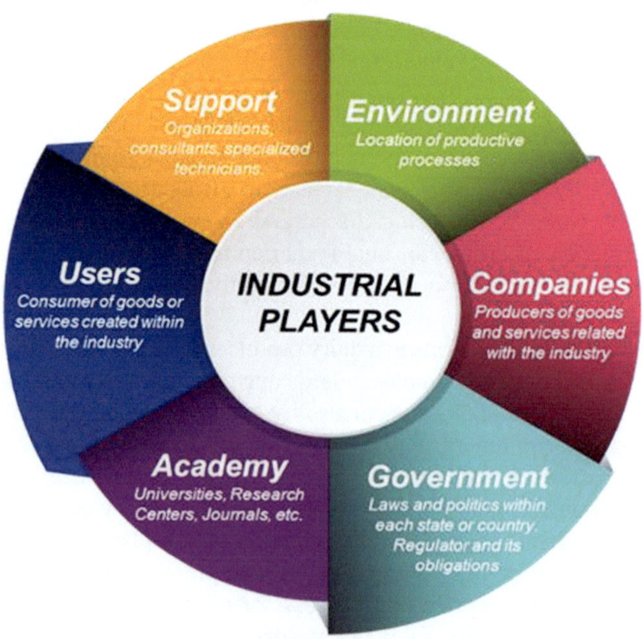

Figure 20.1 Industrial players map – and example of an industrial sample.

and others. Support from banking institutions can be critical for small companies that do not have the financial strength to pay for advisory, research, upgrades to machinery and equipment, facility expansions, or the introduction of cutting-edge technologies to improve productivity, efficiency, competitiveness, and sustainability.

To provide this level of support, banking institutions must expand beyond funding services (financial structures with expanded terms, competitive rates, etc.) and have developed an active interaction between customers and specialists in areas like regulation, management, environmental affairs, productivity, and technology. Produbanco has developed a program called SME Talks: '*Charlas PYMES*.' This program provide a physical place to gather SME customers with specialists (consultants, regulators, etc.) with information about all relevant new trends, legislation, risks, and opportunities from a globalized world. This educational website has been developed at a national level, serving all branches of Produbanco in Ecuador and is visited by more than 700 customers per year.

20.4 BANKING AND THE SUSTAINABLE VALUE PROPOSITION

In developed countries, the banking sector has pioneered the inclusion of analysis and measurement of environmental and social impacts in credit portfolios. Banks in the United States began accounting for such impacts as early as the 1980s, while in the 1990s European institutions created green financing vehicles for businesses with positive environmental impacts. These models served as examples for banks in other parts of the world to actively support corporate sustainability. Since that time, all banks have begun including new variables in their credit analyses, such as the impact of climate change and other environmental and social risks. They had to update their basic lines of analysis considering new trends and business models developed in the market (digital platforms, renewable energy generation and storage, etc.) that consider stakeholder concerns about social and environmental issues. They also began using new tools like digital platforms that have given customers and other stakeholders the ability to make purchasing decisions based on a company's behavior. Today, people around the world want to know and approve of the production processes associated with the goods and services they consume. For example, many developed markets require certifications that explain traceability in all the production process of food. Certifications like Global GAP Crops, Rainforest Alliance, Fair Trade, UTZ, etc. allow authorities and consumers alike to determine whether a product was developed using good environmental and social practices.

20.4.1 Interaction with multilaterals, consultants, governments, and academies

In Latin America, the banking sector's first steps toward increased support for sustainable use of water by industry resulted from interactions with development banks such as the International Monetary Fund, the World Bank, the InterAmerican Development Bank (IDB), and certain specialized funds (e.g., Green Climate Fund). Academic institutions, specialized consultants, and regulatory entities with expertise in environmental and social impacts also played a supporting role. This collaboration resulted in the development of the following:

- A system of control and analysis for social and environmental risks (ESMS);
- Financial solutions that promote the development of projects, processes, or activities with positive impacts (e.g., Produbanco's 'Green Lines' program); and
- Transfer of knowledge to the portfolio of clients in industries and sectors that have a potentially greater impact on the environment or that involve high social risks.

Examples of industries with elevated environmental risk include palm oil, slaughterhouses, metals, mining, petroleum, sugar mills, and pharmaceuticals. Increased environmental risks result from such factors as use of hazardous materials and production of hazardous wastes, and intensive use of resources (land, water, and energy). Enterprises with high social risks include those which include hazardous processes.

These industries include metals, mining, construction, petroleum, and gas. Processes that constitute a high risk to health, are characterized by overexploitation of workers, or involve child labor (e.g., agricultural production of some crops in poor areas), must be categorized as a high social risk.

The financial system, and specifically banks, now have the capacity to catalyze the creation and transmission of relevant information on legal and regulatory issues, as well as knowledge and technologies applied to production processes (see Figure 20.2). Companies that operate in a certain industry now have more information about the minimum environmental and social requirements associated with that industry. This can lead to the implementation of more efficient technologies and processes that are socially responsible and environmentally friendly.

The majority of improvements relating to more sustainable use of water by industry are created in developed countries: more than 80% of related patents are issued in these markets. By contrast, application of these improvements in emerging markets like Latin America is complicated by several factors, including:

- Limitations in data collection, administration, and analysis;
- Informality in the administrative management of companies;
- Inadequate availability of technical education, resulting in a labor market with limited training;
- A limited number of regulators to monitor operations; and
- High implementation costs (e.g., technical studies and the costs associated with machinery and process changes).

These limitations challenge all sectors in emerging markets, including government agencies and regulatory institutions, as well as the companies themselves. As a result, affected stakeholders need to build initiatives together to overcome these limitations. For example, there are significant gaps between regulatory body requirements and the technical capacities of operators. As regulations are tightened to reflect best practices, there is a need for local education and training to produce staff capable of using water in ways that will meet regulatory requirements (e.g., in relation to the water quality of effluent from industrial processes). This pressure is greatest on small businesses, such as those involved in the extraction and exploitation of raw materials (e.g., mining), agriculture, and aquaculture, which do not have sufficient financial, technical, or administrative resources to implement these practices or staff trained to use them.

In Ecuador, enterprises need to obtain an environmental license in order to start or maintain operations. Where a company does not have the capacity to meet all goals, it may be allowed to establish action plans to address gaps. Large companies with the financial capacity to contract with consultants or set up specialized teams within the company to develop action plans, initiatives, and new processes and projects. These investments allow them to upgrade or expand their operational sustainably. Due to funding limitations, small and medium-sized enterprises are at a distinct disadvantage in this respect. They lack the means to hire expensive technical consultants, and action plans prepared by less competent consultants may not comply with minimum requirements, resulting in reversals and delays in the company's ability to obtain environment licenses. Such plans could even result in sanctions that could cause the closure of the small enterprise.

In such cases, a bank that functions as an articulate advisor can provide tools to all production levels throughout the entire value chain of each industry. In this way the banking industry can support development of sectors that are strategic for the local and national economy while making them more

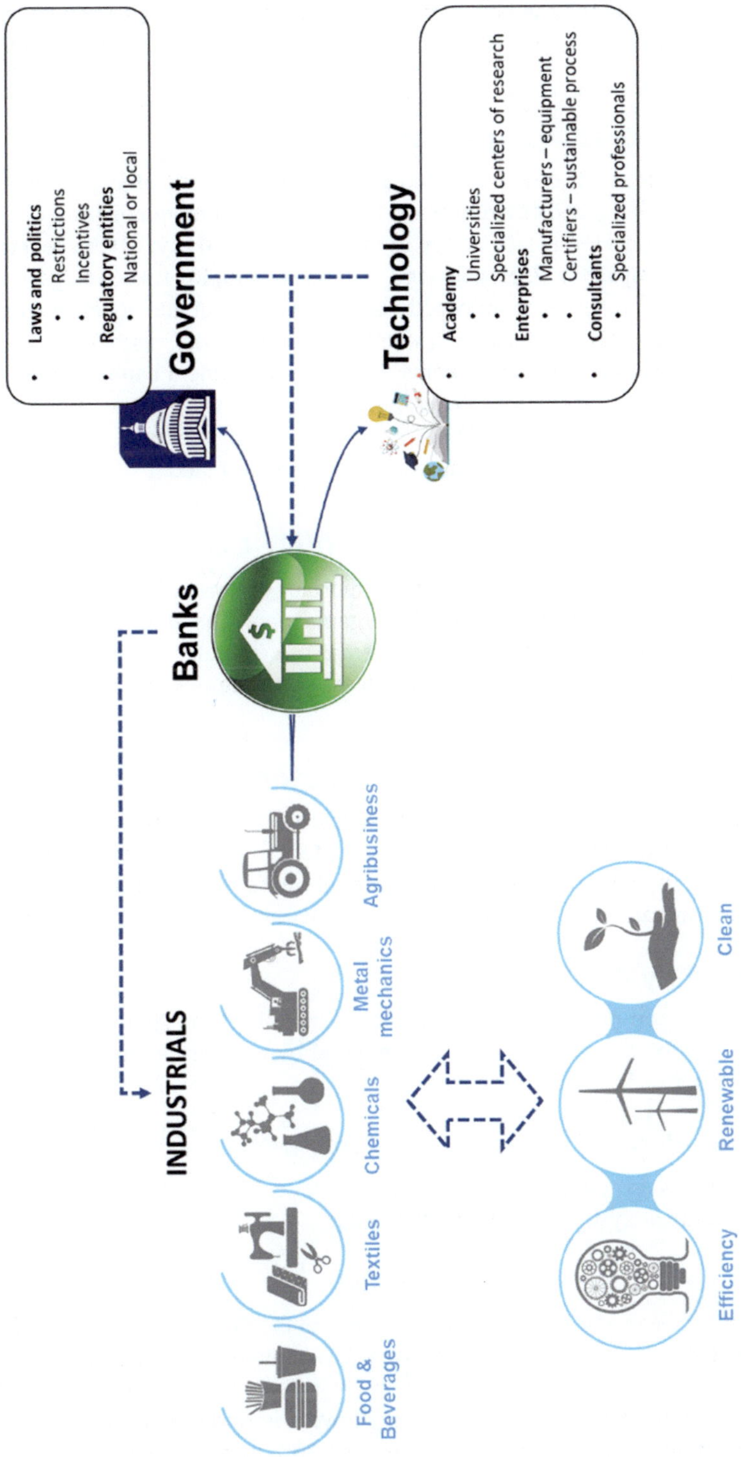

Figure 20.2 Interaction between industrials, government, and developers of new technologies, using the banking sector as a clearinghouse for information.

sustainable in the process. Regulations can be approached in a way that is consistent with local capabilities, accompanied by improvement initiatives and potential funding to achieve adequate environmental and social standards.

20.4.2 Situational understanding of the market

A banking institution can provide to its customers (and even to regulators and other stakeholders) situational understanding of the market's social and environmental issues, based on regulatory baselines. In its analysis it is necessary for the bank to consider: (1) the nature of the economic activity; (2) the geographical location of production; and (3) the company's size (sales and number of employees). Analyses related to these variables allow clarity on the main economic activities developed in the local economy that would present greater potential risks, geographical areas prone to such risks, and the relative capacity of different players to handle these kinds of investments. From this baseline a banking institution can define strategies to address the priority segments, developing tools (measuring and monitoring), processes (attention, analysis, and evaluation), and policies (exposure limits, tolerance, etc.) that will allow an orderly evolution of internal systems and their implementation in the market.

20.5 PROJECTS TO SUPPORT THE EFFICIENT USE OF WATER

Efficient water management programs are particularly important because of the traceability of water as a resource through the entire production process of the organization. The source from which the water is taken (provider and resource arrival mechanism), its internal use (productive and administrative), and its subsequent return to the environment all generate different opportunities for projects and initiatives to use water more efficiently.

20.5.1 Identification of the most relevant sectors in water use

According to AQUAE Foundation data, the primary industrial users of water globally are the paper manufacturers; chemical producers; agriculture and livestock; and the extraction, production and transformation of metals. In emerging countries, exploitation of commodities has been a focus of production; oil, minerals, and agriculture are particularly significant components of their economy.

20.5.2 Industrial analysis process

Once the company's production process has been analyzed, initial efforts should focus on activities that involve the greatest amount of water use or that are regulated by an external authority. Potential projects for the efficient use of water include the following:

20.5.2.1 Water reservoir construction (includes rainwater)

Current construction projects are evaluating the use of rainwater storage tanks in communal areas and rainwater storage and treatment systems for individual houses.

20.5.2.2 Efficient pumping, spraying, and irrigation systems

Drip irrigation systems have been applied to control water use in the floriculture industry (Ecuadorian roses are considered the best roses in the world), and growers have implemented preservative floral boxes to prevent excessive water consumption and use of chemical preservatives during post-harvest activities. Since water used for pumping, spraying, and irrigating flowers is traditionally greater than 0.025 m^3 or 25 liters per stem, consumption can be reduced by more than 70% using these technologies (Figure 20.3).

Figure 20.3 Drip irrigation and other sustainable water use practices have cut water use by some Ecuadorian rose growers by more than 70%. (*Credit: Produbanco*)

20.5.2.3 Wastewater treatment

Adequate treatment of water after it has been used for industrial purposes is one of the most important components of protecting both the environment (i.e., receiving streams) and communities where public health may be impacted by discharged substances. Because of the significant impacts that can result from inadequately treated effluent from industrial processes, projects that support compliance with discharge regulations are a high priority.

20.5.2.4 Water recovery

Water recovery can be achieved through the development of wastewater treatment plants that support the safe reuse of water. The processes of washing floors, hygienic services, and irrigation of communal areas are the most frequent uses for this type of recovered resource. The savings seen from use of reused water for such purposes can reduce the water consumption associated with an industrial process by 5% to 15%.

20.5.2.5 Reduction of water consumption

In the dairy industry (industrial producer), the use of cold water at the end of cheese production often results in the consumption of an additional 0.8–1.0 liters of water per liter of process milk used to make cheese; it also requires more energy to chill the water. In all industries, within its bathrooms, the installation of double discharge toilets and faucet aerators (at a cost of approximately $3–5 USD each) can reduce bathroom water use between 15% and 35% (Figure 20.4).

Analysis done by Produbanco on the savings and efficiencies generated in the sustainability initiatives of the dairy and flower industries of Ecuador indicates that these initiatives have increased the profits these enterprises by 0.5–3.5%. New ongoing income flows have been used to cover debt costs, which in turn

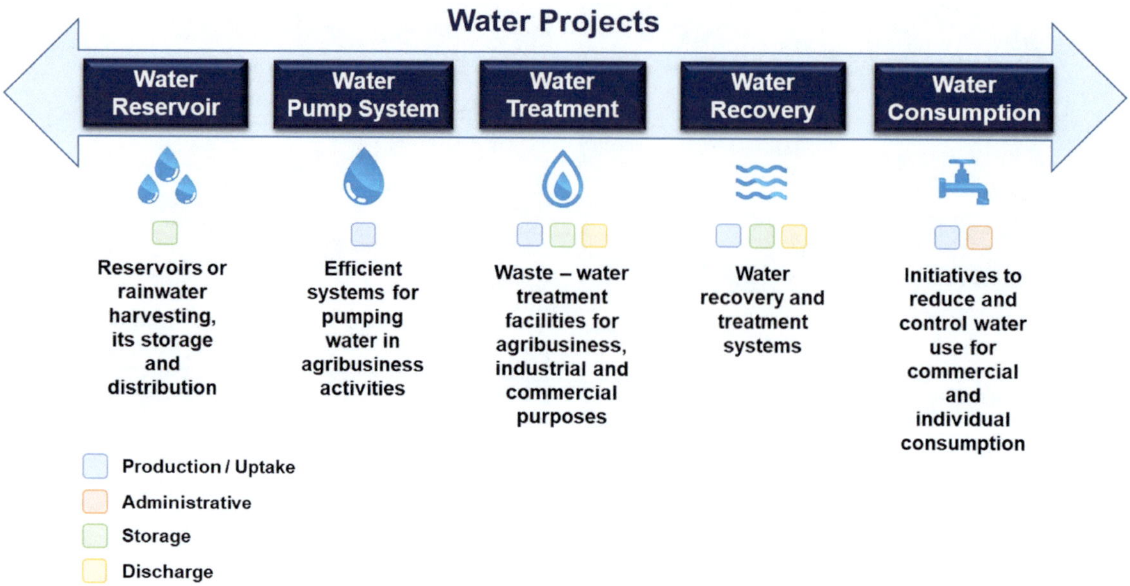

Figure 20.4 Water projects analyzed by their impact on the production process.

have allowed projects to be self-sufficient in paying their obligations. Changes made to address environmental and social concerns have also proven to have economic benefit to these enterprises.

20.6 CONCLUSIONS

Creating business models that involve sustainability principles allows businesses to be in a better position to face the market. Not only does it help them attract and keep consumers increasingly concerned with sustainability, it also generates better relationships with suppliers and regulatory agencies. In addition, efficiency and productivity increase with improved technical processes and better trained staff. A reduction in the perceived level of risk results in higher credit ratings and more positive relationships with the communities where companies operate. In time, we expect that a sustainable approach will become a market necessity, leaving those who have failed to adopt this approach out of the market.

Adopting win–win solutions in order to ensure a better tomorrow is the basis of sustainable development in companies. The banking sector is moving from a spectator to a relevant actor in this process, making a positive impact not only on specific companies, but also the long-term well-being of communities and the environment.

Chapter 21

The Certificado Azul: Peru's innovation for encouraging sustainable use of water by industry

Juan Pablo Mariluz[1] *and Ursula Antunez de Mayolo*[2]

[1]*National Water Authority of Peru, Peru*
[2]*SGS, Peru*

Keywords: Certificado Azul, incentive for good water use, innovate incentive program, Peru National Water Authority, recognition for water efficiency, shared value plan, sustainable water use, water footprint, water and social responsibility, watershed

21.1 INTRODUCTION

Peru, a country with approximately 32 million inhabitants, faces significant water challenges. The country contains 70% of the world's tropical glaciers and its total natural water resources are 2 million cubic hectometers (2×10^{12} m^3), which includes superficial and underground water, per year. The United Nations Office for Project Services describes Peru as a country with 'vast natural resources and rich biodiversity', but also notes that 'years of misuse of water resources by the manufacturing industry, effects of climate change, a growing population, and inadequate agriculture practices have increased water scarcity.' Water distribution is particularly difficult: according to UNOPS (2020) the Peruvian coast is home to more than 65% of the country's population, but has access to less than 2% of its freshwater supply.'

Climate change is affecting Peru's rainfall patterns and the intensity of rainfall, causing extreme hydrometeorological phenomena and increasingly frequent flooding in urban areas; it is also accelerating the process of deglaciation in the Andes mountain range, increasing river discharges, so that freshwater is lost to the sea (Figure 21.1). Water governance in watersheds is often lacking, so that the public, private, and community actors independently compete to meet their own needs. Private businesses are often unaware of water supply and water quality risks in the watersheds where they operate, not recognizing that these could ultimately make it impossible for them to maintain operations and productivity in the catchment. Water costs are low, regulation and public policy related to water use are

DOI: 10.2166/9781789060676_0235

Figure 21.1 Lagoon Lacacocha in the Raura mountain range is an example of the glacier-fed lakes threatened by climate change.

limited, and financing is generally lacking to support water system improvements needed in both public and private infrastructure.

Recognizing these challenges, the National Authority of Peru has found a way to move forward by implementing an innovative incentive program, the Certificado Azul. This program, which provides guidance, encouragement, and recognition for efficient water use by industry, addresses water use at both the facility and the watershed level. This initiative, first implemented by the Peruvian government in 2015, continues to evolve over time, and is under consideration for implementation more broadly in Latin America (e.g., Chile, Mexico, and Colombia). This chapter provides: (1) information on the Certificado Azul methodology; (2) examples of achievements in the area of water supply, water quality, and watershed-level water management; and (3) a discussion of possible next steps to improve on the current program to further support sustainable use of water by industry.

21.2 THE PURPOSE AND METHODOLOGY OF PERU'S CERTIFICADO AZUL

The Certificado Azul recognizes private companies for efficient use of water, encouraging not only sustainable water use within the facilities and operations of private companies, but also social responsibility in the management and sharing of water resources within the watersheds where they operate. Participation in the Certificado Azul program reflects the commitment of a business to modify its own water use and provide leadership in water sustainability efforts through collaboration with public and private entities and communities (including business partners and suppliers).

The program is a voluntary procedure implemented by the National Water Authority of Peru in the framework of the water resources law. The Participation is open to Peruvian businesses that are in compliance with all water regulations, and have no outstanding administrative action against them by the National Water Authority (e.g., penalties or sanctions). The initiative is intended to create synergies between the private sector, public sector, and communities to promote sustainable water use at the basin level. While the initiative has been implemented at a national level, it is notably international in its perspective, reflecting United Nations Sustainable Development Goals (SDGs) relating to water supply and water quality, and internationally recognized methodologies for assessing and improving industrial water use.

The Certificado Azul program helps businesses by improving their economic sustainability by addressing water risks, strengthening their image and brand with customers and suppliers, reducing their operating expenses, and strengthening their relationships with neighboring communities. In addition, companies that have earned the Certificado Azul have a slight competitive advantage when the State tender for the acquisition of goods and/or services needed by the national government. To obtain the Certificado Azul, companies must complete each of the following steps:

(1) Definition of the Water Footprint of a facility's current operations
(2) Development of a plan and commitment for reduction of the company's Water Footprint at the site
(3) Collaboration between the private sector, the public sector, and the community to develop a plan for managing and sharing water resources at the watershed level
(4) Execution by the company of the plans and commitments outlined in the plans adopted for changes at both the site and watershed level
(5) Evaluation of the company's execution of its plans by the Committee of National Water Authority (Water Authority of Peru)

21.2.1 Definition of the water footprint of a facility's current operations

The National Water Authority provides forms as well as guidance on how the Water Footprint Analysis is to be conducted and documented: The methodology used to define the Water Footprint of the company's existing industrial activities at a given site must be consistent with the Water Footprint Network, the ISO 14046 guideline for water footprint analysis, or the AWS Standard developed by the Alliance for Water Stewardship.

- The methods and models used for analysis must be scientifically and technically valid.
- The report on the findings of the Water Footprint analysis must be transparent and consistent.
- The report must be done with guidance and support from a third-party verification entity with sufficient expertise to ensure the validity of findings.

21.2.2 Development of a plan and commitment for reduction of the company's water footprint at the site

The commitment to using less water and/or protecting water quality should apply to direct uses (e.g., in principal activities and processes) as well as indirect uses (e.g., energy consumption, transportation processes and use of fuel, and water use by suppliers). Examples of plans developed by companies that have earned the Certificado Azul are provided below

21.2.2.1 CELESPA

At both the facility and the watershed level, CELESPA, a hydropower company, focused on SDG 6.A, improved efficiency in water use. A water footprint analysis of the company's San Juanito power plant showed that the dominant water use associated with the facility was evaporation from the reservoirs that stored water prior to its use for hydropower production. Evaporation is extremely difficult to address in terms of both technology and investment. However, the company was able to reduce water use related to the facility by installing water monitoring and control of the domestic water use at the residences and dining rooms of the employees who were employed at the plant.

21.2.2.2 CIA Minera Coimolache

The facility-based improvement adopted by CIA Minera Coimolache in relation to its mining facilities related to SDG 6.3, Water Quality. CIA Minera Coimolache operates both gold and silver mines. When rainfall enters open mines, its water quality is degraded by contact with minerals; this effluent needs to be treated before discharge into the environment. The water treatment process requires energy consumption and also results in water evaporation. Installing geomembrane 'raincoats' over open mines kept water from coming into contact with minerals, so that over 100,000 m^3 of water per year no longer needed to be treated.

21.2.2.3 Nestlé

Nestlé addressed the SDG 6.4 water efficiency goal by savings 14,232 m^3 of water per year at its D'onofrio Ice Cream Plant in Lima. The ice cream plant has 25 distribution points for cleaning and maintenance of the facility, and production machines in the plant. The traditional hoses did not allow optimized water consumption, so they were replaced with high-pressure hoses. Procedures for use of the hoses were drafted, and training was provided to employees on how to use the new equipment, which has had the additional benefit of increasing working efficiency, productivity, and safety.

21.2.3 Collaboration between the private sector, the public sector, and the community to develop a plan for managing and sharing water resources at the watershed level

This watershed level plan, which should be developed in collaboration with others drawing water from the watershed, should allow for the needs of the community as well as natural ecosystems. Within catchments, water risks are defined through use of the World Wildlife Fund's Water Risk Filter. Examples of components to be considered in the watershed-level plan include the following:

- Improved availability and accessibility of water resources to different kinds of users in the watershed;
- Improved quality of the water available: treatment plants for potable water and/or wastewater treatment for effluent from domestic, industrial, and municipal users;
- Improved water and sanitation in rural and periurban areas;
- Automated irrigation to reduce water consumption for agriculture;
- Construction and optimization of hydraulic systems that support increased water efficiency;
- Public education on the importance of public education on the importance of the protection and conservation of water resources;
- Implementing systems for monitoring and control of water resources;
- Protection of the watershed and conservation of ecosystems; and
- Funding for water improvements.

Examples of watershed-level contributions made by private companies in connection with the Certificado Azul initiative include the following:

21.2.3.1 El Platanal hydroelectric power plant

CELEPSA CELEPSA invested in concrete coating for irrigation channels s used to transport water from the Caete River to fields used by 232 users associated with agricultural associations in campesina community. Their activities included choosing channels to be coated, identification of sections to be coated, determining the efficiency of the uncoated channels, hiring local workers to do the concrete coating, coating the channels, and determining the effectiveness of the coatings. Over 1,000 meters of irrigation channels were coated in connection with this project, resulting in water savings of 2,900,000 cubic meters of water.

21.2.3.2 CIA Minera Coimolache

Coimolache's water-shed level program related to water use efficiency. Within the watershed focused on the communities of Chencha and Ramirez; both of these communities have livestock farming as one of their major economic activities. Water is used to irrigate the grass pastures used to feed cattle. In this project, eight mini-reservoirs that were coated with geo-membranes were installed, as well as eight sprinkler irrigation systems. The project also included training to residents on how to maintain these water improvements, which benefited 60 people and saved 5,760 cubic meters of water per year.

21.2.3.3 Nestlé

Nestlé donated treated wastewater from a Nestlé wastewater treatment plant for irrigation and sustainable development of green areas in Lima Peru. This was done to improve environmental conditions in Cercado de Lima, which was characterized by a deficit of green space for the population, insufficient water for irrigation of green areas, limited ornamental plants and infrastructure for maintaining them, and insufficient public awareness of the value of upgrading the local environment. Nestlé worked with the National Water Authority (ANA) to obtain permission to reuse treated wastewater for irrigation, coordinated with the community where the treated water would be used, and was able to provide indirect benefits for 20,000 people through the reuse of over 24,000 m^3 of water per year.

21.2.4 Award

If the plans, and the company's execution of the plans, meet the standards set by the Water Authority of Peru, the Certificado Azul is awarded by administrative resolution to the private business that has met the Certificado standards (Figure 21.2).

21.3 ACHIEVEMENTS OF THE CERTIFICADO AZUL INITIATIVE

To date, eight businesses in Peru have been awarded the Certifado Azul. Three have renewed their certifications. Eleven companies are in the process of obtaining certification, and 10 are initiating a water footprint analysis.

Programs initiated in connection with the initiative have resulted in the following benefits:

- 4.3 million cubic hectometers (4.3×10^6 m^3) of water saved;
- 122 million cubic meters of water reused'
- 110 million cubic meters of water treated;
- \$11.2 millions of dollars invested (USD);

Figure 21.2 Delivery of Blue Certificate: NESTLE PERÚ, El Platanal Electric Company - CELEPSA y Mining Company Coimolache.

- Almost 25,000 persons receiving indirect benefit of investments; and
- Over 12,000 persons receiving direct benefits from the program.

21.4 CONSIDERATIONS FOR FUTURE IMPROVEMENT

As the National Water Authority of Peru considers how to build on the achievements of the Certificado Azul initiative, one alternative under consideration is to increase the financial incentives associated with the award, which are currently minimal. While the current financial incentives are sufficient to interest companies that have a long-term perspective, an awareness of water issues, sufficient financial resources to invest in upgrades at the facility and watershed level, and a commitment to positive company image and branding, they are not large enough to make the program appealing to the majority of businesses in Peru. The positive impact of the program can be increased by making the program appealing to a wider range of businesses.

The National Water Authority is aware of the need to create both economic and non-economic incentives for sustainable water use by industry; numerous factors influence industrial water use. Factors where government can play a role include the following:

- Water charges, because unmonitored consumption or low water rates encourage water waste;
- Regulations on the amount of water that can be consumed, and the quality of water released to the environment (with adequate enforcement and penalties for inappropriate use, applying the polluter pays principle);

- Public and private sector awareness of water risks, and the urgent need for water conservation, water resource protection, water equity, and natural resource protection, to protect business interests as well as the community;
- Policies and regulations that support better governance of water resources and water-sharing at the catchment level; and
- Support for alliances between the private sector, public sector, and civil society, prioritizing environmental and social responsibility.

Many positive changes require investments in technology, such as technologies for monitoring and managing water flow and water quality, and to support water reuse. In some cases, private entities have sufficient funding to make these investments. However, additional financing options should be developed to help both public and private entities that lack sufficient resources to make the investments required (e.g., green or grey infrastructure projects). One important aspect will be improved use of information technology, such as analysis of big data and storage of data on the cloud.

It is important not only to do more but to move in these directions as quickly as possible to counteract the pace of population growth, the cumulative impacts of environmental damage, and climate change. Increased collaboration among countries, among different levels of government, and among different sectors of society will be required in order to bring improvements that are commensurate with the magnitude of the challenges before us.

REFERENCE

UNOPS. (2020). 'Combating Water Scarcity in Peru.' Website https://www.unops.org/news-and-stories/stories/combating-water-scarcity-in-peru. (accessed May 5, 2020).

Chapter 22

The culture of water needs to change

Michael Spencer

Monash University, Monash Business School, Melbourne, Australia

Keywords: attitudes, behavioralism, collaboration, culture, drought, management, pollution, professions, stewardship, water

22.1 INTRODUCTION

The need for industry participation in addressing the world's water crises has been well documented. (See, e.g., 2030 Water Resources Group, 2009; Asian Development Bank, 2016; High Level Panel on Water, 2018; International Conference on Water and the Environment, 1992). What is less well documented is how this is going to happen. By and large, not enough has been done to understand the obstacles to water-using industries addressing scarcity, water quality, ecosystem health and water equity issues, or the incentives that might encourage their participation. Isolated examples of collaboration between industry, water agencies, government, and other stakeholders can receive considerable attention but really only highlight the extent to which these parties normally work independently of one another. Despite the urgent need for change, water professionals largely continue to employ the same approaches, despite demand for water growing beyond supply in some regions and population growth, industrialization, urbanization, changing consumption patterns, and climate change creating new challenges.

In this chapter, I argue that change needs to start with the attitudes of water professionals, who must encourage industry customers to participate in solving water problems through the development and celebration of new forms of collaboration. In short, so long as water professionals accept responsibility for solving industry's water problems, they will perpetuate the traditional agency relationship between utilities and customers, reinforcing industry's attitude that water is 'someone else's responsibility.' To start the change process, there is a need to look at the attitudes and culture of the water industry – including how water professionals are trained, recognized, and rewarded – to understand how these

DOI: 10.2166/9781789060676_0243

cultural issues have become an obstacle to collaborative problem-solving. Water professionals need to help companies understand their part in responding to the overall water crisis and seek their participation in addressing the underlying causes. When this happens, it can create a meaningful framework for collaboration between water agencies (suppliers and regulators) and industry (customers) that can deliver measurable outcomes.

22.2 SETTING OURSELVES UP FOR FAILURE

To illustrate this 'attitudinal' problem, let me recount a recent experience as part of an Australian delegation to an unnamed state water ministry in India. The Minister for Water took great pride in his team of engineers who had set up a technically impressive water monitoring network that provided an overview of the complex mix of water resources from rivers flowing through the region. This monitoring system allowed them to control flows and direct water through an expanding network of channels that could eventually divert water from one part of the state to the other. They were sure this would waterproof the state and provide a solid base for economic growth, while allowing them to keep their promise to maintain low water prices.

One member of our delegation asked Ministry staff how they involved the public in their water projects, but the question didn't generate much discussion. Later the delegate recounted his conversation with a local taxi driver. He had asked the driver, whose parents were farmers, about their water use: where they got their water, how it was allocated, how much it cost, etc. He was informed that the water came from the river with few restrictions on its use and at little cost. This was consistent with the Ministry's approach but for delegates, who had experienced the problems associated with severely over-allocated Australian river basins, this meant danger. We understood the personal, social, and economic cost of believing that water available today would always be available in the future, and we knew how the appearance of 'unlimited water' encourages inefficient and excessive use, creating dependencies that one day would need to be painfully unravelled. There was considerable technical expertise in the state, but no desire to build collaboration with customers or undertake behavioral solutions. The engineers wanted to fix water problems the way they had been taught: increasing control through engineering solutions.

In fairness, the problem is not just with engineers and engineering solutions. Other types of top-down solutions also require coordination with stakeholders, including enhanced regulation and cap-and-trade water allocation. For example, in dealing with water pollution public policy generally turns to regulation. Policy-makers set regulatory limits on contaminants in wastewater emissions in the expectation that these will stop pollution. However, as a recent United Nations report found (United Nations Environment Program, 2019), too often environmental regulations fall far short of their intended goals due to inadequate implementation and enforcement efforts. Implementing ministries for environment are often underfunded, and lack the political muscle of economic ministries responsible for exploiting natural resources or economic development. Laws that are not enforced, or not respected, are generally ineffective laws.

These problems are common in both developed and developing countries. Despite its environmental protection laws (established after the Stockholm Conference on the Human Environment in 1972), China's development goals often took precedence over its environmental priorities and until very recently environmental laws were not vigorously enforced, resulting in serious pollution issues. Similarly, in Australian there are examples of water quality rules being overlooked. Some years ago, in preparation for a water stewardship trial, I was gathering information about pollution levels in local rivers for a catchment assessment. Buried in Environmental Protection Agency annual reports I discovered the statement that few Victorian rivers complied with state-wide environmental quality parameters. On closer examination, it turned out several quality parameters specified were not being monitored during a period of severe drought. For others, different metrics were being monitored than those specified in the regulations. It was almost impossible to assess performance, and there appeared to have been little effort to enforce regulations.

If legal enforcement has its problems, so too do solutions such as fixed allocations and market-based pricing through cap-and-trade systems. These essentially direct water to the highest value use. Inevitably, particularly in agriculture, the highest value use is by large corporations that have big budgets and operate efficient systems on a large scale. These larger users have the ability to buy water rights from smaller operators through their economic power. Sale of water by small users will impact towns and communities because the smaller operators employ more people and are the life-blood of their local communities. As the fabric of these communities is impacted by loss of scale (schools shrink, services are centralized, sporting clubs can't field teams), a sense of crisis develops and there is political interference in the-cap-and-trade system. Efficiency (which is at best a limited goal) starts to give way to profligacy and the integrity of the water allocation system is undermined. To be successful, regulatory and cap-and-trade approaches need bottom-up engagement and buy-in by people affected; otherwise they will fail to achieve the purposes for which they were established.

22.3 A KNOWN PROBLEM

Nothing in the above analysis of 'attitudinal' failures in water management would surprise those who have followed global discussions of our water crises. The need for better collaboration was called out over a quarter century ago. The 1992 Dublin Statement on Water and Sustainable Development (particularly Principle 2) spoke of the need for a '*participatory approach involving users, planners and policy-makers at all levels*' (1992). The Statement spurred development of approaches such as Integrated Water Resources Management (IWRM), which recognized the holistic nature of water management and the role of stakeholders, including industry, in achieving change. However, by 2004, when the San Francisco based Pacific Institute published its report on private sector water risks, it found that both corporations and investors were '*unfamiliar with freshwater-related risks and unprepared to implement the suite of measures available to reduce them*' despite the fact that these risks could cost industry billions [USD] (Morrison & Gleik, 2004).

Subsequently, a number of initiatives sought to engage industry in changed behavior towards water management. In August 2007, UN Secretary General Kofi Annan initiated the CEO Water Mandate, and in 2009 the concept of water stewardship was formalised in the global Alliance for Water Stewardship (AWS); both programs are ongoing. Also in 2009, the Carbon Disclosure Project urged investors to take up the challenge of addressing industrial water risks while the 2030 Water Resources Group (2030 WRG) (a collaboration between the World Bank Group and industry) released 'Charting Our Water Future; Economic frameworks to improve decision-making,' a report on the state of the planet's freshwater resources. The 2030 WRG (2009) report drew attention to the emerging disequilibrium between water supply and demand in most parts of the planet, especially in developing countries where the deficit could be as high as 50% by 2030. It concluded that the private sector was '*critical to the transformation of water use in a country*' and that government policy needed to help align industrial behavior with the broader objectives of the water sector (pp. 122–123).

Those views were echoed by multilaterals such as the Asian Development Bank whose 2016 Water Outlook report stressed the need for the water sector to engage with industry. More recently, a UN-World Bank High Level Panel of Water (2019) argued that technical solutions alone could not solve the planet's water problems, and that water-use sectors needed to embrace water stewardship and engage in collaboration around IWRM. There has not been a shortage of urging for the water sector to engage with business and business to engage with water, but the sector continues to grope in the dark for ways to make that happen.

The 2030 WRG report had said solutions that only engage a few central decision-makers would have significantly greater cost than solutions incorporating all available measures. Specifically, the report referred to measures whose adoption would require behavior change from millions of farmers andindustrial and domestic water users. Seven years later, Newborne and Dalton (2016) found there had been little evolution by companies from business as usual on water. Their report, described as 'taking stock' of corporate action on water management and water stewardship, found that, while corporate concern about water risks was evident, companies interviewed in the study struggled to create a business case for action. Companies that did take action were more likely to fund water stewardship programs out of their corporate philanthropy budget. While any action is welcome, as Newborne and Dalton pointed out, charitably funded programs are likely to be more ephemeral than those incorporated into core business models.

WWF (World Wide Fund for Nature) Water Practice Lead, Stuart Orr, has said that the rate of private sector progress on water issues is too slow to mitigate the risks posed by the world's soaring population, rapid development, and urbanization (2018). Sarni and Grant (2018) agree, citing industry failure to adequately recognize and capture the value of an effective water strategy. They argue that a more expansive approach to valuing sustainable water use innovations may provide the traction needed for change: '*an abundance strategy designed to create greater business value and impact, along with societal and brand value*' (p. 99). Newborne and Dalton, hold that collaboration between government water policy and industry leaders is needed to achieve change and recommend connecting private actions with government-led water stewardship plans to improve water governance. Our research in China finds that this sort of collaboration can provide legitimacy for water stewardship projects, a foundation for industry leadership and a framework for government investment in collaborative projects (Spencer & Xu, 2021). However, as I will argue below, meaningful collaboration will not happen without cultural change in the water industry.

22.4 MARKET FAILURE AND BEHAVIORAL FAILURE

The 2030 WRG report (2009) made the case that water showed evidence of ***market failure***; that scarcity was not drawing in sufficient new investment nor changing behavior toward water, and that normal demand and supply market forces '*were not happening for water.*' It argued that appropriate strategies for meeting competing demands for water would not emerge naturally from market dynamics. While market failure occurs on a macroeconomic level, our research suggests that on a microeconomic scale – at the firm level – there is evidence of ***behavioral failure,*** as well. Economists describe behavioral failure as a failure to behave as predicted by rational choice theory (Shogren & Taylor, 2008). The point here is that that reversing market failure through the creation of market-like incentives rests on a presumption that people make rational choices that are consistent and systematic. If industry decisions about water cannot be relied upon to be rational and consistent, then any analysis of what will motivate or incentivise behavior change will need to look at a broader range of potential drivers than economic incentives.

Evidence of behavioral failure is illustrated in results from our recent survey of decision-making about water-use by 30 operating sites in China. Some sites surveyed had already adopted sustainable water practices; others had not but were considering them, some were not considering them, and some did not know. Respondents were asked to rate a range of potential benefits from water stewardship, both tangible and intangible, as well as potential costs such as investment in new infrastructure and audit costs. Using an index of perceived benefits minus costs, analysis found that there was not a strong relationship between whether sites said they were considering or not considering water stewardship and perceptions of net benefits. This suggests the decision to adopt or reject water stewardship is not based simply on a rational perception of costs and benefits. In other words, for the sites surveyed, their perception of net benefits was relevant but not sufficient to motivate them to participate in a water stewardship program. There were other factors, beyond economic considerations, holding back their participation.

In examining the motivations and constraints for participation in water stewardship more closely, we identified government as an important influencer of participation – even more than customers (who were also important). Although this initial research was conducted in China, initial findings from ongoing research in Australia and New Zealand, suggest that the influence of government is also important particularly for industrial water-users. If this is borne out, it would support the proposition that government plays a key role in determining industry's attitudes to water management regardless of the political regime.

In China, our research found that compliance with government requirements was a major driver of investment in water management and an important influence on a facility's participation in water stewardship training. Customer pressure, although present, was not sufficient to ensure commitment to a water stewardship program, nor was the attraction of enhanced reputation. (Here, too, the site's reputation with government and regulators was as important as its reputation with customers.). Thus, while all sites identified 'improved water management in operations' as the primary benefit of water stewardship, sites that opted out of such programs considered 'staying ahead of government regulators' to be their major consideration. This points to the need to connect a holistic water stewardship approach with support from government agencies (water authorities) to engage facilities outside their fence-line. Strong collaboration between government, customers, civil society, and operating sites, as well as preparedness to recognize shared costs and benefits, offers the best chance of success.

22.5 BARRIERS TO ENGAGING GOVERNMENT

Building the sort of collaboration discussed above does not come easily or naturally. The current culture of the water industry as a whole does not appear to value collaborative behavior except as a last resort during a crisis such as drought. On the industry side, there appears to be an attitude of *Government delivers water, I pay a fee and follow the rules*. On the utility side, there is a similar aversion to engagement: interviews with government water and environmental officials reveal their main interest is in top-down technical, regulatory, and engineering fixes rather than the collaborative development of social and managerial solutions. As one interviewee in China pointed out, most employees in government agencies are driven by performance goals such as delivering specific engineering or regulatory outcomes within a defined period of time. Since solutions based on changing management systems are longer term and involve less measurable outcomes, they are unlikely to be recognized by managers who measure shorter-term performance, and employees have little incentive to spend time, effort, and resources on behavior change projects. Not surprisingly, then, in our research with Chinese industrial facilities (Figure 22.1), we found that the most enthusiastic supporter of water stewardship was a water professional at or near retirement who no longer felt bound by the constraint of his performance framework.

This bifurcated attitude towards collaboration is also reflected in the water industry's professional development activities, such as conferences. On the one hand, global conferences (e.g., Stockholm Water Week) feature numerous representatives from customer industries who present on their latest water management innovations. On the other hand, more practically focused 'bread and butter' national, state, and local conferences, (e.g., Australia's annual OzWater meeting, that attracts some 5000 water

Figure 22.1 Water treatment at a textile factory in Zhejang Province, China. (*Credit: Michael Spencer*).

professionals) focus almost exclusively on technical solutions to water problems. For example, an analysis of three days of the recent OzWater 2019 program revealed that of the 261 presentations, panels, workshops, and poster sessions, words such as 'customers,' 'partnership,' 'collaboration,' or 'engagement' were featured in fewer than 20 titles, while fewer than 10 presentations appeared to actually discuss engaging customers in solving water problems. (The Alliance for Water Stewardship was told there would not be sufficient interest to warrant a presentation on water stewardship.)

But conferences are only symptomatic of a broader cultural problem. Water authorities are generally run by male engineers, often with additional business training, surrounded by other technical experts. The fact that it is an engineer who normally rises to the top in a water agency reflects the technical and engineering bias of the agency culture. A quick review of the heads of major capital city water agencies in each of the six Australian states shows that all except one are run by people with an engineering background (often complemented by a business qualification). It is interesting in this context that the one water agency not headed by an engineer has demonstrated more interest in water stewardship than the other five, and the greatest degree of willingness to consider how it might work with others to solve water challenges.

The point here is not to criticize engineers, technicians, and scientists whose skillsets are all vital to solving our water challenges. Rather, it is to emphasize that a different set of skills will be needed for behavior change to take place, and to introduce a broader idea of what constitutes a solution into our collective water management toolkit. It is for this reason that ecological economist Robert Costanza and his colleagues lament what they refer to as the *reductionist paradigm* which assumes the world is infinitely separable into isolated units of knowledge. This idea took hold at the start of the 20th century, stimulated by universities where internal reinforcement systems rewarded specialized work and led to subdivision of knowledge into more disciplines. Each discipline developed their own unique language, culture and ways of looking at the world (Costanza *et al.*, 2015). Or as Donna Meadows observed in her primer on systems thinking: 'Too often, universities are living monuments to boundary rigidity' (2008). A cursory review of university engineering programs reinforces the perception of enforced isolation of the disciplines.

This reductionism transfers from universities to professions that find their thinking constrained by what economist Herbert Simon referred to as *bounded rationality* (Simon, 1986). Quoting *The Professions* (a near century-old study by Carr-Saunders and Wilson) Costanza *et al.* note: '*Every profession lives in a world of its own. The language spoken by the inhabitants, the landmarks familiar to them, their customs and conventions can only be learnt by those who reside there*' (p. 54). Unfortunately, the rapidly developing challenges of managing the planet's natural capital cannot be solved by any one profession and will require the application of a range of disciplines. Water management has been dominated by engineering solutions for millennia – some would argue (e.g., Solomon, 2010) that the origins of organized human social systems lie in the successful management of water – so developing these different skills will require a major cultural shift to change long-held attitudes.

As noted earlier, the problem is not all one-way. Just as water engineers will want to solve water problems using their amazing box of engineering solutions, industrial customers are by and large happy to leave them to it. In our previously cited work on motivations and constraints for uptake of water stewardship in China, managers at water-using sites were asked how responsibility for water and catchment health was spread among various actors (government, industry, agriculture, households, and NGOs). Sites that embraced water stewardship saw responsibility spread among a broader range of actors, while those that were not interested had a narrower view and saw government as primarily responsible for water and watershed health. Early indications suggest this finding will hold with industrial water-users in Australia and New Zealand. In short, water agencies and water professionals must begin to embrace behavioral solutions that involve their customers while industry must accept its role in resolving water challenges. But this change has the best chance of success by starting with water professionals (Figure 22.2).

22.6 OPPORTUNITIES FOR COLLABORATION

Solving the world's water crisis won't happen unless behavioral solutions are embraced (2030 WRG, 2009). If attitudinal issues can be addressed, there is significant potential for building collaboration and shared solutions to water challenges. One of the most successful recent examples of collaboration has been near Shanghai in the Chinese municipality of Kunshan. Here a collaboration between the Environment Protection Bureau, the municipal government, industry, international customers, and civil society groups (World Wildlife Fund China, Alliance for Water Stewardship) engaged in an education and training program (Figure 22.3) that helped water-using sites that helped water-using sites understand water challenges facing the region, identify solutions, and evaluate their potential rewards.

Collaboration was built not just between the organizations, but between key individuals in those organizations. Government and customers engaged water-using sites while the civil society organizations

Figure 22.2 Tour of AWS-certified Ecolab facility in Taicang, China (*Credit: Ecolab*).

Figure 22.3 Alliance for Water Stewardship training session in Shanghai, China. (*Credit: Michael Spencer*).

offered training in water stewardship and pointed sites to best practice and available technical solutions. This created legitimacy for the program and a practical focus on measures that site managers could see would enhance operational performance. After almost two years, the municipal government offered an incentive payment of approximately USD $15,000 for sites that achieved water stewardship certification and multinational customer companies offered further support for the program. This support was intended to assist with certification, project management and technical support. This stage-by-stage engagement process is illustrated in Figure 22.4.

The Kunshan example in China provides a useful model for *cluster-based* water stewardship projects. Cluster-based projects, focused on groups of operating sites, are distinct from projects by large customer-facing corporations for whom reputational risk to their brands may represent a stronger motivation for engagement in water stewardship; for example, Nestlé Waters has announced that it will certify up to 85 sites world-wide to the AWS International Water Stewardship Standard. Research in China tested interest in a range of incentives to encourage participation: direct incentive payments, reduced taxes and levies, access to green finance, and reduced compliance obligations. While the previously discussed 'behavioral failure' indicated that incentives alone were unlikely to persuade a company to participate, incentives did support an internal business case once a site becomes interested in water stewardship. The research confirmed that all participating sites were interested in most forms of incentive or reward. Firms in Kunshan, the first municipality in the world to offer an incentive payment, were generally positive about this payment. In other areas, where there was less familiarity with the payment, firms were slightly more interested in reduced taxes on water or lower environmental levies. Green finance was favoured slightly less than incentive payments and reduced taxes. In Kushan, where the water and environmental compliance burden was high, sites were also interested in receiving relief from compliance reporting as a result of complying with an international audited protocol like the AWS International Water Stewardship Standard (Abdel Al *et al.*, 2014).

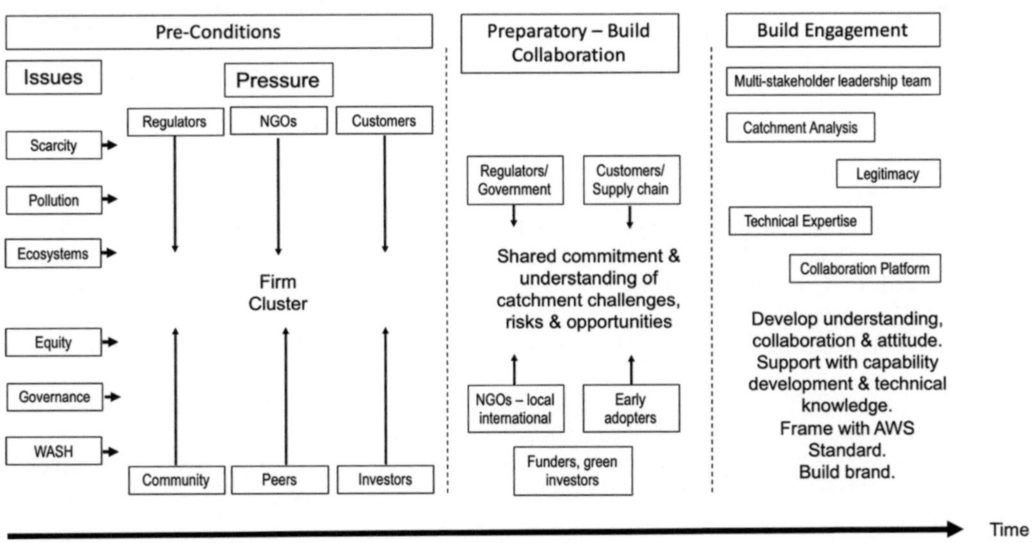

Figure 22.4 Stage-by-stage engagement process for water-using sites based on experience in Kunshan China. Building engagement of different actors before embarking on certification discussions (Spencer (forthcoming)).

In thinking about how this approach could be institutionalized, consideration needs to be given to how impacts can be better measured and tracked, particularly if the program is delivering both public and private benefits. Government agencies need to be able to track public benefits from water stewardship where public funds may be offered to reward companies participating in water stewardship. To justify these payments or relief measures, AWS looked at how the AWS Standard could be linked to the System of Environmental Economic Accounting (SEEA). A system for measuring and reporting environmental performance sponsored by the United Nations and World Bank, SEEA is designed to sit alongside the System of National Accounts used to report national accounting measures such as Gross Domestic Product (GDP). By reporting AWS impacts in a universal reporting framework such as the SEEA, governments could more easily relate water stewardship to a public policy framework and justify investment in the program. The SEEA is increasingly being taken up by national and provincial governments. A pilot project linked to the AWS program in China found that there was scope for linking AWS and SEEA which could provide a basis for measuring, accounting for, and reporting environmental outcomes and prioritising investment in a water stewardship program (IDEEA Group, 2018; Obst & Eigenraam, 2017).

22.7 A FINAL WORD – ACCELERATE CHANGE

In identifying the need for behavior change to broaden the solutions in our global water toolbox, I have given voice to frustration with the current slow rate of progress and spoken of the need to disseminate water management and behavioral solutions beyond the purview of a narrow group of international companies. A key obstacle to these changes is the persistent culture among both water providers (government and private) and water-consuming industries that must shift to make room for expanded implementation of behavior solutions, such as water stewardship. Where this obstacle has been overcome, there is considerable scope for collaboration, especially where a staged approach can build participation in cluster-based water stewardship projects and impacts can be measured and reported in a language government and industry can both understand. If this can be achieved, there is scope to build shared investment in water stewardship programs that can deliver shared public-private benefits. None of this will be easy or quick, and important work remains to be done in the education and training available to future water leaders.

REFERENCES

2030 Water Resources Group. (2009). Charting Our Water Future; Economic frameworks to improve decision-making.

Abdel Al I., Ahmad S., Ballestero M., Bezbaroa S., Cookey P., Dourojeanni A., Galli C., Langford J., Mensink M., Opondo G., Park J. M., Pinerso E., Ruffier P., Witmer L. and Xin H. (2014). The AWS International Standard: Alliance for Water Stewardship.

Asian Development Bank. (2016). Asian water development outlook 2016: Strengthening water security in Asia and the Pacific.

Costanza R., Cumberland J. H., Daly H., Goodland R., Norgaard R. B., Kubiszewski I. and Franco C. (2015). An Introduction to Ecological Economics, 2nd edn. CRC Press, Boca Raton, FL.

High Level Panel on Water. (2018). An Agenda for Water Action; An Open Letter from the High Level Panel on Water. https://sustainabledevelopment.un.org/content/documents/17829Open_Letter_HLPWater.pdf (accessed 5 October 2020).

IDEEA (Institute for Development of Environmental-Economic Accounting). (2018). Application of AWS and SEEA to accounting for water: China industrial park demonstration (Final Report).

International Conference on Water and the Environment. (1992). The Dublin Statement on Water and Sustainable Development. Paper presented at the International Conference on Water and the Environment, Dublin.

Meadows D. H. (2008). Thinking in Systems: A Primer. Chelsea Green Publishing, White River Junction, VT.

Morrison J. and Gleik P. (2004). Freshwater Resources: Managing the Risks Facing the Private Sector. https://pacinst.org/wp-content/uploads/2013/02/business_risks_of_water.pdf (accessed 5 October 2020).

Newborne P. and Dalton J. (2016). Water Management and Stewardship: Taking stock of corporate behaviour. https://portals.iucn.org/library/sites/library/files/documents/2016-069.pdf (accessed 5 October 2020).

Obst C. and Eigenraam M. (2017). Linking AWS and the SEEA: Applying advances in accounting for natural capital to support implementation of AWS. https://a4ws.org/download/seea-final (accessed 5 October 2020).

Orr S. (2018). Foreword. In: Water Stewardship and Business Value: Creating Abundance from Scarcity, W. Sarni and D. Grant (eds.), Earthscan from Routledge, London.

Sarni W. and Grant D. (2018). Water Stewardship and Business Value: Creating Abundance from Scarcity. Earthscan from Routledge, London.

Shogren J. F. and Taylor L. O. (2008). On behavioural-environmental economics. *Review of Environmental Economics and Policy*, **2**(1), 26–44.

Simon H. (1986). Theories of bounded rationality. In: Decision and Organisation: A Volume in Honor of Jacob Marschak, 2nd edn, C. B. McGuire and R. Radner (eds.), Univeristy of Minneapolis Press, Minneapolis.

Solomon S. (2010). Water: the Epic Struggle for Wealth, Power and Civilization. Harper Collins, New York.

Spencer M. (forthcoming). Managing the interface between humans and nature; opportunities and constraints for collaborative governance doctoral dissertation. Monash University, Melbourne.

Spencer M. and Xu Z. (2021). Water stewardship; engaging business, civil society and government in collaborative solutions to China's freshwater challenges. In: Non-state Actors in China and Global Environmental Governance, O. Young, D. Guttman and Y. Jing (eds.), Palgrave McMillan (forthcoming).

United Nations Environment Program. (2019). Environmental Rule of Law; First global report.

Chapter 23

Designer water: One utility's unique approach to industrial sustainability

E. J. Caldwell

West Basin Municipal Water District, Water Policy & Resources Development, Carson, CA, USA

Keywords: designer water, industrial reuse, incentives, public-private partnerships (PPP), West Basin Municipal Water District (WBMWD)

23.1 INTRODUCTION

The West Basin Municipal Water District ('West Basin') is widely known for treating and delivering five types of 'designer water' to a variety of recycled water customers. In order to maximize the use of recycled water in the region, West Basin has adopted a 'fit for purpose' approach to treating wastewater, in which five different treatment processes ensure that the water quality meets the specific needs of each customer. While this approach requires West Basin to meet a variety of stringent water quality standards, it has allowed us to satisfy the needs of our diverse municipal and industrial customer base.

The amount of water Southern California receives from the Sierra Nevada Mountains (via the State Water Project) and from the Colorado River (via the Colorado River Aqueduct) has historically been reduced by factors that include prolonged droughts, varying snow and precipitation, and regulatory limits to protect endangered species in the Sacramento River Delta. By adding locally produced water (recycled, conserved, and desalted groundwater) to its portfolio, West Basin is able to significantly enhance the reliability of its water supply. Currently West Basin manufactures as much recycled water as the volume used by more than 130,000 households annually. However, in order to further reduce its reliance on imported water supplies, it must have customers who can use its 'designer water' in place of existing potable supplies.

Each of the 17 cities and unincorporated portions of Los Angeles County (California, USA) served by West Basin is home to a wide range of water users. Under the guidance of its Board of Directors,

DOI: 10.2166/9781789060676_0255

West Basin's engineering and operations teams utilize the best available wastewater treatment technologies to meet the needs of the communities it serves, whether irrigating public parks and sports fields, injecting recycled water in the ground to augment groundwater supplies, or treating water to specification to serve the region's many industrial water users. This approach allows West Basin to maximize the use of recycled water. The five types of recycled water served by West Basin include the following:

(1) *Irrigation water*: Wastewater filtered and disinfected to California's 'Title 22' standards for turf irrigation and a variety of other non-potable uses;
(2) *Cooling tower water*: Wastewater filtered, disinfected, and nitrified to remove ammonia through a biological process for use in industrial cooling towers;
(3) *Seawater barrier and groundwater replenishment water*: Wastewater filtered through microfiltration and reverse osmosis membranes and disinfected prior to pumping underground to form a barrier against seawater intrusion into local groundwater supplies;
(4) *Low-pressure boiler feed water*: Wastewater filtered through microfiltration and reverse osmosis membranes for use as low-pressure boiler feed water; and
(5) *High-pressure boiler feed water*: Wastewater filtered through microfiltration membranes and passed twice through reverse osmosis membranes for use as feed water for high-pressure boilers.

23.2 ECLWRF: THE HEART OF THE PROCESS

West Basin's Edward C. Little Water Recycling Facility (ECLWRF), located in the city of El Segundo, is the largest water recycling facility of its kind in the United States. ECLWRF is designed to filter and disinfect over 150ML/day (40 mgd) of partially treated effluent from the City of Los Angeles Hyperion Treatment Plant, turning wastewater into useable water. Nearly one-third of that highly treated flow (up to 50 ML/day or 12.5 mgd) can be treated by a combination of microfiltration and reverse osmosis membranes for use in a range of industrial settings, including cooling towers and low- and high-pressure boilers. In addition, nearly 20 ML/day (5 mgd) is treated to potable standards and pumped into local groundwater supplies to protect them from seawater intrusion (Figure 23.1).

The ECLWRF was recognized by the National Water Research Institute in 2002 as one of only six national centers for water treatment technologies, and the only one in the United States producing five different qualities of recycled water ('designer water') to meet the specific needs of West Basin's municipal, commercial, and industrial customers. The ECLRWF also houses a 60,000 square foot solar power generating system that has reduced emissions of carbon dioxide by over 356 tons over one year. These emissions reductions are equivalent to planting nearly 100 acres of trees or not driving 890,007 miles.

23.3 SATELLITE FACILITIES

In addition to the primary ECLWRF, West Basin also owns and operates three satellite facilities:

- The Chevron Nitrification Facility (El Segundo) which treats approximately 20 ML/day (5 mgd) of recycled water from ECLWRF through a nitrification process for industrial applications. The Phase V Expansion Project will provide the facility improvements to accommodate an additional 2 ML/d (0.58 mgd) of nitrified water demand.
- The Juanita Millender-McDonald Carson Regional Water Recycling Plant (Carson) treats 15 ML/d (3.5 mgd) of recycled water from ECLWRF through microfiltration, reverse osmosis, and nitrification to provide water for boiler-feed and cooling tower applications in an adjacent refinery. The Carson Facility capacity will be expanded by 4 ML/d (1 mgd). The treatment capacity of the Carson

Figure 23.1 West Basin's Edward C. Little Water Recycling Facility (El Segundo, CA) treats 150 ML/d (40 mgd) of secondary effluent to non-potable and potable standards for municipal and industrial reuse. (*Credit*: West *Basin Municipal Water District*)

Facility will also be modified to add ultraviolet (UV) disinfection and advanced oxidation processes (AOP) in order to supply water to the Dominguez Gap Barrier to prevent seawater intrusion into local groundwater supplies.

- The Torrance Refinery Water Recycling Plant (Torrance) is located in the Torrance Refining Company facility. This recycling plant nitrifies recycled water to remove ammonia prior to introduction into cooling towers and as boiler feed water. Since 1995, this major industrial customer has used recycled water at a daily average rate of over 20 ML/d (6 mgd).

23.4 FROM CONCEPT TO IMPLEMENTATION

23.4.1 Institutional challenges

Leadership and outreach are two major factors in West Basin's success. West Basin was created in 1947 to reduce groundwater over-pumping by providing new sources of water. In 1948 West Basin joined Metropolitan Water District of Southern California, which sells imported water from the Colorado River and Northern California to more than two dozen cities and water agencies in Southern California. After a severe drought in the late 1980s taxed Metropolitan's ability to supply all its retailer demands, West Basin's Board decided to implement aggressive water conservation and reuse to bolster its reliability and protect against future shortages. West Basin remains committed to being a leader in the water industry by exploring new methods and technologies that enhance reliability in the region's water supply, including seawater desalination.

West Basin's leadership was demonstrated early on in its development of a number of productive collaborations, beginning with its agreement to purchase secondary-treated wastewater from the City of

Los Angeles. West Basin also executed agreements and maintained productive working relationships with its many retail water customers, including the California, Cal-American, and Golden State water companies, the cities of Inglewood, El Segundo, Manhattan Beach, and Lomita, and the Water Replenishment District of Southern California, the regional groundwater manager. Along with the cost of water, these agreements included a range of operational details such as minimum pressures and flows and maintenance of transmission and distribution infrastructure. West Basin's customized agreements with its larger retail industrial customers (including the Chevron, Marathon, and Torrance refineries) stipulated water quality and price and established protocols for communication between the utility and the industrial customer.

Another challenge was financial. West Basin staff obtained a number of federal and state grants and loans to help limit the impact of the project on local ratepayers. However, industrial water users were responsible for on-site retrofitting, including investment in new equipment. Customers were also responsible for the cost of water which varied with the level of treatment, and for low- and high-pressure boiler feed water was actually higher than the cost of potable water.

With respect to outreach, from the beginning West Basin conducted an extensive education campaign to inform the public, business community, and community leaders about the value of using recycled water. Through this effort, West Basin reduced public fears about health risks related to a process some disparaged as 'toilet to tap', and built community trust in the treatment technologies and the recycled water product. In addition to health issues, industrial customers were concerned about the uniformity of water quality, especially since the quality of the effluent from the City of Los Angeles (the influent to West Basin facilities) could vary over time. In response to these concerns, West Basin worked with the customers to set up and maintain communication channels where utility staff could advise industrial plant operators about pending deviations in water quality.

23.4.2 Engineering challenges

Operating a system designed to meet the needs of different communities presents many challenges. In addition to constructing satellite treatment facilities to produce water of appropriate quality for its industrial customers, West Basin has invested in roughly 160 km (100 miles) of purple pipeline (the color designated for non-potable water), with an additional 100 km (60 miles) of pipeline planned. Delivering recycled water further and further from the ECLWRF will also require additional pump stations and satellite treatment facilities, which could have significant cost impacts on system operations.

West Basin's recycled water system was designed and constructed to deliver recycled water to 'high volume' industrial water users (refineries), or anchor customers. Using the pipelines constructed to deliver recycled water to the anchor customers, West Basin was able to build individual laterals to supply additional recycled water to parks, sports fields, and other industrial water users. To facilitate their design and implementation of water reuse projects, West Basin performed water audits for its industrial customers to help them use water more efficiently and to determine if recycled water is available for their specific needs. By reducing water demand by industrial customers, West Basin is able to make more recycled water available for others.

23.4.3 Industrial challenges

The term 'designer water' is part marketing and part technology. West Basin coined the term to reflect the way the utility designs and treats water to meet the needs of its customers. Every community served by West Basin is unique, and so are the industrial users in those communities. As specific problems were encountered, such as the sensitivity of admiralty metals to ammonia, West Basin adjusted its water quality specifications to the needs of its industrial customers. To ensure that these water quality needs are always met, West Basin routinely tests its recycled water for some constituents not required by state health regulations (e.g., silica). There are several factors that impact the utility's ability to meet those specifications, such as a higher influent contaminant concentration at the West Basin treatment facility resulting from reduced wastewater flows into the City of Los Angeles' Hyperion wastewater plant. When changes in recycled water quality occur, West Basin staff communicate directly with the affected customers to ensure that no processes are impacted.

Despite the utility's efforts at outreach and communication, some companies did not immediately embrace the change in their water supply. This caution is understandable, since altering an industrial facility's water supply has the potential to create an assortment of 'unknowns' that might impact plant operations. In those cases, it was often up to the plant manager to demonstrate the leadership required to recognize the benefits of reducing water demand and using water more efficiently.

23.4.4 Lessons learned

Past challenges to implementing the designer water program included limited public acceptance of recycled water (in California) and low demand for recycled water by industry. However, the water industry and trade associations together have done an excellent job of educating the public; where West Basin once had to 'sell' recycled water to a potential water user, the utility is now contacted by customers who want to purchase recycled water. Equally important, the regulatory community has developed trust in the recycled water agencies and their use of technology to safely produce recycled water for potable and non-potable uses.

West Basin's experience demonstrates the value of tailoring the quality of recycled water to maximize its use by industrial customers. Moving forward, the new challenge might be that reduced wastewater flows due to conservation will create a competitive market for plant effluent, forcing some recycled water utilities to look for new sources of water to supply their industrial customers. Currently, several large Southern California utilities – including the Metropolitan Water Districts and the City of Los Angeles – are developing plans to reuse even more water for potable purposes over the next 10 years. In that meantime, West Basin will continue to provide high-quality 'designer water' specifically suited to its industrial customers, helping them save water and energy now and in the future.

Chapter 24

Incentivizing sustainability: How utilities can support industrial water conservation and reuse

Eric Rosenblum

Water Resource Consultant, San Jose, CA, USA

Keywords: cost/benefit, incentives, industrial reuse, public-private partnerships (PPP), recycled water, sustainability

24.1 INTRODUCTION

Water and wastewater utilities around the world have an important and difficult job, protecting the health of their communities and the environment by providing safe drinking water and removing domestic and industrial pollution. Their efforts are usually overlooked, and the attention they do receive is often negative, so they tend to stay out of the public eye. While understandable, this shyness does not advance global progress towards sustainability. Utilities alone cannot be expected to create a truly resilient, renewable water system sustained by a restored, thriving environment ('fishable, swimmable rivers'). A task this large needs the broad support of every sector, including the participation of industry.

How can a utility gain the cooperation of industry? Spencer (2021) suggests that the first step is for utilities to acknowledge that the companies they serve have a role to play in solving local, regional, and national water problems. This approach may be particularly appropriate where companies are already connected to a local utility, involved in the community, and disposed to see themselves as 'good neighbors.' In some areas, however, industries operate more or less independently of local authorities: they pay no water bill and do not treat their wastewater prior to discharge. In that case, the national government may have a more important role to play than the local utility in bringing the public and private sectors together (Nizami *et al.*, 2021).

But even where local government and industry are well-aquainted, their working relationship may be limited to the monthly water bill. How then, can utilities better support and encourage companies to invest in water conservation and reuse? Conservation-based block rate structures, bill inserts advocating conservation, and cash rebates for investment in water improvement are all routinely used to encourage

DOI: 10.2166/9781789060676_0261

sustainable industrial water use. Without some expectation of ongoing support, however, even these incentives may not be attractive enough to entice business into dialog with a government-run agency.

From my experience as former manager of a recycled water utility, this is precisely the challenge many agency employees encounter when they approach industrial water users about switching from drinking water to more environmentally sustainable non-potable water. To be persuasive, a good start is to find some common ground between utility bureaucrats and company managers before showing how protection of public health and the environment can harmonize with industry's desire to earn a profit. With candid conversation and attention to detail, utilities can help even reluctant companies reduce their water impact and 'green' their operations.

24.2 PUBLIC BENEFITS AND PRIVATE COSTS

24.2.1 Water recycling comes to Silicon Valley

Along with several other partner communities, the northern California cities of San Jose and Santa Clara operate and maintain a large 600 ML/day (150 mgd) advanced wastewater treatment plant that discharges into the south end of San Francisco Bay. In 1989, the Regional Water Quality Control Board determined that the highly treated freshwater discharged from the plant was reducing the salinity in the adjacent salt marsh, home to two endangered species – the California clapper rail and the salt marsh harvest mouse. To protect these animals and their wetland habitat, the Board limited the amount of effluent that could be released into the South Bay during the summer months to 450 ML/d (120 mgd) – a 10% reduction in dry weather discharge. Since California was in the midst of a seven-year drought, the cities decided that, rather than simply pumping the effluent further into the Bay away from the salt marsh, it was better to reuse it for irrigation and cooling in this water-short region (Rosenblum, 1999).

24.2.2 Selling used water

Project development commenced in 1994 at an accelerated pace to meet the 1998 flow reduction deadline. The pipeline was designed to serve recycled water to the largest non-potable water users, so a campaign was immediately launched to connect property owners to the recycled water system. In addition to golf courses, parks, and other irrigation customers, several large industrial water users near the pipeline were approached about using recycled water for manufacturing, cleaning, and cooling. Offering factory owners the opportunity to substitute cheaper recycled water for their traditional drinking water supply revealed some important characteristics shared by many companies, such as:

(1) A fierce protectiveness of the profitability of their enterprise;
(2) An instinctive avoidance of risk without reward; and
(3) A deep-rooted suspicion of government.

Addressing these concerns taught us some valuable lessons about why companies are reluctant to participate in public environmental programs, and how they can be encouraged to use water more sustainably. Among the other clues we received were the importance of (1) willingness to help defray connection costs; (2) clarity about risks and benefits; and (3) timely communication and assistance at every step of the planning and implementation process.

24.3 BARRIERS TO SUSTAINABILITY

The rationale most commonly offered by companies as to why they don't adopt water conservation and reuse practices is 'cost.' The price of sustainability, they argue, is more than they can afford.

But companies routinely spend large sums of money developing new products, polling customer focus groups, advertising their wares, and even throwing parties to celebrate their success – activities which, on balance, might seem more costly and less valuable than reducing their environmental footprint. From the perspective of an advocate of sustainable water use, businesses may appear miserly and heedless of the public welfare.

In fact, the issue of 'cost' is often just a convenient box into which a variety of objections are tossed, including worry over market share, care for aged and fragile equipment, fear of regulatory delays, and even an interest in worker health and safety. With time and effort, however, a sympathetic utility representative can unpack these concerns, address them one by one, and demonstrate the ultimate value of sustainable industrial water use to the most reluctant company.

24.3.1 Cost and competition

Public sector personnel often fail to appreciate the extent to which competition influences the decision-making process in business. Utility managers who define success in terms of protecting public health, and utility workers who receive a regular paycheck from a tax-funded institution, may have a hard time understanding the perspective of those whose continued employment depends upon meeting quarterly sales goals.

For one thing, business owners are keenly aware of their margins: the cost of goods sold, the price they sell for, and the profit represented by the difference. This simple equation often determines how companies assess any proposal. That is, if the proposed action incurs a cost but does not increase the market value of the product or boost sales it represents a decline in profits and deserves no further consideration. Projects that can't pass this simple test are rarely implemented.

For this reason, the first step in selling water reuse involves the determination of the return on investment, comparing the project cost to the estimated savings. Along with an even-handed analysis of implementation issues, this comprises the basic business case which is evaluated by managers who decide whether or not to adopt the recommended measures. As noted by several other authors in this anthology, one reason why companies are often reluctant to take steps to conserve or reuse water is that the cost of water is so low that the financial benefits of water savings rarely exceed the project cost (Chesnutt, 2021; Tenuta, 2021).

To make a persuasive business case for sustainable water use, utility managers must be prepared to demonstrate how other advantages outweigh implementation costs, how the cost of improvements can be reduced through subsidies, or both (Rosenblum & Moore, 2021). With respect to our program, this meant that to get the interest of prospective customers we had to discount the price of recycled below the already low cost of drinking water. We also had to assume some of the cost of connection and the burden of installation, and provide additional incentives in the form of publicity and goodwill.

The most common expenses our utility absorbed to encourage the use or recycled water were (1) the cost of connection to the recycled water pipeline; and (2) the cost of re-plumbing the site ('retrofitting') to use recycled water. In order to ensure that potable and non-potable water supplies remain separate, sites using both types of water are required by California law to distinctively label recycled water pipes and to separate the two internal piping systems. (Recycled water users must also provide backflow devices to prevent non-potable water from entering the public potable water system.) We sometimes reimbursed the cost of designing and building onsite pipeline modifications as loans which the customer could repay through an incrementally higher water rate – usually the same rate as potable water, so that the 'discount' paid for the retrofit. In all cases, we paid the cost of extending the pipeline to the customer site, at least until the system was well established and the use of recycled water was generally accepted.

24.3.2 Risk factors

As eager as they are to maximize profits, businesspersons are just as motivated to minimize risk. Risk can't be avoided – it is an inevitable part of growth and change. Experienced company managers understand that the market is fickle and success is cyclic: today's top seller is tomorrow's reject. From this standpoint, companies must constantly re-evaluate their performance and re-calibrate their operations to meet their customers' evolving demands.

By the same token, it is a rare company that takes unnecessary chances. There are enough 'unknowns' in business that it seems foolish to voluntarily assume risk without a clear reward. On the contrary, most companies put a good deal of time, effort, and money into risk reduction, to protect them from unforeseen losses and limit their exposure to the risks they take intentionally for good reason and with their eyes on the prize.

This presented us with both a challenge and an opportunity when discussing with potential customers the costs and benefits of connecting to the recycled water pipeline. On the one hand, switching a manufacturing plant from potable to non-potable water can impact performance, especially those processes sensitive to increases in ionic strength or conductance such as cooling or microchip manufacturing. On the other hand, using recycled water reduces vulnerability to water supply cutbacks during drought – a frequent occurrence in California.

In the first instance it was important to develop (or retain) expertise in industrial water use to properly advise companies about the solutions to potential water quality issues. Without some assurance that these issues are manageable, companies are unlikely to voluntarily change their source of water. The quality of the recycled water we supplied was comparable to drinking water available elsewhere in the Western United States, and significantly lower in salinity than many places in southern California, so no process problem was unsolvable. However, changing source water does alter specific aspects of the pretreatment needed before using water in cooling towers or in applications requiring extremely pure water, like those involved in the production of many computer components. For many customers, the small adjustment to their water treatment regime required was reason enough to reject recycled water out of hand, regardless of the potential savings. As one company manager responded when informed that he could save $100,000 per year using recycled water at his chip manufacturing plant, 'That's fine, but if my production line goes down I'll lose $1 million per day!'

The utility has no way of knowing how fragile a company's manufacturing process may be, whether due to the age of the equipment or exactness of the product specifications. Unless water reuse is mandated by law, the company will decide for itself whether to accept the risk – no matter how minor – of switching to recycled water. A utility representative can help, however, by providing technical information and support to make it easier for a company manager to reach a favorable decision. This could be as simple as offering historic data that accurately describes any seasonal variability in water quality of interest. In one example profiled elsewhere in this book, a water agency maintains an open communications process to notify an industrial customer of any significant changes in water quality (Caldwell, 2021). By contrast, technical support could be as robust as hiring a consultant to advise the company on process impacts. Making a recognized expert on the subject of industrial water treatment available to a potential reuse customer can give them enough of a sense of security to connect to the system. (To encourage a local university to use recycled water, we even provided funding for a graduate student to monitor water quality changes and develop a thesis on the use of recycled water in its cooling system.)

The challenge of addressing operational risk is compounded when a company outsources its water treatment to a third party. Activities like cooling water treatment services are frequently provided on retainer, such that the contractor is paid a set amount monthly to keep the system running within specified parameters, regardless of the effort required. Unless the company directs them otherwise, the contractor has no incentive to modify treatment, even if such a change entails only minimal cost. As a result, we noted that contractors frequently exaggerated the risk of using recycled water, implying that it was incompatible with process operations even when similar facilities elsewhere were using it effectively. By walking through the operational issues step-by-step, we were able to convince one cooling tower service company to acknowledge the feasibility of reuse – and then only after we pointed out that there were other contractors willing to provide the same service at the same price with recycled water.

On the other side of the ledger, a company sensitive to the risk of future water shortages can appreciate the long-term advantage of using recycled water. These advantages are not readily expressed in 'dollars and cents,' so their worth has to be translated into monetary terms. One way to calculate the value of a resilient water supply like recycled water (either municipally supplied or produced onsite) is by estimating the cost of reducing current water usage by 10%, 20%, or 30%. This could include the cost of additional chemical treatment required to reuse available supplies, the expense of retooling a production line to use a waterless system, or the lost revenue of having fewer products to sell. Whatever this cost is, the 'insurance value' of water reuse can be based on likelihood of such cutbacks occurring. One equipment manufacturer we approached figured that it would cost in excess of $100,000 to accommodate a 20% water cutback, a possibility which they estimated was likely to occur at least once or twice over the next 10 years. By factoring into the business case, an 'avoided cost of $10,000–20,000 they were able to justify the expense of connecting to the recycled water system.

24.3.3 Regulatory roadblocks

Industry welcomes regulation as eagerly as cats take baths. Unfortunately, their reflexive rejection of any proposed rule (usually accompanied by dire predictions about the end of all commerce) overshadows the grain of truth in the complaint that some regulations are in fact counter-productive, doing little to improve the environment while burdening business unnecessarily. This was brought home to us repeatedly in our efforts to connect industrial customers to our non-potable water system.

One of the first businesses in Silicon Valley to embrace recycled water was a firm that used over 1 ML of water per day (300,000 gpd) to manufacturer a variety of paper products. The company had already incorporated a number of 'green' elements into their process, including waste heat recovery and reuse of newsprint collected through the community's solid waste recycling program, so it seemed natural for them to use recycled water to process the recycled paper into pulp before pressing it into boxes. However, some of the cardboard they made was bleached, waxed, and formed into boxes that were used to contain frozen vegetables. That put the company in the crosshairs of the US Food and Drug Administration (FDA) which required that frozen food packaging had to be manufactured with potable water.

Essentially, the FDA rule meant that the company would have to abandon a lucrative product line in order to use recycled water. The inappropriateness of this restriction was highlighted by the fact that they proposed to use recycled water primarily in the pulping process, in which the scrap paper gathered from the street was cooked at high temperature before being strained, filtered, and pressed into cardboard. Happily, after only 18 months of correspondence, the FDA reconsidered their position and issued a letter allowing the use of non-potable water in this particular application: the company continues to use recycled water in lieu of drinking water in its pulping process today.

No doubt most rules are enacted to protect the public and the environment against very specific threats to health and safety. But government rule-makers are now challenged to operate in an increasingly complex and interconnected world. In the United States, for instance, this has given rise to the use of 'whole effluent toxicity' tests that measure treatment plant effluent quality by its ability to sustain various forms of marine life, rather than by the concentration of specific contaminants it contains. While this innovation often places an additional burden on utilities, there is no question that it provides a more realistic assessment of the safety of plant discharge than sequential measurement of individual pollutants. As the saying goes, '*You can't solve network problems with linear solutions.*'

Government must strictly enforce industrial regulations to protect the public and the environment from harm, but they must also recognize when 'one size fits all' directives actually prevent companies from moving towards sustainability (Marcus, 2021). Such counter-productive rules reinforce the opinion of some companies that government is incompetent and must be kept well outside the fence-line, and nowhere near the factory floor. The more knowledgeable utilities are about their business processes, the more they can help their industrial customers participate in programs designed to benefit both the public and private sectors.

24.3.4 The clock vs. the calendar

Recycled water regulations can also discourage companies just by the amount of time required to obtain permits, so we continually had to work to speed up the process. The reasons previously cited that prevent utility employees from appreciating the fiscal pressures businesses operate under also make it hard for them to see how companies measure time differently, too. Where dozens of utility workers are busy creating detailed five-year capital plans and proposing biannual budgets, the majority of people in any given business are focused on the next quarter. So when companies and utilities work together, the first hurdle they often face is simply to agree on the meaning of the term 'timely.'

An industrial customer that supplied compressed gases to all the major companies in Silicon Valley opted to connect to the recycled water system as part of its national corporate commitment to sustainability. The process of separation and compression of gases is exothermic (heat producing), and the company used in excess of 600 kL/day (150 kgal/day) of potable water to cool its industrial facility. The utility paid to extend the pipeline some 400 m (1300 feet) to reach the company's fence-line, and the company paid to modify its onsite plumbing to ensure that the recycled water was isolated from the potable water piping. The chief issue for the company wasn't the cost, but the time it took to coordinate with utility staff to construct the extension and to receive permits to connect to the system. In reviewing the process, one study noted that 'the customer's expectation was six months. Ultimately, the project took 18 months to complete, which according to the company's project sponsor, was 'six months too long'' (Bowden & Layton, 2015).

To put it in a nutshell, too often bureaucrats are marking time on the calendar while business people are looking at their wristwatches. The study went on to offer the following recommendation:

> If water providers truly want an exponential increase in recycled water use by industrial customers, they will have to revisit and revamp their existing project planning processes, approvals, and execution to bring a more timely response to businesses requests for delivery of recycled water. At the same time, business customers must come to understand that many regulatory reviews and approvals are outside the agencies' jurisdiction and sphere of influence. The result will be project delays that are, in many cases, unavoidable. (Bowden & Layton, 2015).

To facilitate collaboration between business and government, the study included a chartering tool designed to let all parties put down in writing their rationale for promoting water reuse, and their contributions to the project (Figure 24.1).

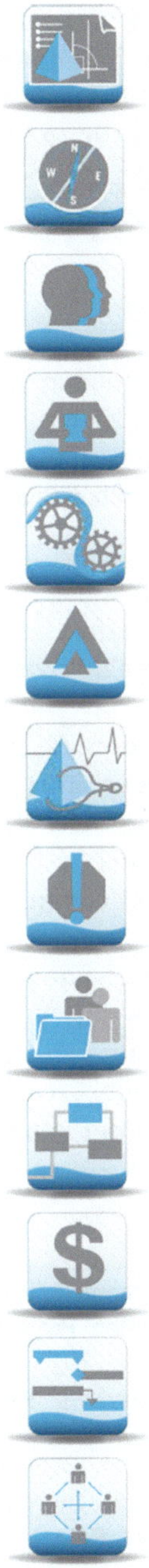

Description
Identify the nature or type of IWR application; e.g., cooling system, boiler system, process; type of plant/industry, location, etc.

Purpose
Identify the business reason for the project (i.e., project drivers/needs analysis). Describe how the project links to major strategies, objectives and drivers.

Key Stakeholders
Identify for whom/by whom this project is being undertaken and other organizations/agencies/entities that will be involved; e.g., regulatory agencies, water district, plant operations, etc.

Decision Maker(s)
Specify for whom the project is being created and list others who eventually may be in the approval loop

Functional Requirements
Identify in general terms what operating conditions the IWR solution must satisfy; e.g., demand/capacity, guaranteed quality range, etc.

Scope/Approach
List and describe in brief the major steps and activities involved in completing the project. For each of the major steps, estimate the amount of effort and time required.

Success Indicators
List the measures – business and operational – by which the project will be considered a success. Example metrics include: amount of fresh water saved, cost factors, median flow rates, operability, etc.

Risk Factors
Identify any and all factors that could undermine the timely and cost-effective delivery of the project; e.g., regulatory and compliance issues, resource availability/bandwidth, plant runtime, maintenance concerns

Resources and Information
Identify the primary project contributors and potential, additional resources or information needed to complete the project on time and within budget.

Organizational Links
Identify other organizations, departments or initiatives that are linked to this project for input / output.

Budget
Is there a designated budget for this project? Can an estimate be developed based on previous projects of similar scope/complexity? Also, funding considerations/ contributor.

Scheduling
Estimate start-to-finish cycle times with consideration given to simultaneous activities and potential critical timing issues. List milestone review and approval dates.

Core Project Team
The Core Team will include organizational leads and support personnel, contractor and vendor representatives, primary contacts for stakeholders, etc.

Figure 24.1 Chartering the industrial water reuse project (Bowden & Layton, 2015).

In short, while corporate incentives reward speed and agility, the public sector is devoted to reliable delivery of clean, safe water to the public unhurried by a profit motive, secure in knowing that operational costs will always be covered by rates. The onus, then, is on utilities to put themselves in the place of business, and respond promptly as though their jobs depended on it. In addition to communicating directly with perspective customers, agency employees can attend meetings of local and state industrial associations where companies will be glad to share their issues and concerns.

24.4 INCENTIVIZING REUSE

As long as the public expresses their desire for a healthy environment, companies will continue to burnish their image by demonstrating a commitment to green principles – putting up LEED [Leadership in Energy and Environmental Design] Platinum buildings, reducing emission of greenhouse gases and using less water. In the meantime, public utilities should remember a few rules about working with private sector companies that will help them develop more productive relationships:

(1) *Be honest about cost*. Recognize the fiscal impact of implementing water reuse at the outset, and candidly discuss options for obtaining grants. In the meantime, subsidize the effort, reduce the scope of the project, or provide financial assistance in the form of a loan payable over time to construct the project.

(2) *Acknowledge negative – and positive! – risks*. Express the negative risks of water reuse including challenges associated with the use of recycled water in cooling towers. At the same time, point out the advantages of using recycled water as a strategy to attenuate the risk of future water shortages.

(3) *Set clocks ahead to Business Standard Time*. As challenging as it may seem at first, respond to all calls and emails from industrial customers in 24 hours or less. Demonstrate your commitment by ensuring that they have your full attention and support whenever it is requested.

(4) *A chartering tool*. As a means of confirming the collective support for implementation of water reuse, consider using something like the 'Chartering Process' described in the WateReuse Foundation report (Bowden & Layton, 2015). By identifying each of the primary partners' roles and responsibilities, the Charter can set the stage for future collaboration. A graphical explanation of the chartering process and a sample charter document are shown in Figure 24.1.

REFERENCES

Bowden J. and Layton R. (2015). 'Evaluation of Historical Reuse Applications and Summary of Technical/Regulatory Issues and Related Solutions for Industrial Reuse Projects' WateReuse Research Foundation Project 12-03. WateReuse Foundation, Arlington, VA.

Caldwell E. J. (2021). Designer water: one utility's unique approach to industrial sustainability. In: Sustainable Industrial Water Use: Perspectives, Incentives, and Tools, C. Davis and E. Rosenblum (eds.), IWA Publishing, London.

Chesnutt T. (2021). The economics of sustainable industrial water use, reuse, and the value of water. In: Sustainable Industrial Water Use: Perspectives, Incentives, and Tools, C. Davis and E. Rosenblum (eds.), IWA Publishing, London.

Marcus F. (2021). Musings of a former regulator: we can do better. In: Sustainable Industrial Water Use: Perspectives, Incentives, and Tools, C. Davis and E. Rosenblum (eds.), IWA Publishing, London.

Nizami A., Ali J. and Nguyen-Koah S. (2021). Government-industry partnership for sustainable water use: Insights from Pakistan (Ch.11). In: Sustainable Industrial Water Use: Perspectives, Incentives, and Tools, C. Davis and E. Rosenblum (eds.), IWA Publishing, London.

Rosenblum E. (1999). Selection and implementation of nonpotable water recycling in Silicon Valley (San Jose area) California. *Water Science and Technology*, **40**(4), 51–58.

Rosenblum E. and Moore B. (2021). Implementing sustainable water use: a roadmap for industry. In: Sustainable Industrial Water Use: Perspectives, Incentives, and Tools, C. Davis and E. Rosenblum (eds.), IWA Publishing, London.

Spencer M. (2021). Why the culture of water needs to change. In: Sustainable Industrial Water Use: Perspectives, Incentives, and Tools, C. Davis and E. Rosenblum (eds.), IWA Publishing, London.

Tenuta E. (2021). Closing the execution gap. In: Sustainable Industrial Water Use: Perspectives, Incentives, and Tools, C. Davis and E. Rosenblum (eds.), IWA Publishing, London.

Chapter 25

Onsite water reuse: A collaborative strategy to manage water

Paula Kehoe and Taylor Chang

San Francisco Public Utilities Commission, San Francisco, USA

Keywords: brewery process water reuse, collaboration, log reduction targets, National Blue Ribbon Commission for Onsite Non-potable Water Systems, non-potable, one water, onsite water reuse, risk-based standards, San Francisco Public Utilities Commission (SFPUC)

Increasing water demands and greater water scarcity resulting from climate change, population growth, and aging infrastructure are threatening water security in communities across the globe. These complex dynamics are compromising access to water and sanitation services in urban, suburban, and rural areas. New approaches are needed to meet these 21st century challenges.

The One Water approach has emerged as a transformative framework for how we integrate the management of wastewater, drinking water, and stormwater. The San Francisco Public Utilities Commission (SFPUC) has taken a One Water approach to its water, sanitation, and power utility services. OneWaterSF acknowledges the utility's change from its traditional role of meeting all demand generated by customers with potable drinking water to collaborating with the community and matching the right resource to the right use. As described below, recent examples of successful applications of OneWaterSF in the commercial/industrial sector include developing regulations and permitting onsite office building water reuse, diversion of dewatered drainage for steam generation, and direct reuse of treated process water in local microbreweries.

25.1 SAN FRANCISCO'S ONSITE WATER REUSE PROGRAM

The SFPUC provides retail drinking water and wastewater services to the City of San Francisco, wholesale water to three surrounding counties, and power to municipal, residential, and commercial customers throughout the city. Rethinking the traditional approach to water resources management, the SFPUC

DOI: 10.2166/9781789060676_0271

recognized the value of buildings collecting and treating blackwater, graywater, stormwater, rainwater, and/or foundation drainage onsite for reuse for toilet flushing, clothes washing, cooling, and irrigation, as it can reduce the potable water demand in buildings by 25–75%.

To ensure the design, installation, and ongoing operation of these onsite water treatment systems are protective of public health, it was necessary for the SFPUC to prepare an oversight and management program for the City in the absence of a national or state regulatory program. SFPUC staff collaborated with other city departments, namely the San Francisco Department of Public Health-Environmental Health (SFDPH-EH), Department of Building Inspection (SFDBI), and Public Works Department (SFPW) to develop a program to regulate and permit the use of treated non-potable water in buildings. Initially in setting up the program, it was critical to understand the existing regulatory authorities and existing water quality standards. Looking to the California Plumbing Code, there were some standards for graywater and rainwater uses, but no guidance on oversight and management.

Recognizing that oversight was needed to be protective of public health, in 2012 San Francisco became the first municipality in the United States to adopt a local ordinance to allow buildings to capture water onsite for non-potable uses. To ensure a successful program, each city department had clear regulatory roles and responsibilities defined in the ordinance. For example, the agency responsible for public and environmental health (SFDPH-EH) took on the role of becoming the permitting agency for onsite water systems, while the Department of Building Inspection (SFDBI) took responsibility for checking plumbing plans and overseeing construction. As a result, a streamlined permitting process was put in place, which allowed the program to expand in 2015 to become a mandatory requirement for new commercial, mixed-use, and multi-family development projects of 250,000 square feet (approximately 23,000 m^2) or greater.

25.2 REUSE OF DRAINAGE FOR STEAM GENERATION IN SAN FRANCISCO

To encourage onsite non-potable water reuse in the industrial sector, the SFPUC also actively partners with the city's industrial customers to reduce their water consumption, as they represent some of the utility's largest water users.

One of these projects, the Energy Center San Francisco (ECSF)-Bay Area Rapid Transit (BART) Water Recovery Project, is a unique example of the SFPUC providing financial assistance for onsite water reuse in an industrial setting. Beneath the crowds of people that visit San Francisco's Union Square every day, millions of gallons of under-utilized water flow directly under the BART train station to the sewer system. ECSF, owned by Clearway Energy, and BART have partnered on a unique project to bring that underground resource to the surface for reuse.

To maintain the structural integrity of its transportation system, BART captures foundation drainage from the Powell Street BART station in a large cistern and pumps it to SFPUC's sewer system. Recognizing an opportunity, ECSF approached BART to divert that water instead for use in the district steam loop. ECSF, an energy company formerly known as NRG, is the sole steam heating system operator in San Francisco: 24/7/365 days a year, ECSF provides steam heating to hotels and buildings in downtown San Francisco. Reflecting their commitment to the sustainable use of energy and water, ECSF spearheaded the project to reclaim the foundation drainage and redirect it to their District Energy Plant located nearby on Jessie Street for use in the district steam loop (Figure 25.1).

ECSF first worked with BART to replace and upgrade the aging sump pumps at Powell BART station that were used to pump the foundation drainage to the sewer system. Next, a new pipeline roughly 1000 feet long was constructed to transport the foundation drainage to ECSF's nearby steam heating plant. At the plant, ECSF installed an onsite water treatment system to treat the foundation drainage to a quality that is

Figure 25.1 Energy Center San Francisco is operating the largest industrial water recycling project in San Francisco, saving 30 million gallons of drinking water per year. (*Credit: San Francisco Public Utilities Commission*)

suitable for use in a district steam heating system. The onsite water treatment system includes a raw water collection tank with a coarse strainer, microfiltration (MF), and closed circuit reverse osmosis (CCRO). The water also undergoes softening to remove minerals that interfere with the process of steam production. This treatment system is unique in that the CCRO allows for 80–90% recovery of the treated water, as compared to 75% recovery from traditional reverse osmosis systems (Figure 25.2).

Commissioning of the project began in September 2018. Currently, ECSF is successfully operating the onsite water treatment system and will continue to monitor its integration with the district steam heating system. In total, ECSF will reduce their potable water usage by 30 million gallons each year, which is equivalent to 30% of their annual water consumption. ECSF's innovative approach began with identifying foundation drainage as a resource rather than what traditionally has been considered a nuisance to be discharged to the sewer. This project is part of larger effort to continue investigating ways to use less potable water in their operations. Similarly, BART is dedicated to achieving its sustainability goals of reducing its impact on energy and water use. Through this joint effort, the SFPUC, BART, and ECSF are modeling the path for successful public-private partnerships in San Francisco.

25.3 PROCESS WATER REUSE IN BREWERIES

The SFPUC also promotes the reuse of water in other industrial sectors such as breweries. Water plays an important role in brewery operations, as it makes up more than 90% of the product and is used for numerous applications within a brewery. Brewers know the value of water, as a typical brewery can use up to seven gallons of water to produce one gallon of beer. Much of this water is used for rinsing bottles and cleaning equipment. This type of water, also known as 'process water', can be collected and reused onsite at the brewery. Treating and reusing process water onsite can help breweries reduce their water footprint by as much as 50%.

Figure 25.2 Reverse osmosis filtration system is used to treat foundation drainage for reuse at Energy Center San Francisco. Similar smaller systems are used to treat brewery process water for reuse at breweries. (*Credit: San Francisco Public Utilities Commission*)

However, breweries receive limited guidance in how to safely reuse process water onsite. In San Francisco, breweries interested in process water reuse looked to the SFPUC for help. To address this gap, the SFPUC developed guidance materials, including pathogen and chemical control strategies for process water to be reused for tank and bottle rinses, floor wash down, boiler feed water, and as a source water for the beer. The guidance includes requirements for source water characterization, source control, treatment, and ongoing monitoring to ensure the water is safe for these uses. The guidelines also ensure the same level of public health protection as the California drinking water standards for chemicals, and is consistent with the risk-reduction goals of the California drinking water standards for microbial pathogens. In order to help local breweries reduce their water footprint, the SFPUC offers financial assistance to breweries implementing process water reuse in accordance with the guidance.

In California, several innovative breweries have implemented process water reuse onsite in order to use water more efficiently. In addition to saving water, brewery process water treatment systems can also reduce the volume and strength of discharges to the sewer system, which can help breweries reduce their sewer bills and comply with local regulations. Some examples include Seismic Brewing Company in Santa Rosa, California that is recycling 95% of process water generated onsite for applications such as cleaning and boiler feed water, Stone Brewing Company in Escondido, California, and Lagunitas Brewing Company in Petaluma, California.

25.4 SAN FRANCISCO LEADS NORTH AMERICA COLLABORATIVE

San Francisco continues to play a significant role in shaping the national discourse around onsite water reuse in North America. As the chair of the National Blue Ribbon Commission for Onsite Non-potable Water Systems (NBRC), the SFPUC has been leading a collaborative of public health regulators and water and wastewater utilities from across North America since 2014 to further promote the installation of decentralized, onsite water treatment systems. The NBRC, in partnership with the US Water Alliance, Water Research Foundation, and WateReuse Association, has been creating the tools for state and local jurisdictions across the country to adopt consistent water quality standards, governance approaches, and institutional frameworks.

The NBRC tackled one of the primary challenges to widespread adoption through its participation in research on water quality standards for onsite non-potable water systems. The research led to the *Risk-Based Framework for the Development of Public Health Guidance for Decentralized Non-Potable Water Systems* report, which established risk-based water quality standards for onsite water systems. Utilizing similar methodology as is employed in potable reuse and drinking water regulations, the report developed performance-based log reduction targets for the treatment of pathogens that align with the Water Safety Plan approach promoted by the World Health Organization.

More recently, the NBRC has spurred significant progress in advancing policies and regulations for onsite water reuse across North America. The NBRC's document *A Guidebook for Implementing Regulations for Onsite Non-potable Water Systems* has helped several states within the United States, including Colorado, Hawaii, Minnesota, Washington, Oregon, and California implement oversight programs for onsite systems. Additionally, the NBRC's report *Making the Utility Case for Onsite Non-potable Water Systems* is helping utilities and other stakeholders understand the benefits of onsite reuse, how utilities have addressed potential challenges, and best practices for the ongoing operation of these systems. In early 2020, the NBRC completed a training manual on how best to design and permit onsite non-potable water systems that meet the risk-based water quality standards.

25.5 THE NEED FOR NEW WATER MANAGEMENT

Reimagining our urban water systems demands collaboration, consistency, cross-cutting ideas, and active utility leadership. It requires innovation in governance and institutional capacity building, as well as embracing new business models. Utilities can play an important leadership role by actively promoting new water resources management, fostering partnerships with the community, and building an enabling environment for water supply diversification.

Collaborating with local, state, and national partners is a critical piece of reaching consistency in approaches across the country. San Francisco's examples of innovative industrial water reuse projects also show that collaboration can result in breakthroughs that transform water management. When faced with the complexities of water management in the 21st century, onsite water systems are helping communities build upon their centralized infrastructure and increase the resiliency of their water and sanitation systems. Such onsite systems reflect not only the potential of this specific resource management strategy, but also the capacity and responsibility of utilities to support more sustainable use of water by industry.

Chapter 26

Challenges in regional collaboration

Cornelis (Niels) Groot

Dow Benelux BV, Environmental Technology Center, Terneuzen, The Netherlands

Keywords: collaboration, partnerships, public, stakeholders

26.1 WHY COLLABORATION WITHIN A REGION?

When industries dependent on reliable freshwater face structural water stress, they must develop long term strategies to minimize the risk of an interruption in their water supply. This is especially true of industries in the petrochemical sector that use water predominantly for steam production and cooling. Different approaches can be chosen to create a more sustainable water supply, depending on the local situation (Figure 26.1). For example, water stress may occur at the source due to limits on the availability of water, its quality, or the presence of competing users. Similarly, discharge of used water may be limited due to concern about the impact of its discharge into nearby water bodies on water quality, temperature, or other environmental factors.

When it comes to reducing water stress, industries should generally expect to 'reduce, reuse, and recycle' their own water, internally, before looking to 'share' their water issues with others. Nevertheless, there are situations when the best solution for all stakeholders requires regional collaboration with both public sector and private sector partners. In that case, industry managers must work closely with others both inside and outside the company to solve many different kinds of problems, both technical and political, even taking a leadership role as required. And while the challenges are significant, the rewards can be great. This was the situation faced by The Dow Chemical Company in Catalunya, Spain and in Terneuzen, the Netherlands, as described further in the following account.

26.1.1 First things first: reducing the footprint

A measure with great impact is to minimize the footprint of the chemical facility by redesigning or optimizing the chemical process itself so that its net water demand is lower (m^3 of water per kg of

DOI: 10.2166/9781789060676_0277

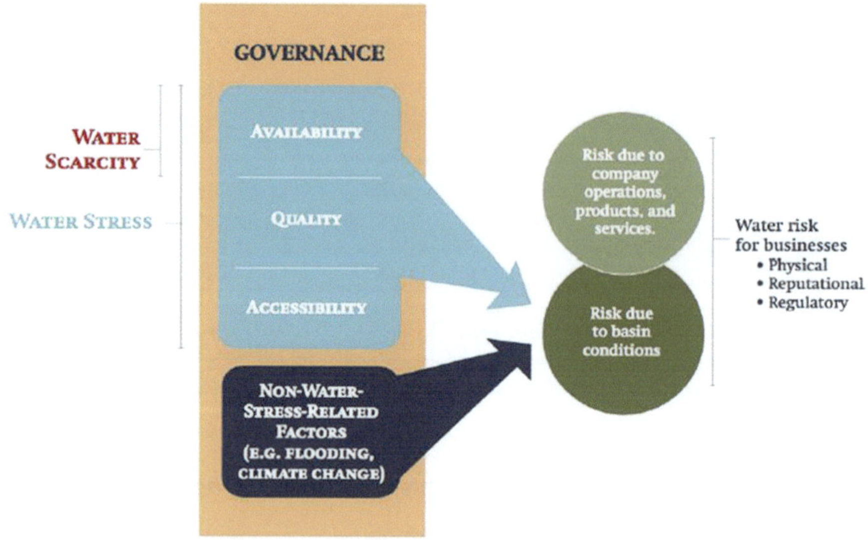

Figure 26.1 Factors determining water stress.

product produced). Significant water savings in cooling, cleaning, and product processing can usually be found in a step-wise method:

(1) Look for ways to *reduce* the amount of water required to perform a process;
(2) *Reuse* water purged from one process in another process where it can be applied directly without further treatment; and
(3) *Recycle* wastewater from a process by treating it to the quality required for reuse.

Reducing steam or cooling demand by interchanging heat, and improving cooling tower operation by increasing cycles of concentration (thereby reducing discharge) can often be accomplished quickly or with a minimum of technical support. It is also possible to reduce the water footprint just by standardizing maintenance and cleaning activities (assure that methods and procedures are optimized and leveraged at the operational levels). As an example, at the Dow facility in Freeport (USA, Texas) Dow saved four million m^3 of fresh water annually through applying Nalco Water's 3D TRASAR™ technology for cooling tower operation. There may also be opportunities to exchange heat between different plants within a chemical complex although the distance between facilities and the complexity of infrastructure connections may ultimately determine the overall cost and feasibility of the project.

Reusing wastewater from one process in another without significant quality enhancement is another attractive option for reducing the water footprint. A classic example here is the return of steam condensate as feed water for medium pressure steam boilers (up to 35 bar). (At higher pressures – 100 bar or more – the condensate may require additional polishing to remove minerals and ions to avoid corrosion and the formation of deposits in the steam/condensate circuit.) Finally, process water can be recycled for reuse at several places within the site's water envelope, after additional and dedicated treatment.

26.1.2 From internal to external

While the aforementioned measures are all within the fence-line of the facility, the first approach should always be to 'clean one's own kitchen first,' and only then search for alternative or additional 'outdoor' measures. Implementing on-site measures first will reduce the demand for freshwater make-up, by reducing the water used in products, losses to the atmosphere (steam, evaporative cooling), and water discharged to a receiving stream. To do more, however, requires collaboration with external stakeholders.

Few large chemical sites possess their own dedicated water sources. Unless they have access to unregulated ground water or have pre-existing surface water rights, chemical facilities are dependent on jointly managed regional sources (e.g., rivers and lakes, coastal (sea) water, groundwater), or on water supplied by public or private utility companies. As a result, facility owners have at least a shared responsibility for solving water supply issues within that region due to their substantial demand for freshwater. In some cases, the government entity managing the watershed actively includes the industry in solving regional problems. In other cases, it is up to the industry to take the lead. These two ideas are discussed further below.

26.2 TERNEUZEN

Dow's Terneuzen site demands 12 million m^3/y of freshwater make-up – about three times more than the water use of the entire region of Zeeuws-Vlaanderen (1200 km^2, 100,000 inhabitants). Water in this region is supplied to residential, commercial, and industrial customers by the water company Evides. Since the 1960s, the water company has built and maintained reservoirs with a total capacity of 6 million m^3 to store river water from the Biesbosch region, a preserved wetland area situated 120 km distance northeast of Terneuzen. To supply Dow's Terneuzen plant, Evides currently provides 6–7 million m^3/y of water collected from Belgian polders in the south under a long-term contract. The plant also receives roughly 1.5 million m^3/y of high quality effluent from the Terneuzen City wastewater treatment plant (WWTP), which leaves a gap of 4 million m^3/y of demand, currently supplied with freshwater from the Biesbosch area.

26.2.1 Early partnerships set the stage for future collaboration

Dow Terneuzen now aims to eliminate its industrial use of freshwater from the Biesbosch in anticipation of increasing pressures from food production, salinization due to sea-level rise, and more severe periods of droughts due to climate change. Dow's intention is to replace its Biesbosch intake by reclaiming and deploying additional regional water sources. To accomplish this, Dow will build upon and beyond existing relationships (established to realize the reuse of effluent from the City of Terneuzen WWTP) in order to create and use additional alternative regional supplies.

The collaboration chiefly involved Dow Terneuzen, Evides Industriewater, the regional Water Board, and the City of Terneuzen. At first, all parties agreed to the reuse the City's effluent, and for two years Dow Terneuzen recycled the WWTP's secondary clarified effluent by directing it to feed Evides' reverse osmosis (RO) membrane plant. At that point, an opportunity arose for enhanced cooperation based on the parties' evolving needs:

(1) In order for the Evides RO plant to continue to operate economically, the WWTP effluent quality had to be improved. Levels of residual organic material and nutrients (e.g., ammonium and phosphate) in the effluent began fouling the RO membranes, despite microfiltration pretreatment to remove the suspended material. This resulted in frequent chemical cleaning, reduced membrane life, higher cost, and lower treated water recovery.

(2) The Water Board determined that WWTP capacity should be expanded by 20% to accommodate future demand.

In response to these issues, all parties agreed to work together to expand WWTP capacity through construction of a membrane bioreactor (MBR) facility. This facility consisted of a dedicated bioreactor to metabolize organic and nitrogen-containing compounds followed by ultrafiltration membrane modules to remove solids, replacing the microfiltration unit at Evides' RO plant. As a result, the treatment plant gained its desired additional capacity and the effluent quality was significantly enhanced to improve operation of the Evides RO facility. Each of the partners recognized the benefit and the full-scale installation was realized ahead of time.

In the absence of such significant drivers, however, the collaboration to replace Dow's freshwater supply proved to be more challenging. Lacking such clearly defined benefits, Dow's previous institutional partners – Evides, the Water Board, and Terneuzen City – were reluctant to move forward quickly. Joint projects with other industries in the area were made difficult by the long distance and varied landscape between facilities, and collaboration with local farmers was complicated by the lack of a common voice that might allow individual farmers to work together with Dow. Farmers also struggled with the fact that in general they cultivate low value crops, while having more water might allow them to move to higher value crops. However, they could not afford the improvements that would make the water available, including the infrastructure for irrigation to reach the many farmers scattered within the region.

Dow Terneuzen might still have developed its own water supply strategy despite these obstacles, dealing with individual partners on a one-to-one basis. However, this posed the risk that the final outcome might satisfy the needs of industry but be suboptimal for the region as a whole. When the regional water planning process is left to 'market players' only, economics tend to become dominant and other values will receive less than desired emphasis. In addition, public support for the decision may decrease, which is negative for all partners at the end. As a result, Dow made every effort to engage the province and government parties in the role of initiator and process coordinator.

26.2.2 A new regional initiative

As pictured in Figure 26.2, the case of Dow Terneuzen was initiated by Dow with the prime objective to explore and use alternative water sources in the region for the reasons stated above. However, from an early stage, other stakeholders and potential parties in the field were invited not only to cooperate, but also to co-create synergy for the good sake of others like agriculture, nature, and other industrial users. So, farmer associations and nature groups have been involved from the beginning to tap into Dow's strategic approach and concrete plans. Other parties have been:

- the regional Water Board ('owning' water ways, dikes, and ditches, 'controlling' groundwater levels, and owning/operating the City's WWTP)
- the municipality of Terneuzen (owning the City's sewer system, but also being responsible for spatial planning)
- the province of Zeeland (responsible for fresh water strategy for that part of the country)
- Evides Industriewater (water supply for city and industry, and owning some vital infrastructure)
- HZ University of Applied Sciences (regional university with specializations in water management and technology, resilience to climate change and spatial planning)
- other industrial users like Yara (fertilizers), Cargill (food), ELSTA (cogen powerplant), ICL-IP (fire retardants), and Heros (solid and liquid waste handling).

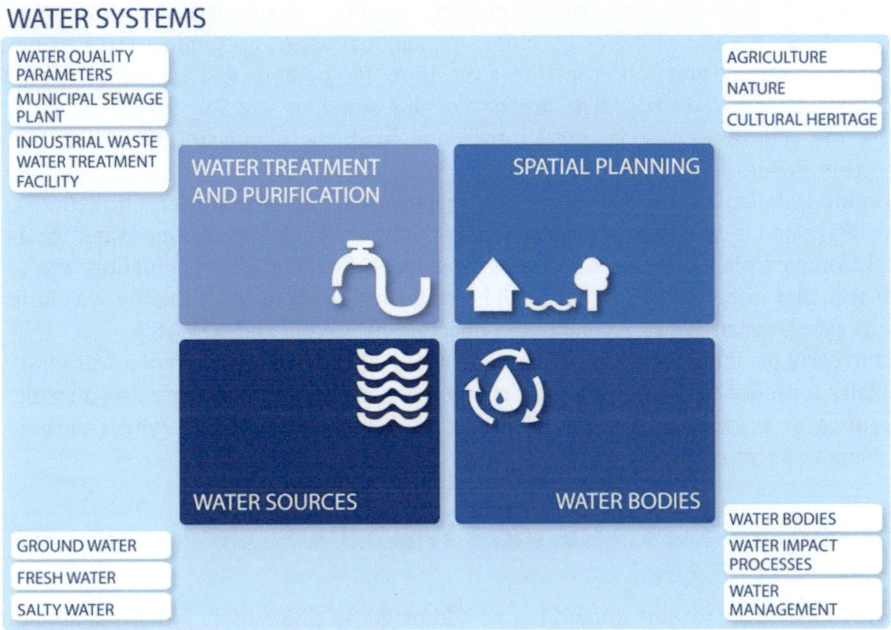

Figure 26.2 Building blocks for integrating different functions of water in a regional setting (regional Impact project by HZ University of Applied Sciences – 2016).

In addition to those mentioned above a few others have been kept informed over the course of the process, for example, Rijkswaterstaat (owner of the nearby major canal and locks) and North Seaports, the regional harbor entity.

Fortunately, the other parties successfully turned the 'Dow-only' scope into a broader strategic approach, called 'Robust Watersystem Zeeuws-Vlaanderen' (Figure 26.2). This approach is committed to the development and implementation of a long-term sustainable water system for the region, optimally providing water for all, including farmers, industry, citizens, recreation, and nature. This regional collaboration is able to identify all needs and receive each stakeholder's perspective, bringing them all together in a short- and long-term project portfolio and a road map for regional implementation.

One major lesson learned from this process is that governmental bodies should be in the driver's seat when searching for regional development to help avoid the risk for sub-optimization. That said, some attention should be paid to the cultural differences between government and industry in order to avoid discomfort or even frustration. For example, while industry typically has a relatively short line for decision-making and a strong focus on getting things done, public parties have a well-established hierarchical structure with multiple procedures to follow and a complex decision-making process. As a result, public and private timelines may diverge significantly. By acknowledging these practical differences in the beginning of the collaboration process, one can avoid unpleasant surprises and unmet expectations further down the road.

26.3 CATALUNYA

The other example of industry and government collaboration is illustrated by events in the region of Catalunya (northeast Spain). There the authorities mandated a regional partnership among a number of public parties, research institutes, and private companies in order to preserve the water level in the river Ebro, while sustaining the use of water by cities, recreational areas, and industries. During a period of extended drought, water abstraction from the river Ebro for potable and industrial use exceeded the outflow, resulting in a dry river bed. The urgency of this situation was felt by all parties, so it seemed appropriate for the authorities to convene a consortium with the responsibility to create a sustainable water management system for the region.

The consortium installed a 19,000 m^3/day water reclamation plant to treat water from the Tarragona and Vilaseca WWTPs, blend it with Ebro river water and supply it as make-up cooling water for the Tarragona Petrochemical Complex plants. Treatment provided included flocculation, coagulation, and filtration, salt removal with RO, and ultraviolet disinfection. The water reclamation plant facility was built by Veolia, owned by ACA (the Catalan Water Agency) and operated by Veolia and AITASA.

In contrast to early planning efforts by Dow Terneuzen, the process in Catalunya was top-down, driven by clear mandates for individual entities to participate in the collaboration process. As a result, there was a common incentive at a sufficiently high level for all parties to cooperate which ensured there was widespread awareness and support.

26.4 COLLABORATING TO REDUCE THE IMPACT OF INDUSTRIAL WATER USE

Dealing with regional water scarcity should not be addressed by single users independently, but a broader collaboration is needed between multiple stakeholders. Setting a common agenda requires consensus between parties with different cultures, work practices, and business models. Financing of required infrastructure also needs to be resolved to raise integrated water management to a sustainable level. With this in mind, it's interesting to see how industry's role can be optimized with the help of other regional stakeholders to facilitate more sustainable industrial water use.

When a direct common and overarching incentive for collaboration (as in Catalunya) is not apparent, special emphasis has to be given to defining common and mutual interests. Industries are typically focused on long-term uninterrupted water provision for reliable operation, while public entities need to ensure equal access to water for all stakeholders to maintain quality of life for the community as a whole. Once a common approach has been agreed, the collaboration process can begin to address the many other collaborative issues like economics and financial terms, contract clauses, quality guarantees, response to unexpected situations, etc. Only when each partner recognizes and knows its own interest ('what's in it for him or her') can the process accelerate to execution.

First of all, *industry* should define its own position within the region and clearly articulate its ambitions and boundaries. A long-term policy for water usage should be established and a sound business environment should be created to support investments in water-related projects. Internally, industry should encourage a multidisciplinary approach to develop more sustainable production processes with substantially lower water footprints ('reduce/reuse/recycle'). In parallel, industry should reach out to other industrial parties to collaborate 'over the fence' in non-competitive areas.

The national and international *industry associations* within the various industrial sectors can help by collecting and publicizing best water management practices among their industry members. Collaborating with other associations in other sectors, these industry groups can also help identify opportunities for 'over the fence' collaboration among industry clusters (such as food processing and agricultural sectors, and (petro) chemical and steel or mining sectors). Finally associations can play an important role in reaching out to policy-makers to bring across industry's ambitions and limitations.

Government agencies (including policy-makers, legislators, and regulatory officials) can promote effective implementation of improved industrial water use strategies by establishing consistent, fair, and transparent policies and regulations that support long-term industry commitments. Clear guidelines should be provided for effluent discharge quality; strategies can be adjusted but drastic changes ought to be avoided. For example, the European Water Framework Directive specifies national permit limits that require member states to reach agreed upon river basin water qualities by 2027. Water abstraction rights should also be clearly defined to reduce competition among different end-users (urban areas, agriculture, nature, and industry) for access to surface water, groundwater and recycled water (i.e., treatment plant effluent). Finally, officials can stimulate multi-stakeholder collaboration by setting regional objectives and providing financial incentives for the implementation of integrated water projects. Regional policy-makers can foster an open environment for collaboration by engaging multiple stakeholders in goal setting and providing financing through public subsidies or private investments.

The role of *financial Institutions* can have a game-changing impact. The value of water availability at sufficient quality for society as a whole is not reflected by its price alone but also comprises social and demographic aspects. Hence, pay-back times for water-related projects are usually much longer than those in which industry typically invests. Related project benefits not tied to the price of water, like ecosystems improvement, are difficult to monetize and don't figure in business cases. By collaborating with policy-makers, financial institutions can create criteria for financing regional water infrastructure projects that stimulate sustainable water management among multiple parties.

Other entities can also support the development of an effective regional collaboration by helping to define a regional scope that best fits the needs of all stakeholders or by working to create an effective organizational structure and associated governance. For instance, *water and wastewater utilities* have working relationships with both freshwater source suppliers and end-users and can collaborate with parties in all sectors, including policy/law-makers, urban areas, agriculture, industry, financial institutions, and NGOs. Their *professional associations* in turn can stimulate the exchange of information, including best practices and lessons learned in regional watershed management. *Academics* and *consultants* can also contribute their knowledge. Over the past two decades universities and research organizations have generated countless technological advances, such that even a country as small as the Netherlands has over 1000 water tech companies. Many research projects are financed through public-private partnerships, which in turn have led to university spin-offs emerging technology incubators that have piloted new technologies on a meaningful scale. By working with industry to better understand the technical, economic, and social aspects of sustainable water use, these partnerships can help prepare professionals to better address the water needs of society today. In addition, the concerns of environmental and social *groups* should be heard and incorporated in a region's long-term water plan, as they often articulate the 'non-market' value of water to enhance biodiversity. To take one example, in the Breda region of the Netherlands a 'bottom-up' planning approach was initiated by a consortium of local farmers and nature communities who successfully convinced both the municipality and the Water Board to dedicate resources for its implementation. Related to this effort is the role of the public itself – the individual civilians, neighborhoods, communities, and protest groups whose voice must be heard. Crucial for success is to inform the public early in the planning process, to invite their input and suggestions, and to allow them to directly interact with all parties (including industries).

26.5 MAPPING THE COLLABORATION AREA

As the above analysis makes clear, effective communication and trust are critical ingredients in a successful regional strategy. If industry is to reduce its water footprint further by participating 'outside the fence-line', a

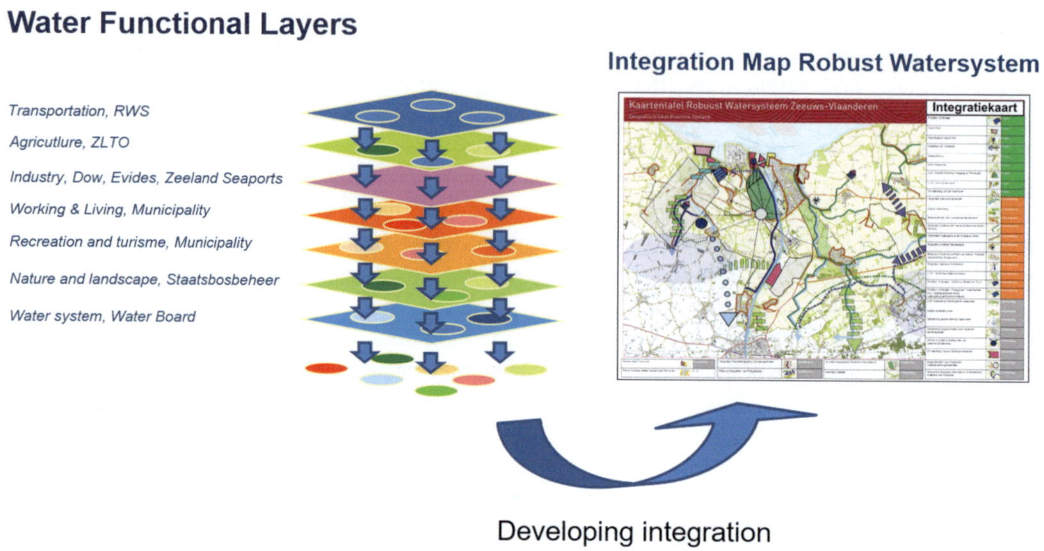

Figure 26.3 Mapping various stakeholders related to water functional layers.

strategy must be in place that clearly represents industry's interest and its impact on regional water use. When industry's issues align well with regional needs, communication of the expected benefits will make these relationships easier to establish. When the advantages of collaboration are less obvious, communication and trust become even more crucial. Certain headwinds are to be expected during the process, and all stakeholders should be asked to collectively develop alternatives that benefit the region as a whole.

A determining factor in the collaboration process is the initiating organization. Typically many different 'players' within a region each play a role in the intended collaboration. When a governmental body like a municipality, province, or (like in the Netherlands) a Water Board is the initiator of regional collaboration, the basic governance is institutionalized and public support relatively easy to obtain given the fact that decisions are taken and strategies are developed on a political level. When an end-user is the initiator – especially an industry – the common interest is not obvious and stakeholders are likely to be more cautious in communicating their real objectives and drivers. For an industry to create a successful collaboration, it must carefully map the organizational structure and identify each participating institution, authority, association, and individual, especially competitive users from agriculture, industry, recreation, and nature (e.g., NGOs). For each of these, individual representatives should be identified as speaking partners in the collaboration process to understand their positions and learn their internal approval processes. One example of such a map is shown in Figure 26.3.

26.6 CONCLUSIONS

Based on Dow's experience at various water-stressed locations in Europe, industries can reduce their water footprint beyond what can be achieved within their own fence line through regional collaborations. Where such collaborations don't already exist, they can be created in a stepwise approach:

(1) Describe the industry's strategy for sustainable water use, clearly articulating its impacts, both positive and negative, on others in the region.

(2) Engage with public authorities at both the strategic and implementation levels to ensure the industry's strategy fits within the region's approach, including legislation, planning, economic development, and environment protection.
(3) Involve a range of participants with a stake in the regional water system, including agriculture, recreation and tourism, academic institutions, environmental groups, NGOs, other commercial and industrial companies, and the public itself.
(4) Challenge the participants to agree on a process with defined owners and leads. When the goal is to implement a regional watershed objective, a public authority may more effectively lead the process until a formal management structure is established. When participant benefits are less clear, industry can lead the development of its plans in concurrence with regional development;
(5) Align project expectations by understanding how timelines and outcomes are addressed in each participant's culture, and encourage each participant to be sensitive to the wishes and contraints of the others parties. Provide an overview showing how parties can support industry efforts towards more sustainable water use.
(6) Develop a business and financing model as soon as project outlines are defined and encourage authorities and financial institutions to create a feasible economic foundation for executing regional improvements.

While there is no 'one size fits all' solution, by following these steps a company can overcome the most common barriers to successful industry-led regional collaboration.

Part 3
Tools

The 'Tools' section offers information on techniques and technologies available to support sustainable industrial water use. These include management techniques, diagnostic models, water balance software, and other analytical tools; water treatment technologies (including energy and resource recovery), water monitoring equipment, and staff training practices to ensure that tools are appropriately used; communication and outreach approaches that nurture public support for sustainable water use; and government policies and regulations that promote sustainable business behaviors. While industry-specific technologies are cited in some cases, the focus of this section is on issues relevant to all types of industrial water use and tools that can be adapted for widespread application.

This brief 'tasting menu' of tools and techniques is designed to suggest to readers the abundance of resources available, and to offer them some links to more comprehensive information on specific tools.

27. **Tenuta** *Closing the execution gap: How industry can lower its water use and help tackle global water scarcity*
    The section begins with an analysis of the gap between the sustainability goals that many companies now profess to embrace and the actual performance of industry in achieving those goals. Making changes up and down the supply chain, from sourcing to shipping, is hard work and can be daunting for even the most well-intentioned firms. Emphasizing this point, **Tenuta** cites a 2017 survey revealing that while three-fourths of companies with revenues over $1 billion had water conservation goals, fewer than two in ten dedicated the resources needed to achieve them. He refers to this as 'The Execution Gap', and traces its source to an inability to determine the real value of water. In many areas (e.g., Brazil), water is both cheap and scarce, putting industry at odds with social and environmental objectives. Knowing the true value of water, he claims, allows companies to make a realistic business case for sustainable water use. To address this problem, global hygiene and water service company Ecolab developed its Water Risk Monetizer and Smart Water Navigator programs to calculate the market value of water in any location and

DOI: 10.2166/9781789060676_0287

determine which different water-related solutions will yield the best return. He acknowledges that water problems are best addressed at the watershed level, and predicts that future analytical tools developed by Ecolab and others will employ artificial intelligence and machine learning, to help operators detect previously invisible trends and fix problems in advance.

28. **Easton** *Water stewardship*
First defining the term 'water stewardship,' **Easton** introduces a variety of tools and techniques to support the practice. Water stewardship, he offers, involves understanding the impacts of one's water use on others and the natural environment, and requires working cooperatively with regulators, other water users, and conservation groups to ensure a healthy and sustainable water environment for all. As a result, true stewardship 'goes beyond the factory walls to include the relevant surface and/or groundwater catchment.' He goes on to describe the different types of infrastructure associated with surface and groundwater use and wastewater treatment and discharge withdrawal. The Water Footprint is a tool often used to estimate a company's impact on the catchment, calculating total water use and consumptive (net) water use so the water steward can consider both in the context of the specific geography where water is used. The effective water steward also takes into consideration a range of non-monetized benefits including regulatory relationships, employee morale, brand reputation, and customer support.

29. **Sym and Wade** *The AWS standard: A common language for the global water stewardship community*
The Alliance for Water Stewardship (AWS), already referenced by previous authors, is a multi-stakeholder membership framework whose AWS Standard has been adopted by many of the world's largest corporations. As explained by Sym and Wade, the AWS Standard is designed to promote good water governance, sanitation and hygiene, and sustainable water use. It does this through cooperative collection and interpretation of data, development of an appropriate plan, implementation at the site level; and transparent evaluation of performance. After an organization has decided to prepare a water stewardship plan, an important component of the planning process is to conduct a water risk assessment that gathers data on current practices, and maps operations and supply chains across the process. Through certification, sites can make credible claims relating to their water stewardship performance. This is done through the use of independent third-party verification, which is done by AWS Accredited Conformity Assessment Bodies.

30. **Laporte-Bisquit** *WWF water risk filter: Assess, respond & value water risks*
To conduct the water risk assessment, many businesses use the World Resources Institute (WRI) Aqueduct and Aqueduct Water Risk Atlas tools, either alone or in combination with the World Wildlife Fund (WWF) Water Risk Filter. As described by Laporte-Bisquit, the Water Risk Filter is designed to assess the water risk in a specific area, with the goal of determining how best to manage those risks. It assesses the physical, regulatory and reputational risks faced by a company due to its location in a specific river basin (basin risks), as well as the impacts of its operations (operational risks). German food producer EDEKA incorporated a variety of high-resolution mapping tools into the Water Risk Filter to identify risks in its global supply chain and make appropriate investments in water conservation. Clothing giant H&M used the Water Risk Filter to evaluate data from over 1100 sites within its value chain to provide specific local water stewardship recommendations, and is also using the tool to work on water stewardship projects in

high priority river basins in China and Turkey. A recent enhancement of the Water Risk Filter was the development of the Water And ValuE (WAVE) tool used to justify investment in sustainable water projects through the construction of a business case.

31. **Diaz** *Water footprint: A sustainability tool for industries*
The water footprint is another approach that can be used by companies to assess their water use. Diaz compares two popular water footprint tools – WFN's Water Footprint, and the ISO's Water Footprint, which includes lifecycle analysis. As she describes them, the WFN tool has the advantage of simplicity, separating the types of use into direct consumption (blue water), water required to dilute pollution (grey water) and agricultural irrigation (green water). By comparison, the ISO tool, while more complicated, allows the practitioner to capture the impact of upstream supply chain production as well as consumer use of the product to give a more comprehensive picture of overall water impact. She provides examples of each analytical approach applied to the beverage industry to demonstrate their use. Diaz concludes that although both methods can help improve water management, regulation is still necessary to encourage sustainable water management.

32. **Tomei** *Best available techniques as a sustainability tool for industrial water management and treatment*
Another technique used widely to support sustainable water use by industry at the facility level is the application of 'Best Available Technology' (BAT) standards, which companies can use to help select appropriate treatment for wastewater prior to discharge or reuse. Tomei spells out the purpose and approach of the program implemented by the European Union (EU) which requests its member states use the BAT as a reference for writing industry permits that achieve environmental objectives. The BATs adopted by the EU address the potential for both upstream pollution prevention and downstream pollution removal. Using the example of the chemical industry, Tomei explains that the standards apply not only to the selection of treatment technology, but to the design of process systems as well. As she illustrates in detail, the BATs for water management in the chemical industry consist of applying closed loop circuits to maximize recycling and reduce water discharge; optimizing the washing processes; avoiding direct contact cooling systems; employing closed circuits in vacuum generators, replacing water or steam jet pumps with dry systems; and assessing the possibility of adopting techniques for the treatment of exhaust gases at low water consumption.

33. **Bjerre** *Applying the UN sustainable development goals as a framework for corporate sustainability*
Reinforcing the importance of metrics, Bjerre suggests that even targets as far-reaching as the United Nations Sustainable Development Goals (SDGs) can be addressed if companies translate them into operational objectives. He contends that a company's ability to implement sustainable practices can be nurtured, encouraged, and facilitated by a national climate of concern about the environment, and a sense of responsibility to do one's part to solve collective problems. Individual leaders can make a difference by directing their companies to adopt sustainable practices, setting an example for others both within and outside their own industries. Once the commitment to sustainability is made, he points out, it takes a company-wide effort to clarify the goals and see that all sectors of the company are involved in the program. To accomplish this implementation step, it helps to be very methodical and inclusive, seeing that the goals and the intermediate steps are clearly

articulated, and that all employees are involved in the process, since the implementation plans adopted will impact their working lives.

34. **Rosenblum and Moore** *Implementing sustainable water use: A roadmap for industry*
Once a company has evaluated the impact of its water use, it can consider its water management options. **Rosenblum and Moore** prescribe a series of detailed steps that industries can take to reduce water use through conservation or wastewater recycling and reuse. Industrial facilities are complicated systems, but by comparing the quality of water required by various processes to the quality of wastewater they produce, facility managers can identify opportunities for reuse and design appropriate treatment systems to recycle wastewater. The authors also offer suggestions on how to develop a robust business case for industry investments in sustainable water use, and how to present this business case effectively to gain approval for projects.

35. **Groot and Davis** *Staff training: An integral component of sustainable water use by industry*
Groot and Davis remind companies that even when they adopt more sustainable water practices and technologies, they will only reap the rewards when the staff responsible for implementation are educated about their purpose and trained in their operation. Further, they point out that this training must take into account the attitudes and skills of staff (including language, literacy, and computer proficiency) as well as their different roles and responsibilities. The importance of investing in staff training was discussed in relation to the San Francisco Public Utilities Commission, a US public water and wastewater utility, which provided training to staff, contractors, and permit holders working on watershed lands owned by the utility. The training was designed to protect both the quality of water in the reservoirs and the wildlife and vegetation within the watershed. A training handbook covered topics including proper use of chemicals that could harm wildlife; how to avoid damaging sensitive and endangered species when mowing vegetated areas; and how to raise or lower reservoir water levels without disturbing native vegetation. In another example, Dow Chemical developed a three-tier training program for its employees: one for line staff (self-study materials), one for technical staff and engineers (made available on-line through the company's global E-Learning platform), and a third for business and operational leaders, designed to prepare them to further upgrade production processes in the future. These training programs include modules relating to water systems, water quality for industrial applications, water technologies, and wastewater treatment.

36. **Etgen *et al.*** *The business value of water education to corporations*
As many authors representing industry have noted, successful water stewardship is based on an understanding of the character of the local watershed and the people who live there. Believing that this success depends equally on the ability of the community to understand water issues, the nonprofit Project WET Foundation (Water Education Today) helps companies connect with their communities by offering education programs for non-technical audiences – including adult stakeholders, their children, and even company employees. As related by Etgen et al., Project WET developed classroom educational materials on water conservation and hygiene on behalf of Newmont Goldcorp, owner and operator of Peru's Yanacocha goldmine, the largest in South America. After consultation with local teachers, Project WET constructed programs designed to encourage the community to engage more meaningfully with the company by teaching about water as a finite but renewable resource and its importance to health and hygiene. Project WET also created interactive exhibits for the regional Museum of Water and Earth and a mobile

exhibition to reach rural areas. Working with Ecolab, Project WET developed the 'Clean and Conserve' curriculum which to date has provided training in water conservation and hygiene to more than 8 million students and teachers, worldwide. Clothing manufacturer Levi Strauss (LS&Co.) engaged Project WET to provide water education to the public by first training their own employees to become 'water ambassadors.' Their training culminates in teaching water conservation, hygiene, and sustainable water practices to children whose parents work in the LS&Co. supply chain.

37. **Greig and Rathjen** *Promoting sustainable industrial water use: Scotland's Hydro Nation at home and abroad*
Just as companies can manage water more wisely when they engage with people in the regions where they operate, Greig and Rathjen, maintain that public utilities can operate more efficiently when they respond to the needs of their customers first. By their account, Scottish Water's commitment to making measurable progress towards sustainable water use through their 'Hydro Nation' initiative not only encourages water conservation at home, but also includes a robust international element in which Scottish Water exports successful practices to countries in the developing world. Domestically, Scottish Water decided to allow companies to compete for the opportunity to supply water to industrial water users, with the goal of reducing customer costs while providing higher levels of service, including greater support for industrial water conservation. Scottish Water provided clear goals, guidelines, and boundaries for the program; more than 60% of industrial water users participated, saving billions of liters of water per year. Internationally, they built on their nation's historical relationships with countries like Tanzania and Malawi to provide technical assistance and support for improved water governance, asset management, and technical capacity. In one notable case, in conjunction with other Scottish and German agencies they worked with global distiller Diageo's Serengeti Breweries Ltd and Tanzanian ministries, universities, and NGOs to address water and climate-related risks and opportunities facing the barley supply-chain and develop a model for others.

38. **Tomei and Angelucci** *Biological wastewater treatment as an opportunity for energy and resource recovery*
Shifting focus to specific water treatment technologies, Tomei and Angelucci note that advances in biological treatment make these technologies more useful than ever in rendering all types of industrial wastewater suitable for reuse. Historically, biological treatment by anaerobic digestion has been employed to produce methane. This technology has been updated by combining it with membranes and expanded granular sludge bed reactors that maximize contact with the substrate and increase energy yields. Using a different approach, microbial fuel cells have the potential to revolutionize energy production from waste, though presently only at pilot scale. Multiple-phase bioreactors can concentrate toxic substrates in the nonaqueous phase, gradually transporting them to the liquid phase so bacteria can treat wastes previously too toxic for biological methods. Further research will ultimately allow industry to recover both resources and energy from wastewater, making industrial processes more sustainable.

39. **Spanjers** *Extreme industrial effluents: Opportunities for reuse*
Spanjers also takes up the challenge of treating 'extreme' industrial effluents, including wastewater that is highly corrosive, toxic to biological treatment, and contains high concentrations of inhibitive or refractory contaminants. These types of effluents resist easy adaptation into the 'circular

economy,' but with proper selection of treatment methods they can be rendered suitable for reuse. In fact, some methods are actually more effective at treating extreme wastes at higher concentrations. To help industries find the right solution, Spanjers provides an overview of available technologies and suggests their appropriate applications. For instance, he suggests that high salinity wastewater with a low concentration of organics can be readily treated by reverse osmosis (RO) to the point that chemicals can be more readily recovered from the concentrated reject brine stream. Conversely, as discussed by Tomei and Angelucci, wastes with high concentrations of organic matter and low salinity can be treated through anaerobic membrane treatment. While no one method is capable of treating all wastewaters, given the range of technologies available virtually every industry can become more sustainable by recovering constituents from waste and reusing water.

40. **Stefanakis** *Promoting sustainability in the oil industry: The benefits of using constructed wetlands for oily wastewater treatment*
Evidence suggests that industries are adopting, however haltingly, the principles of 'the circular economy,' a concept borrowed from nature in which every waste becomes raw material ('food') for other products. As this includes the use of treatment technologies that render industrial wastewater suitable for reuse, it seems appropriate for the treatment methods themselves to incorporate natural processes. As Stefanakis demonstrates, natural systems can efficiently and effectively treat industrial wastewater and provide a range of environmental benefits. They generally consume less energy, need fewer chemicals, and – once properly designed and constructed – require less operator attention than other biological or physicochemical methods. By his account, constructed wetlands have been successfully utilized to treat oil field waste in Oman, replacing deep well injection, a high energy demand process with many environmental drawbacks. As a case study, he describes one of the world's largest constructed wetlands (490 hectares). Its beds are planted with common reeds from nearby areas, propagated at an onsite nursery. Oil from the wellfield is pumped to the wetlands, where most of the oil content is separated by hydrocyclone, without the use of additional chemicals. It then flows by gravity through the wetlands which remove the remaining oil. Treated water is evaporated in beds where salt is collected and purified for industrial use. (The primary tradeoff is the amount of land required to provide adequate treatment.) The facility started its operation in December 2010 with a treatment capacity of 45,000 m^3/day. Nine years and three expansion phases later, treatment capacity has reached 175,000 m^3/day, which is more than two-thirds of the oilfield's wastewater.

CONTRIBUTOR SURVEY RESULTS: TOOLS

As part of the process of preparing this anthology, the editors asked each author to complete a survey reflecting their views on sustainable use of water by industry. The survey gave authors the opportunity to report on a wide array of tools currently available to support more sustainable use of water. These included risk assessment tools; sustainability standards; technologies; and processes (such as collaboration, regulation, and staff training) that support more sustainable water use.

Contributors listed a wide array of analytical, management, and engineering tools they use to manage water more effectively. Water risk analysis, water footprint, and life-cycle programs, like those described by **Laporte-Bisquit** and **Diaz**, were cited by several academic and consultant authors as a means of assessing the influence of facilities on local water supplies and quantifying the impact of their industrial processes. Geographic mapping software was used to trace the flow of water at watershed scale and process modeling software was used to help industries understand the flow of water at their facilities. As detailed in chapters by **Easton** and **Sym and Wade**, those surveyed frequently mentioned the Alliance for Water Stewardship's 'Water Stewardship Standard' as an effective management tool to help a company move from analysis to planning and implementation. Confirming the point made by **Bjerre**, the United Nations Sustainable Development Goals were also acknowledged as important by both industry and utility managers as they helped focus a company's attention on 'big picture' goals. Consultant contributors highlighted the value of water sensors to monitor water use in a facility, and industry, utility, academic, and consultant contributors all drew attention to the great number of advanced technologies, like those mentioned by **Tomei and Angelucci** and others, now available to turn industrial wastewater into 'fit for purpose' water suitable for reuse.

All noted the need to reduce the lag between the development of new technical solutions and their widespread adoption in practice. This is akin to the problem discussed by **Tenuta** which he labeled 'the execution gap.' As one industry respondent observed, 'The technology necessary to recycle wastewater and reduce freshwater use in the textile industry by 90% is commercially available today but has not been fully deployed and scaled across the entire industry.' It was also noted that technical solutions must be properly integrated into the operation of the facility. According to one consultant contributor, 'Rarely does one technological approach solve an entire problem. Rather, the real creative solution comes about when multiple technologies are applied, each having its contributing niche to the whole solution.' The same might be said of the analytical and management tools, now readily accessible to help companies assess their water use and develop water management plans from the catchment to the factory level. That is, the techniques are available, but managers must relate them appropriately to the specific conditions at their individual facilities.

The articles in this chapter describe the multiple tools now available to support sustainable use of water, and also provide practical suggestions for when, where, and how to use the tools.

Chapter 27

Closing the execution gap: How industry can lower its water use and help tackle global water scarcity

Emilio Tenuta

EcoLab, St. Paul, MN, USA

Keywords: business, climate, Ecolab, industry, technology, water, water management, water scarcity, water stewardship, water stress

27.1 THE NEW NORMAL: CHALLENGES AND MEGATRENDS

When we think of global water scarcity, we may think of major cities facing 'Day Zero' when the taps run dry. Recent examples include Sao Paulo, Brazil (population 12 million) which came within days of running out of water; Chennai, India (7 million) which had to resort to shipping in drinking water on special trains; and Cape Town, South Africa (3.8 million) which seriously considered towing in icebergs from Antarctica.

While these local crises are alarming, the reality is that on a global level chronic water scarcity (Figure 27.1) has already become 'the new normal'. Two billion people now live in water-stressed areas while almost 850 million – one out of every nine people on earth – have no access to safe drinking water (UN Development Programme, 2019; UN Environment Programme, 2016). According to projections, if humanity keeps using water at the current rate, the world will face a 40% freshwater deficit by 2030, while global demand for water is estimated to increase 40–85% depending on the sector by 2050, due to population and economic growth (2030 Water Resources Group, 2009; World Business Council for Sustainable Development, 2020). Overshadowing all is climate change, the water effects of which are expressed in phenomena like droughts, excessive rains, receding glaciers, and sea level rise.

27.2 THE EXECUTION GAP

The challenge to companies is simple: to limit their water use by using smart water management practices. Unfortunately, the majority of companies seem to fall into an 'execution gap' where their desire to meet this challenge is not matched by their ability to use water sustainably.

DOI: 10.2166/9781789060676_0295

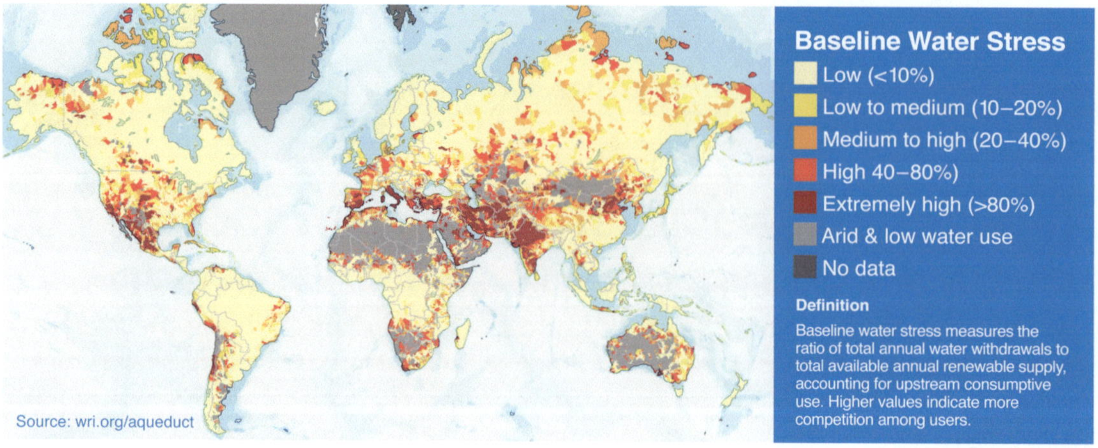

Figure 27.1 Water scarcity affects all parts of the world, including high-, middle- and low-income countries and regions. (*WRI Aqueduct*, 2020)

In a growing number of regions, companies understand the challenges of water scarcity because they are already feeling its effects. One luxury hotel chain in Pune, India, was forced to truck in water at a significant cost when the municipality cut its water supply drastically in response to a drought. Even in places with ample water, industries are reaching the limits of available supplies. For example, when a well-known food producer added a string cheese line to its plant in upstate New York – a water-rich region known for cheese-making – on some days it consumed up to 80% of the water supply of the nearby town, to the point that residents were banned from watering their lawns and washing their cars.

Given the extent of their 'water risk', one might expect businesses to be scrambling to mitigate their exposure. In fact, good intentions abound: Ecolab's 2017 survey (with GreenBiz) revealed that 75% of companies with revenues over one billion dollars have corporate water-reduction goals. However, the survey also revealed that 82% of these companies lack the capabilities to achieve them (Ecolab, 2017). As a result, over the past four years, corporate water use has been trending up, not down, and according to the Carbon Disclosure Project almost 50% more companies reported higher water withdrawals in 2018 (Carbon Disclosure Project, 2018; Greenbiz, 2017). Again in 2019, 88% of companies surveyed said they intended to tackle water issues over the next three years, but half had no plans for how they would make that happen (Ecolab, 2019a). In short, the gulf between corporate water goals and actual water use represents an execution gap between intentions and accomplishments.

27.3 BRIDGING THE GAP: PRACTICAL STEPS TOWARD SMART WATER MANAGEMENT

Although achieving corporate water targets may seem daunting, there are specific actions companies can take to make real strides towards sustainable water use. As shown in Table 27.1 below, these actions are directed at solving problems related to: (1) the under-valuation of water; (2) the site-specific nature of water use; (3) the challenge of identifying appropriate technologies; and (4) the need to look 'beyond the fence-line' for the most efficient, collaborative solutions. The following sections describe solutions industries have developed and implemented to address each of these problems.

Table 27.1 Challenges to smart water management and available solutions.

Problem	Solution	Available Tools
Water is underpriced, which means it is hard to find sufficient ROI for water-related investments.	Determine the full value of water	Ecolab Water Risk Monetizer (Ecolab) WRI Aqueduct (WRI)
Across-the-board, corporate level goals fall short because of water's inherently local nature	Set context-based goals; focus on shared water challenges at the local level Develop a water-energy footprint assessment to understand site-level water issues.	Ecolab Smart Water Navigator (Ecolab)
Users aren't aware of technical solutions to help them reduce, reuse and recycle water.	Deploy the right technology to manage the needs of critical water circuits.	Ecolab Smart Water Navigator (Ecolab)
Water is a shared resource, but not enough companies look beyond the fence-line for collaborative solutions.	Develop mature water stewardship practices and adopt appropriate measurements to support continuous improvement	AWS Standard 2.0 (Alliance for Water Stewardship)

27.3.1 Determining the real value of water

Determining the real value of water to a business is the key to making the business case for improved water use, which is often a prerequisite for implementing any smart water management practices. The 'real value' must be determined because, in general, the price paid for water as a commodity doesn't begin to reflect the full value of water to an industry.

Furthermore, the market price of water does not always reflect its availability. According to the 2017 Global Water Intelligence Tariff Survey, a cubic meter of water in water-rich Amsterdam cost \$6.25, while it cost only \$1.35 in water-stressed Sao Paulo (Global Water Intelligence, 2017). The truth is that water in Sao Paolo is scarce, but that scarcity isn't factored into its price. Since few companies will invest in a resource they see as plentiful and inexpensive, when water is scarce but cheap, a traditional approach to calculating return on investment won't justify much investment in water conservation and reuse.

To capture the real value of water in 'dollars and cents' Ecolab developed the Water Risk Monetizer, a publicly available online tool designed to describe the full value of water to an enterprise's facilities, not only in terms of its cost but also its impact on productivity and the risk of shortage (Ecolab, 2019b). For instance, a dwindling water supply may mean a higher water bill and/or production interruptions due to lower water quality and availability, which may also pose process and product quality challenges. Furthermore, in response to shortages government may impose usage restrictions while communities may turn against companies they perceive as selfish water-guzzlers. Once these risks are factored in, water is no longer a low-value proposition.

Companies enter information into the Water Risk Monetizer describing local factors including water availability, quality, pricing, and environmental impacts, for one or more of their facilities. Based on its publicly available methodology, the tool then generates a dollar figure representing water risk for each location, helping decision-makers understand which of their facilities are at risk, so they can allocate investment dollars more effectively. As an example, Microsoft used the Water Risk Monetizer to model water use at their San Antonio data center. The full, risk-adjusted value of water to that facility was 11 times higher than its retail price reported on the water bill.

This higher value calculated by the Water Risk Monetizer demonstrated a significantly elevated risk level to the facility, prompting Microsoft to invest preemptively in using recycled instead of potable water at its San Antonio facility, where it is now saving 58.3 million gallons of potable water and more than $140,000 USD annually – and it has lowered its impact on the surrounding community (Ecolab, 2019c). Such examples are increasingly important as cooling a typical data center takes millions of gallons of water per year, data centers must be located near users (to combat the challenge of latency), and with the growth in data traffic, in the coming years thousands of new data centers will be needed, many located in water-scarce regions.

27.3.2 Setting context-based goals

Since all water use is local, water solutions must be local, too. This is where corporate water policies often stumble. Too often, water goals are set in similar ways to carbon reduction goals, but water is not carbon. It doesn't matter *where* you reduce carbon emissions since CO_2 emitted anywhere in the world has the same impact on the Earth's atmosphere. Watersheds, on the other hand, are not interconnected, so saving water at your plant in Los Angeles won't make more water available to your plant in Beijing. Whether you are by a river or a coastline, in a floodplain or a desert, where you are ultimately determines how you should use water. Precipitation patterns and soil composition, even the built environment and degree of pollution all factor into the determination of optimal water use.

Given that variability, broad-brush corporate water goals are insufficient to guide action. The solution is setting context-based goals formulated for specific facilities, taking into account the local challenges in the watersheds in which they operate. To facilitate this process, the publicly available Ecolab Smart Water Navigator allows companies to sort and filter their facilities by location, water stress, and current management practices, so they can identify the actions that will generate the best return on investment (ROI) in locations where these actions will make the greatest difference (Ecolab, 2019d).

Based on a 13-question assessment, the Smart Water Navigator places each facility on the '*Water Maturity Curve*', which visualizes how well it manages water (Figure 27.2). For instance, a facility at the beginning of its journey is 'Untapped', while one with fully mature practices is 'Water-smart'. For each facility, the tool generates a practical guide to get to the next level. Once users implement the proposed steps, they can repeat this process until they reach the top level.

Water-smart: Smart, circular water management is fully embedded in your site's decision-making and operations. You are actively working on water issues with the surrounding community.

Exploratory: You have mastered water conservation and your site is deploying pilots for circular water management. You are reaching out to other water users in the surrounding community.

Linear: Your site is primarily focused on water conservation with successful water reduction pilots in place.

Untapped: Your site has not yet adopted smart water management practices.

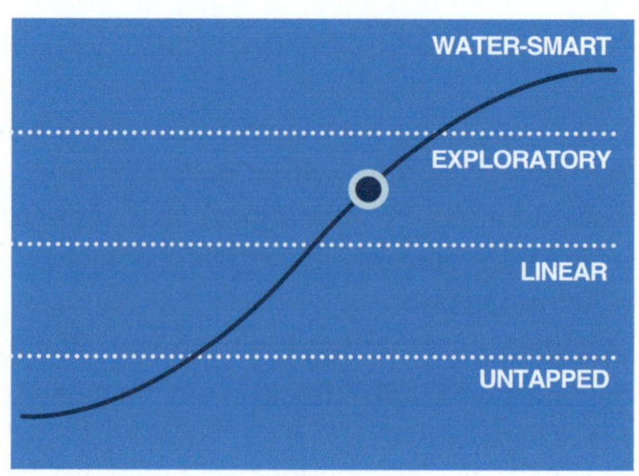

Figure 27.2 The water maturity curve. (*Source: Ecolab, 2019d*)

By way of illustration, Digital Realty, a large data management company, used the Smart Water Navigator to analyze each of its water-cooled facilities in different regions. They identified 14 'at-risk' locations where improved water management would have the greatest impact. Based on that holistic approach, the company is now employing enhanced water stewardship and water management technologies to improve its water resiliency, strengthen its monitoring and measurement capabilities and reduce water consumption.

27.3.3 Using technology to its full potential

Once the real value of water is known and context-based goals have been set, it's time to design and implement technical solutions to meet the targets. This was the situation at Archer Daniels Midland (ADM), one of the world's largest food and agriculture companies, when they adopted ambitious sustainability goals that included reducing water usage 15% by 2018, and energy and greenhouse gas emissions 15% by 2020.

Starting in 2012, ADM worked with Ecolab to review water use at a number of locations and determined that they could make the greatest strides by improving water quality monitoring and control technology to better manage their cooling towers, boilers, and wastewater treatment facilities. ADM adopted Ecolab's 3D TRASAR™ technology which uses ultraviolet light to continuously monitor chemical concentrations and contaminants in water, so they can be corrected in real time. This helps reduce scale formation, corrosion, and biofouling in installation such as cooling towers, boilers, and chillers. As a result, the equipment works more efficiently, requiring less water and energy. It also helps plants to reuse and recycle water.

By implementing similar technology in over 200 individual projects around the world, ADM saved 2.3 billion gallons of water, equivalent to the annual drinking water needs of 7.95 million people. And because water must be pumped, heated, cooled, and treated, it also saved 159,000 million British Thermal Units (MMBTUs) of energy and reduced CO_2 emissions by 70 million pounds.

This improved water quality monitoring and control technology was also used in the hotel in Pune, India, which was able to replace the drinking water in its cooling system with wastewater treatment plant effluent, as well as the Microsoft data center in San Antonio, which was better able to maintain the quality of its recycled water for reuse.

Even more improvements in water management can be achieved when big data is deployed to analyze performance data in real time. Ecolab3D, a new, cloud-based digital platform, supports analysis of multiple information streams, including water quality data, financial data, lab results, and other third-party information. These richer data sets can be combined with advanced analytics, including artificial intelligence and machine learning, to allow operators to detect previously invisible trends, signal potential problems, and fix issues before they arise.

27.3.4 Developing mature water stewardship practices

Water stewardship is the capstone of a sustainable corporate water approach. It is among the criteria to reach the Ecolab Smart Water Navigator's 'Water-smart' top level as well as a requirement for many sustainable water use certifications. In short, since watersheds are the basic unit of water distribution, mature water management must happen at the watershed level to be truly sustainable. In addition, to addressing water use 'inside the fence-line' of their facilities, companies must collaborate with stakeholders in their watershed – communities, farmers, and even other industries.

While there is no 'one-size-fits-all' solution, a logical first step is for each company to learn about other stakeholders' concerns, sharing knowledge and best practices to develop solutions to shared problems. On a local level this can involve working with farmers around a facility and engaging with the local water district or town government. An example of this type of engagement on a broader scale is the California Water Action

Collaborative (CWAC, 2019). Founded in 2014, CWAC brings together dozens of non-governmental organizations, agricultural producers, investors, and global companies to address challenges to the state's water supply, including drought, population growth, aging infrastructure, and climate change.

Stakeholders can even share process water. For example, Ecolab's plant in Garyville, Louisiana, draws water from the Mississippi, pre-treats it, and sends 30% of the treated water to a neighboring plant that makes superabsorbent polymers for diapers. The polymer plant returns the used water to Ecolab for treatment and discharge back to the Mississippi, cleaner than when it started.

Water stewardship also means rebuilding the natural environment's capacity to purify and supply water, using 'nature-based solutions' to restore hydrological features including rivers, flood plains, wetlands, aquifers, and related ecosystems. Ecolab's Garyville plant has worked with The Nature Conservancy for many years to restore the upstream Loch Leven wetlands which feed into the Mississippi River. The project aims to restore 10,000 acres of wetlands, add 12.1 billion gallons of water storage capacity and recharge the underlying aquifer, offsetting the Garyville plant's total consumptive water use. In a similar project, Ecolab worked with the World Wildlife Fund to restore wetlands along the Yangtze River near its plant in Taicang, China, just north of water-stressed Shanghai. Both the Garyville and Taicang plants are certified under the stringent Alliance for Water Stewardship (AWS) Standard.

27.4 CONCLUSION: BECOMING WATER-SMART

Companies can thrive in a water-scarce world by adopting water-smart tools and practices. Business alone can't be expected to stave off a global water crisis: government has a role to play in formulating smart policies, providing modern infrastructure, and creating a regulatory framework that rewards sustainable water use practices. But, since business uses 20% of all water globally and up to 59% in certain regions, the corporate sector can make a very significant impact.

It is within the power of individual businesses to reduce the amount of water they use to produce their products, saving water while reducing energy usage and greenhouse gas emissions. Wherever companies use more water to produce comparable products (Figure 27.3), opportunities exist for them to close the

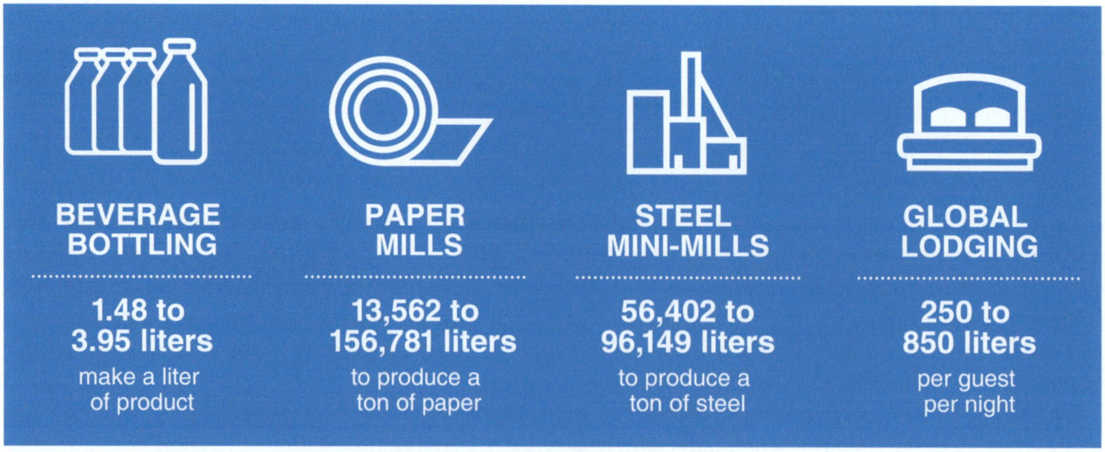

Figure 27.3 Wide performance disparities exist within industrial sectors indicate opportunities for progress in water management.

execution gap, become water-smart, and enhance their economic bottom line while protecting communities and the environment.

REFERENCES

California Water Action Collaborative. (2019). www.cawateraction.org (accessed 14 October 2020).

Carbon Disclosure Project (2018). Treading Water: Corporate Responses to Rising Water Challenges. CDP Global Water Report. https://www.cdp.net/en/research/global-reports/global-water-report-2018 (ccessed 10 December, 2019)

Ecolab. (2017). Survey: Business Water Management Efforts are Still a Drop in the Bucket October 23, 2017. https://www.ecolab.com/news/2017/10/survey-business-water-management-efforts-are-still-a-drop-in-the-bucket (accessed 10 December 2019)

Ecolab. (2019a). Ecolab-GreenBiz Survey Finds Corporations Still Struggle to Translate Water Goals into Action on the Ground (March 21, 2019) https://www.ecolab.com/news/2019/03/ecolabgreenbiz-survey-finds-corporations-still-struggle-to-translate-water-goals-into-action-on-the (accessed 10 December 2019)

Ecolab. (2019b). Water Risk Monetizer. www.waterriskmonetizer.com (accessed 10 December 2019)

Ecolab. (2019c). "How Dry is Texas?" https://www.ecolab.com/nalco-water/stories/how-dry-is-texas (accessed 10 December 2019)

Ecolab. (2019d). Smart Water Navigator. https://www.smartwaternavigator.com/ (accessed 10 December 2019)

Global Water Intelligence. (2017). https://www.globalwaterintel.com/global-water-tariff-survey (accessed 5October 2020)

Greenbiz. (2017). The State of Green Business. https://www.greenbiz.com/report/state-green-business-2017 (Accessed on 10 December 2019)

UN Development Programme. (2019). Goal 6: Clean water and sanitation. UN Sustainable Development Goals. https://www.undp.org/content/undp/en/home/sustainable-development-goals/goal-6-clean-water-and-sanitation.html (accessed on 10 December 2019)

UN Environment Programme. (2016). Half the World to Face Severe Water Stress by 2030 unless Water Use is Decoupled from Economic Growth, Says International Resource Panel. UNEP News, 21 March 2016. https://www.unenvironment.org/news-and-stories/press-release/half-world-face-severe-water-stress-2030-unless-water-use-decoupled (accessed 10 December 2019)

World Business Council for Sustainable Development (2020). https://www.wbcsd.org/Programs/Food-and-Nature/Water

WRI (World Resources Institute). (2020). Aqueduct Water Risk Atlas. Physical Risks Quantity: Water Stress. Website https://www.wri.org/applications/aqueduct/water-risk-atlas/#/?advanced=false&basemap=hydro&indicator=bws_cat&lat=29.99300228455108&lng=-77.87109375&mapMode=view&month=1&opacity=0.5&ponderation=DEF&predefined=false&projection=absolute&scenario=optimistic&scope=baseline&timeScale=annual&year=baseline&zoom=3 (accessed 5 October 2020)

2030 Water Resources Group. (2009). Charting our Water Future: Economic Frameworks to Inform Decision-making. McKinsey.com. https://www.mckinsey.com/~/media/mckinsey/dotcom/client_service/sustainability/pdfs/charting%20our%20water%20future/charting_our_water_future_full_report_.ashx (accessed 10 December 2019)

Chapter 28

Water stewardship

Peter Easton

Easton Consulting, Brussels, Belgium

Keywords: barriers, business case, catchment, incentives, industry, plain language, reputation, safe yield, sustainability, water stewardship

28.1 INTRODUCTION

Among industries with an interest in using water more sustainably, there is increasing awareness of the benefits of water stewardship. Water stewardship is a management framework that takes into account economic, social, and environmental impacts of water use and calls for both assessment and collaboration at a catchment level. The Alliance for Water Stewardship (AWS, 2019) defines water stewardship as:

> The use of water that is socially and culturally equitable, environmentally sustainable and economically beneficial, achieved through a stakeholder-inclusive process that involves site and catchment-based actions.

More broadly, water stewardship means doing much more than just regulatory compliance and protecting the water supply for your business. It includes the need to understand the impacts of your water use on others and the natural environment, and requires working cooperatively with regulators, other water users, and conservation groups to ensure a healthy and sustainable water environment for all. Geographically, the focus goes beyond the factory walls to include the relevant surface and/or groundwater catchment.

Many tools have been developed and promoted for implementing water stewardship, but to date it remains practiced by a minority of businesses. This chapter examines incentives and barriers to its further adoption and suggests ways to promote wider take-up by industry.

While much literature on the concept of water stewardship is written from a global perspective, the basic need is for a business case that makes sense to the business itself – its ability to ensure a reliable supply of the

DOI: 10.2166/9781789060676_0303

water needed to support its enterprise over the long-term and be protected from water-related risks. This perspective requires a deeper understanding of risks to the local water catchment and other water users.

At present, many businesses lack an awareness of the nature and scale of water-related risks, and the potentially high cost of not addressing them. They often assume that others (e.g., regulators and water suppliers) will sufficiently monitor and manage these risks, so they avoid what they perceive to be the high cost of studies and expert support. The low and sometimes non-existent cost of water leads to a disincentive for water efficiency. Water-related concerns compete for attention with other pressing sustainability issues with typically higher direct cost implications, such as energy, packaging, waste, and chemical emissions. Businesses often also find the language of water stewardship to be overly complex and abstract.

28.2 THE INDUSTRIAL WATER CYCLE

28.2.1 Water origins

Before an industry can mitigate or manage risks, it must first understand the location, nature, and vulnerabilities of its source water. In general, industrial water is obtained from either surface water or groundwater. Surface water is withdrawn through an intake from rivers and lakes while groundwater is abstracted from boreholes (water wells). Transport is by canal, pipeline, or tanker. Alternative and less common water sources include: desalinized water; recycled wastewater (after treatment); dewatering withdrawals, such as from mining operations; and captured rainfall (rainwater harvesting).

Historically, surface water was the dominant source due to its accessibility and abundance. Rivers brought a constant supply of freshwater to cities and carried waste away. Groundwater was generally used only when surface water was scarce or absent, typically from natural springs or hand dug wells. During the 20th century, drilling, exploration, and pumping technology advanced enormously, enabling access to much deeper groundwater horizons, from tens to hundreds of metres below ground. Huge quantities of freshwater previously inaccessible became available in many regions with limited surface water. This stimulated population and economic growth and expansion of agriculture in previously water-scarce regions. Unfortunately, this has also led to overexploitation of groundwater in many places with falling water levels that, in turn, may impact surface water bodies.

Surface water and groundwater have different advantages and risks. Surface water is typically easier to find and transport but must almost universally be treated to achieve drinking water quality and is sensitive to seasonal and year-to-year variability in abundance. By contrast, good groundwater reserves (aquifers) require more effort and cost to identify, to assess their reliable ‘safe yield’ and to install an appropriately designed borehole and pumps. However, once established, a groundwater source is usually better protected from pollution and is of naturally high quality due to the natural filtration properties of aquifers. That is why bottled natural mineral waters, which must be safe to drink without treatment (according to European and CODEX international standards) are sourced exclusively from groundwater (FAO, 2011). Many aquifers have a high storage capacity to buffer seasonal and year-to-year variability in freshwater recharge. On the down side, once groundwater becomes polluted, it is more difficult to reverse. There are exceptions to these attributes. While some regions are exclusively dependent on groundwater (e.g., 99% of Denmark’s public water supply) and others exclusively on surface water, on a global level both sources are commonly used.

‘Safe yield’ – a long-established principle in water supply management – is the rate that water can be sustainably withdrawn without causing significant negative impacts on water levels, flows, and quality, for other users and the natural environment. All water bodies are vulnerable to overexploitation when their 'safe yield' is exceeded. One of the most important responsibilities of a water professional is to determine the 'safe yield' of a water body and to ensure abstractions do not exceed it.

In general, hotter, drier regions are more vulnerable to water scarcity than cooler, wetter regions. There are exceptions, however, as some arid regions are underlain by abundant aquifers recharged in distant wetter zones, while some wetter regions with limited surface and groundwater storage capacity are vulnerable to drought. These exceptions highlight the importance of understanding the local situation, not just broad patterns and assumptions. Global water risk tools, such as WRI Aqueduct (WRI, 2019) and the WWF Water Risk Filter (WWF, 2019) are useful for identifying regional patterns and risks but have limitations in data granularity and detail at local level. For site and catchment specific assessments, such tools should always be supported by local data and expert studies so as to avoid missing important risks or investing in mitigating the 'wrong' risks.

28.2.2 The water supply

The water abstraction point is a critical and vulnerable component in water supply. Its location, design, age, and condition, should be known and routinely inspected and maintained. Older boreholes can fail due to corrosion, collapse, clogging, or contamination. A borehole that is not sufficiently protected and secure at the surface is vulnerable to accidental or malicious acts of contamination. Abstraction points have a finite life, needing to be refurbished or replaced eventually.

Water may be provided by an external supplier (municipal or private water supplier), or a business may have its own private water sources. For an external supplier, it is usual for the user to pay a regular fee based on volumes. The supplier develops, manages, and protects the water resources, and treats it to drinking water quality. A good supplier manages its water sources and catchments responsibly, minimizing the risks to its customers and the environment.

Across the world, there is a large variation in the rigor and reliability of external water suppliers. An underinvested and poorly managed municipal supply can present a range of risks such as supply interruption (for minutes, hours, or days); contaminated supply; or low pressure. Capacity limitations may restrict business growth of a customer. If a business considers the risks too high, it can apply additional treatment barriers for greater protection and may develop its own private water sources as a replacement or back-up.

There are a number of reasons why a business may develop its own water sources. The site may be remote from available municipal networks, it may require more water than the municipal capacity, or it may be more cost effective. A business should not be complacent and assume that 'the supplier knows what they are doing'. Find out. The water user should engage with the supplier to learn how they manage their risks and ensure the supplier understands the needs of the business.

28.2.3 Wastewater treatment and discharge

Following its use, water can take a number of routes. Some may be incorporated within the manufactured product (e.g., beverages); it may be re-used within operations (or even supplied to a third party for use). Some water may be 'lost' via evaporation or leakages. Most used water is discharged as wastewater, with or without treatment, to an onsite or external/municipal facility. In some cases, the used water may be returned to its original water body with minimal change (e.g., cooling water).

For a business managing its own wastewater, it should ensure it is treated to a safe and compliant standard before discharge to an appropriate facility or water body. Inadequately treated wastewater may pollute the natural environment and the water supply of others. This is also a regulatory and reputational risk. It is usually assumed that wastewater sent to a municipal facility will be safely managed. However, where government regulations are weak or non-existent, the user should treat its wastewater to international minimum standards prior to discharge to the environment.

28.2.4 The water catchment

Every water source has a catchment representing the geographical and physical zone from which the water originates. Events in the catchment can impact water availability or quality for the user. For surface water, the catchment is represented by the land area that drains towards a central water body, usually a principal river (in which case the catchment is commonly called the river basin or watershed) (Figure 28.1). For groundwater, the catchment is defined by the geological structure of the aquifer, recharge zones, and groundwater flow patterns.

For surface water, the whole catchment is vulnerable to pollution from such events as chemical spills, agricultural runoff, and poorly managed wastewater. For groundwater the risk from surface pollution depends on the nature of the soil and rocks overlying the aquifer. Those overlain by low permeable rocks such as clay and shale are better protected (Figure 28.2). Aquifers are most vulnerable at their recharge zone – where they are replenished with freshwater. Groundwater is also at risk from sub-surface

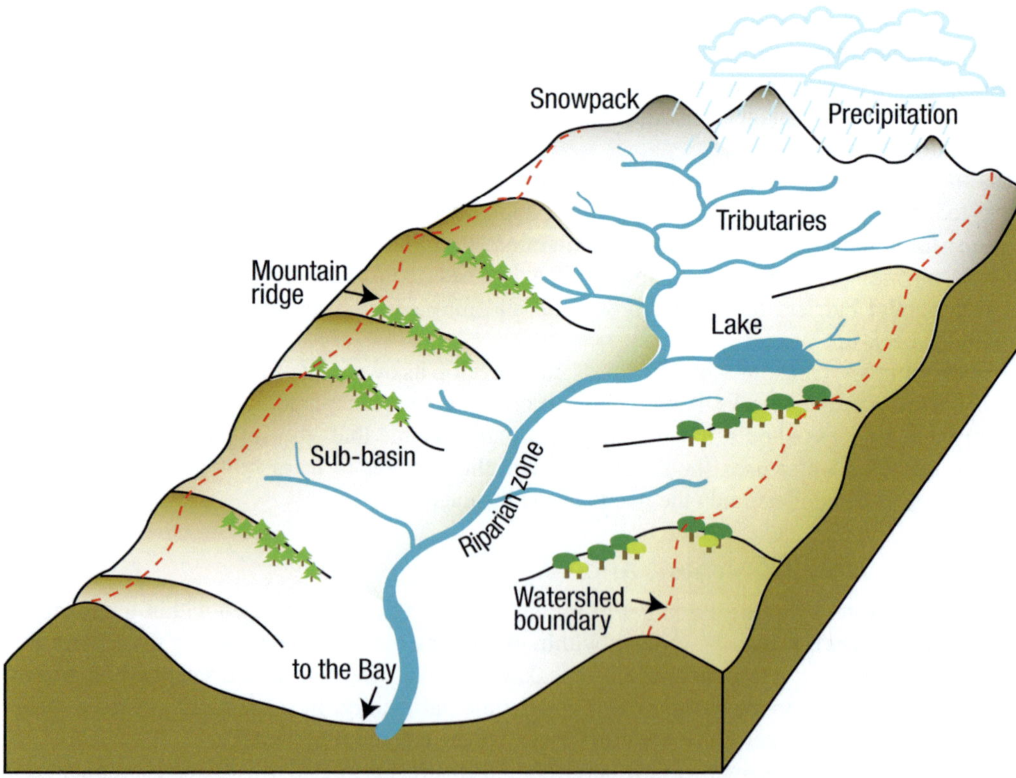

Figure 28.1 Surface water catchment. Note: For precipitation within the dashed red lines, some flows towards the river, some percolates down to groundwater, some evaporates and some is taken up by vegetation. (*Credit: The Watershed Project* https://thewatershedproject.org)

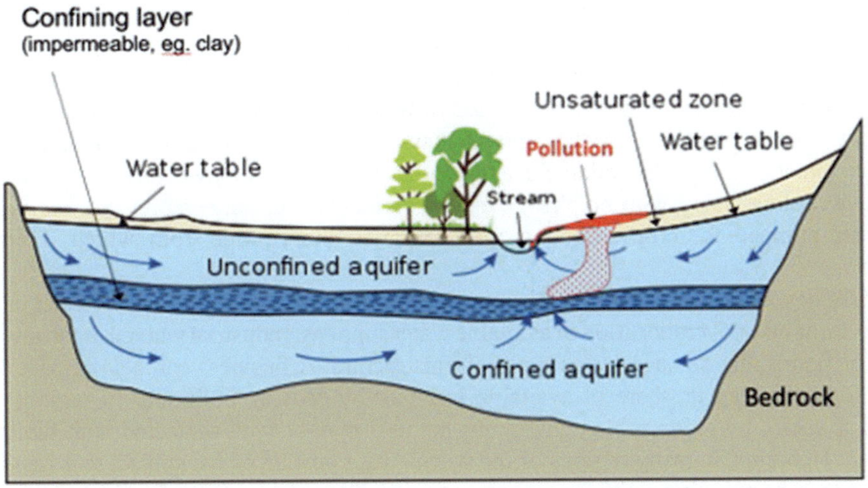

Figure 28.2 Vulnerability of water bodies to surface pollution. Note: Deeper groundwater in a confined aquifer has good protection from surface pollution. (*Credit: Peter Easton adapted from United States Geological Survey diagram*)

activities and structures such as boreholes, tunnels, and sewer systems when any of these are poorly constructed or in a poor condition.

Surface and groundwater bodies are often interconnected. Pollution can cross between the two bodies and overabstraction from one can impact on the other. For example, over pumping of groundwater can cause groundwater-fed wetlands or streams to dry out. It is important to know the catchment of a water source, the land use, and activities on it and the risks these present to the water user, as well as the risks the water user may present to others and the natural environment. Good knowledge of the catchment forms the basis of effective stakeholder engagement and collective action.

28.3 WATER IN THE INDUSTRIAL SUPPLY CHAIN

Industry accounts for approximately 20% of all human water use when considering direct operations and manufacturing. This includes water for processing, washing, rinsing, cooling, and as an ingredient in manufactured products, such as beverages and cleaning fluids. For many industries, however, much more water is used 'upstream' in the supply chain than directly in manufacturing. This is especially the case where agricultural products are used as ingredients and materials (such as for food, drinks, and textiles), since agricultural water use makes up 70% of all human water consumption, the majority for irrigation. Other parts of the supply chain where water use may be significant include the manufacture of primary materials and components, for packaging and for energy.

A common approach to assessing the total water used to make a product is the 'water footprint' method (Hoekstra *et al.*, 2011). This calculates, for example, that a cup of coffee requires 140 litres of water, most of that for growing the coffee plant. However, such numbers should be viewed in context. Despite the often very large numbers, much of the water footprint is not lost or 'used up', but is 'borrowed' and returns to the water cycle, becoming available for reuse. The standard method also includes the rain-fed component of crop water use and an additional component for pollution impacts. Evaporation from land and vegetation (rather than the oceans) is in fact an ecosystem service for inner continental areas, contributing to more than 50% of essential rainfall (Keys *et al.*, 2016). The critical issue is not the crude total of water use, but to avoid overexploitation in a given location, and for water to be responsibly returned to the water cycle.

In order to understand the scale of impacts and to reliably prioritize actions, it is critical to differentiate between total and net water use. Where a proportion of water withdrawn is returned to the local water cycle in good quality, the net water use is the amount that is not returned locally, or remains polluted. The difference can sometimes be substantial; for example, power stations may return to the environment up to 95% of their cooling water, with no change except an increase in temperature. Similarly, a proportion of groundwater pumped for crop irrigation may return to the aquifer from which it originated, via soil infiltration.

External risks beyond its direct control can impact the ability of a business to access water in a catchment. These include droughts or overallocation of available water supplies. Industrial water use can also be impacted by pollution or floods that damage water supply infrastructure. A business can also impact the supply of others by using more than its share of available water or by causing pollution. In regions where water resources are protected by enforceable regulations, the user may be well protected with limited additional action required. However, in many regions of the world, water resources are subject to a free-for-all where excessive water use or polluting activities are uncontrolled. To address them effectively, a 'water steward' needs to understand the issues and risks relevant to the specific geography where water is used.

28.4 INCENTIVES TO INDUSTRIAL WATER STEWARDSHIP

For many businesses, protecting themselves from water risk will be the principal incentive to start the journey towards water stewardship. A strong and stable business provides employment and supports a strong economy through taxes, investments and purchases. This is of particular importance in the developing world, where economic growth and stable businesses are vital to improving the conditions and wellbeing of much of the population. When a business carefully considers its own water supply risks it is taking the first step to understanding the surrounding water environment. Conversely, a water user who does not manage water sustainably will likely and ultimately find its own business negatively impacted.

28.4.1 Need

Most manufacturing businesses need water to operate and produce, so for a business to grow, its water supply must be sustainable over the longer term, and be obtained without causing negative impacts that would damage its future access or reputation.

28.4.2 Regulation and enforcement

A responsible business wants to avoid a breach of regulations, prosecution, and fines. A water risk assessment, conducted in conjunction with a water stewardship program, will help identify a risk of breaching regulations, such as through pollution, overabstraction, or the manufacturing of unsafe food or beverage products.

28.4.3 Cost savings

The cost of water to a business (which may include purchase fees, pumping, and treatment) is typically small compared to other operational costs, such as energy, materials, and salaries. Also, the cost of infrastructure and its maintenance poorly correlates with the volumes used. Therefore, direct cost savings from water efficiency are often a weak incentive to practice water stewardship. A stronger motivation is to avoid unexpected and potentially large costs of supply interruption or contamination. Such events can result in expensive production delays, and in the worst case, business failure. The knowledge gained through learning about your water supply and the resulting water stewardship plan will help protect a business from such risks. Where risks are shown to be low, this improves business confidence for owners, employees, and investors.

28.4.4 Investment and reputation

There is growing pressure on all businesses, especially large and high profile ones, to operate more sustainably. Water stewardship is already an important, and growing, component of many corporate social responsibility initiatives including GRI Standards (GRI, 2019), GlobalG.A.P. (2019), and ISO 26000 'Social responsibility' (ISO, 2019). These efforts are supported by specific water-related initiatives like the CEO Water Mandate (UN Global Compact, 2019), CDP-Water Survey (CDP, 2019), and the AWS Water Stewardship Standard (AWS, 2019). There is also a corresponding trend among investors to invest in businesses and funds with a high ethical and sustainability rating many of which include water as a component of corporate responsibility. The ratings are defined by a number of ethical indices, including FTES4Good and the Dow Jones Sustainability Index (DJSI). Leading companies compete to be recognized by these indices and to rise up the rankings.

Many companies are sensitive to reputational risk. Civil society or 'non-governmental organizations' (NGOs) put constant pressure on businesses to do better on environmental and social matters, and will sometimes 'name and shame' businesses they believe to be indifferent or irresponsible. Some NGOs go into partnership with companies to support social and environment improvement projects in order to engage in a direct dialogue with the companies and to receive financial support for sustainability-related projects. The company in turn gets a reputational benefit by partnering with a respected organization and receives its expert guidance on improving its policies and practices.

The news media can also influence companies to be better water stewards, not only by naming and shaming but also by highlighting best practices. For example, the BBC documentary series, Blue Planet II (BBC, 2019) influenced lawmakers in the UK and Europe to introduce tougher regulatory requirements with regard to plastic waste.

28.4.5 Employee morale

In the modern world, the majority of people care about environmental and social impacts. In their work, most people want to feel they are doing a job that is not harmful and which contributes to a general good, not just through economic development, but also by helping to maintain healthy societies and environment. This makes people happier and more content in their lives and work, whichever level they are at in an organization, from factory worker to CEO. This goodwill attitude of the employees helps to motivate a whole business to do better on sustainability issues, regardless of the core business benefits. Of course, the strength of this driver is strongly dependent on the motivations and support of senior management.

28.4.6 Customers and consumers

More and more businesses are being expected by their downstream customers to meet minimum sustainability standards. This trend reverberates throughout the supply chain with each company expecting compliance by the next supplier up the chain. At the same time, consumers are becoming more discerning regarding the social and environmental impacts of the products they buy. Consumers may be influenced negatively by the news media, especially when a big story critical of a company breaks. Some products carry logos linked to good practices and some carry direct claims. Businesses may communicate to their customers and the public on their good practices through advertising, sustainability reporting, their websites, and social media.

28.5 BARRIERS TO WATER STEWARDSHIP

While many global companies are committed to water stewardship, adoption of these practices has been relatively slow. Along with the incentives identified above, there are also a number of barriers businesses commonly encounter that discourage them from doing more.

As noted earlier, one major incentive for global companies is their desire to protect their corporate reputation and respond positively to the scrutiny and pressure they receive. Many smaller businesses, however, do not feel such pressure and do not feel the need to address water stewardship until they experience a water-related impact directly. In the absence of such risks, many companies also assume that 'someone else' is managing water responsibly so no further action on their part is warranted – in effect, a disincentive to involve themselves in water stewardship. As explained earlier, not all municipal water supplies are secure and low risk.

While many small businesses aspire to better water stewardship, they feel they do not have the time or funds to do more than the regulatory minimum. However inclined they may, the more costly, complex and time-consuming water stewardship appears to be, the less inclined they are to commit to it. For those businesses, it is most important for them to see a business case for water stewardship that shows a return on investment in terms of cost savings, improved profitability, and reduced risk. Those wishing to promote water stewardship, therefore, could do more to make the business case while ensuring demands are not excessive compared to other business priorities.

Finally, some companies may find it difficult to find a place for water stewardship among a range of other social initiatives. Reducing energy and CO_2 emissions are commonly top of the environmental agenda for many businesses, and along with waste elimination and air pollution reduction, water management must compete with gender equality, consumer health, and other issues as well. Businesses must adjust their priorities to the changing pressures from regulators, NGOs, investors, customers, and consumers. Advocates of water stewardship must be prepared to continue to raise the profile of water as a top priority and the key to sustainability and make the business case.

The key steps to water stewardship

The following steps are focused on the practical aspects of water stewardship, for which accurate and reliable knowledge is the starting point. The ultimate aim is a tangible reduction in water risk for the user, for others in the catchment and for the natural environment. The key practical steps to effective water stewardship are as follows:

(1) Know how much water you use, your future needs, and your water sources.
(2) Know your water catchment – its extent and boundaries, other water users, and water in the natural environment.
(3) Know your stakeholders, their water needs, concerns, and challenges.
(4) Assess and understand the water risks to your business, to your catchment, and to others.
(5) Identify opportunities to reduce risks and address shared water challenges.
(6) Develop a water stewardship plan to cover monitoring, stakeholder engagement, and short- and long-term actions.

Communicate the principles and benefits of water stewardship and influence others by your example, your actions, your achievements, and the benefits gained.

28.6 SUMMARY AND CONCLUSIONS

As the pressure on global water supplies increases, the pressure increases for businesses to commit to water stewardship to protect themselves and others around them. While there are good incentives, there remain a number of barriers to wider take-up. A significant barrier is the lack of awareness of the risks including for the many who rely on an external water supplier. Another is the perceived cost of water stewardship when competing with a range of other social and environmental issues businesses are expected to address. Promoters of water stewardship must ensure that in addition to raising awareness, they present businesses with a clear business case in plain language and practical guidance on the required actions. At the same time, they should not make expectations excessively onerous or costly, keeping in mind it is probably better to have 10,000 good water stewards than 100 excellent ones. Effective water stewardship starts with accurate and reliable knowledge about the water user and their catchment and results in tangible risk reduction and benefits for the water steward and their surrounding stakeholders, communities, and natural environment.

REFERENCES

AWS (Alliance for Water Stewardship). (2019). www.a4ws.org (accessed 1 October 2019).

BBC. (2019). Blue Planet II www.bbcearth.com/blueplanet2/ (accessed 5 October 2020).

CDP. (2019). CDP Water https://www.cdp.net/en/water (accessed 1 October 2019).

FAO. (2011). CODEX Standard for Natural Mineral Waters, CODEX STAN 108-1981, amended 2011.

GlobalG.A.P. (2019). The Worldwide Standard for Good Agricultural Practices. www.globalgap.org (accessed 1 October 2019).

Global Reporting Initiative, GRI. (2019). www.globalreporting.org (accessed 1 October 2019).

Hoekstra A. Y., Chapagain A. K., Aldaya M. M. and Mekonnen M. M. (2011) The Water Footprint Assessment Manual: Setting the Global Standard, Earthscan from Routledge, London, UK.

International Standards Organization (ISO). (2019). ISO 26000 Social Responsibility www.iso.org/iso-26000-social-responsibility.html (accessed 1 October 2019).

Keys P. W., Wang-Erlandsson L. and Gordon L. J. (2016) Revealing invisible water: moisture recycling as an ecosystem service. *PLoS ONE*, **11**(3): e0151993. doi: 10.1371/journal.pone.0151993

UN Global Compact. (2019). CEO Water Mandate https://ceowatermandate.org (accessed 1 October 2019).

WRI (World Resources Institute). (2019). Aqueduct Water Risk Atlas, www.wri.org/resources/maps/aqueduct-water-risk-atlas (accessed 1 October 2019).

WWF (World Wildlife Federation). (2019). Water Risk Filter, created by WWF, KFW-DEG. https://waterriskfilter.panda.org/ (accessed 1 October 2019).

Chapter 29

The AWS Standard: A common language for the global water stewardship community

Adrian Sym and Sarah Wade

The Alliance for Water Stewardship (AWS), North Berwick, Scotland

Keywords: Alliance for Water Stewardship, AWS, multi-stakeholder, sustainability, water stewardship

29.1 WATER STEWARDSHIP: AN INTRODUCTION

Globally, our water challenge is becoming ever-more evident, and increasingly, governments and other organisations are turning to the private sector to help address the challenges we face. The UN Global Goals (Sustainable Development Goals) help focus corporate sustainability work, and water stewardship plays a clear role in achieving SDG6 (clean water and sanitation for all) but also SDG3 (health), SDG5 (gender quality) and SDG12 (responsible production and consumption). The Alliance for Water Stewardship (AWS), through our multi-stakeholder membership network and the AWS Standard, offers a common language through which all water users from government, business and civil society can come together and collaborate to achieve our shared goals.

In practice, water stewardship is about shifting the focus from individual ways in which water is 'managed' to more collaborative approaches rooted in the context of a catchment (or watershed). Good water management by farms, factories and other water-using sites can go some way in ensuring non-domestic water users reduce their water impact. However, it is not enough to simply improve efficiency and concentrate on site water use. To guard against unintended or perverse outcomes from well-intentioned actions, sites need to ensure that they understand the shared water challenges impacting communities and nature in their local area. This means working beyond their fence line, encouraging others to take positive steps, and supporting good water governance.

Water stewardship is based on the principle that sustainable management of a common pool resource can be enhanced by engaging users of that resource in transparent and inclusive processes. Water users need to know where their water comes from, and how their use, and use by others impacts

DOI: 10.2166/9781789060676_0313

upon the continued availability and quality of the water on which they, and others, depend. Water stewardship processes aim to ensure that everyone who is reliant upon the same water source can meet their needs and work collaboratively based on trust and a common vision. By doing so, water users actively support public policy that achieves fair access to water for all. This is the essence of water stewardship.

The rest of this chapter provides an overview of the AWS International Water Stewardship Standard ('AWS Standard') and its global system before a deeper dive into water stewardship through four case studies from around the world.

29.2 THE AWS STANDARD SYSTEM

AWS is a global, multi-stakeholder membership-based organisation. Our mission is to ignite and nurture global and local leadership in credible water stewardship that recognises and secures the social, cultural, environmental and economic value of freshwater. We do this through our global membership and the International Water Stewardship Standard, or AWS Standard. Multi-stakeholder governance is core to AWS, and our members come from the private and public sectors, and civil society. AWS Members adopt a shared definition of water stewardship and endorse a common approach to robust, independently verifiable site and catchment-level water stewardship: the AWS Standard. We define water stewardship as 'the use of water that is socially equitable, environmentally sustainable and economically beneficial, achieved through a stakeholder-inclusive process that includes both site- and catchment-based actions'.

Since the AWS Standard was first launched, it has been implemented in a wide range of countries and industries. In line with best practice, in 2016, a public review and revision of the AWS Standard V1.0 was initiated to take on board feedback from implementations around the world. At the end of this process, in early 2019 AWS Members were presented with the final version of the AWS Standard V2.0, and they voted overwhelmingly to introduce it as the new version of the AWS Standard.

29.2.1 The AWS Standard 2.0

The AWS Standard is structured around five steps which take the user on a journey to assess their water risks and opportunities and then act on them (as shown in Figure 29.1). The process is built around a cycle of evaluating impact and striving for continual improvement, with communication and disclosure being a core part of the process to improve transparency. Within each of the five steps, there are a range of criteria with their own related indicators which must be met to be in conformance with the Standard. Some indicators are 'advanced indicators' through which sites are motivated to increase their performance beyond core certification. Sites that accumulate enough points through achieving advanced indicators may achieve gold or platinum certification.

There are five outcomes that the AWS Standard contributes towards:

1. Good water governance
2. Sustainable water balance
3. Good water quality status
4. Protected Important Water-Related Areas (IWRAs)
5. Safe water, sanitation and hygiene (WASH) for all

Each criterion contributes towards one or more of the five outcomes. Comprehensive guidance is provided to support implementers through the process.

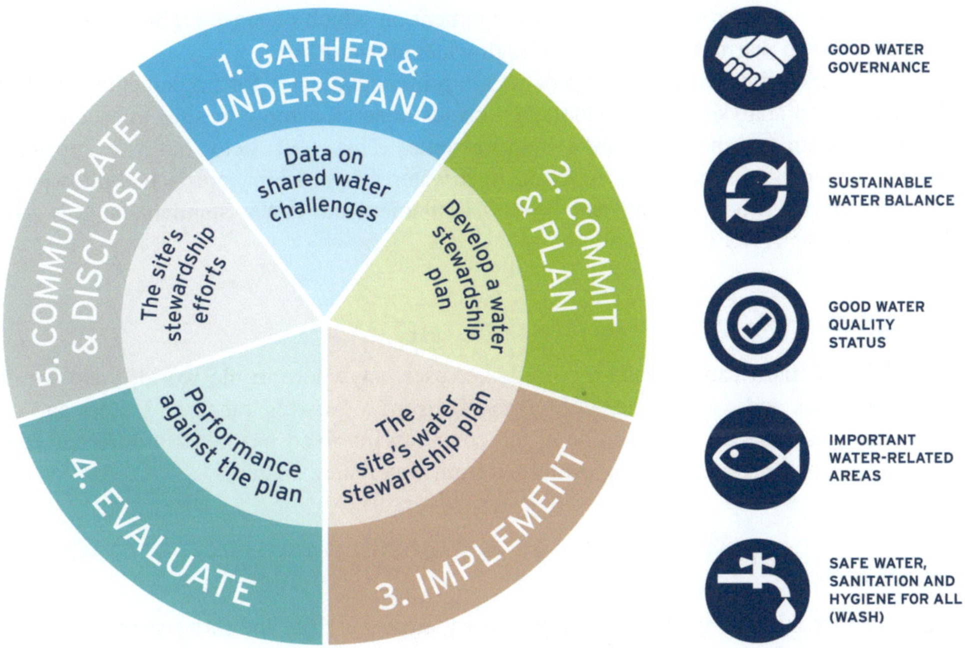

Figure 29.1. The AWS Standard, showing the five steps and the five outcomes (AWS, 2019).

The Standard is responsive to site and catchment context. As a result, it can be implemented by any water user, anywhere in the world. Through certification, sites can make credible, robust claims relating to their water stewardship performance. This is enabled through the use of independent third-party verification.

Throughout the implementation process, there are a range of services available to support sites. Training provides a good grounding in the concept of water stewardship, as well as insight into implementing the AWS Standard. A global network of Professionally Credentialed Individuals can support an organisation through implementation. In addition, membership of AWS provides access to the latest knowledge and innovations on water stewardship, as well as access to peer support and learning.

29.2.2 Water stewardship in action

For industry, a water stewardship journey typically begins with acknowledgement that water efficiency and site-level water management are not enough to mitigate water risks, and that opportunities exist through sustainability-related activity. At this stage, internal stakeholder engagement is vital to gain insight into how the company currently engages on water, and where potential risks and opportunities lie. Once there is sufficient internal buy-in to explore a water stewardship approach, the next step is to gather data and information on current practices and processes, and to map operations and supply chains across the business. This information is used to conduct a water risk assessment. There are a range of tools available which can be used to do this, and many businesses use WWF's Water Risk Filter or WRI's Aqueduct, or a combination of the two. They also engage with experts to understand local water challenges and gather contextual water-related information for their sites to build a picture of their water risks. The water risk assessment highlights where a company has clusters of its operations or supply chains facing high water risk, which then enables the businesses to take a judgement on which of those high-risk sites are valuable enough to the business to warrant investment through water stewardship.

The priority sites identified through this process are prime candidates for implementation of the AWS Standard. An initial gap analysis can be undertaken on just a small sample of sites to ascertain how their current practices relate to the Standard's requirements. This will help build a greater internal understanding of the potential activities required in a water stewardship approach. Following this, the business can then decide when, where, and how to implement the AWS Standard, and as part of this process, they can pursue AWS Certification.

29.3 WATER STEWARDSHIP AROUND THE WORLD

Because the AWS Standard can be used by any water user, anywhere in the world, there are emerging examples of approaches and uses of the AWS Standard from a growing range of sectors and countries. In each of the four case studies discussed here, the company involved has taken a different approach to addressing water risk and realising the opportunities from water stewardship.

29.3.1 Ingham's: catalysing a water stewardship community in the Western port biosphere, Australia

Through water stewardship, companies critically assess their water risks and turn risk into opportunity at a site and catchment level. Ingham's, Australia's largest integrated poultry producer and a major water user, faced water scarcity and public scrutiny at the height of the nationwide Millennium Drought. Further compounding their troubles was a large fire in 2012 that destroyed most of their site in Somerville, Victoria. Ingham's leadership saw these business risks as an opportunity to rethink the way the company approaches water management, prompting them to join the growing global water stewardship movement.

The Somerville plant, situated in the UNESCO Mornington Peninsula and Western Port Biosphere Reserve, became a global pilot site for the AWS Standard. Ingham's rebuilt Somerville to include an advanced water treatment plant that enabled them to reuse 65% of their onsite water use. In the wider catchment, they began collaborating with a diverse group of stakeholders, including the Western Port Biosphere Reserve Foundation, water service providers, government agencies, local government, and commercial and private landholders.

Ingham's demonstrated the range of uses of the AWS Standard not just for large water users, but also smaller users. This prompted other sites in the catchment to take water stewardship action alongside Ingham's, catalysing a community of water stewards in the Biosphere. The community comprises 28 sites actively practising water stewardship and growing. Through the collaborative, stakeholder-inclusive nature of water stewardship, site-level improvements are commonly assisted by public funding for achieving catchment outcomes that benefit all water users in the area. An example of this is the on-site waterway and wetland rehabilitation work undertaken with support from the waterways agencies. It improves water quality in the catchment, helps stabilise tributary banks and reduces sediment inflow. Ultimately, this protects the internationally significant Ramsar wetlands and Yaringa Marine National Part in the Western Port.

In 2015 Ingham's Somerville plant became the second site in the world to be certified against the AWS Standard (Gold-level) and through their continued efforts in championing water stewardship, the plant became the first site in the world to achieve AWS Platinum Certification in 2018. Ingham's is now implementing water stewardship across their sites in Australia and New Zealand, with four currently AWS Certified (Sommerville, Murarrie, Te Aroha and Bolivar).

Through their commitment to transparent, inclusive stakeholder engagement, Ingham's water stewardship activities have stimulated others to follow their lead. This approach has not only benefited the business, through recognition via awards from McDonalds and KFC and becoming a Banksia finalist, but it has also ensured that Ingham's isn't acting alone, creating a legacy through the Western Port Biosphere water stewardship community.

29.3.2 EDEKA and Netto Marken-Discount: responding to supply chain risks through collaboration

EDEKA is a supermarket with 11,000 stores throughout Germany. Since 2009, they have had a partnership with WWF, and in 2012, freshwater became one of the main workstreams for EDEKA and its subsidiary Netto Marken-Discount. Food retail is arguably one of the riskiest businesses when it comes to water due to the complex agricultural supply chains that our food systems rely on. Much of the food and drink consumed globally is produced in countries facing water scarcity, pollution and poor water governance. EDEKA and Netto are reducing their supply chain water risks by increasing transparency, engaging with suppliers and collaborating with other water users.

EDEKA and WWF have created an internal web-based tool based on WWF's Water Risk Filter to enable EDEKA and Netto to evaluate water risks. When the tool identifies suppliers situated in a water risk hot spot (due to water quality, quantity or governance issues), those suppliers are engaged to undertake water stewardship activities to mitigate those risks.

One important response to clusters of suppliers facing significant water risks is water stewardship capacity building and strengthening local networks to share knowledge and provide peer support. EDEKA and Netto have encouraged suppliers to attend AWS Training to increase their water stewardship expertise. In Colombia, the region around the Río Sevilla is an important source of bananas for EDEKA, but water scarcity, flooding and conflict between water users creates water risks. In response, EDEKA has worked with local partners to initiate a multi-stakeholder water stewardship platform with representatives from the banana, coffee and palm oil sectors as well as local and national authorities, communities and civil society. The platform provides a constructive space for discussion, knowledge sharing and collaboration, and acts as a hub for collective projects in the basin. In 2020, 11 privately-owned banana farms involved in this project became the first in the world to achieve AWS Group Certification, a highly significant achievement (AWS *et al.*, 2020).

Standards play an important role in reducing supply chain risks in the agricultural sector, and EDEKA and Netto have several projects underway supporting suppliers to implement the AWS Standard in South America and Spain. Iberesparragal, a citrus farm in southern Spain owned by EDEKA supplier Iberhanse achieved AWS Gold Certification in 2018 and was the first AWS certified agricultural site in Europe. The positive experiences from this first farm have led to over 12 more farms engaging in the project in Spain. EDEKA and Netto have found that supply chain cooperation helps create a business case for producers that may otherwise be unwilling to invest in water stewardship activities.

As an AWS Member since 2016, EDEKA have played an important role in communicating their experiences and lessons learnt via the AWS Global Network, encouraging others to engage in water stewardship.

29.3.3 Nestlé: leveraging a global commitment to achieve maximum impact

In June 2018, Nestlé Waters CEO Maurizio Patarnello announced that all bottled water factories would be certified with the AWS Standard by 2025. 'Water is one of the most critical sustainability challenges facing society and our business', said Patarnello. 'We are 100 percent committed to safeguarding water resources for future generations. By pledging to certify all our Nestlé Waters sites to this publicly recognized, credible water stewardship standard, we demonstrate how we positively contribute to water resources where we operate for the shared benefit of all'.

At Nestlé, the AWS Standard is used as a key reference to guide in the deployment of local water stewardship activities. It is an eye-opening process to identify areas where improvement is needed, it provides a common language in engaging and discussing with our stakeholders, and finally it offers a robust third-party auditing/certification. In particular, AWS promotes a mindset and tools to engage with stakeholders, sharing knowledge on water, to understand collectively the potential water challenges and therefore act collaboratively to address them with relevant and effective solutions. 'We are convinced that achieving certification allows us to communicate better about our good practices. It enables us to move away from self-declarations of good water stewardship', says Carlo Galli, Nestlé's Technical Director for Water Resources and Head of Sustainability at Nestlé Waters. 'Instead, we have a credible third-party process with conformity assessment bodies coming to our sites and screening our practices deeply'.

Nestlé Waters initiated the AWS journey at Nestlé, leading the way to a broader deployment of the Standard within the group. Indeed, water is important and has always been a 'natural' management priority for Nestlé since its beginnings, more than 150 years ago. Farmers need water to grow food, factories need water to operate, and consumers need water to prepare and consume Nestlé's products – water touches every part of Nestlé's value chain.

Therefore, Nestlé is committed to contributing to sustainable water resources management. While they continue to ensure best water use efficiency internally, they need to collectively address, at pre-competitive level, the water challenges that are shared with other users. At company level, Nestlé has crystallized both internal and external focus elements into a comprehensive water stewardship strategy – the 'Caring for Water' Initiative. Caring for Water is their internal platform to deliver the water stewardship approach and principles of AWS everywhere within their operations.

29.3.4 Apple: moving beyond site-based water management to catchment-wide water stewardship

Apple joined AWS as a Member in 2018, but their engagement and activity on water started much earlier than that. Since 2013, Apple's Clean Water Program has provided support and education that enables suppliers to conserve water, find ways to reuse or recycle water in manufacturing operations, and ensure that the water discharged from supplier facilities is as clean as possible. Apple has seen the benefit of stakeholder engagement, a core component of water stewardship, with the number of suppliers participating in the Clean Water Program growing from 13 to 116 in the last five years. In 2018, participating suppliers saved more than 60 million cubic litres of freshwater, bringing the total cumulative savings to 21 billion gallons. Crucially, as the program has scaled, Apple has worked with participating suppliers to expand efforts from conservation to water stewardship.

True water stewardship requires industrial users to go beyond their own facilities, and work with partners in government and civil society to protect water supplies on a basin-wide basis. Recognizing this, Apple began to engage with AWS, and joined as a member in 2018. Through this engagement, Apple is taking their water management and supplier engagement activities beyond the factory and into the community and wider catchment. Through the AWS Standard, Apple's suppliers are beginning to extend their efforts outside of their factories into the communities and catchments in which they operate.

In 2018, efforts to expand beyond conservation to stewardship began with suppliers in the Kunshan region of China. Participating suppliers committed to stricter water stewardship efforts, extended water programming beyond their factories, and engaged their surrounding communities. In 2018, two Apple suppliers received AWS certification, including the first gold certification ever to be awarded in the electronics industry and more suppliers have followed in 2019 and 2020.

As the program continues, suppliers committing to extend efforts beyond their factory walls will establish a comprehensive water management plan, have a deeper understanding of internal and external water risks associated with their operations, promote better water management practices and awareness throughout their own supply chains, and ultimately help improve the water resources in their region.

These experiences from around the world demonstrate the range of contexts within which the AWS Standard can be used to provide a framework for a consistent approach across a business, and the variety of benefits and positive impacts this approach can lead to at a site, catchment and corporate level.

29.4 LOOKING AHEAD: THE FUTURE OF WATER STEWARDSHIP

Water stewardship is now firmly established and accepted as a key part of efforts to ensure the sustainable use of water resources. While huge progress has been made through leadership like that illustrated in the previous section, the fact remains that the vast majority of water use is by smaller companies or farmers within global value chains. Ensuring water stewardship reaches these users remains a critical challenge.

One way to achieve this is to ensure that water stewardship actions and learnings are better connected to policy priorities like Integrated Water Resources Management (IWRM) and climate resilience. Similarly, municipalities and other forms of local government have been largely absent from water stewardship discussions, yet have huge potential as levers of change, either through motivation or legislation. Effective participation by municipalities can also bring water stewardship to the attention of the general public. Experience from other sustainability standards and initiatives shows that this is a crucial route to scale.

Specific to AWS, we will work in different ways to grow water stewardship activity worldwide. Harnessing the expertise of the AWS membership is critical if we are to grow participation in water stewardship, ensure that the benefits are widely accessible, and make sure that water stewardship drives positive change where it is most needed. One of our key roles is to enable peer support and knowledge sharing across a wide range of sectors and geographies.

Another role for AWS is to deepen our collaboration with other water and sustainability initiatives, helping to provide a clear and accessible pathway to engage in water stewardship. At a strategic level there is growing international cooperation focusing on key sectors and framed by the UN Global Goals and, increasingly, climate resilience.

More practically, AWS and other organizations need to work together to develop the range of tools and support that reduce the barriers of entry to water stewardship. One example is in the agricultural sector where we are working with GLOBALG.A.P. to develop an AWS 'add-on' to enable GlobalG.A.P. certified sites to improve their water stewardship activities beyond the farm and achieve AWS Certification. At the corporate level, we collaborate with organisations such as the CEO Water Mandate, the World Business Council on Sustainable Development (WBCSD) and others to raise awareness of water stewardship and make the links between site and catchment-based action and corporate level commitment required to tackle our shared water challenges.

The common thread through all of these is collaboration: only by working together across sectors, geographies and interests can we ensure the potential of water stewardship can be realised.

REFERENCES

AWS (2019). The International Water Stewardship Standard. Available from https://a4ws.org/.

AWS, Dole, EDKEA, Tecbaco, South Pole and WWF (2020). Case Study: The World's First AWS Group Certification. Available from. https://a4ws.org/

Chapter 30

WWF water risk filter: Assess, respond & value water risks

Ariane Laporte-Bisquit

World Wide Fund for Nature, Water Risk Filter Project Manager, Berlin, Germany

Keywords: business, finance, risk assessment, stewardship, tools, water risk

30.1 INTRODUCTION

With worsening water security across the globe, it is widely recognized that the private sector has an important role to play in safeguarding and sustainably using our shared freshwater resources. WWF is a strong advocate for responsible private sector engagement on water issues and a pioneer in the water stewardship space. Within this context, WWF developed a practical online tool, the WWF Water Risk Filter, to raise corporate awareness on water risks and drive effective water stewardship action.

Launched in 2012 in partnership with the German Development Finance Institution DEG, the WWF Water Risk Filter was one of the first online water risk assessment tools for companies and investors to assess water risks in their operations, supply chain and investments. However, the water risk assessment landscape has evolved considerably since. Not only has the use of water risk tools become more commonplace – contributing to a massive growth in awareness of water risks amongst the private sector – but new data and demands from users have also emerged. In response, WWF undertook a major upgrade of the WWF Water Risk Filter tool and a new version 5.0 was launched in 2018. The enhanced tool not only helps users to assess water risks using the best available data but also enables users to evaluate financial impacts (Value section) and identify contextually-appropriate responses (Respond section) to their unique water risk profile. In 2020, the tool was expanded to align with the Task Force on Climate-related Financial Disclosure (TCFD) recommendations by providing new climate and socio-economic pathway-based scenarios for 2030 and 2050.

This chapter aims to provide a comprehensive overview of the three key sections of the WWF Water Risk Filter – (1) Assess section, (2) Respond section and (3) Value section – along with practical examples of how

DOI: 10.2166/9781789060676_0321

it has been used by companies. As the tool is continually evolving, some important future developments and plans will be highlighted throughout this chapter.

30.2 ASSESSING WATER RISKS

Designed to be easy to use by non-water experts, the WWF Water Filter is a free online tool that enables companies and financial institutions to conduct comprehensive water risk assessments. More than 400,000 sites have already been assessed by over 6,000 users from a broad range of sectors – including food and beverage, textile, retail, mining and finance.

The WWF Water Risk Filter's risk assessment framework uses the well-recognized categorization of corporate water risks according to three risk types: physical, regulatory and reputational – as defined by the CEO Water Mandate in collaboration with multiple organizations (UN Global Partnership, 2018). Moreover, the majority of other water risk tools have also adopted the same risk assessment framework, which helps to ensure consistency and aligned approaches to water risk assessments.

In general, water risk tools assess the physical, regulatory and reputational risks faced by a company due to the nature of the river basin in which its sites are located, referred to as basin risks. However, a company's risk exposure also depends on its performance and potential impacts on the basin, referred to as operational risks. The WWF Water Risk Filter was designed to assess the physical, regulatory and reputational risks for both basin and operational water risks as illustrated in Figure 30.1. (Key considerations and guidance points for companies and financial institutions conducting a water risk assessment are addressed in the report, 'Right Tool for the Job' (WWF/WBCSD, 2020) which is available online at https://d2ouvy59p0dg6k.cloudfront.net/downloads/right_tool_for_the_job_1.pdf.)

The WWF Water Risk Filter basin and operational risk assessment framework has a three-level hierarchy: (A) risk types, (B) risk categories and (C) risk indicators. This hierarchical framework ensures a more comprehensive coverage of all aspects of the three corporate water risks types: physical, regulatory and reputational risks. The following paragraphs describe in detail the basin and operational risk assessment framework of the WWF Water Risk Filter.

30.2.1 Basin risk assessment

The WWF Water Risk Filter's basin risk assessment tool evaluates three types of risk – physical, regulatory and reputational risk – which are further subdivided into 12 risk categories. These categories are measured according to 32 separate indicators to evaluate basin risk. (The number increased from 20 to 32 in version 5.0 to provide a more diverse range of best available data sets and to ensure better understanding of basin risk exposure.) The next paragraphs describe the three risk types and associated four risk categories (total of 12 risk categories) of the Water Risk Filter's basin risk assessment framework.

(1) **Physical risks**: Companies can be exposed to a range of different physical water risks, which are classified in the tool into four categories relating to water scarcity, flooding, water quality and ecosystem services status.

- **Quantity – Water scarcity**: Lack of water can result in significant impacts to a business such as production disruption, higher operating costs and growth constraints. Recognizing that there are many models and approaches to assess water scarcity, the tool contains several well-recognized global data sets under this category, including the WRI Aqueduct 'Baseline Water Stress' (Hofste *et al.*, 2019) and the WaterGAP3 'Water Depletion' (Brauman *et al.*, 2016);
- **Quantity – Flooding**: Flood events can impact businesses' operations directly, as well as across their value chain, by closing operations, disrupting supply chains, increasing capital costs and

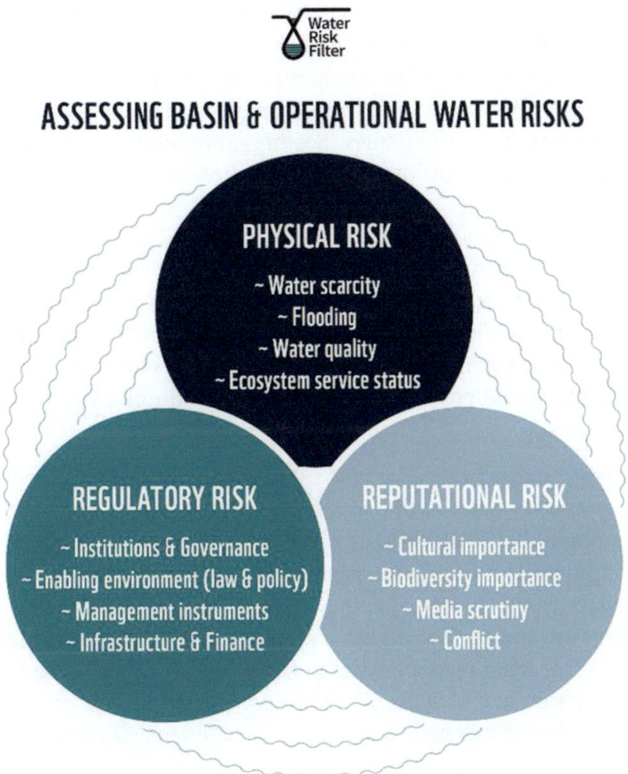

Figure 30.1 WWF Water Risk Filter Risk Assessment: 3 types and 12 categories of basin and operational water risks. (*Source*: WWF (2020a, b) 'Water Risk Filter 5.0 Methodology' Download available at Water Risk Filter website https://waterriskfilter.panda.org/en/Explore/DataAndMethod. Accessed October 5 2020).

reducing revenues from lower sales and/or output. The tool's flood occurrence risk is based on historical data from the University of Colorado's Dartmouth Flood Observatory (Brakenridge, 2019);

- **Water quality**: Poor or degraded water quality can result in sites having no or limited access to water of good enough quality for their activities and performance, increasing operating costs (e.g., treating incoming water), and increasing regulatory and reputational scrutiny. The tool's Water Quality Index is based on the latest global data sets developed by the World Bank (Damiana *et al.*, 2019);
- **Ecosystem service status**: The degradation of ecosystem services can result in sites having restricted access in the long-term to the quantity and quality of water that they need for optimal performance of the sites. For example, the mapping of the world's free-flowing rivers (Grill *et al.*, 2019), which can also be used to determine the degree of rivers fragmentation, was used in the tool as a proxy for assessing ecosystem services status.

(2) **Regulatory risks**: Most businesses thrive in a stable regulatory regime. Thus, regulatory water risk is heavily tied to the concept of good governance. The WWF Water Risk Filter adopted the framework and integrated the data sets used by UN Environment for monitoring countries' progress towards achieving UN Sustainable Development Goal (SDG) 6.5.1 on implementing

Integrated Water Resource Management (IWRM) (UN Environment, 2018). Regulatory risks are assessed according to four categories aligned with SDG 6.5.1 framework:

- **Enabling environment**: companies face higher regulatory risks when operating in countries with no or weak legal frameworks, policies and management plans on freshwater resources;
- **Institutions & governance**: companies face higher regulatory risks when operating in countries with no or weak presence of official forum/stakeholder platforms and institution for integrated water resource management;
- **Management instruments**: companies face higher regulatory risks when operating in countries with no or a low level of management instruments in place as well as monitoring data available; and
- **Infrastructure & finance**: companies face higher regulatory risks when operating in countries with no or a low level of financing for water resource management and infrastructure.

(3) **Reputational risks**: Although a company's reputational risks are primarily linked to its water management performance, there are some basin pre-conditions that make reputational water risk more likely to manifest.

- **Cultural diversity**: water is a social and cultural good, therefore it is critical to understand that businesses face reputational risk due to the importance of freshwater for local communities and indigenous people in their daily life, religion and culture;
- **Biodiversity importance**: companies operating in biodiversity rich areas are exposed to higher reputational risks;
- **Media coverage**: businesses face higher reputational risks when operating in countries or regions with higher global or local media coverage reporting on water-related issues; and
- **Conflict**: WWF partnered with RepRisk (https://www.reprisk.com/) which collects data on documented negative incidents, criticism and controversies that can affect a company's reputational risks.

The 32 global risk indicators are based predominantly on publicly available external, peer-reviewed global data sets. Each original data set is first classified into the WWF Water Risk Filter's risk score categories ranging from a 1-to-5 value: risk score level 1 represents no or very limited risk while risk score level 5 represents very high risk. The data sets are subsequently aggregated at the basin or country level. This process creates a series of basin risk indicators out of the raw basin data sets.

In addition to the global water risk data sets, WWF has developed and integrated several higher-resolution data sets into Water Risk Filter 5.0 to provide country-specific local risk indicators for conducting risk assessment at a finer scale. More specifically, the tool currently has local data sets available for the following countries and regions: South Africa; Brazil; Colombia; Chile; Europe Region; Great Britain; Spain; Hungary; and Greater Mekong Region (Thailand, Vietnam, Laos, Cambodia). Additional local data sets will be integrated into the tool in the coming years.

As new data sets are constantly emerging, the WWF Water Risk Filter's underlying global and local data sets are reviewed and, as appropriate, updated on an annual basis with latest available data. Detailed information on the WWF Water Risk Filter risk assessment framework and latest underlying data sets are available online in the WWF Water Risk Filter's methodology documentation (WWF, 2020a)

Based on the WWF Water Risk Filter basin risk assessment framework, in 2020 the tool launched new climate and socio-economic pathway-based scenarios of water risk for 2030 and 2050. This latest development aims to enable companies and investors to perform scenario analysis to better understand future water risks and opportunities, as recommended by the Task Force for Climate-related Financial Disclosure (TCFD, 2020).

To tackle the complexity of water risks, the Water Risk Filter scenarios comprehensively cover all types of water risks, ranging from TCFD-focused acute physical risks (e.g. flooding) and chronic physical risks (e.g. scarcity, water quality, and ecosystem services status) to the less commonly explored regulatory and reputational water risks, which can cause significant potential impacts if overlooked. Ultimately, the outputs of scenario analysis should be used to evaluate whether the company's current strategy is adequately resilient or will need stronger resilience planning, considering the future risk levels. The WWF Water Risk Filter Brief (WWF, 2020b) provides an overview of the new WWF Water Risk Filter scenarios and guidance on how scenario analysis can help companies and investors to understand future water risks and build resilience in an uncertain future.

30.2.2 Operational risk assessment

A site's operational risk exposure is based on the nature of its activities and water management performance. In the WWF Water Risk Filter tool, operational water risks are assessed by completing a short or detailed operational risk questionnaire developed by WWF through an iterative stakeholder consultation process in order to capture the most important aspects of operational water risk. The framework of the operational risk questionnaire is aligned to the basin risk assessment framework and contains a similar three-level hierarchy comprised of: (A) risk types, (B) risk categories and (C) risk indicators. The detailed assessment questionnaire contains 22 risk indicator questions whereas the rapid assessment questionnaire consists of only 10, both covering all three risk types: physical, regulatory and reputational.

Table 30.1 provides an overview of the WWF Water Risk Filter's short version operational risk questionnaire framework. While the short version questionnaire will provide a rapid operational risk assessment, the higher the quality of input data, the better the quality of the assessment output will be. Therefore, users are encouraged to complete the full version questionnaire for more comprehensive operational risk assessment results. Recognizing that different sectors face different types of operational risks, efforts are currently underway to develop tailored versions of the operational risk questionnaire for specific sectors: agriculture, textile & apparel, forestry, and pulp and paper products.

30.2.3 Case study: how EDEKA uses the WWF water risk filter to assess water risks across their supply chain

Since 2009, WWF has a transformational partnership in place with EDEKA, the leading German food retailer. As a first step in EDEKA's water stewardship work with WWF, the WWF Water Risk Filter was used to assess water risks at a global scale across their supply chain. More specifically, the physical, regulatory and reputational water risks for over 2,300 own-brand products were analyzed.

In addition to a global water risk assessment, EDEKA and subsidiary Netto Marken-Discount conducted finer scale assessments using the high-resolution data sets for Spain in the WWF Water Risk Filter as illustrated in Figure 30.2. EDEKA and WWF have a joint project with citrus fruits growers in southern Spain and the high-resolution data sets provided a more detailed understanding of the local water risk exposure faced by growers.

The results of continuous risk assessments with the WWF Water Risk Filter at global and local level helped EDEKA and Netto Marken-Discount to:

- increase their engagement with their suppliers on water risks and stewardship;
- identify hotspots where they could focus their efforts to reduce high water risks by implementing on-the-ground Alliance for Water Stewardship (AWS) projects; and
- increase customer awareness by including water risk assessments in the approval process for WWF co-branding of their organic own-brand products.

Table 30.1 Three-level hierarchy of the short version operational risk questionnaire framework: (A) risk types, (B) risk categories and (C) operational risk indicator.

Risk Type	Risk Category	Operational Question/Risk Indicator
Physical Risk	(1) Physical: Water Scarcity	O1. In which ways does the site use water?
		O2. How important is the current and future use of water quantity and quality for operating/processing at this site?
	(2) Physical: Quality	O3. Is it necessary to treat/purify the water the site withdraws before its use in operations?
		O4. Is it necessary to treat/purify the water the site withdraws after its use in operations?
		O5. What is the potential impact of the site's operations on downstream water quality in terms of physical, chemical and biological parameters?
Regulatory Risk	(3) Regulatory: Enabling Environment (Policy & Laws)	O6. Relative to other water users in your local catchment (~50 km radius), does this site face heavy water-related regulation and legal enforcement?
	(4) Regulatory: Institutions and Governance	O7. Is the site always in compliance with legal waste water quality standards?
Reputational Risk	(5) Reputational: Community Conflict	O8. Relative to other water users in your local catchment (~50 km radius), would you consider the site a large water user/polluter?
		O9. Relative to other water users in your local catchment (~50 km radius), is the company associated with the site a recognized brand (to the local public)?
		O10. How would you describe this site's general water management/stewardship maturity?

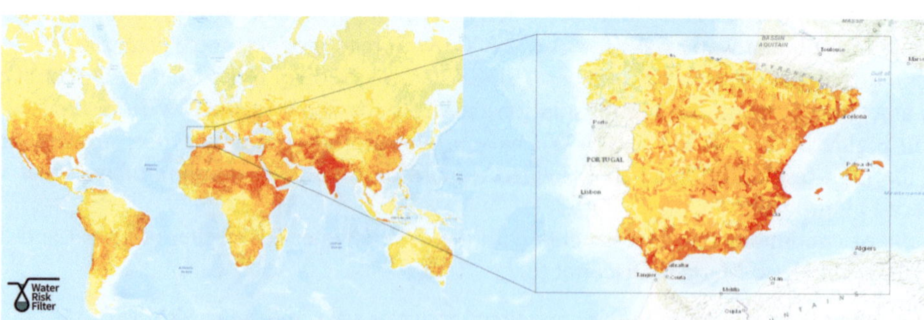

Figure 30.2 WWF Water Risk Filter Global Risk Map (left) and WWF Water Risk Filter High Resolution Map for Spain (right). (*Source*: WWF (2020a, b) Water Risk Filter Maps, 'Map Region: Global' and 'Map Region: Spain'. Accessible at Water Risk Filter website https://waterriskfilter.panda.org/en/Explore/Map. Accessed October 5 2020).

30.3 RESPONDING TO WATER RISKS

The Respond section of the WWF Water Risk Filter tool was borne out of WWF's experience of working with companies who were either requesting further guidance on how to proceed from their risk assessment results, or who in some cases were selecting responses to their water risks that did not align with or match their water risk exposure. Thus, the Respond section was developed with the aim of helping to guide companies towards taking contextually appropriate response actions given the water risks that they are exposed to.

By dynamically linking risk assessment results, the WWF Water Risk Filter's Respond section provides users with recommended response actions to address their unique water risk exposure. It enables users to quickly identify and prioritize response actions that adequately mitigate their water risks and help them set contextually appropriate targets. While WWF recognizes that global data sets are not always sufficiently accurate to be prescriptive in dictating response actions, the WWF Water Risk Filter's Respond section offers a strong starting point for guiding contextually appropriate actions, which can then be further refined and informed based on local data, knowledge and expertise.

Recognizing the existence of well-established water stewardship frameworks, the WWF Water Risk Filter's response actions are based on and aligned to leading frameworks and standards such as the Alliance for Water Stewardship, CEO Water Mandate's Water Stewardship Toolkit, Ceres Aqua Gauge, CDP Water Security and UN Sustainable Development Goals 6. This enables users to better understand and take action within the broader water stewardship ecosystem.

The online component of the WWF Water Risk Filter's Respond section offers users a customized and ranked list of recommended actions according to their unique water risk profile. All actions are classified under 10 categories, which are a hybrid of CEO Water Mandate's Water Stewardship Toolkit, CDP's Water Security Questionnaire, Ceres' Aqua Gauge and WWF's experience. Furthermore, users can filter and select actions according to different criteria of interest. For example, actions can be filtered according to the implementer (actions to be implemented by site managers at the site level vs. actions to be implemented at the corporate level) or the company's water stewardship maturity level (actions for beginners vs. well established companies). Table 30.2 provides examples of WWF water risk filter's recommended response actions.

The water risk assessment space has focused primarily on understanding water risk exposure. However, the enhanced WWF Water Risk Filter enables both companies and investors to understand not only water risk exposure, but also account for response in a more comprehensive manner by understanding how strong corporate response is as well as how appropriately aligned their responses are to exposure. By taking such a risk-response approach, companies and investors can adopt a more nuanced approach to water risk that also helps identify opportunities for value creation.

While water risk remains the key driver to mobilize corporate action and is the main focus of the Water Risk Filter tool, WWF aims to drive a new narrative on water stewardship to capitalize on opportunities, collaboration, resilience and value creation (WWF, 2018a). By leveraging its risk assessment and respond section, the WWF Water Risk Filter is seeking to evolve to provide new ways for companies and investors to identify water-related opportunities and invest in bankable projects as part of WWF's Bankable Water Solutions initiative (WWF, 2018b).

30.3.1 Case study: H&M group

H&M Group and WWF have been working together on water stewardship since 2011. Developing a comprehensive understanding of water risks across H&M Group's operations and suppliers using the WWF Water Risk Filter tool was a critical first step.

Table 30.2 Examples of WWF water risk filter's recommended response actions per water stewardship response categories.

Water Stewardship Response Category	Example of Response Action
(1) Water Awareness & Internal Capacity	Review or conduct a formal study on future water resources scenarios including water supply and quality resulting from higher demands within the basin and how it may affect the company's operations and value chain
(2) Strategy & Business Planning	Calculate how water can affect the site's financial value and integrate into decision-making related to opportunity identification
(3) Collective Action	Engage with peer companies in regionally specific water-related benchmarking
(4) Disclosure & Reporting	Publicly disclose the site's efforts to address shared water challenges
(5) Water Governance	Join a water-related forum (sectoral or intersectoral) as for example: CEO Water Mandate, AWS, WBCSD, ICMM (mining), SAC (apparel), BIER (beverage), etc.
(6) Operations, Performance Measurement & Management	Install or upgrade water treatment systems at the site to treat water to the necessary levels (e.g. reverse osmosis)
(7) Policies, Standards and Plans	Set performance standards and goals through publicly available water policy/statement that align to the water-related SDGs
(8) Risk Awareness	Assess the energy risks (including price increases, brownouts, blackouts, etc.) to the site of shared water challenges in the region (consider the water risks facing the regional energy grid)
(9) Stakeholder Engagement	Identify stakeholders, their water-related challenges and the site's sphere of influence
(10) Value Chain Engagement	Collect and monitor data related to effectiveness of suppliers' water management practices

More recently in 2019, water data from over 1,100 sites within of H&M Group's value chain was used with WWF Water Risk Filter's Respond section to provide both specific local water stewardship recommendations for individual sites as well as recommendations to inform internal regional water stewardship strategies. In addition to this, WWF is working with H&M Group to create a customized sector-specific questionnaire and set of bespoke sector-specific response actions that are tailored to the Apparel and Textiles sector.

Moving beyond assessing to responding to water risks, H&M Group is working actively in water stewardship projects in high priority river basins in China and Turkey. For example in the Taihu Basin in China, WWF, H&M Group and other international partners are providing trainings to increase their knowledge of water risks, regulatory changes and impact reduction opportunities. Moreover, this project has a strong component focused on collective action: textiles brands participation leverages their business partners and stakeholders towards the transformation of the supply chain/textile sector. H&M Group is encouraging other textiles brands to join global industry efforts on water stewardship to work 'beyond the factory fence' by helping to build multi-stakeholder solutions to water issues and support

Figure 30.3 Based on WWF's Water Stewardship Framework, H&M adopted a five-step water stewardship strategy as described in this diagram. (*Source*: WWF (2015) WWF Water Stewardship Framework)

stronger water governance in high water risk regions. Figure 30.3 illustrates the five-step water stewardship strategy adopted by H&M.

30.4 VALUING WATER RISKS

The financial impacts of droughts, floods, more stringent water regulations or the loss of social license to operate can be significant for businesses, reaching up to US$38 billion in financial loss based on CDP's 2018 Global Water Report (CDP, 2019). Investors and businesses are increasingly seeking ways to analyze the potential financial impact of water-related risks. Building on the WWF & IFC co-developed framework on the relationship between water risk, the value of water and water stewardship (WWF/IFC, 2015), WWF developed a database to catalogue all existing valuation tools and approaches that have emerged over the last 15 years. Launched in November 2019, the Valuing Water Database, embedded in the WWF Water Risk Filter, helps users navigate the array of valuation tools and approaches in order to identify the most suitable tool for the job. As explained in the series of WWF Reports 'Linking Water Risk and Financial Value' (WWF, 2019), WWF made the decision to develop its own valuation tool to address some of the gaps and limitations amongst existing valuation tools. In 2018, Water Foundry joined the effort to co-develop the Water And ValuE (WAVE) tool.

By combining data from the user, the WWF Water Risk Filter, CDP and more detailed probability data, WAVE aims to evaluate how water risk exposure currently affects financial value, and may potentially affect it in the future (CDP Global, 2019). The WAVE tool leverages Water Risk Filter data on basin and operational risk exposure, and accounts for site vulnerability and response to calculate financial impacts organized around a series of costs and revenue categories outlining WAVE's logic model (Figure 30.4). Whilst some user financial information is required to be entered, the WAVE tool is powered by CDP's Water Security database with the objective to draw upon corporate water disclosure

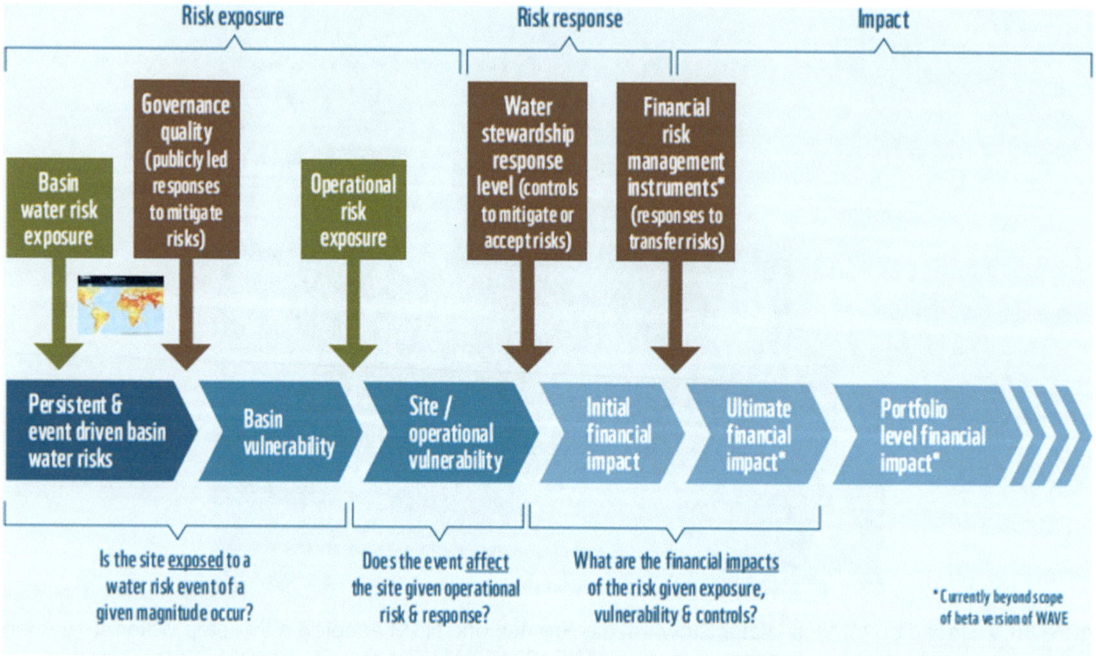

Figure 30.4 Process of translating water risk to financial impact in the WAVE Tool. (*Source*: WWF, 2019)

information to: (1) identify linkages of key water risk events to different types of financial impacts reported by companies and (2) inform financial impact magnitude ranges.

One of the key benefits of WAVE is that it can help users build a clear business case for the need to invest in mitigating a company's water risks. Too often the argument that water procurement is a low cost is used to justify a lack of action on water issues. However, water risk events such as droughts and floods can have significant financial impacts on capital and operational expenditures. WAVE offers a tool for asset-level water risk evaluation that can convert risk exposure into financial impacts suitable for financial analysis purposes. Moreover, the future goal is to enable WAVE to more explicitly handle scenarios to enable users to stress test information as recommended by the TCFD.

30.5 CONCLUSION

WWF's Water Risk Filter tool has evolved considerably since it was first launched in 2012. By shifting beyond risk assessment, WWF's Water Risk Filter has developed into a tool that can support private sector efforts to mitigate the rising tide of water risks and challenges.

The tool's evolution is a continuous journey – with new data availability arising out of new technologies, momentum on the TCFD initiative and WWF's new narrative on water stewardship on value creation – the Water Risk Filter will continue to expand and grow to help drive corporate water stewardship to the next level.

The tool provides a business case for more sustainable use of water by highlighting the risks of inaction. With this innovative tool, WWF will continue to mobilize private sector engagement and aims to raise the level of business sector commitment to supporting a transition to a water secure world for all.

REFERENCES

Brakenridge G. R. (2019). Global Active Archive of Large Flood Events. Dartmouth Flood Observatory, University of Colorado based on the Dartmouth Flood Observatory University of Colorado.

Brauman K. A., Richter B. D., Postel S., Malsy M. and Flörke M. (2016). Water depletion: an improved metric for incorporating seasonal and dry year water scarcity into water risk assessments. *Elementa Science of Anthropocene*, **4**, 000083.

CDP Global. (2019). Global Water Report – Treading Water: Corporate Response to Risk Water Challenges (2019). https://6fefcbb86e61af1b2fc4-c70d8ead6ced550b4d987d7c03fcdd1d.ssl.cf3.rackcdn.com/cms/reports/documents/000/004/232/original/CDP_Global_Water_Report_2018.pdf?1554392583 (accessed 5 May 2020). CDP is a not-for-profit charity that runs the global disclosure system for investors, companies, cities, states and regions to manage their environmental impacts. For more information, please visit: https://www.cdp.net/en/water.

Damiana R., Desbureaux S., Rodella A., Russ J. and Zaveri E. (2019). Quality Unknown: The Invisible Water Crisis. (Washington, DC: World Bank). doi: 10.1596/978-1-4648-1459-4

Grill G., Lehner B., Thieme M., Geenen B., Tickner D., Antonelli F., Babu S., Borrelli P., Cheng L., Crochetiere H., Ehalt Macedo H., Filgueiras R., Goichot M., Higgins J., Hogan Z., Lip B., McClain M., E., Meng J., Mulligan M., Nilsson C., Olden J. D., Opperman J. J., Petry P., Reidy Liermann C., Sáenz L., Salinas-Rodríguez S., Schelle P., Schmitt R. J. P., Snider J., Tan F., Tockner K., Valdujo P. H., van Soesbergen A. and Zarfl C. (2019). Mapping the world's free-flowing rivers. *Nature*, **569**(7755), 215–221.

Hofste R. W., Kuzma S., Walker S., Sutanudjaja E. H., Bierkens M. F. P., Kuijper M. J. M., Sanchez M. F., Van Beek R., Wada Y., Galvis Rodríguez S. and Reig P. (2019). Aqueduct 3.0: Updated Decision Relevant Global Water Risk Indicators. Technical note. World Resources Institute, Washington, DC.

TCFD (Task Force on Climate-related Financial Disclosure). (2020). https://www.fsb-tcfd.org. (accessed 4 November 2020).

UN Environment. (2018). Progress on integrated water resources management. Global baseline for SDG 6.

UN Global Partnership/Pacific Institute. (2018). The CEO Water Mandate: The Types of Water Risk: The Many Ways Water Challenges Can Affect Your Business. https://ceowatermandate.org/posts/types-water-risk-many-ways-water-challenges-can-affect-business/ See also https://ceowatermandate.org/university/101-the-basics/lessons/assessing-your-business-water-risks (accessed 5 May 2020).

WWF. (2018a). Water Stewardship Revisited: Shifting the narrative from risk to value creation (2018). https://d2ouvy59p0dg6k.cloudfront.net/downloads/wwf_waterstewardship_brief_web_final.pdf (accessed 5 October 2020).

WWF. (2018b). WWF's Bankable Water Solutions Initiative. Seizing the Water Opportunity. https://d2ouvy59p0dg6k.cloudfront.net/downloads/seizing_the_water_opportunity_report_pages_2.pdf (accessed 5 October 2020).

WWF. (2019). Linking Water Risk And Financial Value Part III: New Valuation Tool and Database. http://d2ouvy59p0dg6k.cloudfront.net/downloads/linking_water_risk_and_financial_value___part_iii.pdf (accessed 5 October 2020).

WWF. (2020a). Water Risk Filter methodology. https://waterriskfilter.panda.org/en/Explore/DataAndMethod (accessed 27 October 2020/).

WWF. (2020b). Water Risk Filter Brief (2020) Water Risk Scenarios: TCFD-aligned scenarios to help companies and investors turn risk into resilience.

WWF and International Finance Corporation. (2015). The Value of Water: A Framework for understanding water valuation, risk and stewardship (2015). https://commdev.org/wp-content/uploads/2015/05/The-Value-of-Water-Discussion-Draft-Final-August-2015.pdf (accessed 5 October 2020).

WWF & WBCSD. (2020). Right Tool for the Job: Tools and Approaches for Companies and Investors to Assess Water Risks and Shared Water Challenges. https://www.wbcsd.org/Programs/Food-and-Nature/Water/Resources/Right-tool-for-the-job (accessed 27 October 2020).

Chapter 31

Water footprint: A sustainability tool for industries

Leandra Diaz

Servicios Ambientales S.A., La Paz City, Bolivia

Keywords: circular economy, climate change, public-private partnerships (PPP), water compensation mechanism, water management from industries,

31.1 INTRODUCTION

The decreased availability of water resources, in quantity and quality, represents a great threat to humanity as a whole, as well as to the global economy (UN, 2015). Rapid growth of population coupled with the growth of the Global Gross Domestic Product (GDP) encourages more people than ever to aspire to a higher standard of living, producing rapid urbanization and industrialization (UN, 2017; IDB, 2018). Added to these increases are the effects of climate change, and the result is overwhelming pressure on global water resources.

Unsustainable water use and poor management of water resources, resulting in a lack of suitable quality water, also constitutes a significant constraint on industrial productivity. It is anticipated that in 2030 the planet will face a global water deficit of 40%, while the global demand of water for industrial production is projected to increase by 400% by 2050, more than in any other sector (UN, 2015). As a consequence, industrial water management is more important than ever, including water use, distribution, and the pollution generated by all facets of production.

There is an urgent need to reinforce the management of water basins to guarantee the regeneration of the resource and to avoid potential threats to development and economic growth. Companies must measure their impact on water to encourage conservation and to compensate cities for impacts. They need to improve water management by implementing water replacement projects in water basins, conservation of water recharge areas, etc.

This chapter compares two "water footprint" indicators used worldwide to assess and analyze industrial impact on water resources: (1) the Water Footprint Network water footprint tool; and (2) the International

DOI: 10.2166/9781789060676_0333

Standard (ISO) water footprint tool which includes the use of Life Cycle Analysis (LCA). Case studies of the application of each method are presented which revealed practical opportunities to reduce impact on the water environment. In addition, this chapter explores *alternative measurement frameworks* that allow industries to take a more participative role, from regulations to elaborating policies and strategies. In many cases the *municipal governments* take the lead in alliance with the private sector, which may participate through chambers of industry and commerce. A successful case study of collaborative participation among these key actors is also presented.

31.2 CONTEXT

Under the premises '*something that has not been measured, cannot be managed*' and '*something that has been measured, can be improved*', methodologies have been develop to provide indicators of water use, consumption, and other effects. Among the most popular methodologies worldwide are:

- *Water Footprint Network's Water Footprint (WF-WFN)*: Based on the methodology of the Water Footprint Network, this tool allows industries to analyze their direct impact on water quantity and quality during the production stages in which they directly utilize water; and
- *International Organization for Standardization's Water Footprint (WF-ISO)*: Based on the International Organization for Standardization's ISO 14046, this tool enables companies to trace their use of water more broadly throughout the supply chain, and helps to identify key points where timely and convenient actions could mitigate their impact.

31.3 ORIGIN, METHODOLOGIES, AND CASE STUDIES

31.3.1 Water footprint network's water footprint

In 2002 Arjen Hoekstra and Chapagain (2006), a researcher at the UNESCO-IHE Institute for Water Education, developed a water footprint methodology while working to measure the total volume of water consumed and contaminated related to the production of specific goods and services (including 'non-obvious' water uses). (The usual measure used by industries to measure their impact did not consider other impacts such as water sources, the volume of water that is no longer available in the basins, or the impacts of pollution.) Hoekstra's method reflected the total amount of water used, lost, and contaminated per unit of product produced. Given the importance that the methodology was gaining, the Water Footprint Network was created in 2008.

This methodology came under more scrutiny when these water footprint numbers began to be commonly applied as a means to gauge the relative water impact of various agricultural products. Previously, the cost of water did not routinely factor into the evaluation of agricultural exports, and the need to replenish overdrafted supplies was not given particular importance, even in countries with a high water deficit. As an example, a WF study of Kenya's Naivasha basin (a major flower exporting region) showed that to improve water management the price for water had to increase to reflect its actual cost to the community. Only then could the region consider adopting water replenishment projects and various regulatory measures (Mekonnen & Hoekstra, 2012).

31.3.1.1 WF-WFN methodology

The WF-WFN calculates water impact as the sum of (1) losses through evaporation or incorporation, (2) the direct or indirect contamination of the water, and (3) agricultural water remaining in the soil or incorporated into crop material. The information visualized through the WF-WFN allows the user to analyze the

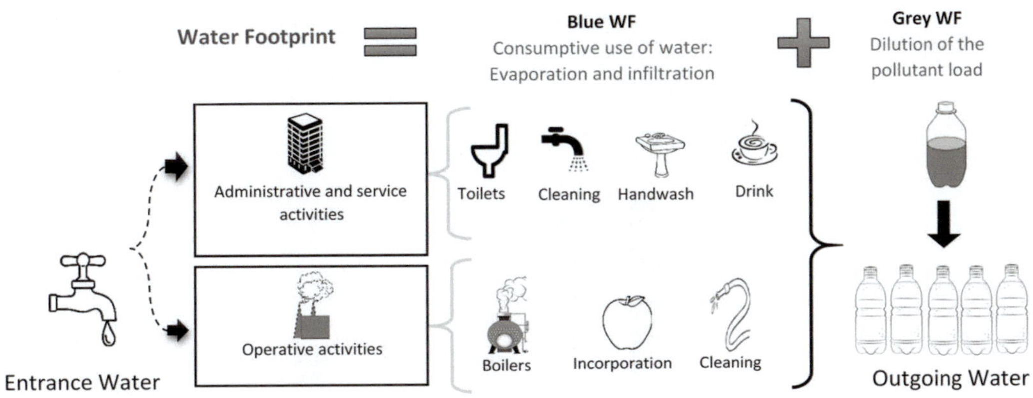

Figure 31.1 The Water Footprint Network's Water Footprint Analysis. (*SASA, 2019*)

environmental, social, and economic implications of the use of water in different geographical areas by calculating the total water footprint as the sum of these three types of water impacts, labeled Blue, Gray, and Green, and defined further as follows:

Blue Water Footprint is an indicator of *consumptive use* of freshwater (surface or underground), including water that evaporates, that is incorporated into a product, that does not return to the same flow zone, which is returned to another catchment area or to the sea, or that does not return in the same period. (An example of the last is water removed during a dry period and returned during a rainy season.)

Gray Water Footprint is an indicator of *pollution* expressed as the volume of water needed to dilute an activity's pollutant load to the point where the quality of the water remains above local regulations and the ecological requirement of the basin. (The lower the regulatory limit, the more water is required to dilute the pollutant loading). By expressing the impact of water pollution on local supplies in this way, the WF-WFN methodology can convey the total water impact of an activity or enterprise as a number – the 'water footprint'.

Green Water Footprint is an indicator of the volume of water that remains in the soil, on the surface, or is incorporated into the vegetation. This footprint measure is only calculated for agricultural crops, defined as any vegetation produced through anthropogenic activities. (Natural vegetation is not considered to exert a 'green' water footprint.)

Figure 31.1 shows the calculation and the analysis that is carried out in an industrial process for the analysis of the water footprint, from the point that the water enters the facility, throughout its use, and the generation of residual effluents.

31.3.1.2 Case study: WF-WFN management in a production facility of beverages

Objective: By determining how production of beverages directly impacts water, recommend steps to reduce those impacts and demonstrate to clients the company's commitment to sustainability (The name of the company has been withheld for reasons of confidentiality; the study was prepared by Servicios Ambientales SA.)

Scope: Measure the WF-WFN that results as a direct consequence of the beverage production based on a one-year study of the production of isotonic drinks, soft drinks, energizers, and water in the company's main facility.

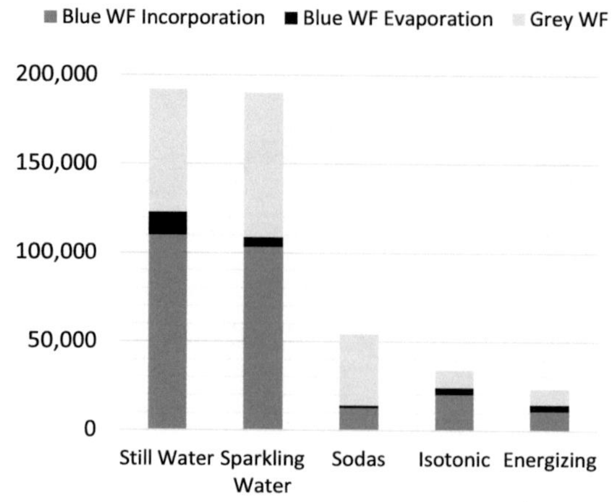

Figure 31.2 Total WF-WFN of evaluated products (m^3). (*SASA, 2019*).

Results: The Company's beverage production was found to represent 7% of the total water footprint of industry in the city where the plant is located. Due to the larger volumes produced (see Figure 31.2), production of bottled water represents 77% of the total WF-WFN. On the other hand, as shown in Figure 31.3, the Gray Water Footprint of soft drinks is three times greater than the other produced drinks since these beverages generate most of the contamination. Isotonic drinks and energizers account for most of water losses due to evaporation, more than double the other products.

Conclusion and Recommendations: Based on this water footprint analysis, reduction proposals focused on the production of sparkling and still bottled water, since together they account for 80% of the whole annual production and represent the dominant share of the total water footprint. Both water types are sold in reusable glass bottles, so the gray water footprint could be reduced by improving efficiency in the bottle washing process and by using biodegradable detergents, which are easier to dilute. In addition, the evaporation losses that occur during the purification of the drinking water could be reduced through the

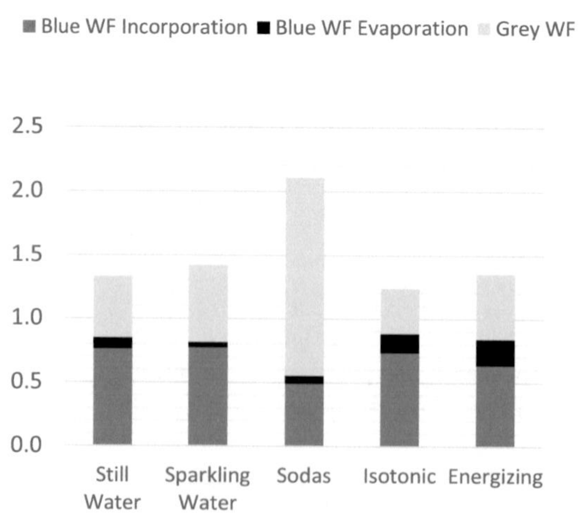

Figure 31.3 Liter WF-WFN per liter of product ($l/l_{produced}$). (*SASA, 2019*)

implementation of a vapor recovery system. The production process of isotonic drinks and energizers could also be improved by adding a vapor recovery system that would recycle water from the continuous filtration process as washing water. An alternate approach to reducing the Gray Water Footprint from production of soft drinks would be to construct a residual effluent treatment plant that would improve the load of organic materials that are generated and emitted.

31.3.1.3 Footprint of cities

The WF-WFN methodology was applied to water use in 14 cities in Latin America in the Footprint of Cities Project, and the impact of different sectors was determined (CAF, 2012–2018). Results are shown in Figures 31.4 and 31.5. Based on the WF-WFN analysis, on average the industrial sector contributes 6% to the total water footprint of the evaluated cities, mostly due to the impact of the water pollution it generates (To calculate Gray Water Footprint, BOD5 and COD were selected as the best indicators of the quality of urban effluent). Although the industrial sector contributes significantly less to the total water footprint than the residential sector, a greater percentage of its contribution is due to 'gray water' (as shown in Figure 31.5) compared with residential use, so reducing industrial pollution may lower water pollution overall.

31.3.2 International organization for standardization's water footprint

The Water Footprint-ISO (ISO 14044) is part of the ISO's environmental management suite of standard methods. It preserves the concept of the WF-WFN methodology, but extends the scope of the measurement of water impacts throughout the supply chain in greater detail. The LCA is '*the collection and evaluation of inputs, outputs and potential environmental impacts of a product system through its life cycle*' (ISO 14040). This analysis considers all the stages of the productive process of a product from obtaining the raw material, transport, its transformation, distribution and generation of waste, taking into account the consumption of resources and generation of impacts, at each of these stages. This methodology is aligned to the measurement of impacts on climate change with the accounting of greenhouse gases (ISO 14067) and the impact on water in terms of quantity and quality (ISO 14046). Use of the LCA became widespread during the 1970s when, in response to the rise in world oil prices, conservation advocates looked for ways to measure the potential to save energy for a given product (Raúl, 2008).

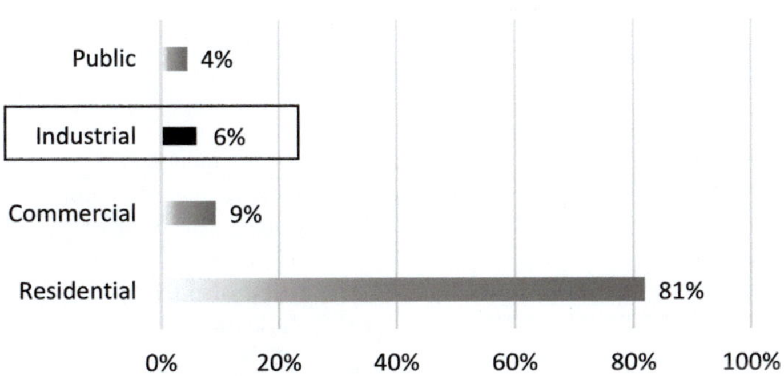

Figure 31.4 Industrial sector contribution to the Water Footprint of the Cities.

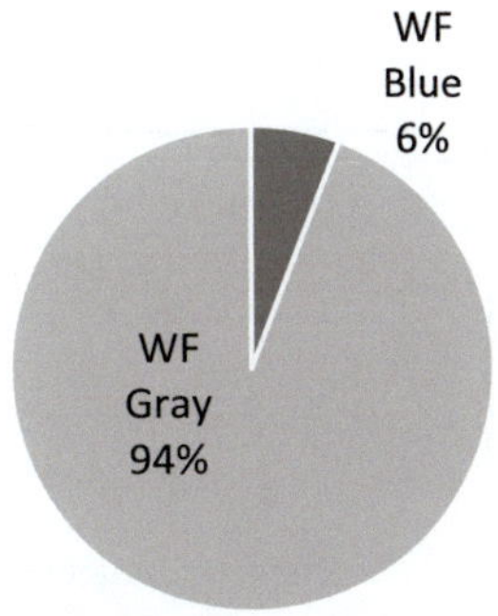

Figure 31.5 Industrial sector footprint types. (*Source: Footprint of Cities Project*)

31.3.2.1 Water footprint-ISO methodology

By carefully accounting for the impact of water use in all steps of the supply chain, the WF-ISO methodology allows the user to compare the water footprint of the same product produced by different methods or with raw materials from different supplies, as well as between similar products manufactured of different materials. As a result, this information can be used in the design and development of products, continuous improvement of processes, strategic planning, green marketing, as well as in the development of policies. The LCA accounts for the potential environmental impacts throughout the life cycle of a product from the acquisition of the raw material, passing through the production, utilization, final treatment, recycling, until its final disposition (that is, from the cradle to the grave) including the use of resources and the environmental consequences of emissions and discharges. As shown in Figure 31.6 below, The ACV ('*A*nálisis del *C*iclo de *V*ida completa, or life cycle analysis) is composed of a number of phases: (1) definition of objectives and scope; (2) inventory analysis; and (3) impact evaluation.

The flow lines in the Figure 31.6 go both ways, forward and backward, because the LCA is an iterative process: that is, as information is acquired about the system studied, the goals and scopes can be modified, adding or suppressing environmental impacts. This means that an LCA can be adapted to the availability of information for its implementation.

The life cycle approach and the methodologies described in the ISO 14040 and 14044 standards can interact with other environmental management tools, such as environmental impact studies and/or environmental audits, and also support other standards such as ISO 14046 (water footprint). This is illustrated in Figure 31.7, where the calculation for each type of impact indicator quantifies the type of substances that specifically cause this impact. Phosphorus, for example, causes eutrophication of freshwater while NOx, NH_3, SO_2, and hydrogen ions result in terrestrial acidification. These tools rely on extensive databases based on field studies that measure the impact on the production of different types of materials. Among the most well-known software systems are SIMAPRO and ECOINVENT.

31.3.2.2 Case study: Analysis of WF-ISO at a soft drink bottling facility

Objective: Evaluate the WF-ISO in the production chain of returnable (RB) and non-returnable bottles (NRB) and promote the consumption of the more footprint-friendly product through massive publicity/advertising campaigns. (This study was prepared by Servicios Ambientales SA for the unnamed company under terms of confidentiality.)

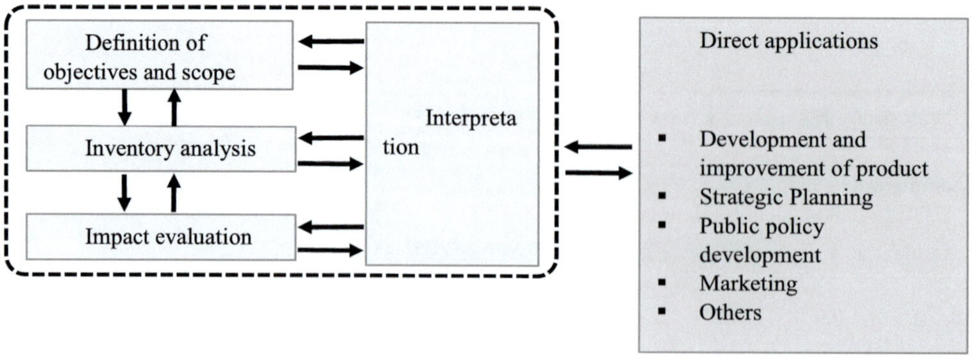

Figure 31.6 LCA phases. (*Source: IBNORCA, 2007*)

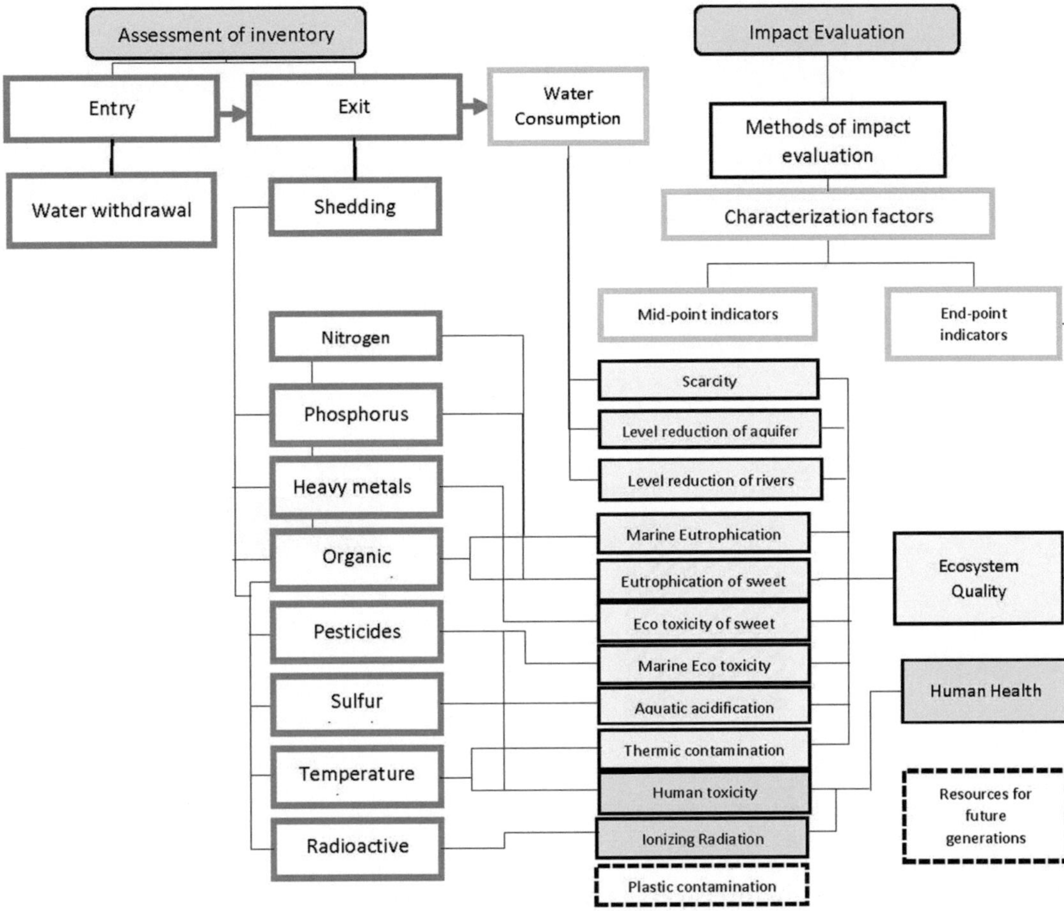

Figure 31.7 Water Footprint-ISO analysis. (*Based on CADIS, 2017*)

Scope: Perform an LCA of the two different bottle types from the procurement of raw materials to the final packaging ('cradle to door') for a timespan of one year.

Results: The production steps and activities were analyzed and are presented in Figure 31.8: Considering the water consumption in all production stages, NRBs have an impact 17% greater than the RBs, mostly due to the use of virgin resin which accounts for 57% of the water footprint. This is followed by the production of the preform which accounts for 21% of the water footprint.

The impact of raw material procurement for the RBs is reduced considerably due to the fact that the bottles are used at least 15 times in the span of 10 years, meaning that fewer new bottles have to be produced. The largest water impact of the RBs can be found at the washing stage. Here, for every liter of bottled soft drink, the washing process requires 0.6 liters of cleaning water. This stage represents 61% of the water footprint, but has to be carried out to ensure the safety of the product. Detailed information about the impacts of the two bottle types can be found in Figure 31.9.

Conclusions and Recommendations: Based on the data, the consumption of RBs should be promoted due to their lower WF-ISO impact. However, it is also recommended that cleaning efficiency should be

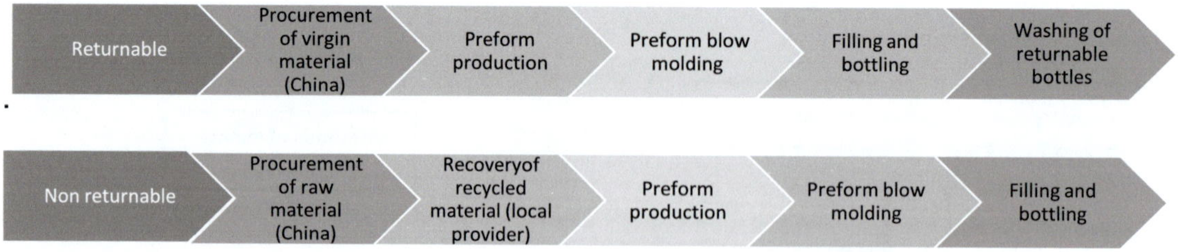

Figure 31.8 Production steps of non-reusable and reusable bottles. (*Source: SASA, 2017*)

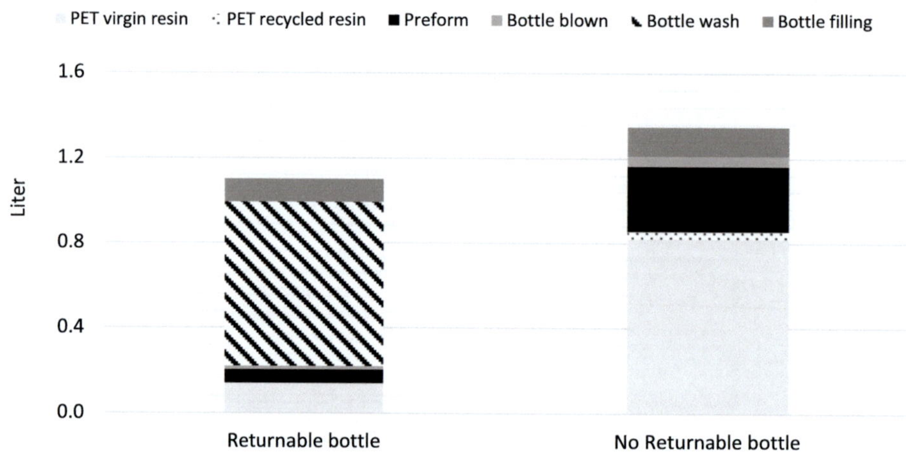

Figure 31.9 WF-ISO of the two bottles per consumption. (*Source: SASA, 2107*)

improved by using biodegradable detergents and reusing the cleaning water for other sanitation purposes. Also, it should be noted that since the LCA was only conducted from 'cradle to door' (not 'cradle to grave') there may be other impacts from product use not accounted for here. One example would be the waste generated by consumers after drinking the water contained in NRBs.

31.4 WATER MANAGEMENT BASED ON THE MEASUREMENT OF THE FOOTPRINT

After measuring the water footprint and implementing actions to reduce it, there may still be a gap between a product's overall water impact and true sustainability (i.e., a neutral water impact). To account for this gap, other tools have been created that suggest additional measures to attain true water sustainability. The following is a case study of the implementation of a successful municipal project involving an alliance between the industrial sector and local government.

31.4.1 Case study: mechanism for footprint management of the city of Cuenca, Ecuador

The municipal government of Cuenca allied with the Chamber of Commerce to implement a pilot project designed to comply with the goals of both the Carbon Footprint and WF-WFN (This study was prepared by Servicios Ambientales S.A. company). Called a 'mechanism of footprint management in alliance with the private sector', it incorporates measurement, reduction, and compensation strategies into industry management policies and businesses models so that they contribute to the fulfillment of footprint reduction goals established by the municipal government. The mechanism aims to make incremental improvements in business operations, promoting voluntary interaction between public and private sector participants within an overall framework of commitment to the goal of footprint reduction and water conservation.

As part of the project an online tool was developed that allows businesses to measure their footprint and to identify possible reduction activities. These activities will help businesses not only to reduce their use of resources, but it can also reduce costs, improve their production processes, and present them with possible technological improvements. Thanks to the progress that participating industries can achieve, they become eligible for international certifications that allow them to be more competitive in foreign and local markets. Furthermore, they identify compensation projects that will secure the conservation of water sources and the sustainability of their business.

31.5 BOUNDARIES, CHALLENGES AND LESSONS LEARNED

It is clear that there is an urgent need to manage water resources sustainably, and to promote these actions at all levels including communities and academic and non-profit institutions as well as private companies, industries, and municipal and regional governments. The WFN and ISO water footprint analysis tools allow industries to measure their impact on water at various stages of the production, so they can take timely action to create greater efficiency, reduce cost, and the lower their impact on all resources, promoting a 'circular economy'. However, widespread use of these tools will not automatically result in the sustainable use of water by industry. While some companies fail to manage their water impact due to a lack of knowledge of the water footprint methodologies, others may simply be waiting for regulations or other government mechanisms to require them to do so.

If a company decides to conduct a water footprint analysis, it is crucial for the management staff of the company to demonstrate their commitment not only to collect valid data, but also to follow-up on the findings gained from the evaluation process. The footprint evaluation is usually done in cooperation with industry technicians, but involvement by decision-makers is key: the greater their participation, the higher the possibilities that the findings from the water footprint analysis will result in measurable improvements. Only at these higher levels will an equilibrium be found between industrial production, which is intended to produce goods and services desired by consumers, and the problem of water sustainability due to productive activities.

To reach a balance, governments must incorporate holistic and sustainable water management strategies and policies. It will be necessary to consider all stages of the water life cycle, from the catchment sources, to the water distribution, the demand of different sectors and the general contamination. Only through a holistic approach will it be possible to guarantee the renewal of water resources. Where water costs are far below the true value of the water consumed, and the cost of water replenishment and wastewater decontamination are not considered, it will be difficult for municipal governments to obtain the funding required to implement best practice projects or better infrastructure works.

Since the benefits of replenishment actions may take decades to become fully visible, it is important to institute sustainability actions in the short-term to ensure the provision of water for following generations, including availability of water to support industry.

REFERENCES

Centro de Análisis de Ciclo de Vida y Diseño Sustentable (CADIS) (2017). Basado en la norma internacional ISO 14046: Huella hídrica – Principios, requisitos y Guía, publicada en 2014.

CAF. (2012–2018). Proyecto Huella de Ciudades. iniciativa del banco de desarrollo de América Latina -CAF- en alianza con la Agencia Francesa de Desarrollo –AFD; y Alianza para el Clima y Desarrollo – CDKN; facilitado por la Fundación Futuro Latino Americano –FFLA; e implementado por SASA. The Footprint of Cities Project is an initiative of CAF (Banco de desarrollo de América Latina) in alliance with CDKN (Climate and Development Knowledge Network), AFD (French Development Agency), FFLA (Fundación Futuro Latinoamericano), and SASA (Servicios Ambientales SA.).

The 14 cities in the project include: Paz, Santa Cruz, Cochabamba, Tarija and El Alto (Bolivia); Quito, Loja, Santa Cruz de Galápagos, Guayaquil and Cuenca (Ecuador); Recife and Fortaleza (Brazil); Cali (Colombia); and Lima (Peru).

Hoekstra A. K. and Chapagain A. Y. (2006). Water footprints of nations: water use by people as a function of their consumption pattern. *Water Resources Management*, **21**(1), 35–48.

IBNORCA. (2007). Bolivian Norm NB- ISO 14040 Environmental Management – Life cycle analysis - Principles and reference framework. Instituto Boliviano de Normalización y Calidad.

IDB. (2018). Macroeconomic Report for Latin America and the Caribbean 2018. IDB-AR-161, Interamerican Development Bank. Department of Research and Economics. The average annual growth of the Global GDP was 3.5% in the period 1960 to 2018.

ISO 14040. (n.d.). ISO 14040: Environmental management – life cycle assessment – principles and framework. International Organization for Standardization, 2006.

ISO 14044. (n.d.). ISO 14044: Environmental management – life cycle assessment – requirements and guidelines. International Standards Organization, 2006.

ISO 14046. (n.d.). ISO 14046: Water Footprint – Principles, requirements and guidelines. International Organization for Standardization.

ISO 14067. (n.d.). ISO 14067: Greenhouse gases – Carbon footprint of products – Requirements and guidelines for quantification. International Standard Organization, 2018.

Mekonnen M. M., Hoekstra A. Y. and Becht R. (2012). Mitigating the water footprint of export cut flowers from the Lake Naivasha Basin, Kenya. *Water Resources Management*, **26**, 3725–3742.

Raúl C. V. (2008). Historia ampliada y comentada del análisis de ciclo de vida (ACV). Escuela colombiana de ingeniería, Programa de Ingeniería Industria.

SASA. (2013). Gestión de la Huella Hídrica en una Planta de fabricación de bebidas en Sudamérica. Servicios Ambientales S.A.

SASA. (2017). Huella de Agua de botellas retornables y no retornables de una embotelladora en Sudamérica. Servicios Ambientales S.A.

UN. (2015). Report of the United Nations on water resources in the world 2015. United Nations. World Water Assessment Programme.

UN. (2017). World Population Prospects: The 2017 Revision, Volume II: Demographic Profiles. (ST/ESA/SER. A/400): United Nations. Department of Economic and Social Affairs, Population Division. It is expected that the population will increase by 22% in the period 2015–2050.

Chapter 32

Best available techniques as a sustainability tool for industrial water management and treatment

Maria Concetta Tomei

Water Research Institute, C.N.R., Via Salaria km 29.300, CP 10, 00015 Monterotondo Stazione, (Rome), Italy

Keywords: BAT Reference Documents, BATs for water management, BATs for wastewater treatment, Best Available Techniques, Chemical industry, Industrial water reuse, Minimization of water emissions

32.1 INDUSTRIAL WASTEWATER: MANAGEMENT AND TREATMENT

The past few decades have been characterized by industry's increasingly intense concern for the environment. While government agencies mandate stricter limits on wastewater discharge, companies struggle to brand themselves as 'greener' and more sustainable than their competition. As a result, industrial wastewater management strategies have focused on treatment methods that are both economical and environmentally sustainable. Specification of Best Available Techniques (BATs) can help industries, decision-makers, and regulators address these concerns by highlighting technological solutions that provide both highly treated effluent for discharge and high quality water for reuse.

EU member states are requested to use BATs as reference points to attain environmental quality objectives and establish environmental permit conditions for large industrial installations. These European BATs are defined by 'BAT Reference documents' (BREFs) prepared by Technical Working Groups organized in compliance with the Industrial Emissions Directive (IED) (2010/75/EU) and including a variety of stakeholders such as regulators, industrial companies, and environmental non-governmental organizations (NGOs). As listed in Table 32.1, EU BREFs address a comprehensive range of environmental issues in a variety of industries including emissions in air, land, and water (http://eippcb.jrc.ec.europa.eu/reference).

DOI: 10.2166/9781789060676_0343

Table 32.1 Best available techniques reference documents (BREFs) and conclusion documents (BATC) developed under the IPPC directive and the IED.

Industry	Code	Document
Ceramic Manufacturing Industry	CER	BREF (08.2007)
Common Waste Water and Waste Gas Treatment/Management Systems in the Chemical Sector	CWW	BATC (06.2016);BREF
Common Waste Gas Treatment in the Chemical Sector	WGC	MR (09.2017 and 03.2018)
Emissions from Storage	EFS	BREF (07.2006)
Energy Efficiency	ENE	BREF (02.2009)
Ferrous Metals Processing Industry	FMP	BREF (12.2001)
Food, Drink and Milk Industries	FDM	BREF (08.2006)
Industrial Cooling Systems	ICS	BREF (12.2001)
Intensive Rearing of Poultry or Pigs	IRPP	BATC (02.2017)
Iron and Steel Production	IS	BATC (03.2012)
Large Combustion Plants	LCP	BATC (07.2017); BREF
Large Volume Inorganic Chemicals – Ammonia, Acids and Fertilisers	LVIC-AAF	BREF (08.2007)
Large Volume Inorganic Chemicals – Solids and Others Industry	LVIC-S	BREF (08.2007)
Manufacture of Glass	GLS	BATC (03.2012)
Manufacture of Organic Fine Chemicals	OFC	BREF (08.2006)
Non-ferrous Metals Industries	NFM	BATC (06.2016); BREF
Production of Cement, Lime and Magnesium Oxide	CLM	BATC (04.2013); BREF
Production of Chlor-alkali	CAK	BATC (12.2013); BREF
Production of Large Volume Organic Chemicals	LVOC	BATC (12.2017); BREF
Production of Polymers	POL	BREF (08.2007)
Production of Pulp, Paper and Board	PP	BATC (09.2014);BREF
Production of Speciality Inorganic Chemicals	SIC	BREF (08.2007)
Refining of Mineral Oil and Gas	REF	BATC (10.2014); BREF
Slaughterhouses and Animals By-products Industries	SA	BREF (05.2005)
Smitheries and Foundries Industry	SF	BREF (05.2005)
Surface Treatment Of Metals and Plastics	STM	BREF (08.2006)
Surface Treatment Using Organic Solvents (including Wood and Wood Products Preservation with Chemicals)	STS	BREF (08.2007)

(*Continued*)

Table 32.1 Best available techniques reference documents (BREFs) and conclusion documents (BATC) developed under the IPPC directive and the IED (*Continued*).

Industry	Code	Document
Tanning of Hides and Skins	TAN	BATC (02.2013); BREF
Textiles Industry	TXT	BREF (07.2003)
Waste Incineration	WI	BREF (08.2006)
Waste Treatment	WT	BATC (08.2018); BREF
Wood-based Panels Production	WBP	BATC (11.2015); BREF

Source: European Commission, Joint Research Centre, Directorate B – Growth and Innovation Reference documents under the IPPC Directive and the IED http://eippcb.jrc.ec.europa.eu/reference/ Accessed September, 2019.

32.1.1 The BAT concept

The concept of BAT originated in the 1970s, and was officially recognized in the European Union's (EU) directive concerning integrated pollution prevention and control (IPPC) (96/61/EC; Council directive, 1996). BATs were defined as '*the most effective and advanced stage in the development of activities and their methods of operation which indicate the practical suitability of particular techniques for providing in principle the basis for emission limit values designed to prevent and, where that is not practicable, generally to reduce emissions and the impact on the environment as a whole*'. To better explain the acronym:

'Best' i.e. most effective strategy in achieving a high general level of protection of the environment as a whole industrial sector, under economically and technically viable conditions, taking into consideration the costs and advantages, whether or not the techniques are used or produced inside the Member State, as long as they are reasonably accessible to the operator.

'Techniques' include both the technology used and the way in which the installation is designed, built, maintained, operated, and decommissioned.

An effective simplification of the BAT concept can be synthetized as: 'BAT application means that the operator has to use the very best possible way that can be economically justified to protect the environment from the industrial emissions'.

32.1.2 Procedure for BAT implementation

The following steps are required for BAT implementation:

- Identification and characterization of the wastewater streams
- Identification of the technologies applicable to the case (referring to the BREF(s), site-specific factors and even innovative technologies not covered by BREF)
- Compilation of an inventory of emissions for each option
- Estimation of the environmental effects
- Evaluation of the technique that offers the highest level of protection for the environment as a whole (i.e., short term vs long term impacts, ozone/global warming potential, noise, and vibration, odor etc. – impact on resource use, waste production)
- Evaluation of costs for each alternative
- BAT definition.

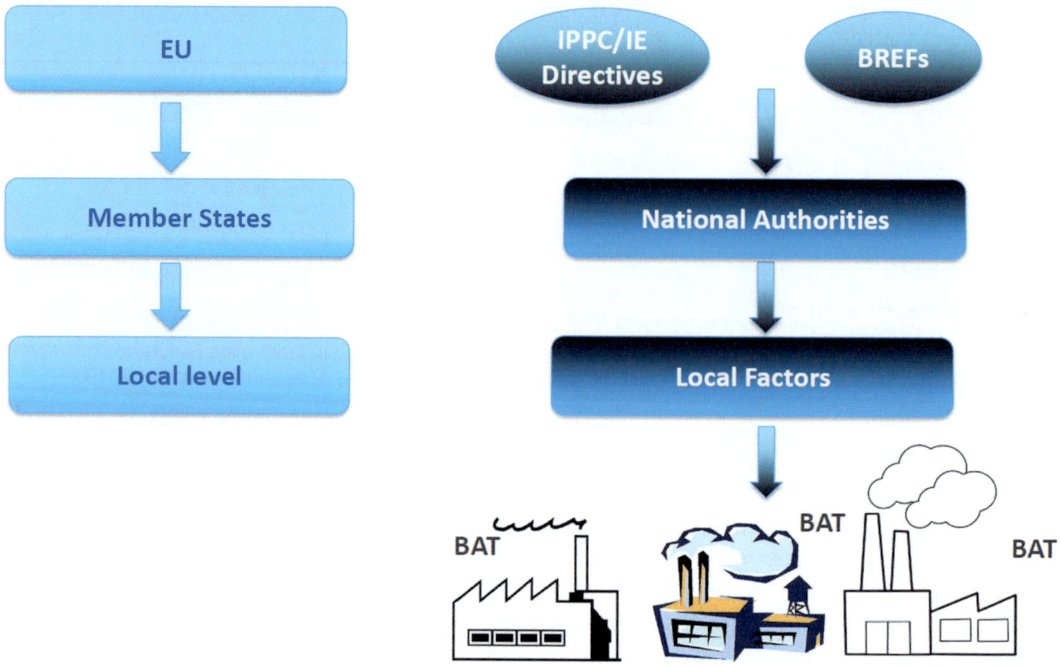

Figure 32.1 Flow chart of adoption of BATs to influence site-specific treatment measures.

It is evident that, besides the technological efficiency, many specific factors determine the BAT selection, and this requires a strong involvement of the industry. For this reason, BATs promulgated by the EU through the IPPC directive must still be evaluated by member countries and adapted for national and local use. On the local level, the site-specific evaluation of the applicability of BATs should be part of the routine management activity and related investments should be included in the budget planning. A basic flow chart illustrating the process of implementing a BAT is illustrated Figure 32.1 and is presented in more detail in the following discussion of applying best available technology in the chemical industry.

32.2 BATS FOR THE CHEMICAL INDUSTRY

The BREF document for the chemical sector (IPPC BREF, 2016) 'Common Waste Water and Waste Gas Treatment/Management Systems in the Chemical Sector' is a general document addressing the management and treatment of produced wastewater in the chemical industry. The Document provides fundamental strategies for the prevention and control of emissions, aimed at achieving a high level of environmental protection irrespective of the specific chemical production process.

In the management and treatment of water from chemical industries, procedures to minimize environmental impact can be implemented with 'upstream' interventions, i.e. with appropriate management strategies within the production cycle aimed at water reuse, (also called 'integrated process actions'), or 'downstream' with the optimum management of wastewater collection and treatment operations.

32.2.1 BATs for water management in chemical industry

Integrated process actions are the optimal operating procedures to reduce the amount of effluents or their pollutant load as they are applied at the production water cycle level, thus exerting a preventative action of environmental protection.

The primary integrated process actions for water management in chemical industry consist of:

- applying closed loop circuits for process water, trying to maximize recycling and reduce water discharge;
- optimizing the washing processes, avoiding those without recycling;
- avoiding direct contact cooling systems;
- employing closed circuits in vacuum generators, replace water or steam jet pumps with dry systems;
- assessing the possibility of adopting techniques for the treatment of exhaust gases at low water consumption.

32.2.2 BATs for collection

The collection system has an essential role in effectively reducing emissions and optimizing the wastewater treatment processes. The basic criterion to follow in planning an appropriate collection system is to separate the streams characterized by different groups of pollutants and avoiding mixing of wastewaters with different contamination levels. The best approach is to separate process water from uncontaminated streams and to segregate the streams based on their organic and inorganic load. Figure 32.2 shows an overview of the technologies available for different pollutants found in chemical industry wastewater.

32.2.3 BATs for treatment

The preliminary phase before planning the optimal strategy for treatment requires the following steps:

- definition of the criteria for management of water emissions;
- assessment of their impact on the receiving water body; and
- adoption of preventive measures to reduce the production of wastewater to be treated.

Once the nature of the produced wastewater to be treated and its impact has been defined and a treatment option has been selected, the remaining steps help select between on-site and centralized treatment for each different pollutant or group of pollutants. Figure 32.3 is a flow chart giving general guidelines for the best treatment route depending on the pollutants in the segregated stream. The diagram is based on the indications of the BAT Conclusions document for the chemical Industry, which, besides the optimal treatment strategy application, strongly supports water re-use.

32.3 CONCLUSIONS

The implementation of BATs in industry can contribute significantly to the reduction of environmental impact and increase of sustainability of industrial activities.

Achieved environmental benefits include reductions in raw material, water and energy consumption, emissions to air or water as well as solid waste, and hazardous waste generation.

It is worth noting that to ensure successful implementation of pollution prevention programs, it is necessary, for each specific case to evaluate environmental benefits, environmental impacts, and cost of implementation associated to their application.

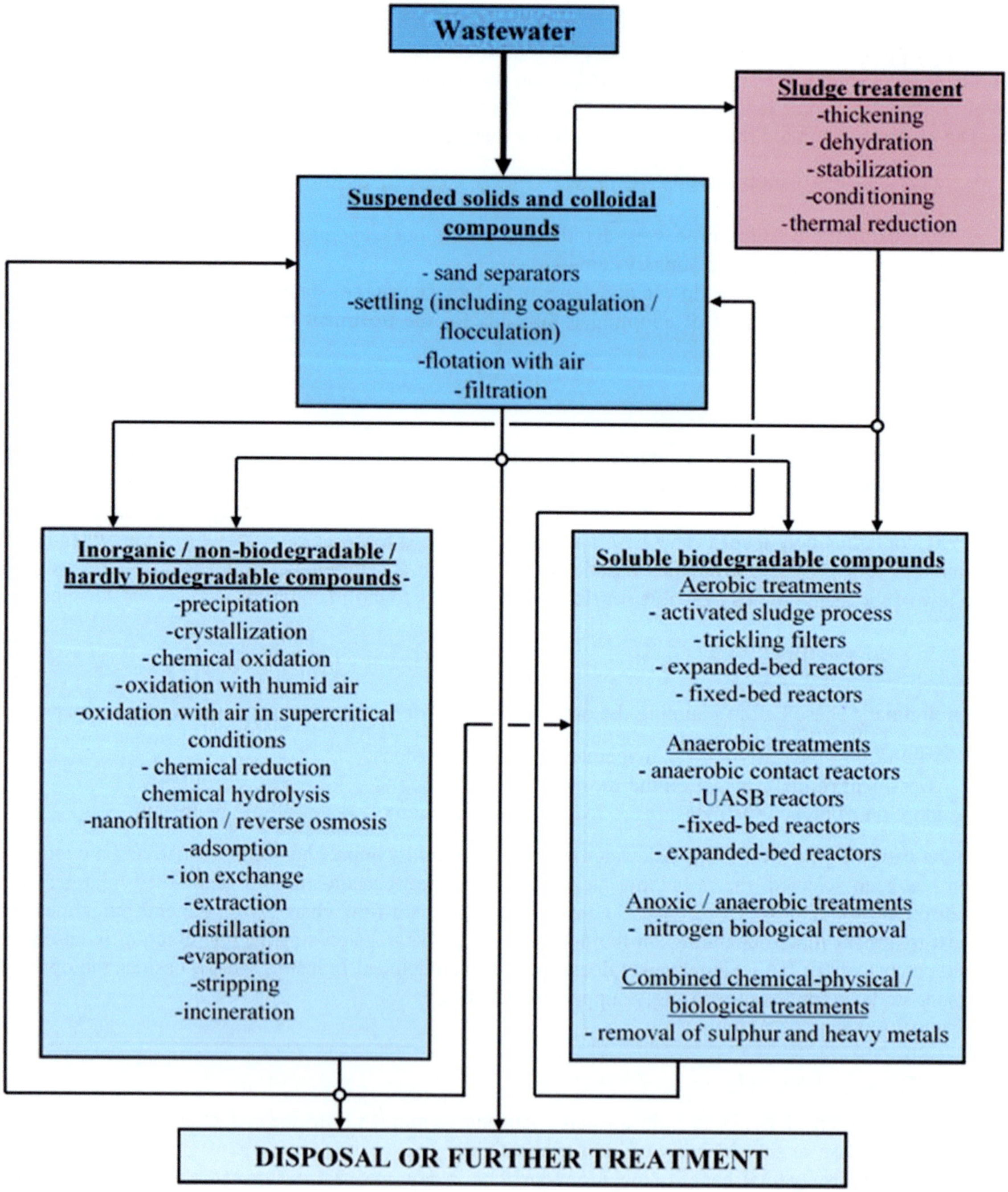

Figure 32.2 Overview of the technologies available for the different groups of pollutants found in chemical industry wastewater.

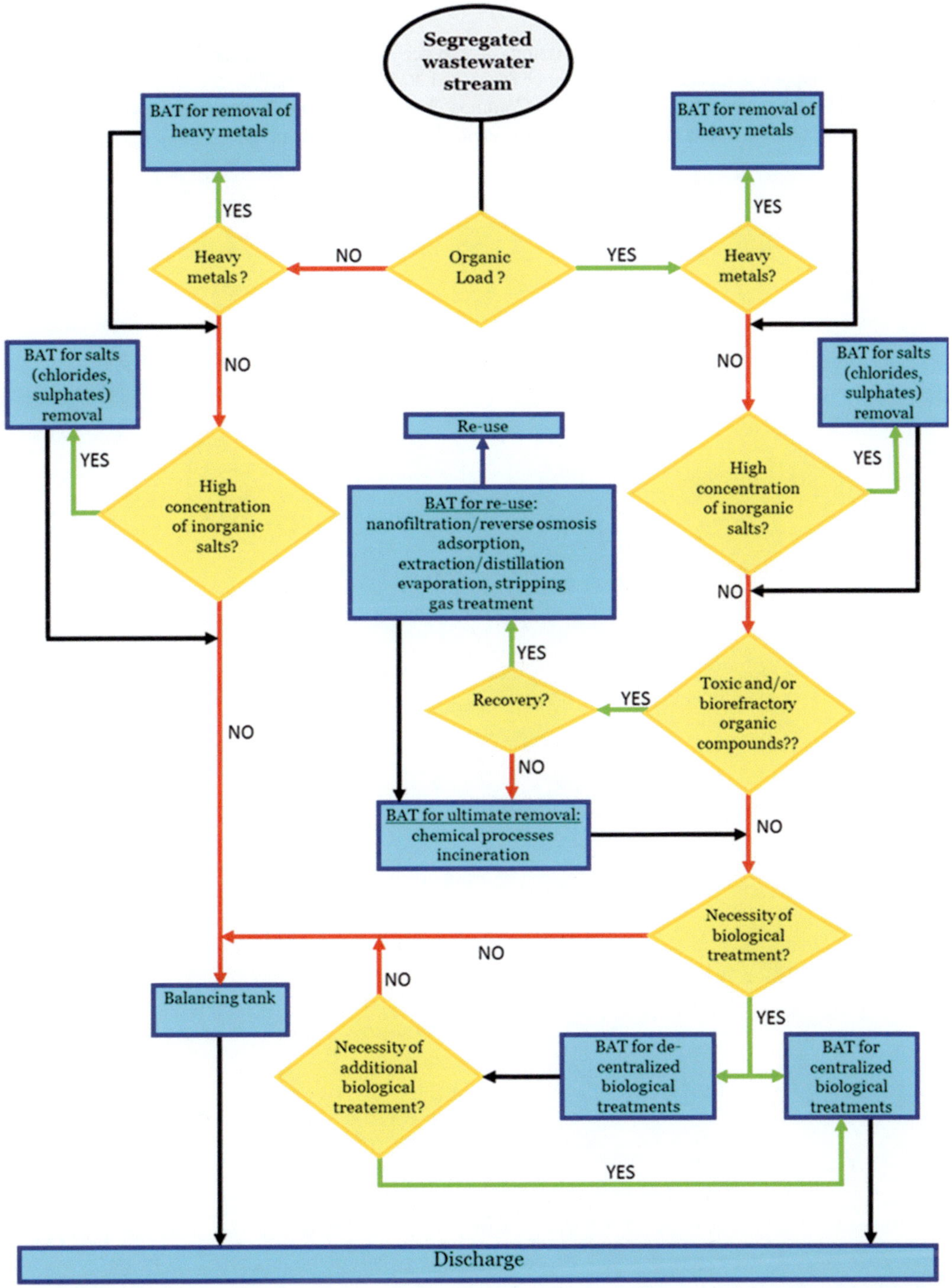

Figure 32.3 Flowchart reporting the selection strategies for wastewater treatment in the chemical industry.

REFERENCES

JRC-EC (Joint Research Centre European Commission). Reference documents under the IPPC Directive and the IED website. http://eippcb.jrc.ec.europa.eu/reference (accessed 30 March 2019).

IPPC-BREF. (2016). Best Available Techniques in Common Waste Water and Waste Gas Treatment/Management System in the Chemical Sector. . European Commission, JRC, Siviglia.

Chapter 33

Applying the UN sustainable development goals as a framework for corporate sustainability

Troels Kærgaard Bjerre

VCS Denmark, Odense, Denmark

Keywords: BHAG, prioritization, SDG, strategy, sustainability, water utility

33.1 INTRODUCTION

The 17 United Nations (UN) Sustainability Development Goals (SDGs) adopted at the UN Sustainable Development Summit in 2015 provide a comprehensive framework for corporate sustainability across all sectors of enterprises (UN, 2020). This chapter explains how VCS Denmark, a large publicly owned water and wastewater utility, incorporated the SDGs into its strategic business plan in order to implement in practice the sustainability goals. It describes how Denmark became a global leader in sustainability which led to VCS Denmark's adoption of the SDGs, the approach the utility followed to bring the SDGs into its organization, the key issues encountered, and some general thoughts about the need for continued organizational development to achieve water sustainability.

33.2 UN SUSTAINABLE DEVELOPMENT GOALS AND CORPORATE SUSTAINABILITY

While the Sustainable Development Goals (SDGs) were adopted only within the past decade, the UN has explored the concept of sustainability over the course of several decades (Caradonne, 2014). Previous efforts are reflected in a series of international agreements, beginning in 1972 with the Stockholm Declaration and continuing through the Brundtland Commission Report (1983–1987), the Montreal Protocol (1987), and the International Protocol on Climate Change (1988). While these documents focused on the role of governments in protecting the environment, the concept of sustainability was ultimately expanded to include ways in which environmental protection could be harmonized with running a profitable business.

DOI: 10.2166/9781789060676_0351

One common denominator for this emerging perspective was a belief that the ultimate purpose of business is something more substantial than making money: *'The promise of business is to increase the general well-being of humankind through service, creative invention and ethical philosophy'* (Hawken, 1993). The concept of *triple bottom line* added to a company's accounting its impact on society and the environment (Elkington, 1998), and was subsequently embraced by businesses as a measure of their 'corporate social responsibility' (CSR). Another useful notion was the *ecological footprint* that gauged environmental impact in terms of the area required to produce a nation's goods and services (Wackernagel & Rees, 1996).

While these tools have given us high-quality information about our impact on the environment, by the turn of the 21st century it was increasingly evident that we were far from realizing the Brundtland Commission's ideal of development '*....that meets the needs of the present without compromising the ability of future generations to meet their own needs'* (Brundtland, 1987). We are only now beginning to address the challenge of sustainably providing for the world's 7.5 billion inhabitants – challenges which will only continue to deepen as the population climbs to an estimated 10.9 billion by 2100.

Described as *'a universal call to action to end poverty, protect the planet and improve the lives and prospects of everyone, everywhere'*, the SDGs were adopted on September 25, 2015 by all UN member states as part of its 2030 Agenda for Sustainable Development (UN, 2020). Where earlier agreements were more limited in scope and geography, the ambitious agenda of the SDGs offered targets in 17 key areas (Figure 33.1) whose accomplishment would help ensure the health and prosperity of every country on earth. The SDGs took effect on January 1, 2016, initiating a 15-year plan to achieve them (by 2030) and setting the UN's course towards a more sustainable future.

Figure 33.1 The United Nations Sustainable Development Goals SDG Compass Guide (2015).

To reach these goals will require sustained and major contributions at all levels – locally, nationally and globally – as well as the active commitment from corporate sectors. Companies must be actively engaged in helping solve the problems addressed by the SDGs, because private enterprise helped to create them in the first place. To secure their involvement, however, business must see value in tackling these global challenges. Since 2015, UN Global Compact has inspired private enterprise to adopt the SDGs as a framework for the performance of corporate sustainability. Together with the Global Reporting Initiative (www.globalreporting.org) and the World Business Council for Sustainable Development (www.wbcsd.org), the UN Global Compact has developed guidelines advising enterprises how to adopt the SDGs. These guidelines has been a source of inspiration in the adoption of the SDGs as a framework for corporate sustainability in VCS Denmark.

33.3 DENMARK'S PATH TO SUSTAINABILITY

Just as decades of dedication to environmental responsibility by the UN eventually led to the adoption of the SDGs, so Denmark's national embrace of sustainability was the product of many years of social and political progress. Climate change led the agenda of Denmark's 2019 Parliamentary Election (dubbed 'The Climate Election'), but only because during the previous 50 years the Danish public, their politicians and their corporate executives all grew used to working together to fight pollution, protect the environment and address the challenges our society faces in general. Danish eco-literacy is taught at home and reinforced in school and at work, making Denmark a prime 'breeding ground' for sustainability.

In many ways Denmark has been a frontrunner in environmental action since 1973 when it passed the Danish Environmental Protection Act, one of the world's most ambitious. This strict regulation of business encouraged the development of both internal and external motivation towards sustainable behavior. Companies like the giant pump manufacturer Grundfos have long embodied Denmark's environmental ethic. It is no accident that their corporate purpose is 'we pioneer solutions to the world's water and climate challenges and improve quality of life for people' Far from 'greenwashing', this purpose reflects the national ethic of environmental protection. The company's corporate vision even includes a classic CSR acknowledgement that '*a high profit level is a means to the Group's continued existence and development - not a goal in itself*' (Grundfos, 2020). Grundfos former CEO Mads Nipper is joined by many other Danish executives like Novozymes' Head of Sustainability Claus Stig Pedersen and pioneering business consultant Prof. Steen Hildebrandt who demonstrated to business leaders that a sustainable business strategy is more than just a philanthropic indulgence.

But the Danish tradition of environmental awareness is not limited to business. On the political front, Mogens Lykketoft and Kristian Jensen were instrumental in putting sustainability – and ultimately the SDGs – at the top of the national agenda. Lykketoft, who held numerous elected and appointed positions in Denmark's Parliament (including Speaker) also presided over the 70th UN session in 2015 when the SDGs were adopted. During that same period, Jensen served as Foreign Minister and later on Finance Minister where he ensured that the SDGs were included in the national action plans by insisting on their relevance, their urgency, and their potential to benefit both individual citizens and corporate enterprises. It was in this context that, in October 2015, VCS Denmark CEO Anders Bækgaard (2020) heard Claus Stig Pedersen testify to Novozyme's success using the SDGs as a framework for corporate sustainability. The time was right for the spark of that presentation to ignite the SDG fire at VCS Denmark.

33.4 VCS DENMARK, CSR AND SDG

VCS Denmark, located in the Danish city of Odense, is the country's oldest and one of the largest water and wastewater utilities. Like most Danish water utilities, VCS Denmark is chartered a municipally owned limited company, operated on a non-profit basis, but with rates structured to fully recover all operating and capital costs. Since its founding in 1853, VCS Denmark's purpose has been to protect and improve public health. This commitment was proved early on when in 1854 a cholera epidemic felled 5% of Copenhagen's residents while their safe water supply protected the citizens of Odense from the water-borne plague. By 1866 sewers were installed and in 1908 VCS Denmark commissioned the city's first wastewater treatment plant. For VCS Denmark, its primary purpose has not been to make money for its shareholders but to create value in a broader sense for the community it serves.

It is no surprise, then, that VCS Denmark's corporate policies already included the concept of sustainability before the SDGs were formally inserted into its Owner Policies in 2018. Since 2007, the company had maintained a number of international certifications for health and safety and environmental practices (ISO 9001, ISO 14001, ISO 22000, and OHSAS 18001) so it was not unusual for them to go above and beyond their basic regulatory requirements.

On the other hand, while the Owner Policies identified sustainable development in general terms, neither the board nor VCS nor the utility's customers knew much about the SDGs to begin with. The decision to use SDGs as inspiration and as stepping stones towards fulfilling VCS Denmark's corporate responsibility originated somewhere else: CEO Anders Bækgaard. For years before coming to the company, Bækgaard promoted sustainability as head of the Danish Water and Wastewater Association (Feilberg, 2020) he brought that same sensibility to his new position. Conditions were therefore ripe at VCS Denmark, and the board welcomed his suggestion to adopt the SDGs as a framework for corporate sustainability.

33.5 ADOPTING THE SDG FRAMEWORK

To adopt the UN SDGs as a framework for corporate sustainability, VCS Denmark first had to evaluate each of the 17 goals and 169 target and identify goals and targets of particular relevance to the company. With goals and targets as broad as '*Eradicate extreme poverty for all people everywhere*' (Goal 1, Target 1) and as focused as '*Improve the regulation and monitoring of global financial markets*' (Goal 10, Target 5) it seemed necessary to determine which SDGs were in fact applicable to VCS Denmark's operations. The results of this initial screening were eye-opening, as the executive board identified 55 targets distributed between 14 goals where the utility might have an impact. This was surprising, as company managers had originally anticipated that its core business first and foremost related to Goal 6, '*Ensure availability and sustainable management of water and sanitation for all*'. The unexpectedly high correspondence between VCS Denmark's operations and the SDGs suggested the company had to take a more holistic approach to sustainability – a conclusion consistent with the logic behind creation of the SDGs.

Within the utility, this initial corporate embrace of the SDGs was characterized by elation and a pioneering spirit. The 'honeymoon period' soon ended, however, as staff confronted the challenge of anchoring the SDGs to the company's core strategies and plans, including its program of CSR. Integrating the SDGs into the actual activities of the company was necessary to avoid reducing them to an afterthought, but the number of potential focus areas made it difficult to know where to start. Where would efforts generate the most value? Where was the company's capacity for action strongest? As American author Patrick Lencioni put it, '*If everything is important, then nothing is*'.

To chart a roadmap for action, in 2017 VCS Denmark convened a cross-organizational team consisting of 10 employees recruited from each of the company's five departments. The project team included four women and six men: several engineers, a legal aide, a geologist, a journalist, a manager with a MSc. degree in Economics who specialized in human relations, and a plumber. Most members had no expertise in the area of the SDGs. A steering committee of senior management team was also established to ensure strong ties between the project and the board of directors. Together with the steering committee, the project committee defined five key characteristics of a successful SDG implementation project:

- **Core Business:** Make the SDGs a real and integral part of the core business rather than a philanthropic appendix of a more symbolic nature;
- **Materiality**: Create a basis for action that makes a real difference: no green-washing;
- **Involvement**: Acknowledge that active involvement of employees is a prerequisite for generating a sense of ownership and commitment in the organization;
- **Anchoring**: Ensure solid anchoring of the SDGs in the company's strategies, plans and responsibilities (CSR) as well as in the corporate culture;
- **Transparency**: Understand that transparency demystifies decision-making and generates better conditions for the sharing of knowledge internally and externally.

As a relatively small enterprise with 200 employees, VCS Denmark did not have capacity to allocate an entire department to the SDG agenda. Instead, it matched its organizational capacities to the process prescribed in the UN's SDG Compass Guide (2015), which divide the the process into five distinct phases as shown in Figure 33.2.

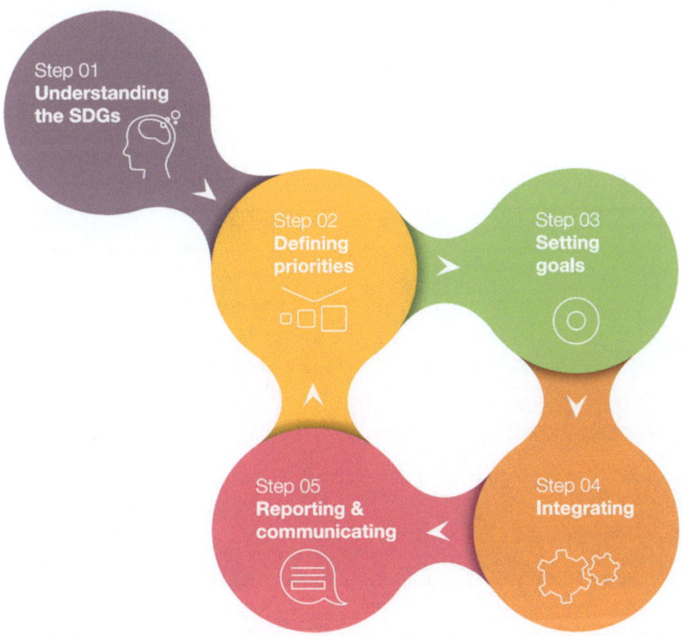

Figure 33.2 The Sustainable Development Goal Compass. (*Source:* UN (2020) *https://www.unglobalcompact.org/library/3101*)

The five phases are described as follows:

Phase I: Understanding the SDGs: Create a shared understanding of the SDGs and the challenges they address to establish a sound basis for the project. This step includes not only the project group and the steering committee, but the entire organization, including top management and the board.
Phase II: Defining Priorities: Map the relationship between the SDGs and activities in the company's value chain to set the stage for selection of goals and targets.
Phase III: Setting: Goals Set business goals that support priority SDGs and related targets.
Phase IV: Integrating: Identify specific aspects of business operation that correspond to selected focus areas and develop an action plan with performance indicators.
Phase V: Reporting and communicating: Establish procedures for assessment and progress reporting.

33.6 ASSESSMENT AND PRIORITIZATION

Although an initial assessment of the SDGs showed that many were relevant to the goals and activities of the utility, it was clear that the organization needed to focus on a smaller number of priority areas. In order to evaluate the SDGs methodically, the organization developed a detailed assessment tool. The tool worked by posing a series of questions designed to identify areas of relevance, gaps, and business opportunities for each candidate target. As shown in Table 33.1, responses were scored numerically, with the higher scores indicating higher priority.

Selection of a methodology for determining priorities is clearly an area where many choices need to be made. Some of the background questions the project team had to answer in developing the tool included:

- To what level of refinement should each SDG be evaluated?
- Does the team applying the tools have the skills and knowledge to make competent assessment?
- Have all the right and relevant variables been chosen?
- How should different variables be rated?

Table 33.1 VCS Denmark's assessment tool for evaluating the SDG targets.

Category/Question	Score
Relevance To what extent is the target relevant to VCS Denmark's activities?	1 = minor 2 = moderate 3 = major
Gap To what extent is there a gap between the target and the state of affairs in the geographical area influenced by VCS Denmark's activities?	1 = small 2 = moderate 3 = large
Business opportunity To what extent would achievement of the target represent a good business opportunity for VCS Denmark?	1 = minor 2 = moderate 3 = major

With regard to the last question, it should be noted that for a publicly owned non-profit company, a business opportunity can be rated based on value creation for owners and customers in a broader sense, on multiple 'bottom lines' or as 'shared value' (Porter & Kramer 2011).

In contrast to this approach, some companies might prefer a more classic 'materiality analysis' where goals are assessed based on whether their accomplishment is material to the company or its stakeholders. Others may choose from among the many SDG assessment tools developed over the last couple of years. Regardless, companies must bear in mind that assessment is not an exact science, and no tool can be relied upon exclusively to produce a clear-cut, conclusive result. The value of the selected assessment tool depends upon the extent to which it offers a new perspective that stimulates discussion about what is important to the company, and helps establish a useful and transparent basis for decision-making.

To speed up the process, the project team added 16 targets to the 55 targets identified by the executive board during its initial screening and limited its assessment to those 71 targets. The result of the assessment is shown in Figure 33.3, where the extension of each SDG icon is proportional to the average score of relevant targets on a SDG level, and the SDGs shown in color were prioritized by VCS Denmark for implementation.

The VCS Denmark board of directors received these results in late 2018 and chose to follow the project group's recommendations, prioritizing the following goals: 6 (clean water and sanitation); 7 (affordable and

Figure 33.3 Results of the SDG assessment by VCS Denmark (VandCenter Syd, 2019). The size of the individual icons reflects the average score of the relevant targets for every goal in the SDG assessment. The five goals appearing in colour represent the five SDGs that VCS Denmark has decided to give high priority (VandCenter Syd, 2019).

clean energy); 9 (industry, innovation and infrastructure); 12 (responsible consumption and production); and 14 (life below water).

33.7 INTEGRATION IN THE BUSINESS STRATEGY

In a planning effort of this type, a key strategic issue is how high to set the goals for achievement. The preparation of a new business strategy for VCS Denmark in 2019 provided an opportunity to incorporate the SDGs into the utility's new strategic plan. It was soon clear that incorporating goals of this magnitude into the strategic plan would require the formulation of ambitious, long-term and inspiring goals whose accomplishment would require a high degree of creativity and innovation. In the term coined by business writers Collins and Porras (1996), BHAGs (Big Hairy Audacious Goals) or 'bee-hags'. Accordingly, during the process of updating its business strategy, company leadership developed BHAGs in five priority areas closely related to the challenges addressed by the prioritized SDGs. These BHAGs are listed in Table 33.2, along with a brief description of the challenges they address. Notice that the BHAGs and prioritized SDGs aren't related one to one. Instead the BHAGs mirror a thematization fitting the activities of VCS Denmark.

Why five BHAGs? Would it not have been more manageable to focus on a few? Maybe – but VCS Denmark adopted the perspective that innovation must be 'multifunctional'. Walking the path of sustainability is a balancing act: we must take into account the complex nature of the challenges we face, and avoid creating 'suboptimal' solutions. Furthermore, we must carry out our new ideas with an eye to the potential synergies that may arise from their interaction with other bold solutions.

Any one of these five BHAGs could undoubtedly be considered unrealistic if the company intended to accomplish it alone. From the outset, however, VCS Denmark has set out to collaborate with its partners and stakeholders including suppliers, customers, local communities, authorities, private and public enterprises, NGOs and knowledge centers. Recognizing the importance of collaboration, the company also realizes that success depends not only on its own efforts, but also on the applied effort of other local, national and global agents as well. The importance of collaboration is further highlighted in the slogan the company chose to label the program: 'Together towards sustainable development'.

33.8 CONCLUSION AND PERSPECTIVES

This chapter has presented an approach used by one utility to apply the SDGs as a framework for corporate sustainability. This approach might not be workable, practical, or even advisable in other organizations. At a minimum, any organization would want to modify the methodological approach to meet its own needs. Nevertheless, one utility's experiences may be helpful, in a broader sense, to those organizations now considering implementing the SDGs. Complying with the intention behind the UN Sustainable Development Agenda means taking the opportunity, as VCS Denmark did, to review and examine established corporate practices in order to apply a holistic perspective to these activities throughout the entire value chain. Efforts like VCS Denmark's inclusion of the SDGs in its updated business plan can even provide an organization with an opportunity to shape future regulations through the establishment of 'best practices'.

Not everyone sees the SDGs as an opportunity to review their established practices and strategies. Some companies merely use the language of the SDGs to enhance their existing organizational branding. Worse yet, the SDGs may be misappropriated as a vehicle for 'greenwashing'. But if the SDGs are taken seriously and applied wholeheartedly, the effort can help transform the organization's business practices,

Table 33.2 'Big Hairy Audacious Goals' (BHAGs) for focus areas corresponding to the five prioritized SDG goals, identified in conjunction with the development of the new VCS Denmark business strategy in 2019.

BHAG	SDG Focus	Challenge
Potable water based on clean groundwater in 2050	Groundwater protection (SDG 6)	Danish groundwater is highly vulnerable to pesticide pollution. Approximately one-third of the abstraction wells in VCS Denmark's catchment areas have been found to have pesticide concentrations exceeding Danish drinking water standards. The national goal of abstracting potable water from only clean groundwater requires massive effort and commitment.
Climate neutral in 2050	Greenhouse gas reduction (SDGs 7, 9, and 12)	Despite major efforts to reduce net energy consumption at water resource recovery facilities (WRRFs), when emissions are assessed from the entire value chain (scope 3 in the GHG Protocol) VCS Denmark's carbon footprint remains high. Major efforts are required to reduce direct and indirect emissions from construction projects and daily operations.
Future-fit, flexible and sustainable infrastructure in 2050	Sustainable infrastructure (SDGs 6, 7, 9, 12 and 14)	In Denmark, climate change is expected to lead to more precipitation, a higher frequency of heavy rainfalls and rising water tables. This poses a major challenge (or threat) as existing water infrastructure is not geared to the new conditions. Major investments in water infrastructure and innovation are required to manage increased and more intense precipitation combined with rising water tables in shallow groundwater aquifers.
The world's most resource-efficient water utility in 2030	Natural resource conservation (SDG 12)	Wastewater constitutes a valuable resource that can be utilized rather regarded as a waste product to be disposed of. Though the transformation is still in its early stages, VCS Denmark has taken the first steps to sustainability through a gradual transition of the wastewater treatment plants (WWTPs) into water resource recovery facilities (WRRFs). Reducing consumption and increasing resource recovery of resources from all of VCS Denmark's construction, maintenance and operation activities will require a substantial and targeted effort to improve resource efficiency throughout the entire value chain.
Satisfactory ecological status in waterbodies in 2027	Wastewater treatment (SDGs 6, 9, 12 and 14)	The aquatic environment in Denmark does not currently meet the EU Water Framework Directive requirements, due to intensive agriculture in rural areas as well as discharge of treated wastewater and sewer overflows in cities and towns. Reducing pollution form nitrate, phosphorous and other contaminants and restoring the aquatic environment requires major efforts and dedication.

strengthen its operation, and provide a competitive advantage ensuring its future relevance and, ultimately, its survival.

Another key lesson learned is that for this process to work, both the goals and the perspective they reflect must enter into the corporate culture. In the words of Peter Drucker: *'Culture eats strategy for breakfast'*. The value of incorporating the SDG agenda must be more than a strategic exercise for top-management or an opportunity to brand the company as 'sustainable'. If employees do not perceive the SDGs as relatable and relevant to their work, they will not be successfully implemented. For this reason, the importance of company-wide communication and training programs cannot be emphasized enough.

In VCS Denmark, the SDGs were presented in the context of the company's core narrative, highlighting its values, its unique history and the corporate raison d'être. Viewing the SDGs from this perspective embeds them as a natural next steps in the company's evolution, while integrating the SDGs into the utility's sustainability plan bolsters the corporate narrative and encourages employee engagement. This happens when employees see their company as a partner in the important work of solving the problems faced by communities, society and humankind in general. It also contributes to creating a workplace that is able to attract the bright minds of the future.

It is also worth noting that VCS Denmark's ability to use the SDGs grew out of Denmark's embrace of sustainability as a national value. Previous UN environmental initiatives were often regarded as 'minimum development goals' for developing world countries, and when the SDGs were released they were also regarded at first as something exotic. In time, however, society caught on to the fact that they involved everyone, the politicians joined in, and many Danish companies took up the challenge. The background of environmental activism in Denmark set the stage for the companies to accept the challenge.

How can this level of support for sustainability be replicated in other countries? What does it take for the idea to seep through the national fabric and manifest itself in the way companies do business? There is no one simple recipe, but it takes the combined actions of many different kinds of people to have an impact. The environmental legislation of the 70 s and 80 s that imposed environmental regulations on Danish companies were important, but so were the ethical and moral values of the public as a whole, the ideas of progressive thinkers in academia and the innovations of the companies themselves.

At some point, sustainable action becomes a reflection not only of outer restriction but also of inner motivation, and corporate priorities extend even beyond environmental concerns to the welfare of society as a whole. This does not happen overnight, but when such values are shared by increasing numbers of individuals and companies, the enthusiasm snowballs. As the Senegalese forester Baba Dioum so eloquently expressed *'In the end we will conserve only what we love, we will love only what we understand, and we will understand only what we are taught'*.

REFERENCES

Bækgaard A. (2020). Personal communication.

Brundtland G. (1987). Report of the World Commission on Environment and Development: Our Common Future. United Nations General Assembly document A/42/427.

Caradonne J. L. (2014). Sustainability: A *H*istory. Oxford University Press.

Collins J. C. and Porras J. I. (1996). Building your company's vision. *Harvard Business Review*, **September–October**, 65–77.

Elkington J. (1998). Cannibals with Forks. New Society Publishers, Gabriola Island, BC

Feilberg M. (2020). Senior Consultant DANVA. Personal communication.

Grundfos. (2020). Corporate Values and Vision. https://www.grundfos.com/about-us/Our%20company/our-values-and-purpose.html (accessed 3 March 2020).

Hawken P. (1993). The Ecology of Commerce. Harper Business, New York.

Porter M. E. and Kramer M. R. (2011). Creating shared value. *Harvard Business Review*, **January–February**, 62–77.

SDG Compass Guide. (2015). The Guide for Business Action on the SDGs. Published by United Nations Global Compact in collaboration with GRI and WBCSD.

UN. (1987). Montreal Protocol on Substances that Deplete the Ozone Layer. UN Treaty no. 26369. Concluded at Montreal on 16 September 1987.

UN Sustainable Development Goals. (2020). The Sustainable Development Agenda. Website. https://www.un.org/sustainabledevelopment/sustainable-development-goals/ (accessed 3 March 2020).

United Nations, Department of Economic and Social Affairs, Population Division. (2019). World Population Prospects 2019: Volume II: Demographic Profiles.

VandCenter Syd. (2019). Sammen om bæredygtig udvikling. Strategi 2020–2024.

Wackernagel M. and Rees W. (1996). Our Ecological Footprint: Reducing Human Impact on Earth. New Society Publishers, Gabriola Island, BC.

Chapter 34

Implementing sustainable water use: A roadmap for industry

Eric Rosenblum[1] and Brian Moore[2]

[1]Water Resource Consultant, San Jose, CA, USA
[2]Aracadis US, Inc., Industrial Water Use Practice Leader, Clifton, NY, USA

Keywords: business case, cost-benefit analysis, industrial reuse, industrial water use, return on investment, sustainability, water conservation

34.1 INTRODUCTION

If industrial water pollution could be eliminated cheaply and easily, no plants would discharge any water: they'd reuse every drop. But implementing sustainable water use takes careful planning and thoughtful engagement of facility staff and management, along with a thorough knowledge of facility operations and technical requirements. All projects that reduce, reuse, and recycle industrial wastewater follow a similar path. Knowing the steps to take and the order to take them can increase a company's ability to move confidently and successfully towards sustainable water use.

In general, projects that improve the use of water in industrial facilities and reduce their overall impact on the water environment must complete the following steps:

(1) Clarify reasons to improve current water use practices, and share goals with internal stakeholders.
(2) Map existing use of water throughout the facility and find opportunities to reduce, reuse, and recycle water;
(3) Identify technically and financially viable opportunities to reduce and directly reuse water. If additional water use reduction is needed, develop options to treat and recycle water considering cost, operational complexity, and other criteria.

DOI: 10.2166/9781789060676_0363

(4) Present water minimization alternatives to corporate decision-makers in a way they can understand and appreciate, and implement the most appropriate projects.

34.2 WHY CONSERVE WATER?

There are many distinct types of industrial water use. In addition to facility-wide uses, like heating, cooling, and cleaning, water is required for many specific applications in product development, manufacture, and testing. As show in in Table 34.1, these include uses like casting and machining (in auto and aerospace), formulation (food and beverage, pharmaceuticals), and dyeing (textiles). Despite these critical uses, water conservation is often an afterthought within industry because water comprises a small percentage of most plant budgets. Industrial water projects need to compete for limited funding with other proposals, but the full breadth of advantages of reducing, reusing, and recycling water are often overlooked and/or not well understood.

Before investing in water projects company managers must recognize the value of water, over and above the size of their utility bill. Depending on the plant location and intensity of water use, these may include:

- improved water availability;
- reduced wastewater discharge;
- protection against the risk of future shortages due to climate change;
- reduced discharge of certain constituents of concern;
- lower water supply and wastewater discharge costs;
- lower costs for purification, heating/cooling water, pumping, onsite wastewater treatment; and
- improved branding as a socially responsible company.

When company managers understand what can be gained by using water more sustainably, they can better communicate these objectives to plant staff and corporate executives alike.

In their report, 'Evaluation of historical reuse applications and summary of technical/regulatory issues and related solutions for industrial reuse projects' Bowdan and Layton (2015) recommended the use of a charter process in which participants were asked to collectively confirm their commitment to the industrial water use improvement project goals and approach. They further recommended that the responsible parties should signify their support for the project in writing, to encourage their continued cooperation.

Another tool to help staff work together to identify water efficiency opportunities is the Water Kaizen Blitz™ (WKB) process. As with the charter process, teams include site personnel who are experts in the manufacturing facility as well as experts in processes and technologies for reducing water use within manufacturing operations. Teams work together using a variety of continuous improvement tools to identify and develop an array of realistic and cost-effective opportunities (Moore & Buzby, 2017).

34.3 MAP WATER USE AND WASTEWATER DISCHARGES

Once water reduction goals have been established, the next step is to trace water demands and wastewater discharges throughout the facility. A facility *water balance* can help identify opportunities to reduce water use and minimize discharges. Due to the diversity of processes and the variety of uses, the task of preparing such a water balance may seem overwhelming. However, it's not necessary to map the quantity and quality of each and every water demand and wastewater discharge. An adequate survey can be performed by measuring major water uses and discharges most likely to result in water saving opportunities, while

Table 34.1 Characteristic water uses in specific industrial sectors. (Adapted from Moore and Buzby 2017. Reprinted with permission. © Water Environment & Reuse Foundation.)

Industry Sector	Facility-Wide Water Uses				Product-Specific Water Uses				
	Heating & Cooling	Advanced Water Treatment	Equipment Cleaning	Casting & Machining	Cleaning	Transport	Formulation	Testing	Painting & Coating
Aerospace	x			x				x	x
Automotive	x	x		x	x		x		x
Chemical and Refinery	x	x	x		x		x		
Flat Glass Manufacture	x				x				
Food & Beverage	x	x	x		x	x	x		
Oil & Gas Exploration						x	x		
Paint & Coatings		x	x				x		
Power Generation	x	x	x						
Pharmaceuticals	x	x	x		x		x		
Textiles	x	x	x		x	x			x

minor water uses (e.g., hand washing, safety showers, small process streams, etc.) can be excluded or grouped together and labelled 'minor uses'. Considering the purpose the water serves can simplify the task of identifying opportunities to reduce water consumption, redirect wastewater for reuse, and recycle treated wastewater throughout the facility.

The water balance should characterize the water *quantity* (volume and flow) and water *quality* (salinity, pH, etc.) of upstream demands and downstream discharges, as well as the effluent from intermediate treatment processes. This data can be used to identify opportunities to reuse water of varying quality in different areas of the facility. A water balance can provide critical information needed to identify major water consumers, reveal high-strength internal waste streams, and determine overall water savings to be expected for a given project. For this reason, development of a water balance is often an iterative process. Frequently an overview of water and wastewater quantity and quality will suggest opportunities to reduce, reuse, or recycle water, suggesting a more detailed analysis. Figure 34.1 shows a water quantity balance for an industrial facility indicating major flows and applications; water quality information could be added based on sampling of flows most readily reused. The same balance can also be conveniently used to show how water use and quality will look after water efficiency projects have been implemented.

In addition to the average quantity and quality of wastewater discharges, it is often necessary to learn how they vary and what factors may impact them in the future. Wastewater flowrates may change on an hourly, daily, and seasonal basis. If no wastewater flow meter exists, a temporary meter may be attached to major discharge lines to record hourly or daily flows. In some countries, local water utilities lend meters at no or low cost. In countries where water supply is unmetered, the facilities will have to implement their own metering programs. In that case, wastewater flows may be estimated based on manufacturing records, or checked against annual sewer data, if available. If wastewater flows vary significantly, it may be necessary to provide upstream storage to stabilize wastewater treatment processes. Variations in the type or number of goods produced may also limit the availability of recycled water, so designers should consider the impact of changes in facility operation on wastewater supply.

The cost of additional treatment associated with wastewater recycling depends on the quality of the wastewater and the water quality needed by the application. While some wastewater quality parameters (e.g., pH, conductivity) are easily measured with online probes other constituents (e.g. BOD, nutrients, heavy metals, and volatile organics) need to be analyzed in a laboratory. As these tests can be costly, it is important to decide which parameters are important for the potential water uses identified. Sampling frequency should also be considered if water quality varies widely. Table 34.2 lists a number of wastewater contaminants that could cause problems in typical industrial applications. (Although the study focused on municipal effluent, these contaminants are found in industrial wastewater as well.) In addition, where membrane processes like reverse osmosis (RO) are used, the wastewater should be tested for oil and grease, COD, TSS, turbidity, hardness, alkalinity, silica, iron, barium, strontium, and particulate density (silt density index).

34.4 IDENTIFY OPPORTUNITIES TO REDUCE, REUSE, AND RECYCLE

34.4.1 Reduce: cut back on water use

Reduce refers to projects that decrease water use by cutting either flowrate or duration of flow. 'Reduce' projects are typically the most cost-effective projects, offering the greatest return on investment (ROI). They can be simple, like the addition of a control valve, or they may involve major process modification,

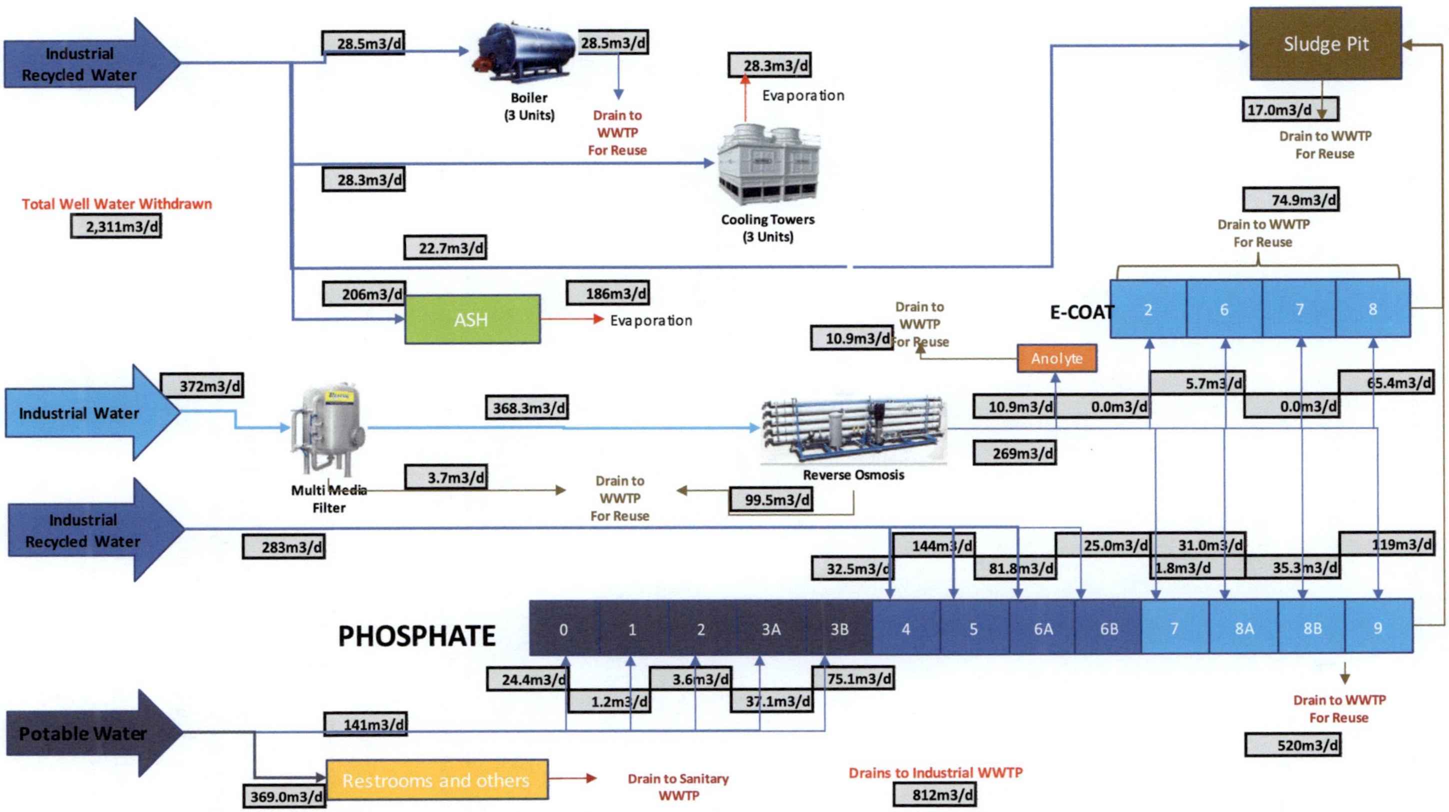

Figure 34.1 Sample water balance. (Adapted from Moore and Buzby 2017. Reprinted with permission. © Water Environment & Reuse Foundation.)

Table 34.2 Summary of recycled water quality parameter and potential industry impacts. (Adapted from Victoria University Institute for Sustainability and Innovation, 2007. Reprinted with permission. © Smart Water Fund.)

Parameter	Health	Corrosion	Scaling	Fouling	Process
Alkalinity		Y	Y		
Ammonia		Y		Y	
Ca, Mg			Y		
Chloride		Y			
Color					Y
Dissolved solids		Y	Y		
Fluoride			Y		
Hardness		Y	Y		
Metals	Y				Y
Microbiology	Y	Y	Y	Y	PD
Nitrate		Y			
Odor (BOD/DO)	Y				Y
pH		Y	Y		
Phosphate		Y	Y	Y	
Residual organics	Y			Y	
Salinity		Y			Y
Silica			Y		
Sulfate		Y	Y		
Suspended solids				Y	Y
Temperature	Y	Y	Y	Y	Y
Turbidity		Y	Y	Y	

Abbrivations: PD = process dependent; BOD = biological oxygen demand, DO = dissolved oxygen.

like substituting dry deposition for wet painting. When investigating opportunities to reduce water, ask the following questions:

- Is the water flowing when the process is offline, or at a higher rate than needed for its desired use?
- For batch processes, could the water be used longer before discharge (e.g. more rinses per batch)?
- Is the line pressure higher than required, and are equipment Standard Operating Procedures designed for water-efficiency?
- Is the flow manually controlled, and if so, do opportunities exist to automate the use of water?
- Are there alternative manufacturing techniques that require less water or do not require water at all?

A thoughtful evaluation of industrial processes reveals opportunities that can be implemented at little or no cost.

34.4.2 Reuse: direct use of untreated wastewater

Reuse refers to projects that direct wastewater from one process to serve as the water supply for another process, without intervening treatment. Industrial facilities typically have a diverse array of water-using operations, each producing effluent of a specific quality and each with distinct water quality needs. Sometimes the effluent from one process meets the quality requirements of another, creating an opportunity for reuse. After 'reduce' projects, direct reuse of water is often the next most cost-effective method of reducing industrial discharges. In some cases, plumbing may represent the greatest capital cost, and the distance between the processes may determine the feasibility of the project.

Washing and rinsing activities (distinct from general cleaning) constitute a significant water demand at many facilities, and discharge from these processes can be suitable for reuse without further treatment. For example, ultrapure water used to rinse computer chips is often appropriate for reuse elsewhere in the facility. As another example, the reject streams from membrane treatment may be suitable for cleaning or cooling. A 'cascade' wash strategy (also termed 'counter-current rinse') may also be employed when products, containers, or equipment are subjected to multi-stage cleaning. In that case, the cleanest water is applied in the final rinse, then collected and re-used in the primary or secondary wash cycles. Waste from the first wash is then either sent to drain or collected and treated for recycling.

34.4.3 Recycle: treating wastewater to be used again

Recycle projects involve the treatment of wastewater from one process for reuse in the same process or elsewhere in the facility. Because of the need for water treatment/purification equipment, recycle projects tend to be the most expensive category of conservation projects. Recycle projects will typically also produce a concentrated waste stream that must be taken into consideration in the planning stage.

It is worth reiterating that an effective discharge minimization program begins with minimizing a facility's water use to protect the value of future capital investments. A project that supplies recycled water to a process will be less valuable if that demand subsequently decreases due to conservation measures. Similarly, if water use in the process supplying recycled water is cut back, the resulting discharge may not be sufficient for the downstream process using the recycled water. Early evaluation and recommendation of water use reduction opportunities is essential to ensure that water reuse and recycling projects will achieve their goals.

Along with aggregate wastewater quality, it can be very useful to sample individual upstream wastewater sources. Such sampling can pinpoint the origin of problematic constituents that may be isolated so that the remaining flow is easier to treat. Most industrial facilities combine wastewater from multiple processes to minimize plumbing costs, combining less contaminated streams that may be suitable for reuse with more heavily contaminated discharges. Segregation of these cleaner streams can simplify design and lower water reuse project costs.

34.5 DEVELOP TREATMENT OPTIONS

Since wastewater can be treated to reach any desired quality, recycled water can be manufactured to suit virtually any type of industrial use. Armed with data on water quality and significant flows, and an understanding of major water uses, staff can select the treatment alternatives best suited to meet identified recycled water demands. Once identified, treatment alternatives can be evaluated with respect to a wide range of criteria to find the project most suitable to accomplish management goals. Along with identified on-site treatment alternatives, project sponsors should consider the use of municipally treated wastewater (if available) or participation in a regional water recycling program. In some countries it has even proved advantageous for companies to provide wastewater treatment for adjacent communities in order to reuse the effluent for industrial purposes.

34.5.1 Matching wastewater quality and use

When the opportunities for direct reuse have been exhausted, a wide range of technologies are available, singly or in combination, to treat wastewater to satisfy remaining applications. A preliminary assessment can be made as to which treatment methods can produce water of acceptable quality by reviewing the wastewater test results to see which constituents might interfere with potential applications, and to what extent they must be removed. Table 34.3 provides general guidance with respect to efficacy of various common types of treatment to remove certain contaminants. Each of these technologies has different space requirements and operational characteristics as well as different capital and operating costs which must be evaluated in order to determine the best option.

34.5.2 Preliminary screening and development of treatment alternatives

Preliminary screening should gauge the basic feasibility of treatment alternatives by considering a range of critical factors. These include the size of the treatment unit, its capacity to handle expected variations in quality and flow, and the ability of plant staff to properly operate and maintain the equipment. At this point one may also develop a rough estimate of project costs sufficient to determine if a particular option is within the available budget. Project cost is largely influenced by: (1) the degree of treatment; (2) the volume of water to be recycled; and (3) the proximity of water demands to the treatment system. As a result, projects will generally be the most cost effective where lightly contaminated wastewater is available to serve a large volume of lower-quality water uses located near the treatment system.

During the development of treatment alternatives, desktop analysis may not be sufficient to determine the suitability of a treatment technology to a particular waste stream. In some cases, information about similar technologies in similar applications can be used. If the waste stream is complex or involves an innovative technology, and if a preliminary assessment suggests that this option is technically and financially feasible, it may be appropriate to perform bench-testing or to construct a pilot treatment system.

34.6 EVALUATING PROJECT ALTERNATIVES

After developing project options, the next step is to select the right treatment method to accomplish the company's goals. To fairly compare treatment alternatives staff must estimate both qualitative and quantitative costs and benefits of the project.

34.6.1 Quantitative costs and benefits

In most cases, quantitative impacts can be expressed in monetary terms so that their relative costs (or savings) can be compared. In addition to project costs and the cost of residuals management, project energy use impacts, utility savings, and costs related to operational complexity can all be considered in a quantitative manner.

34.6.1.1 Direct costs and savings (CapEx and OpEx)

In addition to capital equipment and construction costs, capital expenditures (CapEx) include planning studies, permitting, preliminary design, site preparation, start-up, and other one-time costs. Operating expenses (OpEx) include staffing, chemicals, material replacements, maintenance, solids management, and power. Before determining the cost of the staffing, it should first be determined whether the water

Table 34.3 Treatment alternatives for removal of selected industrial waste contaminants.

Treatment Method	Suspended Solids	Selected Contaminants						
		Oil & Grease		Mono-valent Ions	Dissolved Heavy Metals	Hardness	Dissolved Organics	Target Trace Organics
		High	*Low*					
SEPARATION & FILTRATION								
Bag Filter	◑		○					
Sand Filter	◑		○					
Oil Water Separator		●	◑					
Dissolved Air Flotation	◑	◑	◑					
COAGULATION & CLARIFICATION								
Chemical Precipitation	◑	◑	◑		●		○	
Chemical Softening	◑	◑	◑		●	●	○	
ADSORPION & ABSORPTION								
Granular Activated Carbon	○		●				●	●
Ion Exchange Resin				●	●	●		
Organophilic Resin							●	●
Organoclay	○	◑	●					
OXIDATION								
Ozone			○				●	●
Ultraviolet/H_2O_2							●	●
Fenton's Reagent			○				●	●
MEMBRANES								
Microfiltration	●		◑					
Ultrafiltration	●		●					
Nanofiltration					●	●	○	
RO				●	●	●	●	●
BIOLOGICAL								
Activated Sludge	○		○				●	●
MBR	●		○				●	●

Abbrivations: Treatment Efficacy Symbol Definitions: ○=Poor to Fair; ◑=Fair to Good; ●=Good to Excellent. RO = Reverse Osmosis; MBR = Membrane Bioreactor. (*Source:* Adapted from Tables 5.1 through 5.8, Moore and Buzby (2017)).

recycling equipment will be run by existing staff, new staff, or if third-party operators will be contracted. If staff training will be required for existing staff to use new technologies effectively, training costs should also be taken into account.

34.6.1.2 Residuals management

Membrane treatment and other physical separation processes produce a concentrated waste stream that must be discharged into the sewer system. Local limits may require that these residual streams be neutralized or otherwise treated before discharge. If discharge is not possible, residuals may be managed onsite via evaporation or deep well injection, or by hauling waste offsite for further treatment or disposal. These expenses may add significantly to project costs.

34.6.1.3 Utility savings

A well-designed water recycling project will often result in lower water and wastewater discharge fees. These savings may be calculated based on the estimated use of recycled water and the corresponding reduction in potable water charges, as well as reduced wastewater discharged volumes. In both instances the estimation of cost savings should be checked against utility bills which usually include both fixed and variable costs. One source of savings frequently overlooked is the reduction in the cost of treating water delivered by the utility, since less water is coming into the plant.

34.6.1.4 Greenhouse gas production

Power costs are already included in the OpEx calculation, but energy demand can also be quantified in terms of greenhouse gas (GHG) emissions. If the driver for the project is primarily to meet sustainability goals, this may be a relevant factor to consider.

34.6.1.5 Operational complexity and flexibility

Operational complexity is related to a number of factors, including the variety of treatment processes involved, the number of recycled water applications, and the distance between source, treatment, and end use. For the purpose of comparing alternatives, the cost of complexity can be estimated either quantitatively, in terms of impact on production, or qualitatively, based on the estimated need for additional training or development of standard operating procedures. Since industries are constantly changing, it is important to consider how a recycled water system will serve the facility in the future. A robust treatment train should accommodate anticipated changes in wastewater quality and water demand, with room for expansion provided there is space for future treatment upgrades.

34.6.2 Qualitative costs and benefits

In addition to the quantitative criteria, there are many important factors to consider in selecting recycled water treatment alternatives that are not readily expressed in monetary terms. The relative importance of these factors can often be monetized through contingent valuation or other econometric methods. Alternatively, all factors (including costs and monetized benefits) can be assigned scalar values and compared in the final decision evaluation. For a non-numerical approach, bring stakeholders together (including community representatives outside the company) to discuss and prioritize key aspects of the project, like importance to the environment or contribution to the company's long-term reliability.

34.6.2.1 Permitting complexity

Permits may be required to ensure that a proposed project meets recycled water regulations. Regional or state regulations may come into play when non-potable water crosses local jurisdictional boundaries. The permitting process will become more complicated as the number of parties involved increases, which may in turn increase the length of time required to implement a project (Rosenblum and Anderson, 2006).

34.6.2.2 Reliability and resilience

A significant percent of the value of an onsite water recycling system may be its ability to serve as a reliable 'drought-proof' supply. The benefits of having a reliable source of water can be estimated by asking how production would be impaired by a lack of water and assessing the value of a day's production to the facility. This cost can be used as a benchmark for comparison in the form of the 'no project' alternative. A number of tools are now available to help companies estimate the risk of future water shortages (AWS, 2019). Even if a local utility has published water management plans, they may not forecast shortages. However, some sense of this can be inferred from the size of the surplus they project or the extent to which drought and drought response is addressed in the report.

34.6.3 Assessing costs and benefits

Quantitative and qualitative costs and benefits must be assessed in a manner that facilitates the comparison of project alternatives. Because industrial facilities are under continuous pressure to reduce costs and maximize returns, a financial cost/benefit ratio is often the first thing to be evaluated, typically expressed in terms of either payback period or return on investment (ROI).

A simple payback period calculation can give a rough idea of the project's financial merit. Payback period (Equation 1) simply reflects the time required for project earnings (or savings) to equal the net investment, as follows

$$\text{Payback Period} = \frac{\text{Investment Cost}}{\text{Annual Savings}} \tag{1}$$

A 2-year payback period is an industry benchmark, although the exact payback period threshold will vary between companies. As a project develops, a simple payback period may not adequately reflect its value. ROI represents the total net benefit of a project (rather than its annual return) divided by the total investment cost. A key benefit of ROI compared to simple payback period is that ROI is generally better able to account for the value of a project over time, especially when benefits accrue differently at different points in the life of the project.

Both the payback period and ROI methods are limited to evaluation of quantitative aspects of proposed projects, either market-based factors like CapEx and OpEx costs, water rates, and sewer fees, or qualitative factors that have been monetized like the risk of a water supply shortfall. Where project alternatives are close in cost and distinguished primarily by their qualitative benefits these factors can be reflected in a decision matrix or appraised in expanded accounting methods like "triple bottom line" accounting.

34.6.4 Comparing alternative treatment schemes

Some type of decision matrix is typically required to track both quantitative and qualitative criteria. As noted by O'Brien (2000), 'The failure of cost-benefit analysis to include non-monetary considerations is the main basis for rejecting monetary cost-benefit analysis as a sufficient basis for any environmental, public or private decision making'. A basic decision support tool involves creating a matrix where the various factors identified earlier are converted into selection criteria by assigning scalar values that weight their importance. The translation of qualitative factors into quantitative criteria is a subjective exercise, and its validity depends on the consensus of the participants. It is important to ensure that all factors are fairly considered in order to identify any 'fatal flaws' that might make the project ultimately infeasible. There are a number of software packages now available that support multi-criteria decision analysis of water and wastewater treatment projects (Harris-Lovett, et al., 2019; Patil & Kulkarni, 2014).

The end result of the decision support program is a set of scores for the various projects evaluated. While the relative value of projects is determined by comparing their scores, an equally valuable use of the system is to see which factors contribute to their relative ranking. It can be very useful to understand what circumstances lead to a project being poorly ranked, and to consider how these conditions/circumstances might be mitigated. For instance, if a project ranks low based on its complexity, what steps might be taken to simplify operations? In many cases, the sensitivity analysis is as beneficial as the quantitative assessment itself.

34.7 MAKING THE BUSINESS CASE FOR REUSE

34.7.1 How business decisions are made

Private sector companies will typically have a limited (and often fixed) budget for capital projects each year, and that budget is rarely sufficient to fund all projects requested. To compete for funding, sustainable water projects must emphasize the financial value they provide. All the feasibility studies, conceptual designs, risk analyses, and cost-benefit exercises often culminate in a single 1-hour meeting, where one must be prepared to convincingly answer the following questions:

- What goals does this project achieve?
- Why is this project recommended above the others?
- What is the consequence of not choosing the project?

This underscores the importance of making an effective business case for water reuse and recycling projects which highlights the ROI and explains how it can be implemented without endangering the facility or its operation. It should be detailed enough to identify each major element of the project, but simple enough to be easily understood by non-technical managers. In some cases, convincing a small group of decision-makers to fund a project may be more challenging than designing an effective water treatment system.

34.7.2 Who are the decision-makers?

Industrial plants are highly structured institutions, aligning the efforts of many people with different areas of expertise – managers, engineers, operators, mechanics and electricians, accountants, human resource specialists, and product marketing experts, to name a few. In making a business case for industrial water reuse, project sponsors should be prepared to discuss issues that matter with individuals from different perspectives who will collectively review and approve the proposal. Several key employees whose opinion will determine the acceptance of a proposed project are described in Table 34.4 below. Not all these positions will be represented in every company, and some companies may contract with other companies to provide services for various aspects of facility operation who may be involved in decision-making. For example, if the company has outsourced operation of its cooling towers, a project that changes the cooling tower makeup water

Table 34.4 Company decision-makers and their concerns. (Adapted from Ch. 2. Planning for Water Recycling and Reuse in Industrial Water Recycling and Reuse to Minimize Liquid Discharge © 2020 Water Environment Federation. Used by permission.)

Decision-maker	Responsibilities	Level of Influence	Concerns	Recommended Approach
Owner or CEO	Responsible for outcome of the enterprise and direction for future performance	Ultimate authority, but needs support of managers and line staff to implement projects successfully	Short-term risks and long-term benefits of proposed water use changes	Gain support of line staff and demonstrate long-term competitive benefit
Finance officer or accounting manager	Responsible for the balance sheet; certifies availability of funds	Company-wide; needs to understand fundamental value of proposed project	Uncertainties in cost of proposed projects, potential for overruns	Identify uncertainties; use 'range of costs' to reflect anticipated fluctuations.
Product manager	Selection of products and meeting production targets	Interfaces with manufacturing to ensure sales goals are met	Impact of plant changes on production, profits	Address in advance all constructability issues
Plant manager	Continuous facility operation and achievement of manufacturing goals	Focused on the facility: may have the greatest say in deciding which projects move forward	How projects in one group impact another group's work flow	Discuss project with all work groups, consider
O&M supervisor	Facility performance, management of work groups, plant shutdowns	Must confirm ability of staff to operate and maintain,	Additional staff time or skills required; complexity of system	Consult with staff on implementation issues early and often
Sustainability officer	Guides management about company's impact on environment; may have specific targets to reduce resource use	May serve as the project sponsor or a chief advocate if the project's primary purpose is to reduce the facility's 'environmental footprint'	Importance of environmental improvements; measurable reductions in water use	Explain how investment in the proposed project will help the plant meet company-wide sustainability targets.
Environmental health & safety manager (EHS)	Often called upon to review construction plans to ensure they do not pose a threat to plant personnel	Need to approve the use of any hazardous chemicals that may be required to operate or maintain the proposed treatment technology.	Permits needed; assurance that project will not exceed discharge limits	Minimize use of toxic chemicals; consider storage issues

would likely require input from that company, as well. However, even owners of small and mid-size companies with sole authority over business decisions will want to consider water management investments from several different angles.

34.7.3 The successful business case

A good business case addresses the issues important to each of the decision-makers. It unites the interests of 'the board room and the boiler room' by demonstrating how corporate goals can be accomplished in a way that is compatible with the efficient, uninterrupted operation of the facility. This is especially critical in companies where every team member has 'veto authority' over discretionary process improvements. Certain objections are repeatedly raised in opposition to implementation of industrial water reuse projects (Moore & Buzby, 2017).

- 'The project is too expensive'.
  - Ensure that all less costly options (e.g. reduce, reuse) have been fully vetted.
  - Add the cost of heating and treating water to the utility price to reflect the "true cost of water."
  - Investigate opportunities for government grants and low-interest loans.
  - Look for ways to reduce cost without reducing return, e.g. cut expensive low-yield project elements.
- 'Our current water situation is fine: We don't need this project'.
  - Project a water cost escalator that reflects future utility rates which will only increase in the future.
  - Present the financial impact of water cutbacks and the value of having resilient on-site supplies.

As suggested by the sample questions below, a successful business case will answer these arguments in advance.

- 'We aren't a water company'.
  - If in-house expertise is lacking, hire third-party contract operators or consultants to provide training.
  - Consider partnering with local water utilities to ensure that operational expertise is available.

Finally, it should be noted that when it comes to implementing an industrial water reuse project, timing is everything. A proposal that would be deferred or rejected outright during a rainy year might be accepted in the midst of a multi-year drought. Similarly, during slow business cycles when revenues are down there may be less enthusiasm for process improvements, unless a case can be made for their necessity and the advantage of constructing them when operational demands are less pressing.

34.8 CONCLUSION

A successful industrial water reuse project is the result of the combined effort of many people throughout the facility. To accomplish this level of coordination, the project team must not only perform traditional engineering tasks, but also reflect company objectives, work within budget constraints, adapt to operational sensitivities, effectively communicate the value of the project, and ensure that all stakeholders are included in the process. These key tasks are as follows:

- Clearly define project drivers and objectives and communicate them to all participants.
- Identify and confirm roles of internal and external stakeholders and decision-makers.
- Develop a good water balance, understanding the source of variability.
- Identify 'reduce' opportunities before investigating 'reuse' and 'recycle'.
- Focus on the end-user and understand their water quality requirements.
- Develop a strong business case and communicate the benefits of the project.

ACKNOWLEDGMENTS

An expanded version of this article appears in Chapter 2 of Moore, B. (ed.) Water Reclamation and Reuse to Minimize Liquid Discharge (Water Environment Federation, 2021, First Edition). This article has been revised and adapted expressly for this book.

REFERENCES

Alliance for Water Stewardship. (2019). International Water Stewardship Standard Version 2.0 22.03.2019. https://a4ws.org/download-standard-2 (accessed 21 January 2020)

Bowdan J. and Layton R. (2015). Evaluation of historical reuse applications and summary of technical/regulatory issues and related solutions for industrial reuse projects: Managing industrial reuse projects for success, WRRF Research Report 12-03, WateReuse Research Foundation. The charter tool is explained in more detail in this volume in Chapter 24, Incentivizing sustainability: How utilities can support industrial water conservation and reuse.

Harris Lovett S., Leinert J. and Sedlak D. (2019). A mixed-methods approach to strategic planning for multi-benefit regional water infrastructure. *Journal of Environmental Management*, **233**, 218–237.

Moore B. and Buzby M. (2017). A framework for the successful implementation of on-site industrial water reuse, WRRF No. 14-04, Water Research Foundation.

O'Brien M. (2000). Making Better Environmental Decisions: An Alternative to Risk Assessment, MIT Press, Cambridge, MA.

Patil A. S. and Kulkarni N. J. (2014). Decision support system for waste water management: a review. *International Journal of Innovative Research in Advanced Engineering*, **1**(3), 25–29, special issue.

Rosenblum E. and Anderson J. (2006). The water reclamation matrix: A framework for sustainable urban water use. In: 2nd Leading Edge Conference on Sustainability in Water-Limited Environments, B. Beck and A. Speers (eds.), International Water Association.

Victoria University Institute for Sustainability and Innovation. (2007). Guidance for the use of recycled water by industry. https://waterportal.com.au/swf/images/swf-files/415-001-guidance-for-the-use-of-recycled-water-by-industry_final-report.pdf (accessed 5 October 2020)

Chapter 35

Staff training: An integral component of sustainable water use by industry

Cornelis (Niels) Groot[1] and Cheryl Davis[2]

[1]Dow Benelux BV, Environmental Technology Center, Terneuzen, the Netherlands
[2]CKD Consulting, USA

Keywords: training, workforce

35.1 INTRODUCTION

As anyone who has accomplished this task can testify, it is difficult to change industrial practices. Procedures evolve over time, with the input of dozens of individuals – sometimes hundreds – who continually test and improve methods and materials until they become 'standard operating procedure.' Even in the most highly automated facilities, using water in more sustainable ways requires some level of training. This training involves not only providing information about how to do the old job in new ways, but also explaining why old ways need to be changed, and the benefits to be gained through new practices.

Because staff training is key to the successful adoption of new practices and technologies intended to make more sustainable use of water, this training needs to be carefully designed and implemented, taking into account the existing attitudes and skills of staff (e.g. language, literacy, and computer proficiency), as well as differences in their roles and responsibilities. This chapter reflects two examples of staff training activities (one from a public agency and one from a private company) to prepare their employees to handle water/related tasks more sustainably.

35.2 PUBLIC SECTOR CASE STUDY – SAN FRANCISCO PUBLIC UTILITIES COMMISSION

The San Francisco Public Utilities Commission (SFPUC) is a publicly owned water/wastewater/power utility that provides water services to 2.6 million customers in the San Francisco Bay Area. While the

DOI: 10.2166/9781789060676_0379

water/wastewater industry tends to see itself as a 'green' industry, the actual operations of producing clean potable water and discharging treated wastewater safely back in to the environment involve tasks that can have adverse environmental risks if not carefully handled. Examples include the following:

- Using chemicals to treat water and wastewater that, if discharged to land or water, could harm wildlife;
- Damaging sensitive and endangered species when mowing vegetated areas and maintaining roads and pipelines;
- Raising and lowering water levels in reservoirs in a manner that could disturb native vegetation and wildlife;
- Eroding riverbanks, increasing turbidity, and putting silt into sensitive habitats by discharging stormwater at high velocities;
- Spoiling wildlife habitat during construction of capital projects, and using construction materials with high embedded water and energy costs; and
- Increasing greenhouse gas emissions through the use of energy for pumping and treating water and wastewater and other industrial processes.

Unless employees are trained to be aware of potential adverse environmental impacts, they will tend to manage their work in the way that most expeditiously achieves immediate operational goals, such as treating water and wastewater to meet regulatory standards. As an example, in 1995 SFPUC water treatment plant operators drained potassium permanganate (a water treatment chemical) into a plant floor drain. This was consistent with the practice prescribed in their operating manual; however, at the time of the incident, the manual's directions had been superseded by new rules. As a result, the fish kill that ensued from the discharge of the chemical to the river was a violation of environmental regulations. The organization's internal staff training program had not, in this case, kept pace with changing regulatory standards. This alerted the agency to the need to update the operations and maintenance manuals provided to water treatment operators, and supplement the manuals with additional training. In addition to updating materials specifically related to this incident, the agency also provided more general training to its operational staff on the range of environmental regulations that now applied to their day-to-day activities.

Environmental issues also arose around management of agency-owned watershed land that drained into local reservoirs. Because use of the land had been limited over time to protect the quality of water in the reservoirs, these watersheds were providing habitat for vegetation and wildlife species that had largely disappeared from the urban areas that surrounded them. The laborers, skilled trade workers, watershed keepers, and engineers who performed work that impacted this vegetation and wildlife were sometimes unaware of how their actions could affect environmental resources, and the regulatory guidelines that applied to their work. Managers also needed more information on how to best protect these resources for the future. To address these concerns, the SFPUC prepared Watershed Management Plans for the Peninsula and Alameda watersheds. An important aspect of implementing these plans was to create clear, usable guidance materials for the staff. Watershed Field Manuals were created for use by SFPUC employees, as well as contractors, lessees, and members of the public with access to watershed lands (SFPUC, 2000; SFPUC, 2001). The field manual included information on their responsibilities (to protect watershed resources and water quality), prohibited activities (e.g., vehicle use off designated roads), procedures (e.g., erosion control and protection of vegetation), and maps. SFPUC laborers and skilled trades workers who did maintenance work on the watersheds also received on-site training by a skilled botanist on how to identify sensitive plant species, so that employees could avoid destroying them during the mowing season.

These investments reflected understanding that it was the agency's responsibility to provide employees with the information they needed in order to meet the agency's environmental commitments while meeting the water supply needs of customers.

35.3 PRIVATE SECTOR CASE STUDY – DOW

Dow's water and water treatment assets, like all components of its process units, are managed and operated to assure robust and uninterrupted performance. This requires a fundamental understanding of how the facilities work, including their impact both within and outside the perimeter of Dow facilities. Employees at all levels have to make the right decisions on both an operational and tactical level, so staff knowledge development is considered to be a key aspect of asset management.

Obviously, a great deal of training effort focuses on day-to-day operations. This includes detailed in-plant training materials and manuals to develop an in-depth knowledge of the basic design and maximum capability of unit processes, as well as their performance under extreme conditions. Most of these training resources are in the form of self-study materials that combine theory with in-field plant knowledge. Only after completing the study and passing a certification test is the operator allowed to fulfill certain roles and responsibilities. For example, new employees are assigned training materials suited to their entry level positions, but with an eye towards their subsequent development. Where classroom training is offered it is customized to specific assets and typically facilitated by in-company subject matter experts, although in special cases outside experts are called upon to teach classes.

At an intermediate level (e.g., engineers and technical advisors), training materials typically involve modules available electronically through Dow's global E-learning system. Water-related modules have been developed for water systems, water quality for industrial applications, water technologies, and wastewater treatment to address the background needed to fully understand and support daily operations. These materials do not provide an academic or 'textbook' overview of the field but instead target the assets and practices specifically relevant for Dow. This allows learners to quickly and directly relate theory and background to their daily tasks.

At a tactical level, a limited number of modules have been developed to cover water recycling practices and assess risks associated with droughts, availability of water, and water safety. These modules are developed to prepare engineers and middle management staff to initiate middle- and long-term improvements or enhance Dow's sustainable water use. An example of this tactical training is in Dow's on-line E-Learning on 'Assessing Water Supply Risk.'

As a first step, the target audience for this training is defined by Dow's business-aligned 'learning leaders' and manufacturing leadership. To this group, 'water supply risk' encompasses both Dow's long-term strategic sustainability objectives as well as the need to supply water for day-to-day operations at each industrial site. As a result, separate modules were developed for general management and site leadership, and for engineers at a facility level. Although the two courses share much of the same material (70–80%), each has a specific emphasis. The version developed for engineers provides knowledge and understanding of the impact of short-term supply risks and the technical approach to mitigate risk and manage undesired consequences. The version developed for employees in leadership positions has a stronger focus on the sustainable supply of water in the long- term, taking into account external relationships, current and future legislation, and commercial impact.

Understanding these distinct meanings leads to different learning objectives for each target group. An example of learning objectives for the leadership version is included in Figure 35.1.

LEARNING OBJECTIVES

After completing this course, you will be able to:

- Provide definitions for commonly used water terms.
- List the possible sources of water intake.
- Describe a few water risk case studies.
- Describe the different types of risk related to water sourcing for a site.
- Name the risks that your site is facing from water stress.
- Describe the information needed to assess uninterrupted supply needs at your site.
- Name the data needed to evaluate appropriate reservoir volume for a site.
- Describe the range of technologies offered by Dow for water solutions.
- Describe the response on your site to help mitigate long-term water risk.
- Describe the various groups and roles within Dow that address aspects of water topics.

Figure 35.1 Learning objectives for the course 'assessing water supply risks – leadership'. (*Reproduced and adapted with permission from Dow*)

The course content for this module is designed to provide a framework for understanding how water-related choices impact Dow's operations. Figure 35.2 show the characteristics of raw water options in the vicinity of industrial sites.

Where possible the examples to illustrate this conceptual understanding are selected from in-house cases of practice. Figure 35.3 shows how the hydrologic context of a Dow facility in the United States is used to illustrate the assessment of water supply risk due to drought.

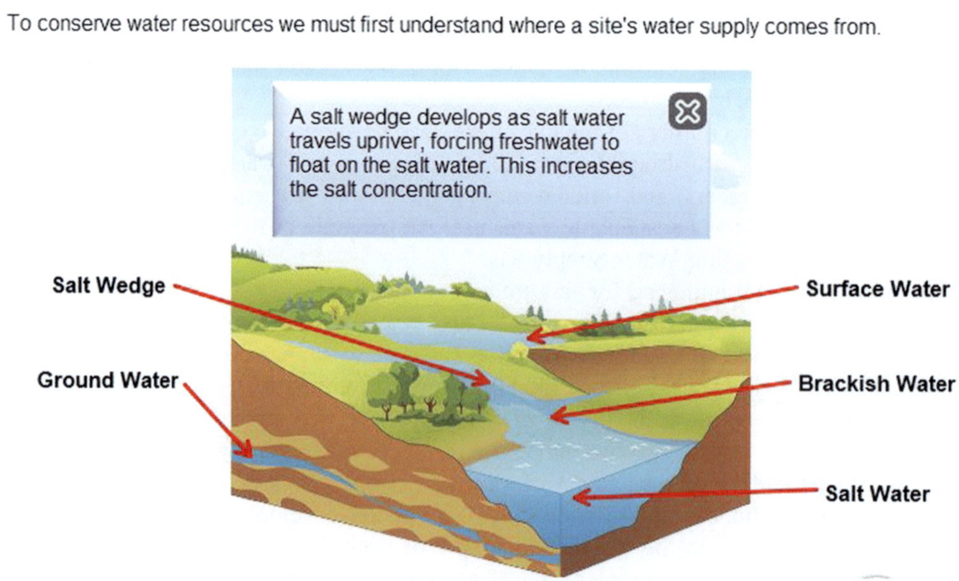

Figure 35.2 Raw water options and characteristics for water supply to industry. (*Reproduced and adapted with permission from Dow*)

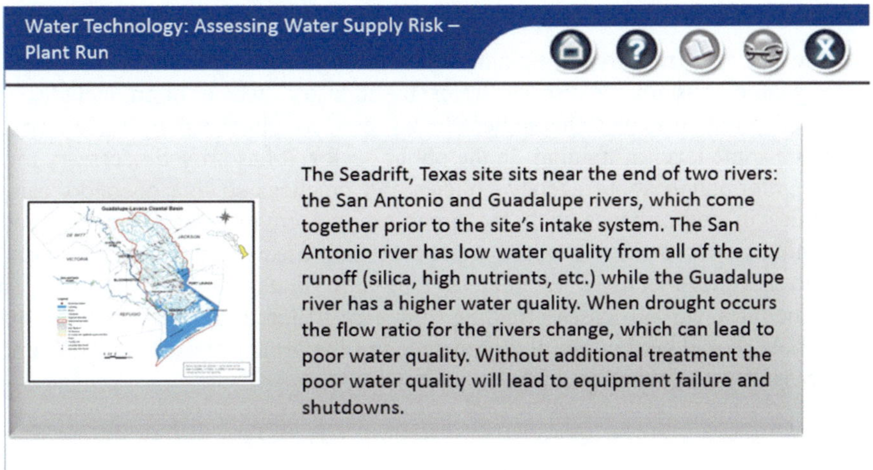

Figure 35.3 Practical example to illustrate water supply risks. (*Reproduced and adapted with permission from Dow*)

Finally, the learner's understanding of the material is tested by means of an interactive feature, as shown in Figure 35.4. The learner can drag and drop answers and receives feedback with identification of correct and incorrect answers.

At the end of the course the learner is required to take a test, typically consisting of 8–10 questions, interlinked with the course's learning objectives. The employee must receive a 90 or 100% correct score before receiving credit and certification in Dow's global learning system.

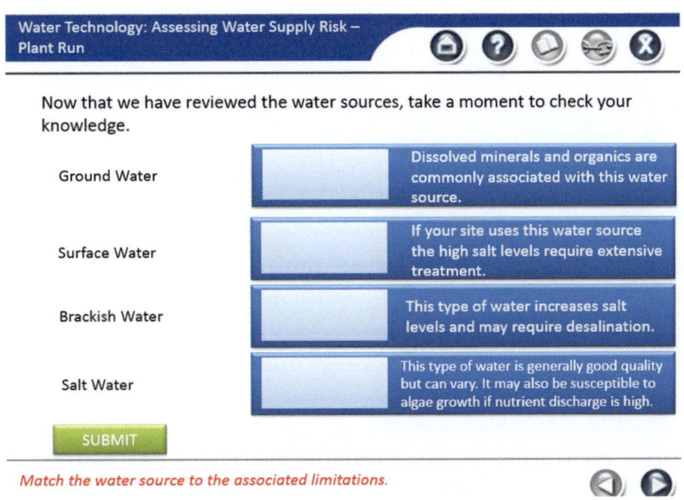

Figure 35.4 'Drag and drop' tables allow learners to check their understanding by filling in the blanks. (*Reproduced and adapted with permission from Dow*)

35.4 CONCLUSION

The road from noble concept to real-world implementation is long and complicated. This is certainly the case in achieving the goal of sustainable use of water by industry, which often includes the research, development, and adoption of new technologies. In the end, as illustrated by these two case studies, these changes require some level of training. In the public sector, it has proved necessary to provide both broad and detailed information about agency policies and practices so line operators can successfully carry out their duties in conformance with the organization's environmental goals. Similarly, in the private sector, training on industrial water and water treatment assets can range from basic tutorials on daily operations to advanced courses to address long-term issues and opportunities supported by distance-learning modules. In all cases, training must be tailored to the learner's daily working environment, materials should be thoughtfully prepared, and the organization should document when the employee has successfully completed training.

REFERENCES

SFPUC. (2001). Peninsula Watershed Management Plan: Final Environmental Impact Report. https://www.sfwater.org/Modules/ShowDocument.aspx?documentID=4343 (accessed 1 July 2020)

SFPUC. (2000). Alameda Watershed Management Plan: Final Environmental Impact Report https://www.sfwater.org/Modules/ShowDocument.aspx?documentID=4344 (accessed 1 July 2020)

Chapter 36

The business value of water education to corporations

John Etgen[1], Nicole Rosenleaf Ritter[1] and Will Sarni[2]

[1]Project WET, Bozeman, MT, USA
[2]Water Foundry, Denver, CO, USA

Keywords: climate, corporate sustainability, distance learning, employee engagement, learning games for kids, Project WET, STEM, sustainability education, water education, Water Education Today

36.1 INTRODUCTION

Businesses today are increasingly concerned about water as a risk to business continuity, growth, and brand. These water risks facing businesses impact the economic development, social well-being, and ecosystem health of the entire community, and other key stakeholders – employees, government officials, community leaders, and non-governmental organizations (NGOs) – are all involved in the process of finding solutions. As a result, businesses are investing in both internal and external communication efforts to engage with stakeholders on improving a basic understanding of water concepts and the value of corporate water stewardship programs.

It is essential to understand the critical role engagement with stakeholders plays in successful corporate water stewardship strategies. Water is a 'Wicked Problem' in that it is (among other things) difficult to define, unstable, and socially complex with many interdependencies (Sarni, 2017). As such, it can't be solved by any one stakeholder. Collective action programs start with effective internal and external communication efforts. These communication efforts range from reaching out to communities within a watershed, corporate workforce, NGOs, water professionals, and others to driving alignment on understanding water-related risks and solutions.

Frequently, discussions about water risks and water stewardship strategies are left to technical experts. However, water education – especially when interactive learning methods designed to appeal to nontechnical audiences are employed – can help a range of stakeholder groups gain 'water literacy.'

DOI: 10.2166/9781789060676_0385

Water literacy forms the basis for stakeholder understanding of the importance of water stewardship and sustainability. When stakeholders understand how water 'works' – its unique properties, the water cycle, the nature of watersheds, and other water basics – then a more representative group of individuals and organizations can play a meaningful role in decisions about water that impact their communities.

The non-profit Project WET Foundation (Water Education for Teachers) has been developing and implementing water education programs and curricula that nontechnical audiences can understand for more than three decades. Project WET grew out of efforts to help water management officials explain the links between land use, groundwater, and human health to rural farmers and ranchers in eastern South Dakota. With the aid of nontechnical terminology and innovative teaching methods, local stakeholders learned the connection between a detected increase in cancer cases and common agricultural practices impacting the quality of water in the Big Sioux Aquifer, their sole drinking water source. Once they understood the importance of proper use and management of fertilizers and pesticides and the need to avoid improper chemical storage, they changed their behavior. That led to improved water quality and a decrease in human health problems.

Since those early days, Project WET has grown exponentially, training hundreds of thousands of teachers to provide hands-on, interactive water education to audiences across the United States and in more than 70 countries. Partnering with some of the world's largest corporations, Project WET educates children and youth as 'tomorrow's water stewards' and works with the water stewards of today through an adult stakeholder education program, The WaterCourse.

The following case studies illustrate how customized, accessible water education that facilitates employee engagement and community relations can help businesses implement impactful water stewardship strategies.

36.2 CASE STUDY: NEWMONT GOLDCORP

Headquartered in Colorado, Newmont Goldcorp is a publicly traded mining company with 19,000 employees and 18,000 contractors in Australia, Canada, Ghana, Peru, Suriname, Mexico, Argentina, Dominican Republic, and the United States. Named four times by the Dow Jones Sustainability World Index as the mining industry leader, the company's stated purpose is 'to create value and improve lives through sustainable and responsible mining' (Newmont, 2019a). In 2018 Newmont Goldcorp engaged Project WET to assist with community water education in the vicinity of two of its South American mines.

36.2.1 Peru: building local capacity to understand water from a science-based perspective

Located near the city of Cajamarca in northern Peru, Newmont Goldcorp's Yanacocha operation is South America's largest gold mine. In cooperation with the Peruvian Ministry of Education and SEDACAL, a local water company, Newmont Goldcorp established a foundation, Asociación de los Andes de Cajamarca (ALAC), to promote sustainable development in the Cajamarca region.

The company believed the community could engage more meaningfully on water issues if its members understood that water is a finite but renewable resource. Project WET was selected to help build capacity in the local community for understanding water from a scientific perspective, and to teach children the skills needed to become environmental stewards. Project WET also designed interactive displays for the Museum of Water and Earth which opened the same year.

The process began by showing trusted local leaders – specifically teachers and ALAC facilitators – how to use local, readily available materials to transform their classrooms into fun, exploratory environments (Figure 36.1). Sixty-four teachers were trained to teach their students how – and why – to conserve water, as well as the importance of water to health and hygiene (Newmont, 2019b). Teacher surveys conducted in the first quarter of 2019 indicated that not only were students effectively integrating the material, they were also changing the way they used water: 'Teachers observed that even if students got distracted playing with the water during handwashing and forgot to turn off the faucet, they quickly remembered the importance of water conservation and adjusted their behavior accordingly.'

ALAC facilitators have continued to work directly with teachers, while Project WET staff provide online support. Moving forward, teachers have asked to include parent education, demonstrating the value not only for their students but for the larger community as well. Meanwhile, Project WET also assisted with the creation of interactive exhibits in the Museum of Water and Earth, a regional experiential learning center housed in ALAC's offices (Newmont, 2019c). In conjunction with the Ministry of Education and the National University of Cajamarca, Project WET developed 11 interactive displays and experiments designed to teach children and adults about water, geology, space, and other fundamentals of natural resources (Figure 36.2). Museum docents (mostly students from the National University of Cajamarca) guide visitors through the exhibits which encourage questions and discovery.

The learning modules have also been adapted into a mobile exhibition to reach rural areas, through a partnership with Peru's National Council of Science, Technology and Innovation (CONCYTEC). According to Newmont Goldcorp, the museum supports their efforts 'to expand the local community's

Figure 36.1 Local educators at a training workshop in Cajamarca participate in a whole-body learning activity, emphasizing the need for cooperation and balance among different water user groups. All of Project WET's activities focus on hands-on learning. (*Credit: Project WET*)

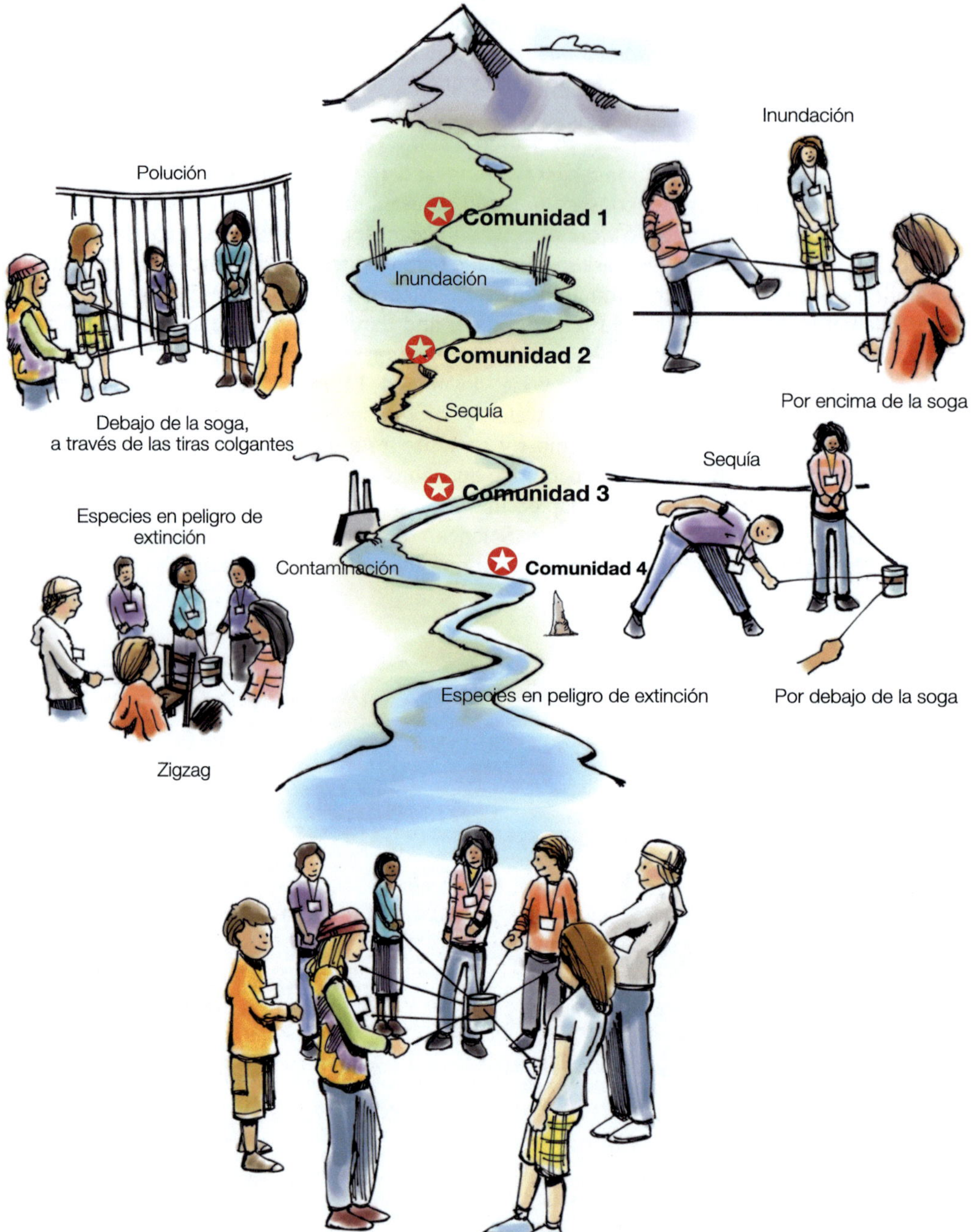

Figure 36.2 Project WET provides both training and customized materials to help educators set up water education activities in their own learning environments, as with this sample page in the guide created for educators in Peru with support from Newmont. (*Credit: Project WET*)

water management capacity through the development of water education programs for current and future leaders.' Their work with Project WET and other stakeholders also aligns with the UN Sustainable Development Goals to ensure access to water and sanitation and to advance goals through strong local, national and global partnerships (SDG-6 and 17).

36.2.2 Suriname: combining water infrastructure with education to promote conservation and water quality

In contrast with the Yanacocha mine's mountainous location, Newmont Goldcorp's Merian operation is located in a jungle in northeast Suriname (Newmont, 2019d). The mine has access to abundant surface water from the Marowijne River which forms the boundary between Suriname and French Guiana, but the water quality in the river has been impaired by a lack of wastewater treatment as well as discharges from small-scale mining operations, many of which use mercury to process gold. These factors have also negatively impacted the native Pamaka Maroon community which depends on the river for its water supply.

To better understand local water issues and community perceptions, Project WET staff visited several towns and villages, including some like Langa Tabiki (the main island in the Pamaka river) only accessible by canoe. Accompanied by Newmont Goldcorp staff, they discussed with community members how water education could be most useful to communities that lack safe water. Through the combined efforts of the mining company, the Suriname government and the Community Development Fund (a foundation governed by the Pamaka community) a potable water system was installed on Langa Tabiki following these visits.

The long-term success and sustainability of the water system depends on the ability of the community to understand issues related to water quality protection and water conservation. Given that, Project WET staff plan to return to Suriname in late 2021 to train more officials in the Ministry of Education and to reach more teachers and students with hands-on education about water.

36.3 CASE STUDY: ECOLAB

Ecolab is a recognized leader in water, hygiene, and energy technologies whose products and services are used in food service, food processing, hospitality, healthcare, industry, and oil and gas markets. The company's 49,000 employees work in nearly 3 million separate customer locations in more than 170 countries to promote safe food, maintain clean environments, optimize water and energy use, and improve operational efficiencies (Ecolab, 2019a).

Project WET began partnering with Ecolab in 2014 in conjunction with the company's 'Solutions for Life' global giving program, designed to enhance the company's mission to conserve water and improve hygiene around the world through global philanthropy and employee volunteerism and collaborations with NGOs (Ecolab 2019b). Project WET and Ecolab worked together to develop the 'Clean and Conserve Water Education Program' as the centerpiece of their partnership. Employee volunteers as well as school and community educators are trained to use the program with young people.

36.3.1 Clean and conserve and WaterStar: providing open access to water conservation and hygiene education to reach a wide audience

Project WET and Ecolab launched Clean and Conserve, a suite of educational materials and online training tools available in multiple languages that focuses on water conservation and hygiene education, in 2015 (Project WET, 2019a). WaterStar is the recognition component of the Clean and Conserve program, profiling individuals and groups around the world who successfully use the Clean and Conserve water education materials. Past WaterStar designees have worked in India, Ghana, Pakistan, Peru, and the United States (Project WET, 2019b).

Clean and Conserve aims to raise awareness and provide practical training in water conservation for students and teachers (Ecolab, 2017). Ecolab employees, classroom teachers, scout troop leaders, parents, informal educators, and many others have received the curriculum through a multiyear implementation process. In fact, anyone interested in educating people about water conservation and hygiene can download it free of charge, along with the 'Soap and Water Science' interactive online lesson (on www.discoverwater.org) that teaches children how to properly wash their hands. Project WET has also provided printed materials to local schools using in-person workshops and a train-the-trainer model. To date, Clean and Conserve has reached some 8 million people in 98 countries worldwide.

36.3.2 Employee engagement: giving local employees tools and training to carry out water education

Ecolab encourages its employees to use water education as part of their community outreach. In the United States, an Ecolab employee resource group focused on the advancement of women has used 'Clean and Conserve' to connect with Boys and Girls Club, United Way, STEM magnet schools, and other educational groups. In Argentina, Ecolab employees have worked with Asociación de Amigos de la Patagonia (AAP) to help local students better understand water conservation and hygiene. Clean and Conserve was also the focus of Ecolab Argentina's first Corporate Volunteering Day (Figure 36.3).

As interest in the program has grown, Ecolab and Project WET have partnered to develop a specialized employee training guide that has been translated into Spanish, Chinese, German, French, Japanese, Turkish, and Portuguese (Figure 36.4). Ecolab's support for water education – not only sponsoring 'Clean and

Figure 36.3 Ecolab employees around the world use Project WET activities to teach water conservation and hygiene concepts to local children, as in this school festival in Argentina. This activity helps learners understand the percentage of water on Earth. (*Credit: Ecolab*)

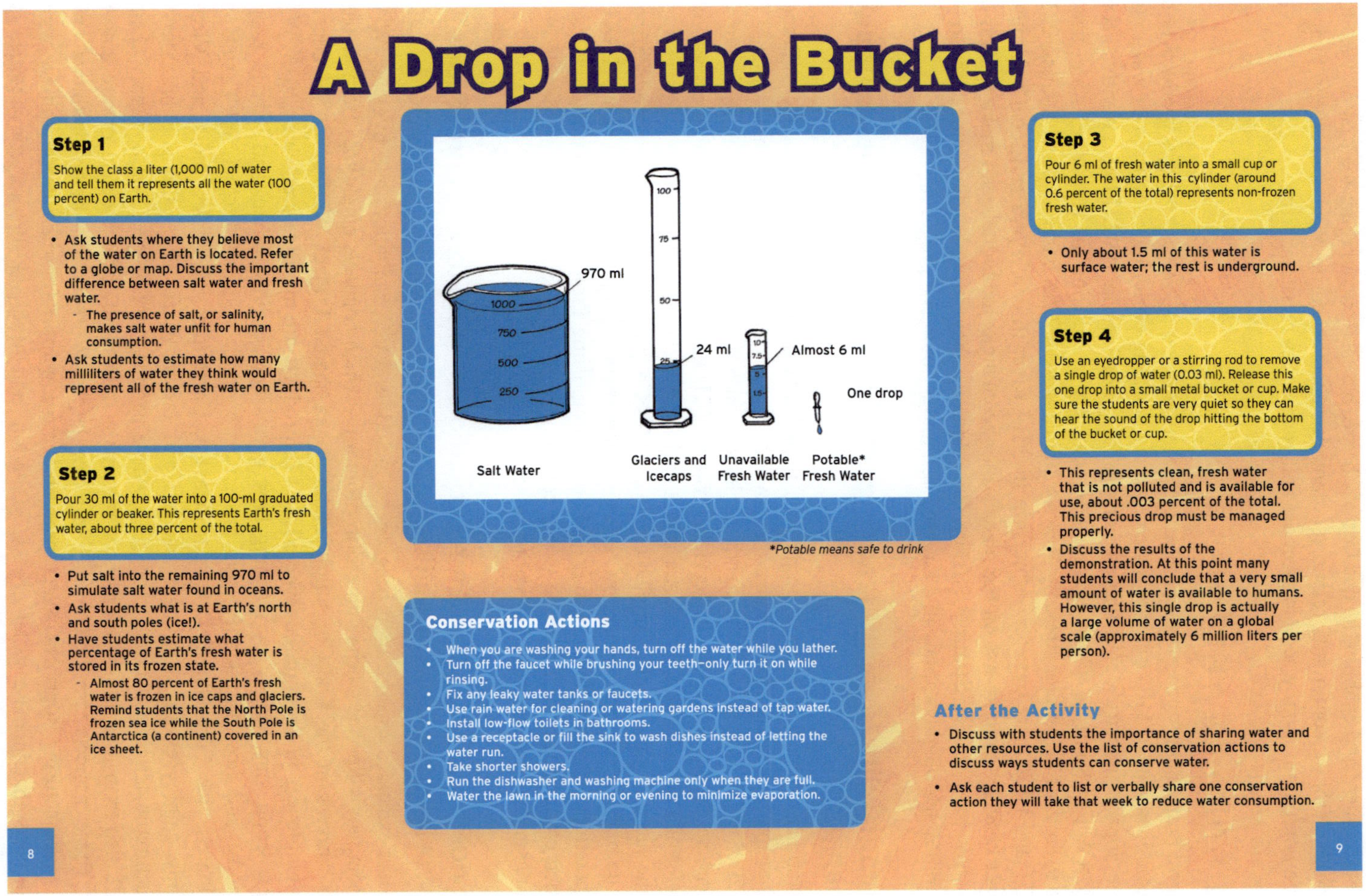

Figure 36.4 Ecolab worked with Project WET to adapt activities designed for classroom teachers into an 'at-a-glance' format that works well for their employees, maximizing their ability to educate young people in classrooms and other learning venues without an intense time commitment. (*Credit: EcoLab*)

Conserve's' development and enabling its use free of charge, but also encouraging their employees to use water education as a tool for community relations – demonstrates how corporations can extend their reach well beyond the boardroom as they seek to promote and enable water sustainability.

36.4 CASE STUDY: LEVI STRAUSS & CO.

One of the largest brand-name apparel companies, Levi Strauss & Co. (LS&Co.) sells its products at more than 50,000 retail locations in 110 countries. The company has long been in the forefront of water issues, establishing the apparel industry's first wastewater quality guidelines in 1992. More recently, LS&Co. released its 2025 Action Strategy, which commits the company to reducing the amount of water used for manufacturing in areas of high water stress by 50% by 2025 against a 2018 baseline (Levi Strauss, 2019a).

Project WET and LS&Co. initiated a partnership in 2014 to teach employees and local communities about the impact LS&Co. clothing has on the planet, and the changes individuals can make in their daily lives to conserve water (Levi Strauss, 2016b). In addition, they brought this education to their employees' families and neighborhoods in a targeted community outreach program.

36.4.1 Sustainability training: teaching employees to share the company's sustainability program

Project WET staff trained the first group of employees at the LS&Co. global headquarters in San Francisco on the use of a customized curriculum to educate youth about the importance of water conservation in 2015 (Levi Straus, 2015). Employees took the hands-on lessons into the community during the company's Community Day, an annual global day of service. Project WET also developed a learning activity based on the 'life cycle' of a pair of jeans, with water use at each step of the process.

Following the success of these initial implementations, LS&Co. pledged to provide water education and sustainability training to all employees by 2020. Today, every employee learns about LS&Co.'s sustainability practices either online or at an in-person training, using Project WET's interactive learning methods. Trained employees in turn become 'water conservation ambassadors,' educating people in their local communities about the importance of saving water.

36.4.2 Service corps: taking water education into factory communities to benefit workers and their families

The company has expanded its education program to include a water, sanitation, and hygiene (WASH) element. The company's Service Corps immersion program – which connects employees from across the company with workers in the LS&Co. supply chain – has shared Project WET's WASH education resources with employees in communities where LS&Co. products are manufactured since 2015. WASH education helps people stay healthy by teaching how germs spread; how, when, and why to properly wash their hands; and how to identify and protect safe water sources. Project WET staff have accompanied the Service Corps on their trips to Mexico, Cambodia, Sri Lanka, India, and Vietnam, training not only employees but also local teachers and factory staff. The training culminates with participants teaching materials in schools to children whose parents work in the LS&Co. supply chain.

Factory staff also train their fellow employees through a train-the-trainer model to ensure long-term sustainability. These projects are tied into LS&Co.'s larger Worker Well-being initiative, where employees of manufacturing suppliers select programs most helpful for their communities, with an emphasis on projects that increase water access, sanitation, and health in supplier communities (Levi Strauss, 2019b). For example, Service Corps volunteers in Cambodia installed handwashing stations

Figure 36.5 LS&Co. employees in the Service Corps are trained to use Project WET activities to teach about water, sanitation, and hygiene in communities where LS&Co. products are manufactured. This education is usually paired with an action project, as shown here, where Service Corps volunteers in Cambodia installed handwashing stations outside a school to enable healthy hygiene habits. (*Credit: Project WET*)

outside a school with the help of a local NGO (Figure 36.5). The installation allowed students who had learned about the importance of proper handwashing through Project WET lessons to have a ready supply of safe water and soap.

36.5 CONCLUSION

As water scarcity and poor water quality impact economic development, business growth, ecosystem health, and social well-being, it becomes increasingly critical for all stakeholders to contribute to solutions. Corporate support for water education is one way the private sector can help give communities a stake in their own water futures.

Water education has its limits, of course. It cannot take the place of coordinated solutions around water and sanitation provision, whether public or private, nor can it replace better and more just governance. And water education is no quick fix – the impacts of educating people about water can take years, if not decades. However, especially when it reaches the children who will one day be charged with solving the world's water challenges, water education is an invaluable investment.

By teaching people about water, corporations can engage and mobilize additional stakeholders into the process along with government agencies, local non-profits, schools, museums, community foundations, and others. As LS&Co. Vice President of Sustainability Michael Kobori said, 'Water is one of the planet's most precious resources, and it is going to take more than just one company or individual to ensure its future' (Levi Straus, 2016b).

REFERENCES

Ecolab. (2017) 'Focus on water: Ecolab supports education program on water conservation and hygiene in Germany.' 2/10/17 https://www.ecolab.com/news/2017/02/ecolab-project-wet-monheim-germany. (accessed October 1, 2019).

Ecolab. (2019a). 'About Ecolab' https://www. ecolab.com/about. (accessed October 1 2019).

Ecolab. (2019b). 'Ecolab's Global Program: Solutions for Life' https://www.ecolab.com/about/corporate-responsibility/community-involvement/giving-programs/solutions-for-life. (accessed October 1, 2019).

Levi Strauss and Company. (2015) 'Dennis Nelson: A modern day pioneer educating youth about water conservation.' 5/4/15 https://www.levistrauss.com/2015/05/04/dennis-nelson-a-modern-day-pioneer-educating-youth-about-water-conservation/. (accessed October 1, 2019).

Levi Strauss and Company. (2016a). 'A look back at our partnership with Project WET Foundation' 3/18/16 https://www.levistrauss.com/2016/03/18/the-perfect-match-a-look-back-at-our-partnership-with-the-project-wet-foundation/. (accessed October 1 2019).

Levi Strauss and Company. (2016b). 'LS&Co., Scholastic and Project WET Team Up to Provide Water Education to 1.5 Million Students' 9/28/16 https://www.levistrauss.com/2016/09/28/lsco-scholastic-and-project-wet-team-up-to-provide-water-education-to-1-5-million-students/. (accessed October 1, 2019).

Levi Strauss and Company. (2019a). '2025 Water Action Strategy' https://www.levistrauss.com/wp-content/uploads/2019/08/2019_LSCO_WATER_STRATEGY_REPORT.pdf. (accessed October 1 2019).

Levi Strauss and Company. (2019b). 'Worker Well-being' Website. https://www.levistrauss.com/how-we-do-business/worker-well-being/. (accessed October 1 2019).

Newmont Gold Corporation. (2019a). 'About Newmont'. https://www.newmontgoldcorp.com/about/ (accessed October 1 2019).

Newmont Gold Corporation. (2019b). 'Newmont Peru – Project WET Program Update' Newmont Blog 8/18/19 https://www.newmontgoldcorp.com/newmont-goldcorp-peru-project-wet-program-update/. (accessed October 1, 2019).

Newmont Gold Corporation. (2019c). 'New Interactive Museum Builds Knowledge About Natural Resource Challenges and Conservation' Social Acceptance Case Study. https://www.newmontgoldcorp.com/new-interactive-museum-builds-knowledge-about-natural-resource-challenges-and-conservation/. (accessed 1 October 2019).

Newmont Gold Corporation. (2019d). 'Newmont Suriname – Project WET Program Update' Newmont Blog 8/21/19 https://www.newmontgoldcorp.com/newmont-goldcorp-suriname-project-wet-program-update/. (accessed 1 October 2019).

Project WET. (2019a). 'The Clean and Conserve Education Program.'. https://www.projectwet.org/cleanandconserve. (accessed 1 October 2019).

Project WET. (2019b). 'WaterStar Recognition Program' https://www.projectwet.org/waterstar. (accessed 1 October 2019).

Sarni W. (2017). Beyond the Energy–Water–Food Nexus: New Strategies for 21st-Century Growth. Routledge, New York.

Chapter 37

Promoting sustainable industrial water use: Scotland's "Hydro Nation" at home and abroad

Barry Greig and Jon Rathjen

Scottish Government, Water Industry Division, Edinburgh, Scotland, UK

Keywords: government, Hydro Nation, international, Malawi, privatization, resources, Scotland, Scottish Water, value, water resources

37.1 INTRODUCTION

Water flows through the heart of Scotland, literally and figuratively. Water is arguably Scotland's chief environmental feature, celebrated in Scottish social life and a critical resource for all sectors of the Scottish economy – manufacturing, energy, agriculture, food and drink, even tourism (Figure 37.1). This case study describes Scotland's approach to water resources management through its 'Hydro Nation' agenda, with particular focus on its introduction of competition in its domestic water supply, and its international activity in support of UN Sustainable Development Goals. By allowing more than 30 government licensed providers to serve non-household (including industrial) water accounts, The Scottish Government has managed to reduce industrial water use by 12%, with commensurate reductions in energy and greenhouse gas production. At the same time, Scottish international efforts in Malawi and Tanzania have supported the development of progressive water management policies and sustainable industrial practices there.

37.2 SCOTTISH WATER: ORIGIN AND INNOVATION

Production and transmission of clean water and treatment and disposal of wastewater is carried out by Scottish Water, a public corporation accountable to Parliament and its Ministers. In addition, for over a decade – since 2008 – licensed providers, now numbering more than 30, have supplied water on a retail basis to business customers, including industrial users, the service sector and public sector, and

DOI: 10.2166/9781789060676_0395

Figure 37.1 Suilven and Cul Mor Mountains seen from Stac Pollaidh, Assynt, Sutherland, Scotland. (*Credit: B. Greig*)

'not-for-profit' businesses, saving water and energy in the process. The evolution of Scottish Water from a single public monopoly into an innovative utility supporting a thriving competitive sector speaks to both the challenges and opportunities in water management today.

37.2.1 Scotland's water environment

With a population just under 5.5 million, Scotland covers an area of 78,000 km^2 including 787 islands. Scotland is generally a wet country with more than 125,000 km of rivers and streams varying from small highland *burns* (streams) to deep lowland rivers and a 220 km canal network. In addition, there are over 25,500 *lochs* (lakes) in Scotland, including Loch Ness, whose volume of 7.4 million m^3 is greater than all the surface water in England and Wales combined.

Water policy in Scotland is set by The Scottish Government in conformance with European Union water and environmental law and agricultural law and policy. The Government is also responsible for Scottish Water, the national water utility, formed in 2002, which supplies 2.5 million households and 150,000 businesses with nearly 1.5 billion litres/day of drinking water through more than 30,000 miles of water pipes and over 200 water treatment works. Wastewater is collected through another 30,000 miles of sewers and treated at nearly 2000 wastewater treatment works before it is returned to the environment (Scottish Water, 2019). But it was not always so.

37.2.2 Consolidation and competition

Throughout the 1990s the organization of water services in Scotland was the subject of debate. For half a century, water supply in Scotland had been steadily consolidating, from some 200 separate water agencies in the 1950s to three regional councils serving southern, northern, and western Scotland (Audit Scotland, 2005). Despite the trend towards unification, however, these areas remained distinct, geographically and demographically, and they each charged different rates for water. The consolidation of these last remaining suppliers into Scottish Water was done in part to establish uniform rates across Scotland.

At the same time, in neighbouring England and Wales, 10 regional water authorities were privatized in 1989 to generate capital and improve efficiency. By contrast, privatization was vigorously resisted in Scotland, despite concerns about the performance of the three authorities and their ability to meet forthcoming challenges. The performance of the three Scottish authorities began to pale beside the gains in efficiency in the newly privatized English utilities, and they lacked the capital to meet the new EU standards for drinking water by the 2004 deadline. In parallel, it was recognized that water rates (collected by local governments) were not covering the cost of service and that significant rate increases were inevitable. (The financial challenge facing the North authority was particularly severe, as it had the smallest, most disparate population with the longest coastline and the least developed systems.) In 2002, the Scottish Executive formed Scottish Water, a single all-Scotland water authority monitored by the independent Water Industry Commission for Scotland (WICS), an independent body responsible for setting rates and charges through transparent analysis.

Since that time, Scottish Water's performance has improved rapidly. More than 99.9% of all water samples comply with relevant standards, and customer service is now comparable with the leading water companies in England and Wales while average household charges are lower by £46 per year. Not only WICS, but all the partners in the Governance structure – the Drinking Water Quality Regulator (DWQR), Scottish Environment Protection Agency (SEPA), and the customer representative body, Citizens Advice Scotland (CAS) – together measure public investment in terms of improving standards, social benefit, economic impact, and emissions reductions. In response to these goals, in 2008 Scottish Water, on top of ambitious legislative targets calling for net-zero carbon emissions by 2045, have committed to the visionary goals of Scotland's 'Hydro Nation' programme, a broad suite of activities designed to promote sustainable management of water resources (Scottish Government, 2012).

37.3 'HYDRO NATION': STRATEGY AND STRUCTURE

As a small, but responsible nation, Scotland has long recognized the principles of Sustainable Development. Following the establishment of the Secretary of State's 'Advisory Group for Sustainable Development' in 1999, the Scottish Government has worked continuously to mainstream the principles of sustainable development across all policy areas. In 2005, the Scottish Executive published a 'manifesto' in the form of a strategy document, 'Choosing Our Future: Scotland's Sustainable Development Plan' which held as its first principle their determination 'to stay the course and bring about the long-term changes in the way we govern, so that Scotland sets an example to the world' (Scottish Executive, 2005). This vision was underpinned further in 2013 by a statutory duty placed on the Scottish Ministers to 'take such reasonable steps as they consider appropriate for the purpose of ensuring the development of the value of Scotland's water resources' (Scottish Government, 2013).

Scotland's 'Hydro Nation' concept can best be understood in this context. As explained in the most recent Annual Report (Scottish Government, 2019), 'Scotland is a Hydro Nation, one that views and manages its water resources responsibly, and sees our relationship and the ways we work with the water environment and industry as inextricably linked to our national identity.' The Hydro Nation aims to **maximize the value of Scotland's water resources** by increasing the entire sector's contribution to the national economy. From this perspective, Scotland's water resources are not limited to its abundant waterways but also include the work of Scottish water experts who advise on government policy and create practical solutions for industry, as well as Scottish Water, now one of the best-performing water utility companies in the UK. The value of these resources themselves is viewed broadly, in terms of both economic and noneconomic worth. While some things, like exports of water technology, can be measured directly in terms of the money they add to the national economy, other things – like clean, good-tasting drinking water and the way Scotland's lakes, rivers, and streams maintain the fisheries and attract tourists – must be measured indirectly. These 'noneconomic' values are often gauged in terms of the price consumers would be willing to pay for these benefits, so an important element of the Hydro Nation programme is to increase public understanding of the value of water bringing solutions to customers in remote areas, generating renewable energy, and reducing or even recovering priority pollutants from wastewater.

The Hydro Nation Strategy is realized in four thematic areas: national; international; knowledge; and innovation. The **national** theme is aimed at preparing and developing the Scottish water industry to tackle a wide range of policy areas, including greenhouse gas reduction, providing water to rural areas, and investing in blue-green infrastructure energy. In addition, the national theme embraces water efficiency as embodied by the initiatives of Scottish Water, and the successful introduction of competition into the industrial water market. The **international** theme was conceived to help export Scottish water knowledge, working with other public sector actors to share their expertise with developing world nations, numbering Malawi, Tanzania, Romania, and India among its partners. These two themes are described in greater detail below. The **knowledge** sharing theme sponsors research in areas including Support Systems for Consumers reliant on Private Water Supplies and Community Engagement Best Practice, and also includes the Centre for Expertise in Water (CREW) which administers the Hydro Nation Scholars Programme on behalf of Government. The **innovation** theme includes the Hydro Nation Water Innovation Service (HNWIS) which has supported over 100 Scottish companies developing water and wastewater technologies by providing market insight and information and connecting them with potential customers. It also includes the Scottish Enterprise Low Carbon Team, which assesses opportunities, including in water efficiency and effluent management in Scottish industries (Scottish Government, 2019).

All Hydro Nation strategies reflect the principles of sustainability, especially evident in the national and international themes. Nationally, the water sector in Scotland is preparing for climate change by reducing its carbon footprint and protecting adaptive resources. The carbon intensity of Scottish Water's water service is

Figure 37.2 A Climate Justice Fund project funded by Scotland's Hydro Nation initiative assists with maintenance of local water systems in Malawi to reduce the time residents must stand in line for clean water. (*Credit: CJF Water Futures Programme*)

among the lowest in the UK sector (mainly due to gravity supply in lieu of pumping), down nearly 13% from 2018 and 41% since reporting started in 2007. The company reached a major milestone in March 2018 when it generated nearly 900 GWh of renewable energy, more than twice the amount used annually. Another carbon-reducing focus of the national agenda is preservation of peatlands, which cover over a fifth of Scotland's land area. By collaborating with local landowners and the Peatland Action Group, Scottish Water is helping to restore nearly 500 hectares of peatland, with a further estimated 1500 hectares identified for restoration in multiple locations.

Internationally, as a Hydro Nation with a global conscience, since 2012 Scotland has been engaged through its Climate Justice Fund (CJF) in water-related projects in Malawi, Zambia, Tanzania, and Rwanda, as well as international projects in India and Pakistan. In September 2015, First Minister Nicola Sturgeon confirmed Scotland's commitment to the newly announced UN Sustainable Development Goals, subsequently supporting international trade opportunities in water technology and exporting Scottish expertise in water governance and management. The drive to achieve the United Nations Sustainable Development Goals sets the global context for much of the work undertaken by the Hydro Nation Strategy internationally, especially in Malawi, a country with which Scotland has enjoyed a long historical connection (Figure 37.2).

37.4 HYDRO NATION NATIONAL: A COMPETITIVE INDUSTRIAL WATER MARKET

Management of water and wastewater in Scotland changed significantly in April 2008 when Scottish retail water and wastewater services were opened to competition. While Scottish Water remains responsible for wholesaler and network operations, business customers can shop among retail providers for the best price and level of water and wastewater service for their organisation. There are now around 30 licensed providers in the market, and since the market was opened to competition all 130,000 nonhousehold customers (i.e., businesses, public sector, charitable, and not-for-profit organizations) choose the retailer that best suits their needs. As well as bringing substantial cost savings to customers, this competitive market is also delivering wide-ranging environmental benefits including greater water conservation.

In allowing multiple entities to compete for retail business, it was believed that competition in this market would achieve both financial and nonfinancial benefits. To gain their patronage, retailers would be motivated to offer industrial, commercial, and institutional customers reduced pricing, bespoke billing options, water efficiency advice, and other incentives. It was also anticipated that licensed providers would pressure Scottish Water on behalf of their customers to operate with improved efficiency and responsiveness, so that competition in the retail market would yield smaller water bills for business, reduced water use by industry, and decreased water abstraction by Scottish Water, resulting in reduced energy use and a lower carbon footprint. Before opening the market, however, certain financial standards had to be assured: (1) wholesale charges would remain the same for all customers regardless of regional service costs; and (2) neither Scottish Water nor its industrial customers would be negatively impacted by the change.

37.4.1 Wholesaler and retailer oversight

WICS, which regulates the wholesale supplier (Scottish Water), was in the best position to license and regulate the new providers competing in the retail market by establishing requirements for both Scottish Water and the private water retailers. Scottish Water was required to publish its prices for wholesale services annually meaning retailers can refer to them as a basis for setting their own prices. Scottish Water was also required to accurately meter water use at all industrial services, so customers can make informed choices about their providers based on their water use. WICS also set up a mechanisms for licensing water retailers and providing appropriate oversight, including requiring evidence of financial stability and demonstration of competency in Scottish Water's central system operations before being granted access to the market.

The operation of the market is overseen by the Central Market Agency which is jointly funded and governed by the players in the market. Today, more than 30 companies from a range of backgrounds hold licences within the Scottish industrial, commercial, and institutional market, including subsidiaries of English wholesale water companies and independent Scottish retailers. '(A list of Licensed Providers can be found on the Water Industry Commission website at https://www.watercommission.co.uk/view_List_of_current_licensees%20.aspx.)'

37.4.2 Encouraging water efficiency

Understanding business customers' need to become more water efficient, retailers have been competing to offer services that help them see how much water they use and when they use it, what they use it for, and how they dispose of it. While these customer relationships did not exist before the introduction of competition, retailers are now offering bespoke environmental solutions along with an increased commitment to leakage reduction and other water-saving measures. They even provide customers with the services of specialists to enhance their water efficiency and further reduce costs. Along with these additional services, retailers also provide specific products to help conserve water, from simple devices that control taps to complete rainwater harvesting systems. Retailers are also offering more tailored waste management and surface drainage services, like the aforementioned water harvesting system and pretreatment of waste before it is discharged to the sewerage system.

37.4.3 Results of competition

WICS continues to monitor overall market shares and switching rates in the nonhousehold market. As a result of the increased choice, over 60% of customers have renegotiated their terms of service and switched to a different licensed provider since market opening. All retailers are required to offer a default level of service and tariff to any customer in Scotland to prevent discrimination and to ensure that no customer could experience any adverse impact from opening the market to competition, regardless of their location or cost to serve. As a consequence of this major shift in emphasis at the retail water supply level, non-household customers have benefited from lower prices and more tailored services industrywide.

The new competition marked the end of a 'one size fits all' approach. While consolidated and electronic billing have reduced administration costs, environmental benefits have also accrued as a result of the marked reduction in the level of water consumed and carbon emitted by nonhousehold customers. In short, these market arrangements have ensured that while the wholesaler is focused on sourcing water, operating pumping and storage assets, and devising low-cost water and waste solutions, retailers can focus on what they do best: providing water efficiency advice and helping customers manage the water and waste water issues they face.

This is reflected in numerous achievements. From 2008 to date, Business Stream, the retail offshoot of wholesaler Scottish Water, one of the largest licensed providers in Scotland, has helped its customers conserve over 38 billion litres of water, saving almost £75 million and removing 66,000 tonnes of CO_2 from the atmosphere. Whisky distillers Chivas Brothers report that, under the Scotch Whisky Association's first sector-wide Environmental Strategy, between 2009 and 2018 they have delivered savings of 20% in energy, 37% in greenhouse gases, and 15% in water consumption across distilling with 98% water recycling or reuse. Glenmorangie distillers went so far as to install an anaerobic digester

Figure 37.3 Water conservation and reuse at Scotland's world-famous distilleries is encouraged by licensed water retail companies that compete for industrial accounts. Pictured above: Dalmunach Distillery on the River Spey, near Caron, Scotland. (*Credit: Chivas Brothers Ltd*)

at their distillery to purify up to 95% of their waste water, and they are working on a ground-breaking programme to manage the remaining 5%. Leading potato producer Albert Bartlett and Sons Ltd collects the majority of their water from the roofs of their facilities; they then reuse the water on-site numerous times following treatment. These accomplishments, which have been recognized under a Scottish Government awards programme that celebrates and promotes water efficiency in industry, reflect the benefits of a regulatory structure that encourages reduced water usage and better waste water management (Figure 37.3).

37.5 HYDRO NATION INTERNATIONAL

The goal of the Hydro Nation International strategy is to share Scotland's knowledge and innovation in a global context. To this end, Hydro Nation Research International (HNRI) coordinates a range of international water-related activities that contribute not only to the Hydro Nation agenda, but also to the United Nation's Sustainable Development Goals, in particular Sustainable Development Goal 6 (SDG6), 'Ensure availability and sustainable management of water and sanitation for all by 2030' (Scottish Government, 2019). The sustainable management and stewardship of water resources is a key feature of the Scottish Government's approach, and the same principles apply to Scotland's activities in sub-Saharan Africa: Hydro Nation projects in Malawi and Tanzania are prime examples.

37.5.1 Scotland and Malawi

Scotland and Malawi have had long-standing historical ties stretching back to the expeditions of Dr David Livingstone, the famous Victorian missionary and abolitionist (World Heritage Centre, 2020). This relationship was reconfirmed in 2013 by legislation in both countries and the establishment of a Government-to-Government cooperation agreement supporting joint work at an official level on water resource management, governance, and legislation. The Scottish Government has conducted this work in part through its CJF delivery partner, the University of Strathclyde which focuses on Malawi's ability to achieve the UN Sustainable Development Goals as they relate to water, wastewater, and infrastructure management. This challenge is readily evident in Malawi, where 10 million people do not have access to adequate sanitation, nearly 2 million people do not have access to safe water, and over 300,000 children under the age of five die each year from diarrhoeal diseases. A safe and effective Malawi water infrastructure is the keystone to population health which, in turn promotes educational achievement and entrepreneurial opportunity, helps address issues such as gender equality and provides an important bridge from poverty to prosperity.[1]

Through the CJF Water Futures Programme, the Scottish Government has helped evaluate the sustainability of over 120,000 rural Malawian water-supplies and supported active communication between the Ministry of Agriculture, Irrigation and Water Development and other ministries, and a host of stakeholders, including industry, NGOs, researchers, and rural communities. This longitudinal approach allows research to assess sustainable development needs while government and community stakeholders develop and evaluate appropriate policy. Initial assessment primarily revealed a lack of capacity to understand and manage the complex nature of groundwater resources which is the primary drinking water supply for more than 80% of its 19 million inhabitants. Malawi's population growth rate of 3% only adds to the urgency of this need, where agricultural development, deforestation, and climate-change vulnerability require a truly integrated approach to water-resource management.

The following initiatives are representative of the work done by CJF and the Scottish and Malawian governments in this arena:

37.5.1.1 Asset management information systems and data collection

The Water Futures Programme has helped the Malawi government collect and access data to improve their decisions as they develop their water policy. New digital tools developed by CJF for mWater now provide real-time access to management information for water resources and supplies across all of Malawi. Malawi staff have collected water infrastructure data at surface water, groundwater, gravity-fed rural and peri-urban water points, and targeted waste and sanitation infrastructure. CJF is near completion of the first National Dataset for Rural Water Supplies in Malawi which will include information about which sanitation facilities and solid waste sites are co-located with water supply points, posing a potential risk of contamination (Addison *et al.*, 2020; Kalin *et al.*, 2019; Kelly *et al.*, 2019; Rivett *et al.*, 2019; Truslove *et al.*, 2020).

37.5.1.2 Capacity building and training

The Programme has delivered training in groundwater resources and rural infrastructure evaluation and management to over 400 government staff across all 28 districts, the 3 regional offices, and at the national Ministry of Agriculture, Irrigation and Water Development. Training has covered such technical

[1]For further reading, see in this volume, Alexander and Cordova, "Alleviating Poverty through Sustainable Industrial Water Use: A Watersheds Perspective."

Figure 37.4 Hydro Nation activities in Malawi include capacity building, as in this example of training on borehole installation and maintenance. (*Credit: CJF Water Futures Programme*)

skills as drilling oversight, hydrogeology, and data collection and analysis. While most training has taken place in Malawi, 38 staff members travelled to Scotland for additional training in team-building and Integrated Water Resource Management (Figure 37.4).

37.5.1.3 Research

Science-based policy must underpin the attainment of SDG6 targets in Malawi. Collaboration between Scotland and Malawi has to date produced 70 co-authored research reports and an array of peer reviewed publications in support of sustainable water management decisions. Topics include borehole forensics, drilling and infrastructure training, water supply contract management, and automated indication and evaluation of facilities in conformance with the UN Joint Monitoring Programme for Water Supply, Sanitation and Hygiene (JMP). In addition, Scotland has provided support for the Government of Malawi Isotope Hydrology facility and Scottish PhD students have worked closely with Malawian partner organisations on fundamental research supporting policy reform.

37.5.1.4 Policy exchange and support

Since the inception of the programme, Scottish and Malawian professionals have shared 'best practices' for sustainable long-term management of water resources. The first policy exchange visit was in 2012, and subsequent annual exchanges have included representatives of District (local) and regional senior staff, who share experiences with a range of Scottish agencies and organizations. In addition to meetings with the Scottish Environmental Protection Agency and Parliament, Malawian representatives have worked closely with various Scottish Water agencies and have been embedded within the international

Organisation for Economic Co-operation and Development (OECD) review of the Scottish Water Industry. These activities supported the establishment in 2018 of the Malawi National Water Resources Authority.

37.5.1.5 Regulatory engagement

Another aspect of this Hydro Nation International initiative is the collaboration between the Scottish Environment Protection Agency (SEPA) and Malawi's new National Water Resources Authority (NWRA), whose mission is to help ensure adequate and sustainable water supplies, prevent pollution of the water environment, manage water catchments, and manage flood risk. Working 'regulator-to-regulator,' the Malawi Scotland Regulatory Partnership has provided regulatory knowledge, advice and guidance to help establish a framework for environmental regulation in Malawi. The Phase 1 goal of this collaboration was to produce a roadmap for operationalisation of the new Authority. Now in Phase 2, this partnership looks to build upon that roadmap in collaboration with the newly formed Malawi Environmental Protection Agency (MEPA) and other stakeholders.

While the scope of these partnerships has related to a range of water uses, all contribute to building Malawi's capacity to encourage and support more sustainable use of water by industry. Strong governance, including the ability to collect and analyse data and create and enforce appropriate policies, is essential for progress in this area.

37.6 SCOTLAND-TANZANIA WATER STEWARDSHIP

Alliance for Water Stewardship (AWS), a Scottish-based NGO, has developed the AWS Standard, a globally applicable framework that helps major water users work collaboratively and transparently to achieve sustainable water management within a catchment context. The Scottish Government Hydro Nation programme supports the annual **AWS Global Water Stewardship Forum** and has also provided funding support for AWS project work in Tanzania.

A new state-of-the-art brewery in Moshi, commissioned in 2012 by Serengeti Brewery Ltd (SBL), has a production capacity of 50 million litres expandable to 80 million litres (Diageo, 2020). When AWS Standard analysis of the new site revealed significant water and climate-related risks to the brewery's barley supply, Scotland's Water Witness International (WWI) stepped up to offer assistance. With funding from Diageo, Scotland's CJF, and the German International Water Stewardship Programme, WWI and Serengeti Breweries Ltd formed a collaboration with Tanzania's Ministry of Water and Irrigation, Ministry of Agriculture, and the District Governments of Arusha, Hanang, and Siha; the Nelson Mandela Institute of Science and Technology; and various NGOs (Shahidi wa Maji Trias, and the Selian Agricultural Research Institute). This group set out to:

- characterize, understand, and address the water and climate-related risks and opportunities facing the barley supply-chain;
- demonstrate the benefits of water and climate risk mitigation in the barley supply chain; and
- develop an approach to reinforcing the supply chain that could serve as a model for others concerned with sustainable supply-chains, corporate engagement on water, and smallholder resilience.

SBL obtains barley from small farmers and other suppliers who face a range of water-related risks, including erratic rainfall, flood and drought events, pollution and catchment degradation, regulatory non-compliance, water conflict, and lack of sanitation and water-related infrastructure. The primary ingredient in beer, barley is highly vulnerable to erratic rainfall, and its lack of availability in dry years presents a very significant business risk to SBL (Figure 37.5).

Figure 37.5 A Tanzanian farmer receives training in conservation agriculture through a programme created and funded by Serengeti Brewery (Diageo) and a number of Tanzanian and international partners, including Scotland's Climate Justice Fund. (*Credit: Water Witness International*)

Analysis of the root causes of these challenges supported the development of six modules designed to improve the water security and reduce risks for small-to-medium sized farm enterprises. Support packages were designed, implemented, and evaluated critical topics that included the following:

(1) conservation agriculture;
(2) climate resilient agronomy;
(3) rights, obligations and empowerment;
(4) weather indexed insurance;
(5) entrepreneurship and financial risk management; and
(6) water supply and sanitation.

Each module contained training, joint analysis and planning, and intervention pilots. Almost a thousand training days were provided to farmers across Tanzania. Key lessons and recommendations from the project can be found on https://waterwitness.org/

37.7 LOOKING TO THE FUTURE

Scotland is making great strides working with industry to understand its needs and capabilities. Opening the industrial, commercial, and institutional water market to retail competition was an important first step, but much remains to be done. State-of-the-art development centres on Scottish Water drinking water and wastewater asset sites are helping to drive forward the introduction of new water and energy reduction technologies to be used at scale by utilities and large industrial water users.

As Scotland moves ever closer to its goal of sustainable water use, it will continue to share its expertise with countries throughout the world.

REFERENCES

Addison M. J., Rivett M. O., Robinson H., Fraser A., Miller A. M., Phiri P., Mleta P. and Kalin R. M. (2020). Fluoride occurrence in the lower East African Rift System, Southern Malawi. *Science of the Total Environment*, **712**, 136260. https://doi.org/10.1016/j.scitotenv.2019.136260 [Open access] (accessed 6 May 2020).

Audit Scotland. (2005). Overview of the water industry in Scotland. https://www.audit-scotland.gov.uk/report/overview-of-the-water-industry-in-scotland (accessed 15 April 2020).

Diageo. (2020). East Africa Brewery Limited website. https://www.eabl.com/en/our-business/our-companies/serengeti-breweries-limited-sbl (accessed 29 April 2020).

Kalin R. M., Mwanamveka J., Coulson A. B., Robertson D. J. C., Clark H., Rathjen J. and Rivett M. O. (2019). Stranded Assets as a key concept to guide investment strategies for Sustainable Development Goal 6. *Water*, **11**, 702. https://doi.org/10.3390/w11040702 [Open access] (accessed 6 May 2020).

Kelly L., Kalin R. M., Bertram D., Kanjaye M., Nkhata M. and Sibande H. (2019). Quantification of temporal variations in base flow index using sporadic river data: application to the Bua Catchment, Malawi. *Water*, **11**(5), 901. https://www.mdpi.com/2073-4441/11/5/901 [Open access] (accessed 6 May 2020).

Rivett M. O., Budimir L., Mannix N., Miller A. V. M., Addison M. J., Moyo P., Wanangwa G. J., Phiri O. L., Songola C. E., Nhlema M., Thomas M. A. S., Polmanteer R. T., Borge A. and Kalin R. M. (2019). Responding to salinity in a rural African alluvial valley aquifer system: to boldly go beyond the world of hand-pumped groundwater supply? *Science of the Total Environment*, **653**, 1005–1024. https://doi.org/10.1016/j.scitotenv.2018.10.337 Strathprints repository version available: https://strathprints.strath.ac.uk/65888/

Scottish Executive. (2005). Choosing Our Future: Scotland's Sustainable Development Strategy. Online at http://www.scotland.gov.uk/Resource/Doc/47121/0020703.pdf (accessed 21 April 2020).

Scottish Government. (2012) Scotland the Hydro Nation: prospectus and proposals for legislation consultation. Online at https://www.gov.scot/publications/scotland-hydro-nation-prospectus-proposals-legislation-consultation/ (accessed 21 April 2020).

Scottish Government. (2013). Water Resources (Scotland) Act, 2013 asp (5). Available online through Legislation.gov.uk at http://www.legislation.gov.uk/asp/2013/5/enacted (accessed 21 April 2020).

Scottish Government. (2019). Scotland: The Hydro Nation Annual Report 2019. Available online at https://www.gov.scot/publications/scotland-hydro-nation-annual-report-2019/ (accessed 21 April 2020).

Scottish Water. (2019). Annual Report 2018/19. https://www.scottishwater.co.uk/help-and-resources/document-hub/key-publications/annual-reports (accessed 19 October 2020).

Truslove J. P., Coulson A. B., Nhlema M., Mbalame E. and Kalin R. M. (2020). Reflecting SDG 6.1 in rural water supply tariffs: considering 'affordability' versus 'operations and maintenance costs' in Malawi. *Sustainability*, **12**, 744. https://www.mdpi.com/2071-1050/12/2/744 (accessed 5 October 2020).

World Heritage Centre. (2020). 'Malawi Slave Routes and Dr. David Livingstone Trail' Tentative List of World Heritage Site Designations. https://whc.unesco.org/en/tentativelists/5603/ (accessed 5 October 2020).

Chapter 38

Biological wastewater treatment as an opportunity for energy and resource recovery

Maria Concetta Tomei and Domenica Mosca Angelucci

Water Research Institute, C.N.R., Via Salaria km 29.300, CP 10, 00015 Monterotondo Stazione (Rome), Italy

Keywords: energy recovery, high-rate anaerobic bioreactors, microbial fuel cells, resource recovery, two-phase partitioning bioreactors, wastewater treatment technologies

38.1 BIO-BASED TECHNOLOGIES FOR ENERGY AND RESOURCE RECOVERY

As raw materials become increasingly scarce, the current linear system of industrial manufacturing – extracting resources, transforming them into products, and discarding the waste – increasingly appears to be inefficient and unsustainable. This approach is now being succeeded by 'the circular economy', a new manufacturing paradigm which closes production cycles by recovering wastes and reusing them as resources. A prime example is the treatment of wastewater, which aside from reuse as a resource of its own (water), is now being treated to recover the chemicals and energy it contains. Figure 38.1 illustrates the potential pathways by which chemicals, metals, nutrients, thermal energy, biofuels, and the water itself may be recovered from industrial wastewater for reuse (e.g., agricultural reuse).

This new circular approach to the reuse of industrial wastewater has given rise to the development of new technologies, especially biological wastewater treatment methods. These tend to be more sustainable and 'eco-friendly' than either physical or chemical methods, in that they use less energy and produce fewer toxic by-products. On the other hand, biological treatment methods can be limited by the high toxicity of industrial wastewater, which is less biodegradable than domestic wastewater. This chapter presents some new and innovative methods currently being developed and applied to biologically recover energy and materials from industrial wastewater.

DOI: 10.2166/9781789060676_0407

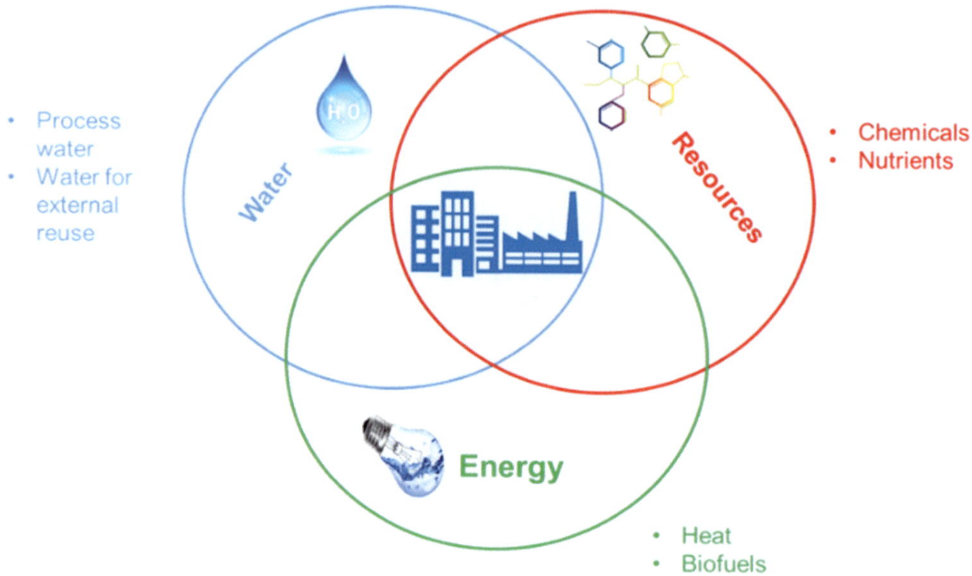

Figure 38.1 Conceptual diagram of recovery of energy and resources from industrial wastewater.

38.2 ENERGY PRODUCTION

38.2.1 Production of methane: anaerobic digestion

The most common way to recover energy from wastewater biologically is by producing methane (CH_4) in biogas as a by-product of anaerobic treatment. Engineered anaerobic systems have been in use for literally hundreds of years. Early designs, applied to organic waste treatment, involved holding waste for relatively long periods of time (on the order of weeks or months) to allow slow-growing anaerobic bacteria to digest organic waste, releasing carbon dioxide and methane in the process. This process is still in use today, for example in nonaerated wastewater treatment and sludge storage lagoons. However, the efficiency of these systems is limited at ambient temperatures, and when feed to the treatment system is diluted.

The design of industrial wastewater treatment is challenging due to the intrinsic characteristics of industrial streams, such as high chemical oxygen demand (COD), extreme pH, and salt content (Massara *et al.*, 2017). Anaerobic systems have been modified to treat various types of industrial wastewater. Enhancements of anaerobic treatment performance are achieved by retaining and concentrating the bacteria in the bioreactor, so that influent comes into contact with a greater amount of biomass, and by increasing the recirculation and settling of the biomass. These high rate anaerobic bioreactors are able to operate at higher efficiencies, because the contact time between the substrate and the biomass (solid retention time or SRT) can be much longer than the residence time of the influent wastewater in the bioreactor (hydraulic retention time or HRT).

Additional improvements have been made by further modifying the equipment and operating strategies. For example, by increasing the concentration of biomass with a filtration unit consisting of a permeable membrane, the anaerobic membrane bioreactor (AnMBR) is able to obtain much higher gas production rates than conventional anaerobic digesters relying on gravity settling. Moreover, the almost complete retention of the biomass allows the production of high quality effluents suitable for reuse. The up-flow anaerobic sludge blanket reactor (UASB) introduces wastewater through the bottom of the tank, allowing

Table 38.1 Methane (CH_4) production rates in different anaerobic treatment systems.

Anaerobic System/Industrial Wastewater	Chemical Oxygen Demand (g/L)	Organic Loading Rate (kg_{COD}/m^3 d)	CH_4 in the Biogas (%)	CH_4 Production Rate, $^a(L_{CH4}/d)$, $^b(L_{CH4}/g_{COD\ removed})$	Scale	Ref.*
Anaerobic membrane bioreactor						
Distillery	22.6	2.06	55	1.54^a	Lab	(1)
Brewery	80–90	28.5	65–80	280^b	Pilot	(1)
Pharmaceutical	4.3	0.025	60	$0.5–0.6^b$	Pilot	(2)
Bamboo	17.2	6		$10.3–13.2^b$	Lab	(2)
Expanded granular sludge bed						
Coal gasification	2.3–2.5	0.63	–	0.27^a	Pilot	(1)
Inverse fluidized bed reactor						
Dairy	1	0.5	–	0.24^a	Lab	(1)
Pulp and paper	1–8	20	–	$0.247–0.283^b$	Lab	(1)
Up-flow anaerobic sludge blanket						
Palm oil mill	95	0.0175	50	7^a	Lab	(3)
Sunflower oil	15.3	1.6–7.8	85	47.6^a	Pilot	(3)

*(1) Massara *et al.* (2017); (2) Musa *et al.* (2018); (3) Ahmad and Ghufran (2019).

it to flow through a thick layer of sludge suspended towards the top, thus maximizing the contact substrate-biomass. The UASB bioreactor also acts as a settling unit; depending on the flux conditions, it can promote biomass granulation. Effective contact between substrate and biomass is also achieved in expanded granular sludge bed (EGSB) reactors with the influent wastewater flowing through the sludge layer at a fast rate, causing it to expand. In the inverse fluidized bed reactor (IFBR), cells are immobilized on low-density particles, to achieve effective biomass retention and high SRTs. Particles are fluidized downwards, that is, the wastewater flows towards the bottom of the tank. Each of these modifications offers advantages that might be adapted to any given industrial setting. Data for biogas production reported in several recent review papers is summarized in Table 38.1 below, which shows energy production rates (in terms of methane production rate) as high as 280 L_{CH4}/g COD removed (for brewery wastewater).

38.2.2 Production of electricity: microbial fuel cells

Another important innovation is the development of the microbial fuel cells (MFCs), which have now produced electricity directly through biological treatment of organic wastewater at both research and pilot scales (Mo & Zhang, 2013). The substitution of a biological catalytic redox reaction for one of the poles in a classic abiotic electrochemical cell allows current to be generated (Santoro *et al.*, 2017). Although MFCs have been widely studied over the last 15 years, applications of these systems are only at a pilot scale for wastewater treatment so far (Mo & Zhang, 2013). Technology limitations of MFCs currently being investigated include high energy loss during the generation, low organic utilization rates, and high capital costs (around 800 times of an anaerobic system). As shown in Table 38.2 below,

Table 38.2 Microbial fuel cells for the treatment of wastewater from various industrial sources.

Anode/Cathode	Industrial Wastewater	COD (g/L)	Working Volume (L)	P_{dmax}, [a](mW/m^2), [b](mW/cm^3)	Ref.*
Single-chambered fuel cells					
Graphite felt	Coal tar	2.01	0.6	4.5[a]	(1)
Graphite fiber-brush/carbon cloth	Coking	3.20	0.3	538[a]	(1)
Graphite fiber-brush anode	Paper recycling	1.46	0.3	501[a]	(2)
Toray carbon/carbon cloth	Swine	8.32	0.03	182–261[a]	(2)
Single-chambered fuel cells with air cathode					
Proofed carbon cloths	Brewery	2.24	0.03	205[a]	(2)
Graphite plates	Pharmaceutical	7.98	0.43	205.6[a]	(1)
Single-chambered fuel cells with granular activated carbon (GAC)					
GAC bed	Textile	2.20	2.5	0.008[b]	(1)
Two-chambered fuel cells					
Graphite plates	Dairy	53.22	2	621[a]	(1)
Carbon graphite	Palm oil mill	2.68	0.1	451.3[a]	(1)
Two-chambered fuel cells with granular graphite					
Carbon rod/graphite flake	Refinery	0.25	0.4	330.4[b]	(1)

*(1) Pandey *et al.* (2016); (2) Pant *et al.* (2010).

maximum power density of different types of MFCs varies widely between of 4.5 and 621 mW/cm^2 depending on their configuration. A threshold value for feasible industrial application of MFCs for energy recovery from organic matter, suggested by Pham *et al.* (2009), is 1.0 kW/m^3 evaluated for reactor volumes >1 L.

38.3 RESOURCE RECOVERY

38.3.1 Membrane treatment methods

Membranes are incorporated into the industrial wastewater treatment process, to provide a 'double optimization', resulting not only in effective removal of toxic pollutants but also in recovery of reusable resources such as heavy metals, salts, and valuable chemicals. As shown in Table 38.3 below, a broad range of recovery strategies has been developed, based on products to be recovered. It should be noted that results for technologies currently under development reflect treatment of synthetic wastewater at lab scale. However, in a recent study (Lu *et al.*, 2019), a pilot membrane bioreactor (working volume 200 L) was successfully used to treat actual brewery wastewater. Biomass consisted of photosynthetic bacteria able to convert substrates into bioproducts used in the agriculture, cosmetic, and health industries, including proteins, amino acids, carotenoids, and coenzyme Q. These valuable resources are directly recovered from the bacteria, avoiding additional operation of sludge treatment and disposal.

Table 38.3 Resource recovery from various industrial wastewaters.

Industrial Wastewater	Resource Recovered	Technology	Reuse	Recovery (%)	Ref.*
Meat processing	Nutrients	Anaerobic membrane bioreactor	N,P	N: 90, P: 74	(1)
Biorefinery	Chemicals	Electrolytic membrane bioreactor	Acetate	96	(2)
Brewery	Chemicals	Photosynthetic membrane bioreactor	Bio-products	0.25–42	(3)
Tannery	Metals	Two-phase partitioning bioreactor	Cr(VI)	100	(4)
Metallurgical process	Metals	Microbial fuel cell, microbial electrolytic cell	Co	100	(5)
Olive processing	Salts	Sequencing batch reactor with ultra- and nano-filtration	Brine	98	(6)
Hypersaline	Salts	Two-phase partitioning bioreactor	Brine	100	(7)

*(1) Jensen (2015); (2) Andersen *et al.* (2014); (3) Lu *et al.* (2019); (4) Mosca Angelucci *et al.* (2017); (5) Huang *et al.* (2014); (6) Ferrer-Polonio *et al.* (2017); (7) Tomei *et al.* (2018).

38.3.2 Microbial electrolysis cells

Microbial electrolysis cells (MECs) require an external source of electricity to produce valuable chemicals (Mo & Zhang, 2013). MEC technology uses microorganisms as a catalyst and the oxidation reaction occurs at the anode with the production of CO_2, electrons, and protons. The electrons then travel through an external circuit to the cathode and combine with the free protons in solution to produce H_2, while CO_2 and protons react to form methane and H_2O. The flux of electrons is ensured by the external power supply (Zou *et al.*, 2017).

Applications of this technology normally produce H_2, CH_4, or other chemicals such as formic acid and hydrogen peroxide (Hua *et al.*, 2019). However, MEC technology is still in its infancy and must overcome serious challenges before practical large-scale applications can be achieved, including reactor configuration optimization, optimization of degradation conditions, and screening dominant strains with good electricity production (Hua *et al.*, 2019).

38.3.3 Case study: two-phase partitioning bioreactors (TPPBs)

One particular challenge of recovering resources from industrial wastewater with bio-based technologies is the toxicity related to these streams: indeed the bacteria able to remove biodegradable pollutants are often sensitive to toxic and/or high concentration of chemicals. In order to overcome this limitation, a unique technology has been developed that concentrates the toxic substrates in a non-aqueous phase (termed partitioning phase, a solvent or solid) and gradually delivers them to the bacteria in the aqueous phase at a tolerable concentration. Release is driven by the bacterial metabolic demand.

These two-phase partitioning bioreactors (TPPBs) have been successfully applied for the treatment of aqueous contaminants (Tomei *et al.*, 2011) and for soil bioremediation (Tomei *et al.*, 2015) with

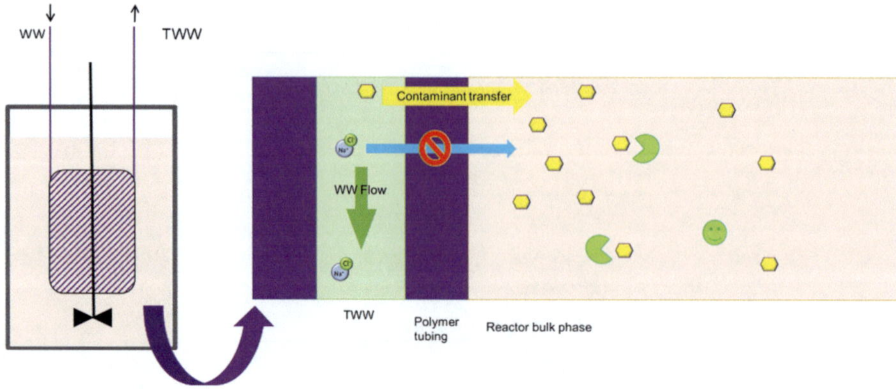

Figure 38.2 Schematic representation and principle of operation of a tubing-TPPB (WW = wastewater; TWW = Treated wastewater).

thermoplastic granular polymers acting as the partitioning phase. These polymeric beads can be tailored to maximize the uptake of specific contaminants to enhance their biodegradation.

38.3.3.1 Tubing TPPBs: Principle of operation

A recent modification to TPPBs has been the development of tubing TPPBs operated as extractive membrane bioreactors (Livingston *et al.*, 1998), by substituting the silicon rubber tubing of their original configuration with polymeric tubing. This technology combines the advantages of extractive membrane systems with the best features of conventional TPPBs because the polymeric composition and affinity can be specifically tailored for the organic contaminant to be removed, thus extending the system applicability to a wider spectrum of compounds. By pumping the influent wastewater through the tube-shaped semipermeable polymer, hazardous organic compounds susceptible to biodegradation can be separated from inorganic (or other ionic) compounds that would interfere with biological treatment, while valuable inorganic materials can be recovered from the effluent. The principle of operation is illustrated in Figure 38.2, below, which shows how biodegradable toxic organic pollutants in the untreated industrial wastewater diffuses through the tubing walls to the biomass on the bioreactor side, where biodegradation occurs. In this case phenols, represented as yellow-filled symbols, flow through the tubing walls while the recoverable constituents in ionic form (such as metals and salts) are left behind in the liquid flowing inside the tubing. The tubing TPPB works in a continuous mode, providing complete separation between the wastewater flowing inside the tubing and the bulk solution containing the biomass, thus protecting the cell-containing aqueous phase from the toxic components present in the industrial stream. Since the biological environment is not affected by the toxic characteristics of the wastewater, tubing TPPB systems are able to treat a variety of 'hostile' industrial wastewater, including high salinity, extreme pH, as well as certain toxic inorganic and organic contaminants.

38.3.3.2 Tubing-TPPB: Applications

Tubing-TPPB systems have been successfully used to treat wastewater from several industrial sources, including phenolic wastewater (Tomei *et al.*, 2016) and fracking fluids (Mullins and Daugulis, 2019). They have also been used for tannery (Mosca Angelucci *et al.*, 2017) and hypersaline wastewater (Tomei *et al.*, 2017, 2018) where tubing bioreactors successfully recovered materials from organic contaminants that were separated from inorganic components. These results are summarized in

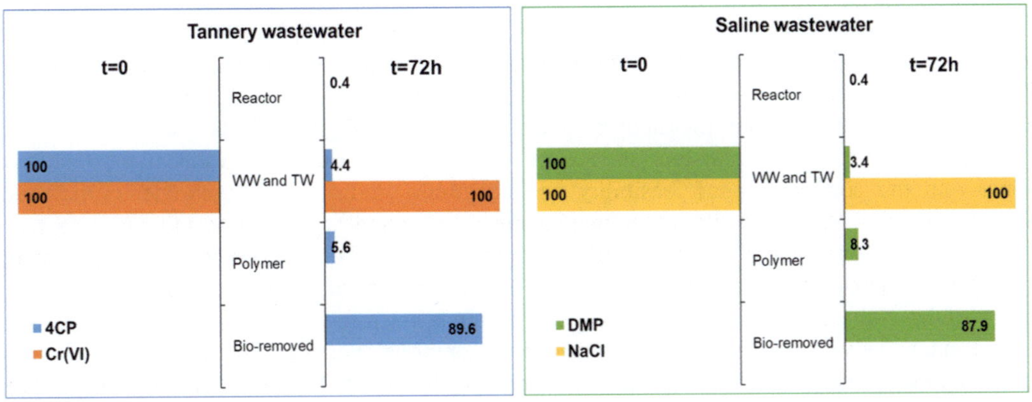

Figure 38.3 Percentage distribution of investigated compounds among phases in the tubing-TPPB.

Figure 38.3, which shows the percentage of 4CP (4-chlorophenol) removed from chromic tannery wastewater, and DMP (2, 4-dimethylphenol) removed from a hypersaline wastewater and the distribution of the contaminants resulting from the mass balance. In both cases, removal efficiencies of target organics were greater than 99%, and biodegradation efficiencies ranged between 88% and 90%. All the CrVI or NaCl was retained either in the wastewater or in the tubing, which absorbed less than 6% and 9% (of the fed amounts) of chrome and salt, respectively for tannery and saline wastewater.

One important advantage of the tubing TPPB process is that inorganics can be recovered without additional pre-treatment. This feature makes tubing-TPPB systems competitive with other resource recovery-aimed technologies as they can be implemented at relatively low cost, simply by adding tubing TPPB modules to existing suspended biomass reactors. A future development will involve the application of anaerobic bacteria to recover energy as well as resources.

The present level of application of TPPBs is at lab-scale and current research is focused on scaling up the technology.

Figure 38.4 Energy and resource recovery concept in industry.

38.4 CONCLUSIONS

To become more sustainable, industrial water management strategies should address the environmental, economic, and social aspects of water use throughout the planning and design process. This is most effectively done by following the 'hierarchy of reuse' illustrated in Figure 38.4 below.

Research activity and scaling testing are needed to achieve the implementation of more sustainable solutions, with the combination of feasibility and economic studies and stakeholder participation in planning and design. A sustainable water management integrated with energy and resource recovery should: (1) consider water and resource supply and demand, (2) enable an efficient recovery of resources by analyzing and maximizing the purity grade, and (3) take account of country regulations.

Review of the available technologies for recovery and reuse in wastewater treatment showed that some of them were extensively investigated and their feasibility demonstrated and this is the case of anaerobic treatment of concentrated streams for energy recovery or technologies for metals and nutrient recovery.

Further work in the development of new technologies like tubing TPPB is required and will be of particular interest to industries in search of pharmaceutical and nutraceutical resources as well as those required to treat toxic or other marginally biodegradable wastewater. By implementing industrial wastewater treatment processes that provide both energy and resource recovery, companies can not only reduce their water demand and recover useful chemicals, they can also more readily meet local and national regulations.

REFERENCES

Ahmad A. and Ghufran R. (2019). Review on industrial wastewater energy sources and carbon emission reduction: towards a clean production. *International Journal of Sustainable Engineering*, **12**(1), 47–57.

Andersen S. J., Hennebel T., Gildemyn S., Coma M., Desloover J., Berton J., Tsukamoto J., Stevens C. and Rabaey K. (2014). Electrolytic membrane extraction enables production of fine chemicals from biorefinery sidestreams. *Environmental Science & Technology*, **48**, 7135–7142.

Ferrer-Polonio E., Carbonell-Alcaina C., Mendoza-Roca J. A., Iborra-Clar A., Alvarez-Blanco S., Bes-Pia A. and Pastor-Alcaniz L. (2017). Brine recovery from hypersaline wastewaters from table olive processing by combination of biological treatment and membrane technologies. *Journal of Cleaner Production*, **142**, 1377–1386.

Hua T., Li S., Li F., Zhou Q. and Ondon B. S. (2019). Microbial electrolysis cell as an emerging versatile technology: a review on its potential application, advance and challenge. *Journal of Chemical Technology & Biotechnology*, **94**, 1697–1711.

Huang L., Yao B., Wu D. and Quan X. (2014). Complete cobalt recovery from lithium cobalt oxide in self-driven microbial fuel cell-microbial electrolysis cell systems. *Journal of Power Sources*, **259**, 54–64.

Jensen P. (2015). Integrated Agri-Industrial Wastewater Treatment and Nutrient Recovery, Year 3. Project Report, 2013/5018. Australian Meat Processor Corporation.

Livingston A. G., Arcangeli J. P., Boam A. T., Zhang S., Marangon M. and Freitas dos Santos L. M. (1998). Extractive membrane bioreactors for detoxification of chemical industry wastes: process development. *Journal of Membrance Science*, **151**, 29–44.

Lu H., Peng M., Zhang G., Li B. and Li Y. (2019). Brewery wastewater treatment and resource recovery through long term continuous-mode operation in pilot photosynthetic bacteria membrane bioreactor. *Science of the Total Environment*, **646**, 196–205.

Massara T. M., Komesli O. T., Sozudigru O., Komesli S. and Katsou E. (2017). A mini review of the techno-environmental sustainability of biological processes for the treatment of high organic content industrial wastewater streams. *Waste and Biomass Valorization*, **8**, 1665–1678.

Mo W. and Zhang Q. (2013). Energy nutrients water nexus: Integrated resource recovery in municipal wastewater treatment plants. *Journal of Environmental Management*, **127**, 255–267.

Mosca Angelucci D., Stazi V., Daugulis A. J. and Tomei M. C. (2017). Treatment of synthetic tannery wastewater in a continuous two-phase partitioning bioreactor: biodegradation of the organic fraction and chromium separation. *Journal of Cleaner Production*, **152**, 321–329.

Mullins N. R. and Daugulis A. J. (2019). The biological treatment of synthetic fracking fluid in an extractive membrane bioreactor: selective transport and biodegradation of hydrophobic and hydrophilic contaminants. *Journal of Hazardous Materials*, **371**, 734–742.

Musa M. A., Idrus S., Man H. C. and Nik Daud N. N. (2018). Wastewater treatment and biogas recovery using anaerobic membrane bioreactors (AnMBRs): strategies and achievements. *Energies*, **11**(7), 1675.

Pandey P., Shinde V. N., Deopurker R. L., Kale S. P., Patil S. A. and Pant D. (2016). Recent advances in the use of different substrates in microbial fuel cells toward wastewater treatment and simultaneous energy recovery. *Applied Energy*, **168**, 706–722.

Pant D., Van Bogaert G., Diels L. and Vanbroekhoven K. (2010). A review of the substrates used in microbial fuel cells (MFCs) for sustainable energy production. *Bioresource Technology*, **101**, 1533–1543.

Pham T. H., Aelterman P. and Verstraete W. (2009). Bioanode performance in bioelectrochemical systems: recent improvements and prospects. *Trends Biotechnology*, **27**(3), 168–178.

Santoro C., Arbizzani C., Erable B. and Ieropoulos I. (2017). Microbial fuel cells: from fundamentals to applications. A review. *Journal of Power Sources*, **356**, 225–244.

Tomei M. C., Rita S., Mosca Angelucci D., Annesini M. C. and Daugulis A. J. (2011). Treatment of substituted phenol mixtures in single phase and two-phase solid-liquid partitioning bioreactors. *Journal of Hazardous Materials*, **191**, 190–195.

Tomei M. C., Mosca Angelucci D., Ademollo N. and Daugulis A. J. (2015). Rapid andeffective decontamination of chlorophenol-contaminated soils by sorption ontocommercial polymers and process modelling. *Journal of Environmental Management*, **150**, 81–91.

Tomei M. C., Mosca Angelucci D. and Daugulis A. J. (2016). Towards a continuous two-phase partitioning bioreactor for xenobiotic removal. *Journal of Hazardous Materials*, **317**, 403–415.

Tomei M. C., Mosca Angelucci D., Stazi V. and Daugulis A. J. (2017). On the applicability of a hybrid bioreactor operated with polymeric tubing for the biological treatment of saline wastewater. *Sciences of the Total Environment*, **599–600**, 1056–1063.

Tomei M. C., Stazi V. and Mosca Angelucci D. (2018). Biological treatment of hypersaline wastewater in a continuous two-phase partitioning bioreactor: analysis of the response to step, ramp and impulse loadings and applicability evaluation. *Journal of Cleaner Production*, **191**, 67–77.

Zou S., Qin M., Moreau Y. and He Z. (2017). Nutrient-energy-water recovery from synthetic sidestream centrate using a microbial electrolysis cell – forward osmosis hybrid system. *Journal of Cleaner Production*, **154**, 16–25.

Chapter 39

Extreme industrial effluents: Opportunities for reuse

Henri Spanjers

Delft University of Technology, Civil Engineering and Geosciences, Delft, the Netherlands

Keywords: biological treatment, high temperature, industrial water reuse, physiochemical treatment, saline effluents, toxic effluents, used process water, water reuse

39.1 PURPOSE AND STRUCTURE OF THIS SECTION

Many industries produce effluents that exhibit extreme properties: high salinity, toxicity, and concentrated inhibitive or refractory contaminants. These effluents may come from a specific process in the company, or in mixed effluent (residual water or wastewater) that needs treatment before disposal in the environment. The trend to separate process effluents and recycle them within individual process operations only increases the extreme characteristics of the residual water. These extreme characteristics can limit the performance of water treatment technologies, jeopardizing further purification and reuse of water, and also opportunities for resource recovery. This chapter focuses on treatment options and reuse opportunities for extreme effluents generated in the manufacturing (secondary) industry.

39.2 INTRODUCTION

Industrial process operations include a wide range of water qualities, from ultrapure feed water to very concentrated process effluents. Moreover, the amount of water used and effluent produced vary considerably. Process water is used for many purposes, including transport, heating, steam production, cooling, cleaning, as a medium for reactions, and as a component in products. Major water consuming industries include:

- food and beverage
- pulp and paper

DOI: 10.2166/9781789060676_0417

- textile, chemical
- pharmaceutical
- oil/gas, and
- energy.

These industries not only consume large quantities of water, they also generate residual waters with a wide range of qualities. In general, industry is a large water consumer and in the chemical industry, for example, direct and indirect water costs have been estimated to reach up to 2% of total production costs. In this context, 'direct' water costs include the cost of water supplied to the facility whereas 'indirect' costs include the costs related to water treatment and transport throughout the industrial process and ultimate disposal of wastewater. Indirect costs may also include maintenance and repair cost due to corrosion, scaling, and biofouling, resulting from insufficient water treatment.

In industries with large water consumption, there is a strong incentive to increase water efficiency and reuse. Reuse of water within industrial operations entails treatment of process effluents to achieve required water quality for process feed water to the various unit operations. This, in turn, results in more concentrated residual waters and brine streams. For example, the use of membrane technology to recover clean water from a process stream by retaining all salts generates a residual saline concentrate. Treatment of residual streams to very high quality, especially streams with high concentrations of contaminants, requires knowledge and expertise of advanced treatment techniques as well as accurate, reliable process monitoring and control.

39.3 EXTREME EFFLUENTS

Extreme effluents are water streams from industrial process operations, which exhibit properties that pose challenges to (further) treatment or discharge into the environment. Extreme effluents can cause corrosion and deterioration of concrete and steel structures and equipment, and can affect the performance of chemical and biological treatment systems, which makes the recovery of resources (water, heat, salt, specific compounds) difficult. Extreme effluents are characterized by one or a combination of the following properties:

39.3.1 High variability

Industrial production processes often follow a daily, weekly, or even seasonal pattern, resulting in strong variations in amount and quality of process effluents. Biological and physicochemical treatment systems are usually designed to process relatively constant hydraulic and contaminant loads, and strong input variations adversely affect system performance. Large fluctuations in amount and quality of industrial effluents may be overcome by using equalization tanks before treatment.

39.3.2 High temperature

Many industrial processes are operated at high temperatures, for example, during distillation and cracking, to dissolve fat, to increase reaction rates or by using steam for sterilization or transport. Although technology exists to recover heat from low grade residual heat, industries often choose to waste the heat along with the process effluent, resulting in hot streams. High temperature can adversely affect the biological material in bio-treatment systems, and fouling, scaling, and corrosion are enhanced.

39.3.3 High salinity

Saline process effluents can be a result of water treatment using ion exchange technology or membrane filtration because these technologies produce saline regenerates and concentrates, respectively. Hydraulic fracturing, or fracking, where water is used to remove oil and gas from shale formation, generates saline effluents containing salts that are released from the deep rocks. Another source of saline effluents are processes where salt is used, like in dyeing textiles which results in production of saline dye bath effluents. Other industries producing saline effluents include tanneries, seafood processing, and chemical industries. Saline waters present a challenge to treatment and recovery technologies. For example, biological and chemical oxidation of organic contaminants from saline water is impeded by high salinity, it is difficult to extract pure salts from mixed brines, and the recovery of water is energy-intensive. Moreover, high salinity may lead to increased corrosion rates of equipment and structures.

39.3.4 Low or high pH

Acids and bases have many applications across industries. For example, acids are used for cleaning steel (pickling) and as a raw material for many chemical products. Bases are used in the paper pulp industry to remove lignin from wood, in the food industry to peel fruit, and for scouring of natural fibers in the textile industry. Both acids and bases are used for cleaning in place (CIP) of the interior surface of pipelines, vessels, filters, and other process equipment without dismantling, especially in the food and beverage industry and the pharmaceutical industry. In a typical CIP operation the equipment is consecutively rinsed with (amongst other solutions) caustic and acid, generating effluents with subsequently very high pH and very low pH. Generally, pH has a substantial impact on physiochemical treatment processes and on the rate of corrosion and scaling. Microorganisms cannot tolerate values below pH 4 or above 9.5, and most biological treatment processes can only operate optimally in a narrow range between pH 6.5 and pH 7.5.

39.3.5 Inhibitive or toxic compounds

Many industries use biocides to control microbial growth in equipment and products. Microbial growth affects the quality of products in the food and beverage, cosmetic, and pharmaceutical industries and must be avoided. Microbial growth (i.e., biofouling) can also affect the performance of process equipment such as membranes, cooling and heating systems, pumps, and valves. In cooling systems biocides are often used to suppress biofouling in order to reduce corrosion and the risk of development of legionella. After their use, biocides end up in the process effluents, and can even pose challenges to biological treatment systems, as these compounds suppress biological growth. Besides biocides, numerous substances used and produced in the chemical industry (including pharma, chemical, textile, etc.) exhibit toxic or inhibitive behavior in biological treatment systems affecting nitrification and the degradation of otherwise degradable organic compounds. If the substances are both toxic (or inhibitive) and biodegradable, then careful design and operation of biological treatment systems may enable successful removal of these compounds. High concentrations of salt in saline effluents or brines also suppress microbial activity in biological treatment systems.

39.3.6 Refractory compounds

Recalcitrant or refractory organic compounds are resistant to biological or physiochemical degradation, and include natural and synthetic (xenobiotic) substances. For example, in the meat processing industry effluents may be generated containing high concentrations of natural humic compounds (also termed natural organic matter, NOM), which are non-degradable. Another example is the production of industrial demineralized water from surface water, which results in a concentrate that has a high humic compound concentration. Although relatively harmless, humic substances render a brownish color to recovered water and salts, which makes them less suitable for reuse. Other naturally occurring refractory substances are found in effluents of the petroleum industry. Many industrial effluents, including pharma and chemical, contain recalcitrant synthetic organic compounds, often termed xenobiotics, that are nondegradable and sometimes toxic in biological treatment systems.

39.3.7 Corrosive compounds

Water can be more or less corrosive depending on its mineral composition. Corrosion is the disintegration of metal through chemical or electrochemical reactions between the metal and its environment. Corrosion is a serious problem in industrial water installations such as boilers and recirculating cooling systems. Corrosion rate increases with dissolved solids concentration (i.e., salinity), although some ions like carbonate and bicarbonate decrease corrosivity. Some ions are extremely aggressive to metals, particularly chloride and sulfate. Chloride can be found in many industrial process effluents, and is, in fact, the major constituent of saline waters and brines, such as those generated during regeneration of ion exchange resins and reverse osmosis filtration. Sulfate in process effluents is a result of the use of sulfuric acid that is commonly used in many industries, especially the manufacture of a wide range of chemicals and in the fermentation industry. Besides dissolved solids, gases such as carbon dioxide, hydrogen sulfide, and oxygen also contribute to the corrosivity of water. A special case is ammonia, present in for example some condensates, which is extremely corrosive to copper alloys. Besides the presence of corrosive constituents, a low pH (acidic) water also accelerates corrosion.

On the other extreme, very pure water (e.g., some condensates) can be very corrosive to most metals when small quantities of carbon dioxide or organic acids are dissolved in it. For example, in water-steam systems, if carbon dioxide in the feed water is carried over via the steam this may result in low pH in the condensate and consequently corrosive conditions. As condensates are generally very pure, they are obviously often considered for reuse. However, small amounts of contaminants can make the condensate unsuitable for direct reuse as feed water or other purpose.

The treatment and reuse of corrosive water poses a challenge to both treatment processes and process equipment. When corrosive effluents are treated for reuse or resource recovery, approaches to reduce corrosivity, include pH correction, de-aeration, addition of scavengers, and corrosion inhibitors.

39.4 MANAGING EXTREME EFFLUENTS

39.4.1 General

Extreme process effluents must be treated for reuse or (less preferably) disposal. Treatment may involve the recovery of water, constituents, heat, or a combination of these. The choice of the technology or combination of technologies depends on the quality of the effluent, the goal of resource recovery and the location of the industry. Table 39.1, below, lists a number of treatment technologies that have been successfully used to treat extreme effluents. Despite the additional capital cost, equalization tanks may be used to balance or moderate fluctuating hydraulic load, temperature, and pH in order to alleviate the impact on the treatment technologies. However, it should be noted that certain treatment and recovery technologies require concentrated effluent streams, and their performance may be negatively affected by lower concentrations.

The following examples, based on real cases, describe the treatment of three different categories of extreme process effluents, technologies, and resources.

Table 39.1 Treatment technologies for extreme process effluents.

Aerobic or anaerobic biodegradation
Chemical and electrochemical oxidation
Ion exchange
Coagulation, flocculation, flotation, sedimentation
Membrane filtration
Electrodialysis
Evaporation
Crystallization
Disinfection
Other

39.4.2 High salt – low organics

The application of zero liquid discharge (ZLD) aims at full water recovery from residual waters. Since many ZLD technologies, such as advanced membrane treatment and thermal evaporation, are energy intensive or costly, application of ZLD can be facilitated where excess waste heat is available and where shortages of (fresh) water sources or limits on discharge of effluents justify the cost of reuse. In this example ZLD strategy is applied in a full-scale petrochemical wastewater treatment plant (WWTP) to maximize the recovery of water for reuse purposes. Figure 39.1 depicts the layout of the treatment plant. After removal

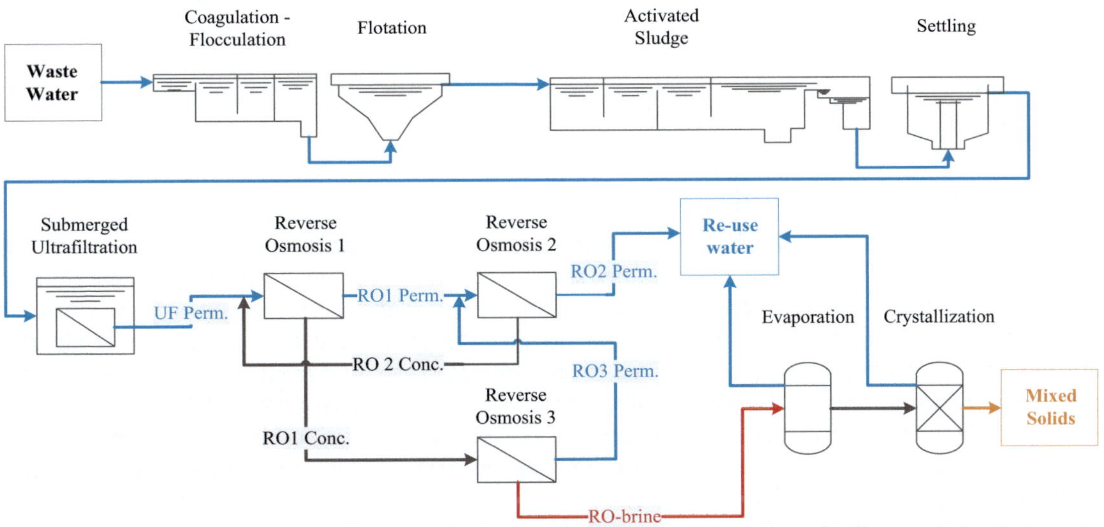

Figure 39.1 Scheme of the ZLD WWTP. The water streams are indicated in blue and black and the RO-brine stream is indicated in red. The produced mixed solids stream is indicated in brown. (van Linden *et al.*, 2019)

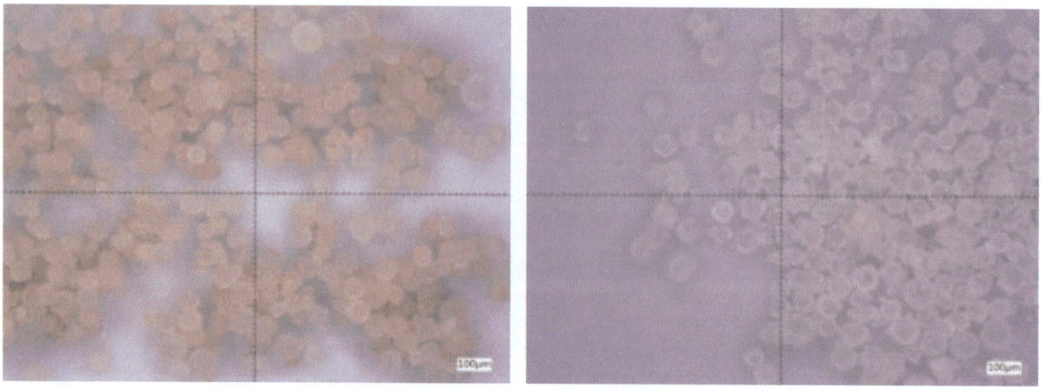

Figure 39.2 Photomicrographs (40× magnification) of the mixed solids (left) and the washed solids (right). (van Linden *et al.*, 2019)

of suspended solids and organic substances by a combination of coagulation/flocculation/flotation and biodegradation, and removal of dissolved solids by reverse osmosis (RO), clean water is obtained that is suitable for direct reuse. The concentrated brine from the RO is further treated by evaporation and crystallization to recover the remaining water, resulting in mixed solids consisting of salts and various organic substances (NOM). The salts can potentially be reused provided the level of NOM is reduced. Investigations are being carried out to test two approaches to separate the salt and NOM. One approach is the separation of NOM from the RO brine by various technologies. Examples are nanofiltration, electrodialysis, and ion exchange, which all effectively separate the salt and NOM. Note, however, that these technologies lead to new waste streams that contain the NOM. The other approach is the removal of NOM from the salts (mixed solids) using the SALEX technology, which involves washing of the salts with a saturated salt solution. This approach also generates a new waste stream, that is, the spent washing solution containing the NOM.

Figure 39.2 compares microscopic pictures of the raw mixed solids and the washed solids. It can be seen that the NOM, responsible for the brown color of the salt crystals, was effectively removed, making the salt suitable for reuse.

39.4.3 Medium salt – medium organics

Many industries generate saline effluents, or brines, as a result of various production processes. The composition of these brines, including impurities such as organic matter, may vary widely depending on the type of industrial process. Although sodium chloride is often the main component, other salts and minerals can be present and may be recovered. In this example, brine from an industrial ultrapure demineralized water production plant is treated to recover water, salt, and magnesium in two large-scale pilots in an industrial area in the Port of Rotterdam. This European Horizon H2020 project (ZERO BRINE, 2019) aims to facilitate the implementation of the circular economy in various process industries. The demin water plant takes water from a lake to produce ultrapure demineralized water for a number of companies in the industrial area. The plant generates two types of residual brine: ion exchange regenerate from the softening units and RO concentrate, which are separately treated in the two different pilots (Figure 39.3). The regenerate contains regeneration salt, that is, sodium chloride, and the hardness ions magnesium and calcium.

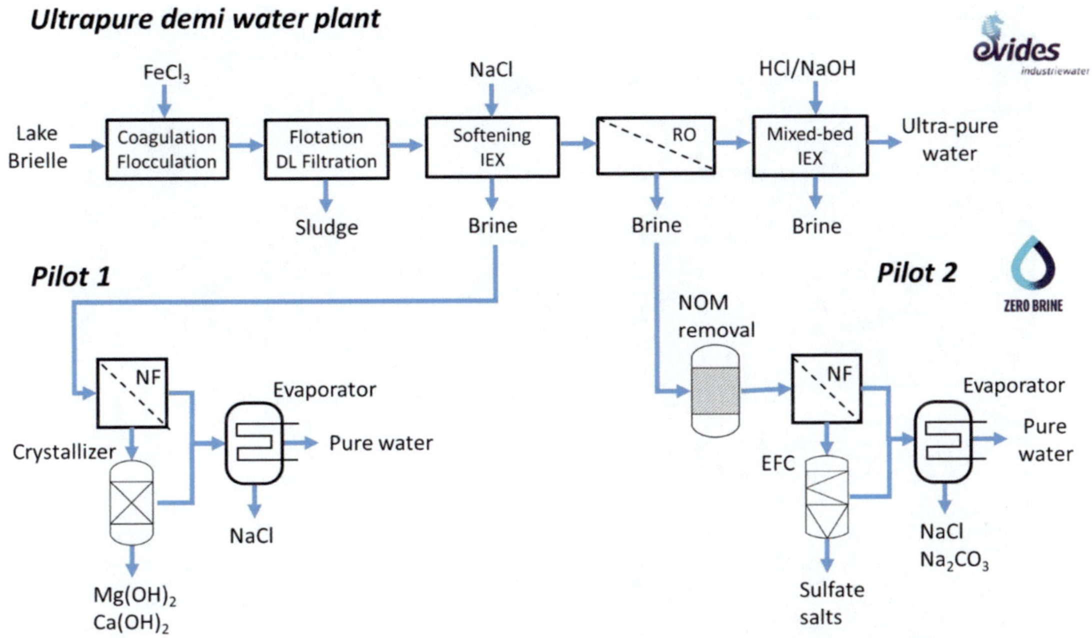

Figure 39.3 Scheme of ultrapure demin water plant generating residual brines, and pilots to recover water, salts, and minerals from the brines. NF: nano filtration, EFC: eutectic freeze crystallization, NOM: natural organic matter.

Pilot 1 consists of an ultrafiltration unit, a crystallizer, and an evaporator, and is designed to recover from the regenerate the minerals magnesium hydroxide and calcium hydroxide, the salt sodium chloride, and pure water. Recovered magnesium and calcium minerals have market value, sodium chloride can be reused to regenerate the ion exchange unit, and pure water can be reused in the demin water plant. Pilot 2 treats the RO concentrate to also recover sodium chloride and pure water, and in addition sulfate salts. For the latter, eutectic freeze crystallization (EFC) is used, and the brine is pretreated by electrochemical oxidation to remove natural organic matter to avoid organic impurities in the salts. In both pilots the evaporator is supplied with waste heat from neighboring industrial processes.

39.4.4 High organics – low/medium salt

The food and beverage, petrochemical, and textile industries, and many others generate effluents that are high in biodegradable organic matter. These effluents are typically treated using anaerobic technologies, which enable the recovery of biochemical energy in the form of biogas. However, the effluents are becoming increasingly associated with extreme conditions, such as the presence of refractory or toxic compounds and high salinity that adversely affect biomass retention or reduce biological activity in bioreactors.

Conventional industrial anaerobic reactors are based on the retention of active biomass by using granulation. Extreme conditions lead to the disintegration of granules and hamper the formation of new granules, resulting in a loss of active biomass. In membrane bioreactors, however, all the biomass is retained by using microfiltration or ultrafiltration membranes, resulting in sustained capacity to degrade

Figure 39.4 Lab-scale anaerobic membrane bioreactor treating saline phenolic wastewater under thermophilic conditions (Muñoz Sierra *et al.*, 2020).

the organic matter, even under extreme conditions. Moreover, since all solids including bacteria are retained, specialized microorganisms are allowed to develop, which are able to degrade refractory substances that might be present in the water. Last but not least, the use of membranes results in a treated effluent that is completely free of solids, which can be relatively easily further polished and reused in the company.

An ongoing research project at Delft University of Technology is developing anaerobic membrane technology for the treatment of saline phenolic industrial effluent at elevated temperatures (i.e., thermophilic conditions). Phenolic effluents, such as those generated in the coal coke industry, have been treated at lab scale (Figure 39.4). Removal efficiencies up to 96% have been achieved for various compounds including phenol, cresol, and resorcinol, at a salinity of 16 g sodium per liter and a temperature of up to 55°C. The technology is now ready for demonstration in an industrial pilot reactor; at the time of writing, a site in The Netherlands has been identified and the pilot is currently under design.

39.5 CONCLUSIONS

Many industries generate extreme effluents, and the increase in the separation of process streams and water reuse creates even more such challenging streams. Technologies are nowadays available to recover water and constituents from extreme effluents and eliminate any remaining impurities. More sustainable use of water by industry can be achieved when companies take the following actions:

- Identify new opportunities to separate process effluents and remove contaminants to recover water and useful materials.
- Avoid diluting concentrated process effluents with dilute effluents, to allow effective recovery of constituents and clean water, respectively.
- Employ proven technologies such as membrane filtration, vacuum stripping, and crystallization used in refinery processes to recover water and useful materials from process effluents.
- Redesign conventional biological water treatment technologies or combine them with new technologies to make them more resistant to extreme conditions. For example, biological treatment is often less chemical- and energy-intensive compared to physiochemical treatment, and contaminants in biological treatment effluent can often be further processed to recover water and materials.
- Use residual, low grade, heat that is often available in many companies to operate thermally driven recovery technologies.

Because of the large variety of extreme effluents across the water intensive industries, there is no one technology that fits all situations; hence any recovery and reuse project must be preceded by a pilot-scale study. However, given the range of technologies available, it is becoming increasingly feasible to make industrial water practices more sustainable by reusing water and recovering constituents from extreme effluents.

REFERENCES

Muñoz Sierra J. D., García Rea V. S., Cerqueda-García D., Spanjers H. and van Lier J. B. (2020). Anaerobic Conversion of Saline Phenol-Containing Wastewater Under Thermophilic Conditions in a Membrane Bioreactor. *Frontiers in Bioengineering and Biotechnology*, **8**, 565311, https://doi.org/10.3389/fbioe.2020.565311.

van Linden N., Shang R., Stockinger G., Heijman S. and Spanjers H. (2019). Separation of natural organic matter and sodium chloride for salt recovery purposes in zero liquid discharge. *Water Resources and Industry*, **22**, 100117.

ZERO BRINE. (2019). Re-designing the value and supply chain of water and minerals: A circular economy approach for the recovery of resources from brine generated by process industries. Project under European Union's Horizon 2020 research and innovation programme. Grant agreement No. 730390. www.zerobrine.eu (accessed 31 October 2019).

Chapter 40

Promoting sustainability in the oil industry: The benefits of using constructed wetlands for oily wastewater treatment

Alex Stefanakis

School of Environmental Engineering, Technical University of Crete, 73100 Chania, Greece

Keywords: circular water economy, constructed wetlands, greenhouse gas emissions, industrial effluent, oil and gas industry, oily wastewater, produced water, surface flow, sustainability

40.1 CONSTRUCTED WETLANDS TECHNOLOGY

Natural processes are widely applied in wastewater treatment. Most of the established technologies and methods are based on these processes (e.g., sedimentation, filtration, biological activity), but they are usually associated with complex and energy-consuming electro-mechanical equipment (Stefanakis, 2020a). The main difference between nature-based solutions and conventional technologies is the use of only natural components and processes for the treatment. Among the various natural treatment systems, such as facultative and oxidation ponds, Constructed Wetlands (CWs) is the most promising sustainable treatment method with increasing worldwide interest. Natural wetlands have been utilized as disposal sites for secondary or tertiary wastewater effluents for thousands of years (Stefanakis *et al.*, 2014). The transition from Natural to Constructed Wetlands was based on the exploitation of naturally occurring processes under a controlled environment for beneficial human and environmental usage. Thus, man-made wetlands are designed to mimic and enhance the functions and operations of natural wetlands. Their excellent treatment performance, the environmentally friendly character, and the lower overall costs placed CWs at the forefront of the scientific and market interest. In addition, they offer the benefits of design flexibility and replicability, low and simple maintenance, low operation cost, and small environmental impact, all of which are consistent with the principles of a 'circular water economy' (Stefanakis, 2019; 2020a).

DOI: 10.2166/9781789060676_0427

40.1.1 Comparison with established treatment technologies

CWs represent one of the most attractive developments in ecological engineering. Their philosophy is based on the decentralized approach in contrast to what had been regarded as the centralized conventional biological treatment methods which dominate the wastewater market. This new approach introduces new parameters and views in wastewater treatment, such as sustainability and overall environmental impact (Stefanakis, 2019). Conventional treatment systems typically include large, end-of-the-pipe facilities and extended sewer networks, which translate to respectively high investment costs not only for their construction but also for their operation. On the other hand, CWs are generally easier to build and simple to operate, while they provide a treatment process robust to flow fluctuations and pollutant concentrations (Stefanakis, 2018). Due to the use of natural materials and processes and the minimum use of electro-mechanical equipment, CWs require only a small energy input since renewable energy sources are used by plants (solar, wind energy) for the treatment processes. At the same time, maintenance needs are low, while global experience from many countries has demonstrated that operational costs of CW facilities can be reduced by more than 90% compared to conventional plants (Stefanakis *et al.*, 2018a). This also translates to a significantly reduced carbon footprint for these facilities. However, the main limitation of CW systems is that they have a larger area demand compared to conventional technologies. Moreover, although issues with odor and/or surface ponding have been reported in the past, this is usually the result of false design and improper construction (Stefanakis *et al.*, 2014).

40.2 CASE STUDY: WETLAND TREATMENT OF OILY WASTEWATER IN OMAN

40.2.1 Wastewater in oilfields

The exploration and production of oil and gas generates a polluted water volume that represents one of the largest industrial waste streams worldwide (Aditya, 2016). This water occurs not only during crude oil recovery, but also during other forms of fossil energy recovery including shale gas, oil sands, and coal bed methane (Jain *et al.*, 2017), and may include water from the reservoir, natural formation water, and water injected into the formation, along with any chemical substances used during the production and treatment processes. Large volumes of oily wastewater (also known as produced water) are generated as an oil production co-product in many countries, while its management imposes a limitation on oil production in many cases. This water stream is typically contaminated with residual hydrocarbons, salts, and various organic and inorganic compounds (e.g., emulsion breakers, chemical additives, solvents, heavy metals etc.). Due to the salt content and the hydrocarbons, oily wastewater cannot be freely discharged to the environment, since it can affect soil salinity and plant productivity and damage the human nervous system (Kim *et al.*, 2013).

40.2.1.1 Established technologies for treating oily produced water

Internationally, the most common management practice of oily wastewater is the deep well disposal (DWD) and in some cases disposal into the ocean (for offshore production activities). A smaller volume is often re-injected into reservoirs to maintain pressure for the oil wells (Arthur *et al.*, 2005). However, these practices pose a significant environmental risk, while they are operationally energy intensive. Several mechanical and chemical technologies have been tested and applied as a treatment step before the final disposal such as membrane filtration (Jain *et al.*, 2017; Munirasu *et al.*, 2016), thermal technologies (Igunnu & Chen, 2014), aerated filters (Su *et al.*, 2007), flotation (Igunnu & Chen, 2014; Saththasivam *et al.*, 2016), and electrocoagulation and electrodialysis (An *et al.*, 2017).

However, almost all these technologies are characterized by high operational and maintenance costs due to the high energy consumption, and frequent mechanical failure, which affects their performance. Mechanical treatment methods, for example, activated sludge and sequential batch reactor, are also energy intensive with high operational costs. Moreover, the implementation of such systems in remote areas in developing countries (where many oil resources are found) often results in inadequate treatment due to the high maintenance cost, lack of local expertise, and poor governance (Daee *et al.*, 2019).

40.2.1.2 Use of constructed wetlands for oily water treatment

Internationally, the technology of CWs has gained attention as a sustainable and cost-effective treatment method. CWs have the advantage of significantly decreasing the capital and, especially, operation costs compared to mechanical systems. They are well established in Europe and North America to treat a wide range of wastewater, such as domestic and municipal wastewaters (Stefanakis *et al.*, 2014, 2019; Wu *et al.*, 2014). The realization of their high treatment capacity enabled their use for the treatment of various industrial effluents (Gholipour *et al.*, 2020; Gomes *et al.*, 2018; Ramírez *et al.*, 2019; Stefanakis, 2018; Wu *et al.*, 2015).

Existing knowledge and experience indicate that CWs can provide a reliable ecological solution for the treatment of water contaminated with petroleum hydrocarbons, additives and phenols (Breuer & Grissemann, 2011; Stefanakis, 2020b; Stefanakis *et al.*, 2016). Wetland systems are particularly appropriate as a remediation technology for oil production fields in remote areas, where land availability is generally high. Currently there are only few wetland systems in various facilities such as refineries, oil and gas fields, and pumping stations (Knight *et al.*, 1999; Stefanakis *et al.*, 2016), located in the USA (Wallace *et al.*, 2011), in Sudan (Saad *et al.*, 2009) and in China (Ji *et al.*, 2007). In general, reports on CWs research and/or applications for oily produced water in the international literature are limited (Alley *et al.*, 2013). One of the largest CW systems worldwide exists in Oman for the treatment of oily produced water from a nearby oilfield (Stefanakis *et al.*, 2018a).

40.2.2 Case study

40.2.2.1 Design and construction

The Nimr oilfield is located in the southern of Oman and produces an oil water ratio of 1:10. Deep wells are used in the area for oily wastewater disposal, an activity that has a high energy demand and that poses many environmental concerns. Hence, there is always a need to reassess the disposal practices and to evaluate other potential methods for treatment and utilization. In 2008, the government oil company of Oman awarded a Design, Build-Own, Operate, and Transfer (DBOOT) contract to the company Bauer Nimr LLC to develop an oily water treatment plant (Stefanakis *et al.*, 2018a). The facility started its operation in December 2010 with a treatment capacity of 45,000 m^3/day. Nine years and three expansion phases later, the current treatment capacity has reached 175,000 m^3/day, a figure accounting for more than 65% of the total oily water generated at that oilfield (Stefanakis, 2020c).

This treatment facility consists of 490 hectares of Surface Flow Constructed Wetland (SFCW) and 780 hectares of downstream evaporation ponds (EPs) (Stefanakis, 2020c). The size of this system makes it is one of the world's largest commercial constructed wetlands (Figure 40.1). The oily wastewater is sent through a pipeline to the plant. First, separation and recovery of most of the oil content takes place in passive hydrocyclone oil separators without the use of energy or chemicals. Then, the water is distributed into

Figure 40.1 Aerial view of the Constructed Wetland facility for oily produced water treatment in Oman. (*Courtesy*: Bauer Nimr LLC)

the SFCW via a long buffer channel without the use of pumps (Stefanakis, 2020c). The water flows through the SFCW by gravity, and the treated clean water flows into the EPs, where water evaporation results in salt formation that can be processed into industrial grade salt (Stefanakis *et al.*, 2018a). The climate in the area is a typical desert climate, with average air temperature in June–July exceeding 50°C. Lower humidity values are usually observed during warmer months (April to October). It is also noticeable that practically no rainfall takes place in the area.

The SFCW beds are sealed with a bottom mineral layer made of locally available clay material in order to reduce the environmental and cost impact of High-Density Polyethylene (HDPE) liner. The HDPE liner is only used in the inlet buffer channel. The wetland beds are planted with various local reed species widely used in SFCWs worldwide such as *Phragmites australis* (common reed), *Typha domingensis*, *Schoenoplectus littoralis*, *Juncus rigidus,* and *Cyperus* spp (Stefanakis, 2020c). All plants were collected from nearby water resources and were propagated at the onsite nursery. The presence of different plant species enhances the biomass production and the resilience of the created ecosystem, which makes this wetland a polyculture (Stefanakis *et al.*, 2018a).

The inlet water is brackish with total dissolved solids concentration exceeding 7000 mg/L (Stefanakis *et al.*, 2018a). Oil in water (OiW) is the target pollutant with an average inlet concentration close to 350 mg/L (in some cases it exceeds even 500 mg/L), while the oily water is low in nutrients, that is, total nitrogen and phosphorus concentrations are lower than 4 mg/L. Each treatment element was selected to maximize the overall effectiveness of the entire system (Stefanakis *et al.*, 2018a). More than 85% of the oil is recovered at the upstream passive hydrocyclones and skimmers. The water with the residual oil hydrocarbons (up to 100 ppm) that flows with gravity to the SFCW cells is biologically degraded there, producing a treated effluent with OiW below the limit value according to the national standards (<0.5 mg OiW/L) (Stefanakis *et al.*, 2018a; 2020c).

40.2.2.2 Advantages

The CW system provides excellent polishing of the pretreated oily water, resulting in complete removal of the oil content in the final outflow. It has been found that the rhizosphere in the wetland system is rich in hydrocarbon-degrading bacteria (Abed *et al.*, 2014), resulting in high rates of oil removal. Additionally, the reed stems act as a physical filter for trapping floating oil, which is subsequently biodegraded by microorganisms growing on the surface of the reed stems, roots, and the soil surface.

Due to the operation of this wetland facility, many high-pressure deep well pumps that are used to dispose the oily wastewater into deep-lying aquifers in the area have been shut down. Considering that this facility is a gravity-based system, the energy demand for the treatment processes is close to zero. Energy is only consumed during the operation for the instrumentation (flow metering), office and accommodation facilities (water supply, air conditioning, kitchen, etc.), and the small onsite reverse osmosis system (Stefanakis *et al.*, 2018a). Compared to the deep disposal wells, the wetland facility uses only 1/50 of the energy consumed including all related infrastructure facilities, which is directly related to significant reduction of carbon emissions. The estimated reduction of carbon emissions reaches 99% compared to the other management options (Breuer & Grissemann, 2011; Stefanakis *et al.*, 2018a). It should be mentioned that it is estimated that this facility alone contributes by more than 5% to Oman's overall Intended Nationally Determined Contributions to reduce emissions by 2% (Stefanakis, 2020c).

It is also worthy to mention that the large CW and the series of evaporation ponds provide a series of ecosystem services, which in combination create a system that offers the most resilience. The whole facility is nowadays a valuable habitat for migratory and resident birds and other wildlife. The SFCW is well integrated in the local environment and accepted by the wildlife, since it provides a comfortable stopover for more than 120 migratory bird species that travel between Asia and Africa (Stefanakis *et al.*, 2018a).

40.3 BEYOND THE OILFIELD

40.3.1 Limitations and challenges

The highly energy-efficient and reliable CW system in Oman provides a free-of-oil treated effluent, while it converted a previously arid and dry desert into a massive new ecosystem and habitat. Such a solution can be adapted not only to warm and dry regions but also to colder climates. However, the space requirement limits the application range of this particular system to regions with adequate space availability, as can be found in the Middle East, in many African countries, and also in North America. It should be noted that the higher space demand is counterbalanced by the significant energy savings and the sustainable character of this nature-based solution. Additionally, as with every biology-based treatment technology, higher salinity content will not allow development and growth of the microbial ecosystem inside the wetland system and will hinder plant growth.

40.3.2 Potential for additional research

The effective and ecological treatment of oily wastewater as demonstrated in this facility in Oman indicates the wide range of additional research that is needed towards not only optimizing the process design but also to better understand and measure its environmental efficiency.

Considering the large volumes of water generated during oil exploration, the effective ecological treatment can and should be combined with further beneficial reuse of the treated effluent. Particularly under hot and arid climates, such as the desert environment, there is the potential to reuse the treated water for irrigation. This practice, however, should be further studied considering that the effluent water remains brackish. A first research study was implemented in this wetland facility in Oman, where a research irrigation field of 22 hectares was established (Stefanakis *et al.*, 2017, 2018a). Due to the relatively high salinity of the irrigation water, salt tolerant plants were tested in order to avoid the need for a costly and nonsustainable reverse osmosis step for desalination. In addition, the crops tested had a related market value, for example, as biofuel and cotton crops. Moreover, research trials should be carried out to test the use of the reed biomass produced in the wetlands for compost production and biogas generation, aiming at closing the loop of materials and waste and promote circularity in this large-scale application (Stefanakis, 2020c).

Another aspect that needs to be studied is the range of ecosystem services provided by CW systems. A first such study was implemented at this massive wetland facility in Oman on the effect of the wetland beds on the microclimate of the area (Stefanakis *et al.*, 2018b). For the first time it was revealed that the presence of the wetland beds regulated the local microclimate: a 10°C decreased temperature value was detected between the wetland body and a radius around the wetland of up to 1 km distance. This study showed the first findings clearly indicating the positive effect of the constructed wetland system on its surrounding environment. Such findings are very helpful to understand the changes in the microclimate due to the wetland's presence. Further studies are needed to determine the impact on all organisms and humans of the temperature difference between a wetland facility and the surrounding area. The reduced temperature and the increased humidity can also modify the biodiversity and affect the energy consumption needed for cooling. Moreover, determining the mean temperature differences is important to communicate to specialists in other disciplines, such as ornithology and ecology, to identify further effects on biodiversity and people living and/or working in wetland surroundings.

40.3.3 Wetland treatment as a paradigm of wastewater treatment for circular economy

Proper and sustainable wastewater management is one of the Sustainable Development Goals (SDG 6) referring to water and sanitation for all. Globally, large volumes of wastewater are still discharged into the environment without proper or with no treatment at all. A main factor for this is typically the related treatment costs and the lack of respective expertise. This is an indication that when choosing a treatment technology, cost-efficiency and simplicity should be among the main decision-making criteria along with environmental performance and reliable technical efficiency. This approach is also enriched by integrating reuse and resource recovery in wastewater management following the circularity principle. The conventional wastewater paradigm views wastewater as a 'waste', focusing only on the treatment effectiveness. Nature-based solutions such as Constructed Wetlands bring a new perspective of integrated and comprehensive wastewater management. This green technology possesses the necessary characteristics envisaged by the new concept of circular economy, that is design flexibility and replicability, low and simple maintenance, low operation cost, small environmental and carbon footprint, use of natural material and processes, among others. In addition, nature-based solutions offer a series of ecosystem services, while their design and operational flexibility allows their installation in different contexts, from the desert environment to the urban environment, as green infrastructure elements. At the same time, the options for effluent reuse, resource efficiency, and nutrient recovery integrate circular economy principles in the water sector, where every 'waste' stream is now considered as a new raw material for new products. Facilities like this large treatment wetland in Oman demonstrate in an emphatic way the technical feasibility, the sustainable character, and circularity options of nature-based solutions under diverse environments and show the way for the transition to a circular water economy.

REFERENCES

Abed R. M. M., Al-Kharusi S., Prigent S. and Headley T. (2014). Diversity, distribution and hydrocarbon biodegradation capabilities of microbial communities in oil-contaminated cyanobacterial mats from a constructed wetland. *PLoS ONE*, **9**(12), e114570.

Aditya R. (2016). Produced water treatment market by application (onshore & offshore), by treatment types (physical, chemical, membrane and others), & by geography. Global Trends & Forecast to 2019.

Alley B. L., Willis B., Rodgers J., Jr and Castle J. W. (2013). Water depths and treatment performance of pilot-scale free water surface constructed wetland treatment systems for simulated fresh oilfield produced water. *Ecological Engineering*, **61**(A), 190–199.

An C., Huang G., Yao Y. and Zhao S. (2017) Emerging usage of electrocoagulation technology for oil removal from wastewater: a review. *Science of the Total Environment*, **579**, 537–556.

Arthur J., Langhus B. and Patel C. (2005). Technical Summary of Oil & Gas Produced Water Treatment Technologies. ALL Consulting, LLC, Tulsa, Oklahoma, USA.

Breuer R. and Grissemann E. (2011). Produced water treatment using wetlands – reducing the environmental impact of oilfield operations. *Paper presented at the SPE European Health, Safety and Environmental Conference in Oil and Gas Exploration and Production*, Vienna, Austria, 22–24 February.

Daee M., Gholipour A. and Stefanakis A. I. (2019). Performance of pilot Horizontal Roughing Filter as polishing stage of waste stabilization ponds in developing regions and modelling verification. *Ecological Engineering*, **138**, 8–18.

Gholipour A., Zahabi H. and Stefanakis A. I. (2020). A novel pilot and full-scale constructed wetland study for glass industry wastewater treatment. *Chemosphere*, **247**, 125966.

Gomes A.C., Silva L., Albuquerque A., Simões R. and Stefanakis A. I. (2018). Investigation of lab-scale horizontal subsurface flow constructed wetlands treating industrial cork boiling wastewater. *Chemosphere*, **207**, 430–439.

Igunnu E. T. and Chen G. Z. (2014). Produced water treatment technologies. *International Journal of Low-Carbon Technologies*, **9**, 157–177.

Jain P., Sharma M., Dureja P., Sarma P. M. and Lal B. (2017). Bioelectrochemical approaches for removal of sulfate, hydrocarbon and salinity from produced water. *Chemosphere*, **166**, 96–108.

Ji G. D., Sun T. H. and Ni J. R. (2007). Surface flow constructed wetland for heavy oil-produced water treatment. *Bioresource Technology*, **98**, 436–441.

Kim K. H., Jahan S. A., Kabir E. and Brown R. J. (2013). A review of airborne polycyclic aromatic hydrocarbons (PAHs) and their human health effects. *Environment International*, **60**, 71–80.

Knight R. L., Kadlec R. H. and Ohlendorf M. (1999). The use of treatment wetlands for petroleum industry effluents. *Environmental Science & Technology*, **33**(7), 973–980.

Munirasu S., Haija M. A. and Banat F. (2016). Use of membrane technology for oil field and refinery produced water treatment – a review. *Process Safety and Environmental*, **100**, 183–202.

Ramírez S., Torrealba G., Lameda-Cuicas E., Molina-Quintero L., Stefanakis A. I. and Pire-Sierra M. C. (2019). Investigation of pilot-scale Constructed Wetlands treating simulated pre-treated tannery wastewater under tropical climate. *Chemosphere*, **234**, 496–504.

Saad A. S. G., Khadam M. A. and Agab M. A. (2009). Biological method for treatment of petroleum water oil content in Sudan. University of Khartoum. Available at http://research.uofk.edu/multisites/UofK_research/images/ stories/research/PDF/BESBC/biological%20treatment.pdf.

Stefanakis A. I., Khadam M. A and Agab M. A. (2009). Biological method for treatment of petroleum water oil content in Sudan. University of Khartoum. Available at http://research.uofk.edu/multisites/UofK_research/images/ stories/research/PDF/BESBC/biological%20treatment.pdf.

Saththasivam J., Loganathan K. and Sarp S. (2016). An overview of oil–water separation using gas flotation systems. *Chemosphere*, **144**, 671–680.

Stefanakis A. I. (2018). Constructed Wetlands for Industrial Wastewater Treatment. John Wiley & Sons Ltd, Chichester, UK.

Stefanakis A. I. (2019). The role of constructed wetlands as green infrastructure for sustainable urban water management. *Sustainability*, **11**(24), 6981. https://doi.org/10.3390/su11246981

Stefanakis A. I. (2020a) Constructed Wetlands: description and benefits of an eco-tech water treatment system. Waste Management: Concepts, Methodologies, Tools, and Applications, IGI Global, pp. 503–525. doi: 10.4018/978-1-7998-1210-4.ch025

Stefanakis A. I. (2020b) The Fate of MTBE and BTEX in Constructed Wetlands. *Applied Science*, **10**, 127. doi: 10.3390/app10010127

Stefanakis A. I. (2020c). Constructed Wetlands for Sustainable Wastewater Treatment in Hot and Arid Climates: Opportunities, Challenges and Case Studies in the Middle East. *Water*, **12**(6), 1665. https://doi.org/10.3390/w12061665

Stefanakis A. I., Akratos C. S. and Tsihrintzis V. A. (2014). Vertical Flow Constructed Wetlands: Eco-Engineering Systems for Wastewater and Sludge Treatment. Elsevier Publishing, Amsterdam.

Stefanakis A. I., Seeger E., Dorer C., Sinke A. and Thullner M. (2016). Performance of pilot-scale horizontal subsurface flow constructed wetlands treating groundwater contaminated with phenols and petroleum derivatives. *Ecological Engineering*, **95**, 514–526.

Stefanakis A. I., Al-Hadrami A. and Prigent S. (2017). Reuse of oilfield produced water treated in a Constructed Wetland for saline irrigation under desert climate. 7th International Symposium on Wetland Pollutant Dynamics and Control – WETPOL, August 21–25, 2017, Montana, USA.

Stefanakis A. I., Prigent S. and Breuer R. (2018a). Integrated produced water management in a desert oilfield using wetland technology and innovative reuse practices. In: Constructed Wetlands for Industrial Wastewater Treatment, A. I. Stefanakis (ed.), John Wiley & Sons Ltd, Chichester, UK, pp. 25–42.

Stefanakis A. I., Charalampopoulos I., Psomiadis E. and Prigent S. (2018b). The thermal regime of a large Constructed Wetland in the desert environment. 16th IWA International Conference on Wetland Systems for Water Pollution Control, September 30–October 10, Valencia, Spain.

Stefanakis A. I., Bardiau M., Silva D. and Taylor H. (2019). Presence of bacteria and bacteriophages in full-scale trickling filters and an aerated constructed wetland. *Science of the Total Environment*, **659**, 1135–1145.

Su D., Wang J., Liu K. and Zhou D. (2007). Kinetic performance of oil-field produced water treatment by biological aerated filter. *Chinese Journal of Engineering*, **15**(4), 591–594.

Wallace S., Schmidt M. and Larson E. (2011). Long-term hydrocarbon removal using treatment wetlands. SPE Annual Technical Conference and Exhibition, 30 October–2 November, Denver, Colorado, USA.

Wu S., Kuschk P., Brix H., Vymazal J. and Dong R. (2014). Development of constructed wetlands in performance intensifications for wastewater treatment: a nitrogen and organic matter targeted review. *Water Research*, **57**, 40–55.

Wu S., Wallace S., Brix H., Kuschk P., Kirui W. K., Masi F. and Dong R. (2015). Treatment of industrial effluents in constructed wetlands: challenges, operational strategies and overall performance. *Environmental Pollution*, **201**, 107–120.

Index